Enabling Indigenous Knowledge Systems in Action Research and Action Learning

Tomé Awshar Mapotse
University of South Africa, South Africa

Emmanuel N. A. Tetteh
Walden University, USA & Saybrook University, USA & Mercy University, USA

Tsebo Kgoto Matsekoleng
University of South Africa, South Africa

IGI Global
Scientific Publishing
Publishing Tomorrow's Research Today

Vice President of Editorial — Melissa Wagner
Director of Acquisitions — Mikaela Felty
Director of Book Development — Jocelynn Hessler
Production Manager — Mike Brehm
Cover Design — Jose Rosado

Published in the United States of America by
IGI Global Scientific Publishing
701 East Chocolate Avenue
Hershey, PA, 17033, USA
Tel: 717-533-8845
Fax: 717-533-7115
Website: https://www.igi-global.com E-mail: cust@igi-global.com

Library of Congress Cataloging-in-Publication Data

Names: Mapotse, Tomé Awshar, editor. | Tetteh, Emmanuel (Emmanuel Nii Ayikwei), DATE- editor. | Matsekoleng, Tsebo (Tsebo Kgoto), 1986- editor.
Title: Enabling indigenous knowledge systems in action research and action learning / edited by: Tomé Mapotse, Emmanuel Tetteh, Tsebo Matsekoleng.

Description: Hershey PA : IGI Global, [2025] | Includes bibliographical references. | Summary: "The objective of this book is to encourage, and influence researchers the world over to apply Indigenous Knowledge Systems (IKS) using Action Research and Action Learning (AR and AL) approaches in their fields of specialization"-- Provided by publisher.
Identifiers: LCCN 2024049256 (print) | LCCN 2024049257 (ebook) | ISBN 9798369335420 (hardcover) | ISBN 9798369351246 (paperback) | ISBN 9798369335437 (ebook)
Subjects: LCSH: Action research--Methodology. | Active learning. | Ethnoscience.
Classification: LCC H62 .E5815 2025 (print) | LCC H62 (ebook) | DDC 001.4/2--dc23/eng/20241219
LC record available at https://lccn.loc.gov/2024049256
LC ebook record available at https://lccn.loc.gov/2024049257

British Cataloguing in Publication Data
A Cataloguing in Publication record for this book is available from the British Library.

Table of Contents

Detailed Table of Contents

Action research (AR) aims to enhance teaching practices through learning and reflection. By viewing teaching challenges as growth opportunities, AR encourages teachers to improve their instructional strategies and foster student engagement. Defined as a cycle of planning, acting, evaluating, refining, and learning (Koshy, 2010), AR supports the development of teachers' professional judgment and contributes to improved educational outcomes. While AR is effective in bridging theory and practice for preservice teachers and enhancing their reflective skills, challenges such as time constraints and limited practical experience can affect its implementation. Nonetheless, AR remains an important tool for professional development and improving educational practices. This chapter seeks to answer the following questions: how teacher professional development is increased through action research, is teacher professional growth developed through a reflective and analytical approach, and is action research suitable as a tool to enhance teacher-researcher competence.

The transformative potential of integrating African Indigenous Knowledge Systems (AIKS) into contemporary research practices. It highlights the limitations of traditional text-based methods and argues for the value of Action Research and Learning (AR/AL) methodologies as a framework for fostering collaboration between researchers and African communities. The paper examines how AR/AL and technology can bridge the gap between embodied knowledge and digital representation, fostering knowledge co-creation and empowering African communities. Ethical considerations are highlighted, ensuring cultural sensitivity and respect for intellectual property rights. The paper concludes by discussing the broader impact of AR/AL on African communities, including fostering knowledge sharing, building research skills, and advocating for policy change. This approach empowers African communities to become stewards of their sustainable development.

Educational stakeholders [Indigenous knowledge System(IKS)] rely on research to make informed decisions that ultimately affect the quality of schooling for their students. Accordingly, the purpose of educational research is to engage in disciplined inquiry to generate knowledge on topics significant to the students, teachers, administrators, schools, and other educational stakeholders. Just as the topics of educational research vary, so do the approaches to conducting educational research in the classroom. Your approach to research will be shaped by your context, your professional identity, and paradigm (set of beliefs and assumptions that guide your inquiry). These will all be key factors in how you generate knowledge related to your work as an educator. This paper discusses the research process from ontological, epistemological and methodological perspectives to explore and understand the social phenomena. The goal of this review is to provide a clear understanding of different research methodologies based on ontology and epistemology.

An inclusive and participative food system is essential to ensure sufficient nutritious food for all while considering sustainability. Agroecological practices require scientific expertise and farmers' familiarity with the local situation, across food systems for sustainable transformations. The co-creation of two knowledge systems—practice and research—rather than a top-down approach, can significantly contribute to comprehensive agroecological evolution while preserving community traditions. The United Nations Food and Agriculture Organization (FAO) has stressed the importance of integrating spiritual elements into the food system. Many valuable practices rooted in indigenous and local knowledge can be observed in diverse communities. Consequently, embracing and learning from these traditions can greatly support the sustainable advancement of humanity. The Dayalbagh community has inventively ensured food security by building a robust and sustainable food system that is both holistic and resilient. Additionally, it upholds social, economic, environmental, and cultural values.

This chapter examines how Action Learning (AL) and Action Research (AR) approaches are being used to decolonize educational systems. In order to create more inclusive and culturally relevant learning environments it emphasizes the significance of tearing down colonial legacies in education and incorporating Indigenous Knowledge Systems (IKS) into curricula. With the help of AR and AL educators and students can co-create knowledge solve real-world problems and advance educational equity. These are reflective participatory methods. The benefits of IKS for diverse learning communities are highlighted as the chapter covers methods for implementing it in educational settings. It also looks at how professional development affects teachers highlighting the necessity of preparing them to act as co-researchers in the decolonization process.

The chapter investigates the transformative potential of coordination of Inborn Information Frameworks (IKS) into instructive systems utilizing techniques such as Activity investigation (AR) and Activity Learning (AL). This approach challenges the bequest of colonialism in instruction by centering inborn points of view, information, and hones within the educational modules' improvement preparation. Activity Investigation and Activity Learning provide collaborative and iterative forms through which teachers, policymakers, and innate communities can co-create instructive activities. These strategies prioritize the dynamic interests of all partners, guaranteeing that activities are responsive to the differing needs and points of view of innate people groups. Teachers can refine their hones to better join IKS in instructing and learning encounters by locking in persistent reflection and adjustment. The chapter analyzes case considerations and best hones from different settings, illustrating the viability of AR and AL in decolonizing instruction frameworks.

This book chapter explores the co-existence of Indigenous Knowledge Systems (IKS) with Technology Education, focusing on their potential to enhance learners' understanding of technology and appreciation for diverse cultural perspectives. Grounded in Constructivism theory, the study involved interviews with three experienced Senior Phase Technology teachers selected for their expertise. Thematic analysis revealed valuable insights into the benefits of incorporating IKS for teachers and learners. We acknowledge challenges such as the dominance of Western viewpoints and the need for better curriculum support. To overcome these challenges, we propose practical strategies like hands-on learning experiences and collaboration with community experts. Ultimately, the chapter emphasizes the importance of making design activities culturally relevant and promoting creative and critical thinking to effectively integrate IKS into Technology Education.

Action Research/Learning means that ideally, the step-by-step process is constantly monitored over varying periods of time and by a variety of mechanisms (questionnaires, diaries, interviews and case studies) so that ensuing feedback may be translated into modifications, adjustment, directional changes, redefinitions, as necessary, so as to bring about lasting benefit to ongoing process itself rather than to some future occasion. These Chapter aiming to describes factors contributing to action research as: a constructive inquiry, during which researcher constructs his or her knowledge of specific issues through planning, acting, evaluating, refining and learning from the experience. It is a continuous learning process in which the researcher learns and also shares the newly generated knowledge with those who may benefit from it. These chapter highlight the distinct formulation features of action research and emphasize the purposeful intent of Ethics of researchers to improve, refine, reform, and problem-solve issues in their educational context.

Action Research develop conceptual structures that took explicit account of the defining social roles and strongly influence the process of any change in the modes of production. Action research was exemplified by the discussion of problems followed by group decisions on how to proceed. Action research must include the active participation by those who have to carry out the work in the exploration of problems that they identify and anticipate. After investigation of these problems the group makes decisions, monitoring and keeping note of the consequences. The group would decide on when a particular plan or strategy had been exhausted and fulfilled, come to nothing, and would bring to these discussions newly perceived problems. These Research elaborates are action research is an emergent process that allows beliefs and decisions to be negotiated as reality and meaning are being. Indeed fair observation demonstrated the efficacy of action research for improving productivity.

The chapter fills the gap in the literature void within Higher Learning Institutions by decolonising the curriculum through utilisation of progressive environmental action research (PEAR) activities to integrate Indigenous knowledge systems (IKS). This qualitative conceptual chapter employs perennialism theory to ground the chapter and the chapter is undergirded by a participatory paradigm. Because of the persistence of environmental issues, the purpose of this chapter is to share or highlight the applicability of PEAR activities in institutions of higher learning studies and close the gap on the death of literature. As such, this chapter will address How Indigenous Knowledge Systems can be integrated into Higher Learning Institutions' studies using progressive environmental action research activities.

Traditional knowledge is the wisdom passed on from generation to generation within a community which forms part of their cultural identity. This community wisdom held by indigenous people is the source of many modern inventions and innovations in various fields of human development. Lack of representation, recognition and protection to this body of knowledge makes it susceptible to misappropriation and misuse. In an effort to create strong indigenous knowledge system it is essential that they are protected through robust legal framework. Intellectual property is one of the tools to guard this traditional knowledge against unauthorized exploitation. This paper aims to study the interface between traditional knowledge and intellectual property. Jabalpur Stone Craft is identified as an expression of traditional knowledge of craftsmen carrying this art for generations. The chapter studies the progression of this craft from traditional knowledge to a commercialized and well-protected piece of art. Even though the craft has been protected as a geographical indication, this protection is limited and focuses on the art itself. This paper studies the traditional knowledge aspect of this stone craft and proposes approaches to protect the underlying wisdom of cultural expressions. To encapsulate, this paper aims at fostering the relationship between traditional knowledge and intellectual property laws through the case study of stone craft of Jabalpur.

The chapter averred that the use of positivism in research in higher education has proved to be of disservice to indigenous communities because of the assumptions inherent in that paradigm. The paradigm assumed that knowledge has to be verified scientifically by using methods derived from modernity and then it must be applied in universal setting. This approach is being challenged by the use of participatory action research which transforms participants from being objects of research to empowered communities who can demonstrate agency in changing their communities. The chapter explores a number of theoretical lenses to discuss the importance of PAR which include communicative theory, ubuntu, decolonial, and education for emancipation by Paulo Freire. The chapter calls for the use of PAR in higher education because of its potential for democracy and empowerment of communities.

This paper addresses disparities in Mathematics teaching between Indigenous and Eurocentric methods in South African primary schools, aiming to improve mathematical performance in underprivileged settings. Through systematic literature review, action research, and action learning, it identifies Afro-centric games

Foreword

Researchers, particularly those working with Indigenous communities, are increasingly drawn to pragmatic, participatory, and action-oriented research methodologies. These approaches—such as Action Research (AR) and Action Learning (AL)—align naturally with Indigenous Knowledge Systems (IKS), which are deeply rooted in practical experience, relational understanding, and close engagement with the natural world.

This book is a timely and significant contribution to the ongoing evolution of research paradigms that center on inclusivity, community engagement, and sustainable transformation. Through chapters in this manuscript, the book offers a unique and dedicated exploration of how Indigenous perspectives, and vice versa, can enrich AR and AL. In doing so, it moves beyond conventional models of inquiry. It challenges the long-standing notion of Indigenous peoples as passive subjects of research. Instead, it celebrates them as active co-creators of knowledge whose wisdom, innovation, and lived experiences are essential to reimagining research modalities.

The editors and contributors have succeeded in marrying Indigenous Knowledge with progressively pragmatic research methodologies in a methodologically sound and ethically compelling way. The book contributes a fresh perspective to AR and AL paradigms. It provides a robust framework for researchers across disciplines—education, business, health, agriculture, engineering, technology, management, public service, the arts, or the social sciences—to integrate IKS into their practice.

What is especially intriguing is how each chapter demonstrates how Indigenous frameworks and paradigms transform the research process. These contributions help reshape research from an extractive endeavor to mutual learning, empowerment, and reflection. In this way, the book engages with a long-standing question—paraphrasing Karl Marx's call in Theses on Feuerbach—by asking: "How can we change our communities and the world through the utilization of IKS in AR and AL, to make it a better place for all human beings?"

The result is a volume that is not only informative but deeply inspiring. It urges scholars, practitioners, and students to rethink their roles—not as distant observers but as engaged participants in co-creating culturally grounded and future-focused knowledge. Henceforth, the chapters offer essential reading for anyone committed to decolonizing research, restoring Indigenous voices to the center of inquiry, and building more just, equitable, and sustainable knowledge systems for all.

Mishack Thiza Gumbo
University of South Africa, South Africa

Emmanuel N. A. Tetteh
Saybrook University, USA & Walden University, USA & Mercy University, USA

Preface

INTRODUCTION

The evolving landscape of research demands methodologies that are innovative and inclusive and contextually grounded in the lived realities, wisdom, and worldviews of communities. This book was conceived out of a growing recognition that Indigenous Knowledge Systems (IKS)—deeply embedded in generations of experience, interaction with nature, and community-based learning—are invaluable to both knowledge creation and problem-solving across disciplines.

The primary aim of this volume is to encourage, inspire, and equip researchers, educators, and practitioners around the world to integrate IKS into their specific fields of specialization through the transformative methodologies of Action Research (AR) and Action Learning (AL). Both AR and AL offer collaborative, reflective, and pragmatic frameworks that align organically with Indigenous ways of knowing, being, and doing. Yet, despite the rising interest in IKS, a significant capacity gap remains in the practical application of IKS through AR/AL within mainstream academic research. This book seeks to close that gap.

By creating a platform for authors to share original research, case studies, theoretical insights, and field experiences, the book brings together 18 chapters from diverse contributors across various disciplines. The content is as varied as the fields themselves—ranging from education, agriculture, environmental sustainability, and health to leadership, community development, business, technology, public service, and knowledge management—but united by a shared purpose: to demonstrate how AR and AL can be used to activate, preserve, and advance Indigenous Knowledge in contextually meaningful ways.

Importantly, this book does not treat IKS as static or historical. Rather, it emphasizes IKS as a dynamic, evolving, and relevant system of knowledge that continues to inform daily life, learning, and innovation in Indigenous communities. The book calls on researchers to draw attention to and investigate the continuity of Indigenous practices, some of which have endured for centuries and remain significant in the present day.

This volume serves as a valuable resource for a broad audience, including academics, researchers, postgraduate students, pre-service teachers, sociologists, agriculturalists, education policymakers, curriculum designers, and community practitioners. It provides concrete examples, methodological insights, and a reflective space for considering how AR/AL and IKS can be mutually reinforcing in shaping research that is not only transformative but also socially and environmentally responsive.

Above all, this book invites a shift—from researching about communities to researching with and for advancing the integration of IKS into AR/AL frameworks across disciplinary fields. It celebrates the role of Indigenous voices in leading inquiry and co-producing knowledge. It is a testament to the idea that change can occur when research is rooted in respect, relationality, reciprocity, and relevance.

We hope this work will spark further interest, collaboration, and pragmatic experimentation with IKS-informed AR/AL approaches in disciplines worldwide, ultimately contributing to more inclusive, equitable, and sustainable systems of knowledge and development.

— The Editors—

CHAPTER OVERVIEW

This book comprises eighteen chapters that collectively explore the integration of Indigenous Knowledge Systems (IKS) into Action Research (AR) and Action Learning (AL) frameworks across a wide array of disciplines. Each chapter contributes a distinct yet complementary perspective to the overarching theme of enabling transformative, community-driven research and practice.

- **Chapter One** introduces the use of Action Research as a tool for enhancing teachers' professional development, demonstrating how reflective practice can empower educators and improve teaching effectiveness.
- **Chapter Two** explores the intersection of AR/AL, Indigenous Knowledge, and digital technology, investigating how embodied knowledge can be meaningfully represented and co-created in digital spaces, with particular focus on empowering African communities.
- **Chapter Three** delves into ontological and epistemological interpretations of Indigenous Knowledge as held and enacted by classroom teachers. It highlights methodological insights for researchers working within diverse worldviews.
- **Chapter Four** presents a case study from the Dayalbagh community, focusing on the sustainability of food systems and how combining expert input with Indigenous knowledge leads to improved food security outcomes.
- **Chapter Five** addresses the decolonisation of education by advocating for participatory action frameworks that empower communities and challenge Eurocentric curricula and pedagogies.
- **Chapter Six** continues this discussion by examining the transformative potential of integrating IKS into education reform, and how AR/AL serve as methodologies for system-wide educational change.
- **Chapter Seven** investigates the coexistence of Indigenous Knowledge in Technology Education, showcasing classroom strategies to enhance student learning through culturally relevant practices.
- **Chapter Eight** focuses on the constructivist nature of Action Research, highlighting how iterative planning, acting, reflecting, and learning shape both the researcher's knowledge and the communities' development.
- **Chapter Nine** offers a foundational overview of Action Research, with illustrative models and reflective techniques showing how AR unfolds as a cyclical problem-solving process.
- **Chapter Ten** looks at integrating IKS into higher education environmental programs, specifically through Progressive Environmental Action Research Activities that make learning more relevant for Indigenous students.
- **Chapter Eleven** evaluates the role of Indigenous Technical Knowledge (ITK) in sustainable agriculture and horticulture, exploring its contributions to biodiversity, food security, and natural resource management.

- **Chapter Twelve** addresses blended and distance education, proposing virtual adaptations of traditional knowledge-sharing methods in response to changes in education delivery modes.
- **Chapter Thirteen** examines how Technology Education teachers incorporate IKS in classroom practice following participation in engaged scholarship and societal impact (ESSI) sessions.
- **Chapter Fourteen** applies Participatory Action Research to explore South African pre-service teachers' experiences and understandings of traditional environmental practices.
- **Chapter Fifteen** critiques the historical exclusion of Indigenous communities in research processes, advocating for genuine community participation in Participatory Action Research.
- **Chapter Sixteen** interrogates the relationship between traditional knowledge and intellectual property rights, using a case study on stone craft in Jabalpur to explore legal and cultural preservation.
- **Chapter Seventeen** explores how higher education institutions can become more relevant by involving local communities in co-creating knowledge through participatory AR and IKS.
- **Chapter Eighteen** focuses on addressing the gaps between Indigenous and Eurocentric methods of teaching mathematics in South African primary schools, proposing culturally grounded solutions to enhance student performance.

Together, these chapters illustrate the power and potential of IKS in reshaping knowledge production, teaching, and research—fostering inclusivity, relevance, and sustainability.

CONCLUSION

This book has provided a rich and interdisciplinary examination of how Indigenous Knowledge Systems can be enabled, activated, and sustained through Action Research and Action Learning. Each chapter demonstrates how scholars, educators, and practitioners can apply IKS in ethically grounded, contextually relevant, and professionally enriching ways.

The editors express deep appreciation to all the contributing authors whose insights, research, and case studies give this volume depth and diversity. The chapters offer theoretical contributions and practical strategies that can be implemented in classrooms, research institutions, and communities across the globe.

Importantly, this book is for academic scholars, teachers, curriculum developers, policymakers, and community leaders. It challenges its readers to rethink how knowledge is constructed and shared, and it encourages a shift towards research that is collaborative, reflective, and responsive to Indigenous realities.

By foregrounding IKS within AR and AL frameworks, this book serves as a beacon for those who seek to transform research into a tool for liberation, empowerment, and sustainable change.

Declaration of Peer Review

This manuscript underwent a rigorous double-blind peer review process for each chapter before publication. Reviewers were carefully selected for their subject matter expertise and independence from the authors, editors, and publisher. Their identities were not disclosed to the authors to ensure objectivity.

Each chapter was evaluated on scholarly merit, originality, methodological soundness, and relevance to Indigenous Knowledge Systems, Action Research, and Action Learning themes. Reviewers provided constructive, candid feedback, which was shared with the authors to guide necessary revisions and strengthen the quality of each submission.

All authors were required to respond to reviewer comments and revise accordingly. Editors provided additional input to ensure consistency, coherence, and academic rigour across chapters. The manuscript then underwent an internal quality assurance review before final publication.

The result is a scholarly and original body of work that contributes meaningfully to the fields of AR, AL, and IKS. All content is free from plagiarism, and every effort was made to maintain the highest ethical and academic standards throughout the review and editing process.

Research Justification

The edited volume Enabling Indigenous Knowledge Systems in Action Research and Action Learning was developed to address an important gap in scholarly literature: the integration of IKS with AR and AL methodologies across diverse fields of study. While Action Research and Action Learning have long been recognized as dynamic, reflective, and transformative approaches, their intersection with Indigenous epistemologies has been underexplored in mainstream academic discourse.

IKS is inherently pragmatic, community-driven, and deeply relational, making it a natural ally to AR and AL. This book brings these methodologies into dialogue, demonstrating how they can work in concert to enhance professional development, promote community empowerment, and preserve Indigenous ways of knowing.

The research presented in this volume is both original and multidisciplinary, offering valuable insights for researchers in Human Sciences, Education, Law, Environmental Studies, Engineering and Technology, Economics and Management Sciences, and Agricultural Sciences. This book targeted the policymakers, curriculum developers, postgraduate students, and community leaders seeking to make knowledge production more inclusive and socially responsive.

Through this work, the editors and contributors have advanced the global discourse on Indigenous Knowledge, contributing to its excavation, preservation, revitalization, and practical application. By embracing lessons from the past, this book offers a vision for the future—one in which research is rooted in cultural relevance, ethical engagement, and transformative purpose. As the book Editor-in-Chief, I unequivocally declare that there is no part of the work that was plagiarised or published elsewhere.

Tomé Awshar Mapotse
University of South Africa, South Africa

Emmanuel N. A. Tetteh
Saybrook University, USA & Walden University, USA & Mercy University, USA.

Tsebo Kgoto Matsekoleng
University of South Africa, South Africa

Chapter 1
Action Research as an Enhancement for Teacher Professional Development

Jolita Horbačauskienė
https://orcid.org/0000-0002-5982-9329
Kaunas University of Technology, Lithuania

Berita Simonaitienė
https://orcid.org/0000-0003-3314-0862
Kaunas University of Technology, Lithuania

Milda Ratkevičienė
https://orcid.org/0000-0003-3924-7281
Kaunas University of Technology, Lithuania

ABSTRACT

Action research (AR) aims to enhance teaching practices through learning and reflection. By viewing teaching challenges as growth opportunities, AR encourages teachers to improve their instructional strategies and foster student engagement. Defined as a cycle of planning, acting, evaluating, refining, and learning (Koshy, 2010), AR supports the development of teachers' professional judgment and contributes to improved educational outcomes. While AR is effective in bridging theory and practice for preservice teachers and enhancing their reflective skills, challenges such as time constraints and limited practical experience can affect its implementation. Nonetheless, AR remains an important tool for professional development and improving educational practices. This chapter seeks to answer the following questions: how teacher professional development is increased through action research, is teacher professional growth developed through a reflective and analytical approach, and is action research suitable as a tool to enhance teacher-researcher competence.

DOI: 10.4018/979-8-3693-3542-0.ch001

INTRODUCTION

Action research (AR) has been recognized as an important tool in teacher education, highlighting its potential to reshape perceptions and foster professional growth through reflection and experiential learning. Many studies (Ulvik et al., 2017; Lattimer, 2012; Yosief et al., 2024) argue that AR is effective as a pedagogical tool in initial teacher education, but at the same time factors like preservice teachers' lack of experience, time consumption, and the specifics of conducting research in school environments could be considered as having a negative impact (Darwin & Barahona, 2019).). Pine (2009) suggests that AR empowers teachers to examine and improve their own teaching contexts, leading to innovative and effective classroom practices. Through reflection, teachers can enhance their decision making and contribute to the creation of engaging, learner-centered environments. Ultimately, action research not only transforms educators' practices, but also deepens their understanding of their professional roles and the contexts in which they work (Kemmis, 2009). Therefore, AR can be used in the education of preservice teachers to develop the pedagogical competences they need for their future work and applying theoretical knowledge in practice.

Our chapter aims to answer the following research questions: How teacher professional development is increased through action research, is teacher professional growth developed through a reflective and analytical approach, and is action research suitable as a tool to enhance teacher-researcher competence.

THEORETICAL STANCE OF THE STUDY

AR has been identified as a relevant approach in teacher education for some time now (Guskey 2002; Ulvik et al., 2017) because it can influence changes in perceptions and attitudes through learning and reflection (Yosief et al., 2024). AR helps teachers interpret and address teaching challenges, encouraging them to view learning difficulties as opportunities for growth rather than shortcomings (Reinhardt et al., 2021). Spencer Clarks et al. (2020) defined AR as a method that empowers and assists teachers in adopting effective teaching strategies by improving the quality of their instructional decisions and actions, ultimately leading to increased student engagement and learning outcomes. Using AR involves taking action, evaluating outcomes, and reflecting on the process. This approach is used to collect evidence that informs and supports changes in teaching practices and therefore improves educational practice. Koshy (2010) defines AR as a productive inquiry in which the teacher builds knowledge on particular issues through a cycle of planning, acting, evaluating, refining, and learning from the process. This ongoing learning experience not only enhances the teacher's understanding but also allows the teacher to share newly acquired knowledge with others. Mertler (2024) notes that AR can be characterized as a method that allows teachers to analyze their own teaching practice as a systematic inquiry by studying assessment, instructional methods, and learners, thus seeking to improve the overall quality of teaching. Bergmark (2020) argues that AR might be understood as a form of professional learning that is intentionally integrated into teachers' daily work. This approach is collaborative, focused on areas identified by teachers, and grounded in both current research and teachers' practical experiences and contextual understanding. Schmuck (2006, p. 28) defines AR as an effort to "study a real school situation with a view to improve the quality of actions and results within it" and provide insights into more effective ways of achieving desired educational outcomes. These definitions highlight the unique characteristics of AR, emphasizing its goal of improving, re-increasing, reforming, and addressing challenges within educational settings, as well as improving teaching practices.

As argued by Crawford-Garrett et al. (2015), action research develops preservice teachers' reflective and analytical skills, particularly in addressing the theory-practice gap in practice and curriculum. This approach reflects a growing awareness of the need to move away from the traditional orientation of imparting pedagogical knowledge and encourages a broader understanding of how knowledge of professional and occupational content is constructed (Cochran-Smith et al., 2012). The increased use of action research in initial teacher education has sparked interest in its effectiveness as a pedagogical tool (Lattimer, 2012). Moreover, AR can play a critical role in professional development of teachers across all educational levels and subject areas due to the relationships between AR and differentiated instruction, data-driven decision making, or progress monitoring of learners (Dana & Yendol-Hoppey, 2019). Research-based education emphasizes that teacher educators should conduct practice-oriented research and train student teachers in relevant research methods. (Smith & Flores, 2019). It also highlights the potential for collaboration between research-active educators and experienced teachers, allowing them to learn from each other and work together to improve educational practices and achieve shared objectives (Caspersen & Smeby, 2023).

Although many studies report positive outcomes for preservice teachers engaging in action research (Lattimer, 2012; Phillips & Carr, 2010; Ulvik, 2014), concerns exist regarding its limitations. These include preservice teachers' lack of experience, time constraints, overlapping research and evaluation requirements, and the limitations of schools as research sites. Volk (2009) argues that action research is often viewed as an academic obligation with little perceived value or authenticity, leading few graduates to continue using it. In general, using action research as an assessment tool can shift the focus from reflection and inquiry to meeting assessment criteria, limiting its effectiveness in developing teacher competences prior to service. (Darwin & Barahona, 2019). Moreover, Darwin & Barahona (2019) argue that despite that preservice teachers are generally prepared to carry out action research, their theoretical knowledge of the process, due to their limited practical experience, may lead to confusion about the technical requirements, especially in relation to cycles and data methods.

Some researchers argue that action research, as a form of practical inquiry, initiates a reform of teachers' practice by allowing them to explore their own situations and contexts in classrooms and schools (Pine, 2009). Reflecting on their practice helps teachers improve it, create innovative teaching and learning environments, make autonomous professional decisions, and increase their excellence and competence. Action research provides teachers with the opportunity to reflect on their practice (Creswell, 2012), as well as to connect to a particular classroom practice (Baan et al., 2020) and individual student learning needs.

According to Kemmis (2009), action research involves changing three aspects: practitioners' practice, their understanding of their practice, and the conditions in which they practice. In this process, each element influences the others in a continuous cycle where each one responds to the others. Reinhardt et al. (2021, p. 441) note that AR is a "meaning-making process for the problems of practice" and argue that AR in teacher education together with research-based teaching strategies could have a twofold outcome for preservice teachers. First, the implementation of AR could serve as a context for better understanding the progress of the learner, and second, research-based strategies could be used to apply differentiated instruction for diverse learners.

Caspersen & Smeby (2023) in their study examined changes in teacher educators' attitudes towards research-based teacher education in Norway within the time period from 2008 to 2021. Their results demonstrate that teacher educators' attitudes have shifted towards a more favorable view of research.

Although this change is in part due to having more research experienced staff in preservice teacher training programmes and institutional changes, the authors argue that the positive outlook on research-based teacher education also appears to be influenced by cultural factors and teacher identity. In Finland research based approach has been an integral part in teacher education, therefore, "all teaching is based on scientific evidence, and theory-based pedagogical thinking is incorporated into teachers' every action" (Puustinen et al., 2018, p. 3), although the empirical results of this study also indicated that preservice teachers valued practical orientation more highly than theoretical elements. In the Lithuanian context action research as a research-based pedagogical approach has been implemented in preservice teacher education for some years now. As Jakavonytė-Staškuvienė et al. (2023) argue, this approach should help in developing highly reflective educators who focus not only on their personal teaching activities but also on their pedagogical practices.

METHODOLOGICAL IMPLICATIONS

Action research is understood as a process of improving educational practice and its methods include action, evaluation, and reflection, i.e. action research is the process of gathering evidence to change practice (Koshy, 2010). Ultimately, in AR, findings emerge as the action develops and unfolds; however, it is not final or absolute, it is ongoing (Koshy, 2010). This means that action research is a quite acceptable and appropriate method, both from a practical point of view, as it empowers teachers themselves to analyze, change, and improve their practice, and from a theoretical-scientific point of view, as it provides valuable information to analyze the processes of education and its change, in order to construct theories for the improvement of education process.

Study setting. There are four different ways to become a teacher in Lithuania and, depending on the route, the duration of the studies varies from one year (60 ETCS) to four years (240 ETCS). We analyzed a programme of professional studies that is a 60 ECTS, one-year programme, which is limited to students with a minimum of a Bachelor's degree. The professional study programme consists of 30 ECTS of professional internship in a school and 30 ECTS of basic theories of pedagogy on the basis of which pedagogical competencies are developed. All teacher education programmes, regardless of the higher education institution and type of programme, teach students to conduct research on pedagogical practice, to analyze, evaluate and summarize in a reasoned way the results of research on performance and change in the classroom.

The methodology of action research was introduced to Lithuanian teachers at the 1999-2000 summer school organized by A.P.P.L.E. (American Professional Partnership for Lithuanian Education). About 200 Lithuanian teachers participated in this project and conducted action research in their schools and classrooms. Doctoral students in Education assisted the A.P.P.L.E. representatives, learning the methodology together with the teachers and advising them during the period as they conducted their action research. As a result of this project, Action Research was also included in teacher training programmes as a compulsory competency for teacher researchers.

However, when it comes to pedagogical studies, it is important to be aware that Lithuanian legislation does not declare the need and necessity to develop students' scientific competences. Contrary to the case of regular first cycle studies, the requirement for the final thesis of pedagogical studies leading to the qualification of a teacher is the integration of theory and practice, with no additional emphasis on scientific competences as defined in the Description of the Group of Educational Studies in Education and Training, "to demonstrate the results of the studies achieved by collecting, systematizing, reflexively and critically analyzing, integrating and applying theoretical, practical, experiential, and empirical information on the objects and subjects of pedagogical activity" (2015, item 40.3.2.1.).

Despite these arguments, it is clear that practical knowledge alone may not be enough for today's teachers. Therefore, as Elliott (2009) argues, a key challenge for teacher educators is to balance their dual roles as educational researchers and practitioners. It is in line with Matos et al. (2023) findings that through the cooperative research-based practices between teacher educators and students or preservice teachers research competences, pedagogical practices and methodological content knowledge may be developed. Moreover, in such case, AR allows for the merging of practice and research, and here, according to Elliott (2009), the role of teacher educators is important, as it is they who have the opportunity to empower teachers to pursue their educational goals by building and testing shared understandings, i.e., to focus research on finding solutions to some of the most pressing problems that teachers face in their classrooms. This means that teacher education must take into account a coherent combination of research and practice, as this ensures informed and timely decision making. Therefore, AR, with its practical applicability and specificity, is one of the most acceptable solutions for enriching a curriculum that focuses on the development of practical pedagogical competencies with scientific knowledge relevant to today's educator.

The final project of the study programme we are analyzing requires students to reflexively assess the change in six competencies related to the teaching profession over the course of their studies. One of the six competences, which is compulsory for all students, is the competency of teacher research. The remaining five competencies can be selected by students from a list of competences for teachers (e.g., DigCompEdu). In their final project, students analyze the research competence in more detail and describe the specific activities that characterize the competence (e.g., designing a research plan, formulating a research question, collecting and analyzing data, making recommendations for improvement, etc.).). It has already been mentioned that 30 ECTS of this programme are devoted to pedagogical internships in schools. The tasks of the pedagogical internship module are designed, supervised, and assessed by the university lecturers. The learning outcomes of the pedagogical internship module state that students will be able to plan, conduct, analyze, and evaluate the development of their pedagogical internship by making appropriate decisions and applying action research methods. Therefore, one of the module's tasks is to plan, carry out action research, and present the results. The aim of this assignment is to develop and strengthen the researcher's attitudes, to understand and learn how to apply action research methodology in pedagogical practice, and to acquire and develop the researcher's competences (e.g., to identify the need for the research, the problem of the research, to collect, carry out, analyze, and present results to others). The action research assignment consists of two parts:

- In the fall semester, students plan an action research (the assignment is called A R design). In this phase, the problem to be solved is selected, the action research question and the planned change are specified, the activities are selected, and the action research activities and timelines are planned in detail. The AR design is an assessed task, students receive detailed feedback and, together with the lecturer, improvements and adjustments are made to the plan. Only after this evaluation can students begin to carry out action research in their internship.
- In the spring semester, students carry out and document AR in detailed diaries. After completion of the AR phase of the action research cycle, students produce a reflective AR report.

The AR assignment requires 65 hours of student workload in the fall semester and 80 hours in the spring semester. A two-contact-hour seminar is scheduled prior to the start of the action research to introduce and discuss the methodology, the assignment requirements, and the action research. The Moodle virtual learning environment also provides instructions for conducting the action research, report and reflection forms, descriptions of the main principles, methods, and examples of action research. Two one-hour tutorials per student are provided, during which interim feedback and comments are given on the action research plan and action research report.

Participants in the study. The primary empirical data in the present study is based on the reports of 54 AR studies of preschool, primary, and secondary (subject) teachers, where the records were relevant to the objectives of this study and to the research questions posed. The data are coded according to this pattern: indication of the year of study of the participant, AR, and number of the report from which the extract is taken. The most illustrative extracts from the AR reports are used in the texts.

Research ethics. The AR was organized and carried out in accordance with all the requirements and ethical principles for this type of research. As Seale (2012) argues, privacy, confidentiality, and data protection, as well as informed consent, are essential in research ethics. In this case, all participants in the action research study were involved voluntarily and were informed of the essential objectives and requirements of the research; the analysis of the data generated by the research aims to ensure confidentiality, privacy, and data protection for both the preservice teachers and the pupils involved in the research.

Study design. Every action research model has essential components: planning for change, taking action to implement the change, monitoring the process and consequences of the change, reflecting on the process and consequences, reacting, observing, reflecting (Spencer Clark et al., 2020).). In this context, the action research was organized in three phases:

1. Planning phase. This phase involves clarifying the research question and the planned change, selecting the modalities of action, and planning the workflow and timeline.
2. Action phase. During this phase, the planned activities were carried out and the research diary is kept.
3. Reflection phase. Throughout the action research, preservice teachers keep reflection diaries and document the beginning and end of action research activities in a report.

Data analysis methods: The data analysis framework is based on qualitative content analysis perspectives, as Azizan et al. (2018) notes verbal or textual data are analyzed applying qualitative content analysis. Shava et al. (2021) argue that quantitative data analysis may not take full account of the many interaction effects that take place in social settings and thus the need to use qualitative data analysis techniques in analyzing educational research data when the results of the analysis can help solve problems, validate existing theories, or develop new theories and models. Čepienė and Vaitiekienė (2019) note that qualitative content analysis allows the researcher to aggregate large volumes of information, grouping research data into subcategories and categories according to content and meaning. Our study analysis is based on the frameworks developed by Feekery (2023), Somekh and Zeichner (2009), and Bergmark (2020). The prevailing domains for our analysis are research form, aims for carrying out the AR, educational solutions, combination theory and practice, collaboration, and critical reflection.

RESEARCH RESULTS

This part analyzes the experiences of preservice teachers in conducting AR during their studies. The analysis of the data aims to reveal how the professional development of the study participants as preservice teachers is manifested through AR. The analysis of the data is presented through six dimensions: research form, aims for carrying out the AR, educational solutions, combination of theory and practice, collaboration, and critical reflection.

Research Form

Somekh & Zeichner (2009) point out that the process of action research can take a spiral (or cyclical) form, focusing on a single research question, or a more complex (multispiral) form, focusing on several research questions simultaneously (Somekh & Zeichner, 2009).). In the case studies presented here, AR is organized as a simple spiral process consisting of planning, action, observation, and reflection. In a cyclical form of inquiry associated with Lewin's methodology, the researcher plans, tests, observes, reflects, and re-tests, sometimes over several iterations, key interventions (Feekery, 2023). When completing the task, students operate within a cycle that includes problem diagnosis, action planning, intervention, and reflexive evaluation of consequences and change.

The peculiarity of action research as a study assignment is that the research is time-limited and not all students have time to replan and repeat the intervention, as, for example, "the activities themselves take a lot of time to devise, to coordinate, to choose the tools" (23VT6).). It is important that students understand and responsibly plan their time resources for the study in the future and that they have experienced what the time costs of the study can be. The challenges of time management are also reflected in the students' experiences: "The research was very interesting for me, it was interesting to see that even a small change was happening. *However, it really should be a long-term work" (22VT4); "In order to extend, to complement this study, more time should be spent on theory search and to look more deeply into the different ways of memorizing words and their application. In the time available for the study, it was not possible to find any sources indicating the effectiveness of different audiovisual and other tools*" (17VT3). The decision of the programme designers to extend the assignment over two semesters was noted and appreciated by the students, which would indicate their awareness of the time resources available for the study, e.g., "*It was very good that it was possible to complete the assignment over a longer period of time, i.e., the assignment was spread over two semesters. The activity requires deeper reflection and implementation, so time is very much needed*" (23VT10). In this case, when there is not enough time for a new cycle, in the reflection phase students have to reflect on how they would continue the research and what other activities they would carry out in a new cycle. It is very important for us, as programme facilitators, that the students consolidate the action research cycle, the sequence of stages, and understand the importance of each stage. In their reflections, they have a recognizable understanding of AR as a cyclical process, and they identify the actions necessary for each stage; e.g. in the first stage, the participants are asked to identify the steps in the process, such as "I developed the ability to identify the problem, to plan and execute the change activity, to notice the change, to recognize the need to adjust the research as it progresses, to reflect on the impact of the activity on the success of the learning, etc. *Planning, acting, and reflecting are the main skills that I have improved during the action research" (21VT10).* In action research, students typically experience and test the whole research process, from formulating questions, collecting and analyzing data, to reflexively evaluating change: "*after completing this assignment, I can see that I am able to plan and carry out an action research study, to collect and analyze data to evaluate my own performance*" (24VT2); "*I can say that the process of "plan -> act -> get the result -> reflect -> act -> ... is becoming almost a daily routine and very necessary for me*" (21VT9); "*the format itself was a new thing, which made me think more before applying the innovation and analyze more afterwards*" (21VT2). Therefore, planning and conducting AR during the course of a study can be useful in consolidating the knowledge and skills needed to improve education and teaching practice in the future.

Aims for Carrying Out the AR

Typically, teachers carry out AR in their own settings to better understand the context of their work and improve their pedagogical practice. According to Somekh & Zeichner, (2009), another reason why teachers conduct this type of research is to create contextualized knowledge that is relevant to the teaching or school practice of more teachers. The facilitators of the teacher qualification programme have a further objective: to enable preservice teachers to carry out AR on their own and to foster an attitude towards basing changes in educational practice on research findings. The students of the programme already have the specific research competencies required for the field of the Bachelor's degree. In these teacher qualification studies, they are trained to conduct AR and to base educational change decisions on their findings (Bergmark, 2020). Although action research is not a controlled experiment and does not show statistically valid causal relationships, it extends the understanding of the phenomenon or situation under study through closer observation, listening, additional questioning, etc., so that normally invisible processes can be described, discussed in detail, and made visible. AR in the classroom follows the same methodological procedures as any other research - problem definition, planning, data collection, analysis, drawing conclusions, etc. However, action research recognizes that the research takes place in the classroom, the researcher is first and foremost the teacher, and the research cannot be allowed to take precedence over the students' learning. What is of interest to us, as programme implementers, is how preservice teachers view the outcome of their action research, the lessons learned, the benefits of this type of research and the change in skills.

The reflections also show signs of anxiety and doubt at the initial stage, i.e. just after being introduced to the assignment of AR. Later on, despite the initial hesitation, the reflective recordings show that the action research was deeply engaging and involving, and that the preservice teachers monitored not only the students' but their own progress as well. For example, "I thought that I did not have enough knowledge, competencies that could help me with this action research. *After defining the field of the change question, creating activities and planning their execution, I felt stronger, but I was afraid. I was afraid I would fail. <...> As the activities progressed, I began to see and feel not only the progress of the children, but also my own. I became convinced that the activities I proposed were effective, purposeful, enjoyable and engaging for the children, and that they gave feedback*" (22VT6). In the AR, preservice teachers identified themselves as researchers and understood the links between research and practice: "*This activity allowed me to put myself in the shoes of a researcher and to see the problem, <...> [in the future] if I find a situation where a particular activity is not going well, I will start asking questions and think about possible solutions*" (21VT5) or "*I think that action research is a great opportunity for every teacher to feel like a researcher in their own*" (22VT10). Preservice teachers recognize a change in their competence and can elaborate on the activities that are part of the competence of a researcher: "*Planning and carrying out the research has broadened my competence in carrying out research: I have* *been learning more about how data should be collected, processed, and analyzed in the classroom. It was the first action research I had done and it made me feel more confident, and I understood why action research is important to achieve better results and to promote students' development*" (21VT8). Another participant notes that "*I followed the processes and requirements of scientific inquiry in my action research. This activity improved my organizational skills (designing, arguing, organizing, implementing, etc.) and my thinking skills (analyzing the problem, the results of the research, selecting activities, etc.). I see the benefits of exercise as going beyond the improvement of my research skills. In particular, by implementing the research activities and observing the students, I had the opportunity to get to know them better (as individuals, as learners or as receivers of information)*" (22VT7). Perhaps the greatest value is the realization that AR is not an end in itself, that its process and results are directly related to the quality of education, to the testing of new strategies, to the teacher's professional development, e.g. "*I have come to realize that action research is a necessary and important part of the teacher's daily routine, contributing to the improvement of the quality of teaching*" (21VT2). A few more participants' records confirm that AR can be a tool for developing pedagogical competences: "*I liked to experience what it means to do action research and to get real results, to reflect on them, to evaluate them, to replan again, and to move on. And I hope for an even better result" (24VT12); "I learned about feedback instruments and their importance, and most importantly, I tried them out in practice. Thus, action research has contributed to my ability as a teacher not only to carry out action research, but also to the development of other pedagogical competencies" (23VT2); "It is an important tool that helps the teacher to observe and improve the content of the lesson or the dialogue with students. It is also a form of inquiry that is not "overloaded" with a variety of mandatory documents, allowing you to concentrate only on what is most interesting and necessary*" (21VT12). In the reports, we find entries (e.g., "After the action research, I can share my experience with others. *For a student whose writing is irregular, I can give tips and tasks to try to fix it*" (24VT5)), which reveal the experience and expertise, the concepts and terms mastered during the research, which gives confidence to participate in professional conversations and to share knowledge and experience with others.

Educational Solutions

Contextual conditions influence the AR, its purpose, participants, and organizational decisions (Somekh & Zeichner, 2009). We analyze ARs that take place in the study process, ARs conducted by preservice teachers under the guidance of university lecturers. This means that the context of these studies is unique, as participants are trained in action research methodology, are consulted, and their research questions and research designs are evaluated, with extensive feedback. In addition, they start their practice in schools by observing the lessons of experienced teachers, using different observation protocols focused on different aspects of the classroom, the lesson, and the educational solution. The requirements of the observation task also guide students in seeing situations that they would like to improve or change by trying a different approach or strategy. Lesson observation becomes an introduction to AR, as it then selects a real, often problematic situation relevant to a particular student, which the preservice teacher seeks to improve during the AR process, while also gaining knowledge from the experience (Feekery, 2023). The observation and design phases of AR focus preservice teachers' attention on clearly identifying the problem, the desired changes, and the possible outcomes in a specific context. These stages also ask to consider the ethical implications of preservice teachers' planned AR and how the results will affect the student participants in the study. Only after these stages do preservice teachers begin the research in the field - at school, where they also have the supervision and support of a school mentor. The context of the study - as the programme is designed to qualify as a teacher - also determines that the AR should be designed to change, improve, or test innovations in educational situations.

The analysis of the preservice teachers' AR reports shows that they recognize that they have broadened their understanding of their role as a teacher, their ways of thinking and acting: "this study has unexpectedly opened up for me a completely different perspective on my already familiar work routine" (21VT15).). Another participant wrote that "*This study showed me what I should change, what I could pay attention to. <...> Now I know for sure what I will do next year. In the summer I will think about the topics for [student projects]. It was difficult to come up with the first time, so I realized halfway through the work that some of the topics [for the projects] were difficult and I could have chosen something more detailed and specific" (21VT9).* Therefore, it can be seen that AR enables students to rethink their actions, to look for other possible ways of doing things, and to look for opportunities in their routines.

Preservice teachers also realized that they too were improving in their AR and their ability to manage the classroom and the educational process in the classroom. One participant stated that "*in terms of my own change, I really think that it has happened, even if it is slight: I have become more coherent in my teaching, I have felt calmer and more confident in the classroom, I have had a clear plan and very clear instructions that the students can see and know, and I have been able to agree with the children more often than not, when they have violated a rule and a consequence has been applied. <...> I feel that my relationship with the sixth formers has become a little stronger, more respectful. I am no longer afraid of them, I can calm myself before the lesson and start it calmly and firmly. So I think my classroom management skills have definitely improved more*" (21VT16). Another quote also notes a change in competence through AR: "I feel a significant change in my perception of students' learning and in my educational competence: *I aim to strengthen my own skills and those of the students by "traveling" through the prescribed steps; I try to get to know the students' capabilities, not to be average, to help them formulate their learning objectives at the beginning of the topic, to help them to monitor and analyze how they are achieving them, to talk about their personal progress" (21VT13).*). It can be seen that the teacher has become more alert, more empathetic, and more responsive to the students and their needs, thus improving her teaching.

As part of the AR, preservice teachers tested new teaching methods and techniques such as "*I learned about feedback tools and their importance and, most importantly, I tried them out in practice. <...> Action research can be a good tool for a teacher to try out new methods in the teaching/learning process*" (23VT2).

Another quote highlights the teacher's efforts to find alternative ways of learning to the previous methods using digital tools: "*I had to come up with a way of changing the way I 'hammered' information into a more effective, more accessible, more comprehensible, and more digitized way for the students. I thought for a long time whether this was even possible... While researching ways to facilitate learning, I came across an interesting programme on the Internet called Anki. So I had to master it and improve my digital competence*" (21VT5). The preservice teachers also saw AR as a good way to look into the reasons behind the success of the learning - the method or other factors, e.g. "*when I noticed more effective methods, I would like to check them out - if they are really that effective, because it raises such questions - if it was not a coincidence, if maybe the students were generally more talkative today*" (23VT4). The AR allowed one to experience the impact of the chosen and specifically applied methods, e.g. "*the children's engagement in reading has increased(I see that showing the film, the illustrations, I get more engagement in reading and discussing the text). After this study, I realized that I will have to use/try at least a few more methods to activate children's reading" (21VT8).* This quote reflects the interplay between teacher thinking and action, the introduction of the chosen new methods improved the teaching and learning situation, which was recognized by the teacher, and this became an incentive to continue the search for methods and the exploration of their impact.

During the AR, preservice teachers explored the reasons that contribute to learning failure and looked for strategies to improve the situation, e.g. "*This study I conducted was aimed at finding out the reasons why pupils do not read or do not fully engage with the assignments. This could be ignored if it did not affect students' grades. It is very unwilling to give a lower grade just because a student does not read the [assignment] conditions, even though he/she knows and understands the subject itself. As a teacher, it is not my aim to write a bad grade. It is much more important to me that the student understands [the assignment] and does the right thing*" (23VT9). Another quote illustrates the teacher's efforts to teach pupils to talk about their feelings, which would help them to recognize and manage situations that can undermine learning success, e.g. "*When comparing the children's openness and motivation to talk about their feelings in the first days of the activity study and in the last days of the activity study, there was a clear difference - while at the beginning there was a lot of "I don't know", at the end of the activity study the students were quickly finding the connections between colors, animals or plants in their feeling*" (21VT3).).

Combination of Theory and Practice

Another important finding of this study is that preservice teachers are looking for links between practice and theory. The programme is professional, with half of the time devoted to professional practice, so there is no requirement in the AR objectives to ground the intervention in theoretical constructs. AR facilitates the alignment of theory and practice because it seeks to test the theoretical issues at stake in practice. On the other hand, AR opens up a space for the creation of new knowledge, perhaps very localized and context-specific, but meaningful for the problems of that context (Feekery, 2023). AR bridges the gap between theory and practice, increasing the teacher's confidence to take on new and untried activities, e.g. "*It is the research that obliges us to find the time and motivation to analyze theory and research and to try to apply it to our own practice. I think that this willingness to look for scientific publications and to analyze them is already a positive development for me as a teacher. I knew in theory that photographs and pictures are important, but after this research they seem more important than before. To extend and complement this study, more time should be spent on theory and more in-depth research into the different ways of memorizing words and their application" (17VT3).* Other quotations identify openness to new methodologies and the study as a condition for testing them in practice: "This study has helped me to better understand the importance of active teaching methods and their impact on students' concentration and general well-being. *I became more open to new methodologies and their application in teaching. I gained valuable experience of how to put theory into practice, which helped me to develop my professional competences*" (24VT7); "*This assignment was particularly valuable for me because it allowed me to put theoretical knowledge and skills into practice in a realistic environment and under realistic conditions*" (24VT10). The experiences of such participants reveal that action research can increase the motivation and engagement of preservice teachers in learning theoretical modules (e.g., developmental and learning psychology, didactics, curriculum development etc.). AR also creates the conditions to delve into theoretical constructs that are studied in parallel, to rethink them, and to relate them to practical educational solutions. AR can enable preservice teachers to read teacher publications that present research findings in a more targeted and professional way, in a simpler language.

Collaboration

Conducting AR is intrinsically related to collaboration (Bergmark, 2020; Feekery, 2023). The analyzed experiences of the preservice teachers reveal that the need for collaboration in action research emerges at very different stages of the AR process and also targets different members of the school community and can influence different processes and individuals.

The need for collaboration that emerged from AR is also inherent in teacher professional development and learning (Bergmark, 2020). The participants' experiences reveal that the AR had a clear positive impact on the development of collaborative competence: "*I also think that I have deepened my professional collaborative competence, as I had a lot of contact with the practice mentor and with the teachers of the different subjects in whose lessons I carried out the activities, both in the planning and implementation of the activities*" (23VT6), and also enabled to make contact and build relationships with other members of the school community and mentors: "*The cooperation with the school administration and with the school professionals, the psychologist, the social worker has intensified a lot in solving the girl's problems.I am very happy with their support and help.*" (23VT10); "*Communication and cooperation with the mentor are also very important. This assignment gave me the opportunity to consult, question, discuss, and cooperate.*" (21VT16). As can be seen, these examples reveal the importance of elements such as thinking, acting, and relating to others (Bergmark, 2020) and the qualitative change during the action research. Constructive and timely engagement with the various parties involved in or likely to benefit from the AR also contributes to the success of the AR itself: "*During the planning of this work, I have learned the importance of collaboration in the planning of the research. It is important to collaborate with all the actors involved in the research, in my case the teacher and the student being researched. It is important to find a way to conduct research that works for everyone so that the change actually happens. It is also very important to take into account the needs of the student*" (21VT7). Therefore, the benefits of communication and collaboration in the context of action research can manifest themselves in a variety of ways.

The study participants' experiences also reveal that they see the importance of collaboration not only during the AR but also afterward. Reflections reveal that preservice teachers have presented what they have learned through action research to others in a variety of ways, including informal presentations in their schools, oral presentations at conferences, and written research reports. This sharing of good practice and ideas can be linked to a desire to continue to improve their practice, e.g., "*It would be good to read more professional literature and talk more with colleagues next time to make sure that the plan and the planned results are in line with the final result*" (21VT4), "*I am also more open to new ideas and willing to collaborate with colleagues*" (24VT6), and to pass on the experience to colleagues, e.g., "*The action research I have done is of interest to other foreign language teachers at the gymnasium. They invited me to a meeting of the methodological group to present the research, the results and the conclusions and to discuss my methods and their effectiveness*" (17VT3). Such preservice teachers expectations and reflections inspire the possibility of action research involving the wider school community.

Critical Reflection

AR in its form encourages and guides reflection on actions, lived experiences, critical analysis of the impact and consequences of actions, and the creation of new actions through reflection. The AR methodology itself contains a tool to help researchers, in this case preservice teachers, to critically reflect on the consequences of their interventions, to better understand their actions and thinking patterns, and to develop theories based on reflection that can inform new practical actions (Feekery, 2023). Preservice teachers kept a reflective journal during the AR, the entries of which formed the basis for the AR report. One part of the report requires a reflexive evaluation of the impact and the result obtained, and asks for a comparison of the result obtained with the expected and planned result. In the case of success, preservice teachers are asked to reflect on whether this is really an achievement due to the impact made or whether other factors are at work. If the AR intervention is not successful, a reflexive reflection is asked about where and what mistakes might have been made and what changes should be made in the action plan when starting a new cycle.

Preservice teachers must consistently write in a reflective journal from the first days of their studies. The students in each cohort see reflective writing as a difficult task at first, but by the time they start to do AR, they are already experienced, so the reflective writing itself is not a problem for them, contrary to what is mentioned in Bergmark's (2000) study. In the reflection section of AR reports, the value of reflection is more often mentioned, e.g. "these reflections help to clearly identify successes and challenges, as well as to identify actions that could improve future activities and personal professional growth" (24VT10); "even the "failures" provided useful information, writing just makes sure that the reflection is meaningful, as new things are revealed" (23VT4); "the reflection helped to summarize the events of the day, the successes and failures, and what should be changed for the next meeting" (23VT5); "the competences of observation and reflection have been very much activated, and the mentor and I have been able to analyze who is reacting, the change we see, and what we should do better" (22VT4).). As can be seen from the quotations, the secondary reflection that the preservice teachers had to carry out already when preparing the AR report encouraged them to reflect more deeply on the intervention and the teaching practice, to look deeper into the problem situation, and to look for ways to improve their teaching practice.

FUTURE RESEARCH DIRECTIONS

AR is an effective and meaningful way to develop teachers' professional growth and competence, as well as their analytical skills and self-awareness. AR projects, when carried out by a larger team, encourage and support collaboration, peer-to-peer discussions about education, its improvement, the value of new solutions or vice versa (Horbačauskienė et al., 2024). According to Khasinah (2013), Kemmis et al. (2014), AR can be a space of dialogue that connects the teacher, the classroom, and the academic researcher, in which cocreation of new theoretical constructs and procedural knowledge needed to improve educational practice occurs.

This study focuses on the analysis of AR as part of a teacher education programme and the added value it creates. The results showed that AR is an effective tool for teacher professional development to consolidate students' theoretical knowledge into practice. This study focused on the situation of the preservice teachers, but in the future it would make sense to dive much deeper into the context and the school community as potential participants in AR. It might be worth to focus this type of future research on the following questions: whether the community is supportive of the idea of AR and the purpose of the study; whether they are willing to collaborate, to engage in the study together, or to become responsible observers of the study; and whether they would be receptive to AR as an alternative tool for teachers' professional development to the traditional professional development courses. Another field of inquiry could also be the leadership of the school leader to promote and initiate AR in his/her institution. In the case of the study presented in this chapter, the students had a university lecturer who provided advice and was a discussion partner in the planning of the student's AR. However, in the case of in-service teacher trainees, it is very likely that support is also needed to initiate and conduct AR. In this case, it is natural for the school leaders to take the support initiative.

CONCLUSION

AR in school and classroom can be carried out for different reasons: to solve problems that have emerged in certain practices or situations, to improve certain existing circumstances, or to introduce additional, previously untried/untested teaching methods or strategies into an ongoing system. In all these situations, AR provides methodological access, increased vigilance, and responsibility of the researcher(s) for both the process and the outcome. In order to achieve quality results, constructive and planned AR is a preferable alternative to subjective, hasty testing of new ways of doing things or spontaneous problem solving. By examining the impact of their own teaching on the educational process and its participants, teachers have the opportunity to assess the strengths and weaknesses of their own teaching, to know the potential of their students, etc., and, therefore, their findings have direct practical implications for further teaching decisions. AR can be seen as a complex situation of gaining new experiences and developing different competences.

This study has shown the importance of integrating AR into the teacher education curriculum as a synergistic tool for the development of the competences of preservice teachers, for the establishment of links between theory and practice, and for the consolidation of the researcher's approach. In order to empower teachers to improve their pedagogical practice, to initiate the necessary changes in the educational process, to introduce new educational methods, and to create new environments, it is recommended that AR be applied in their activities. This method is an acceptable and appropriate tool, as it enables teachers to analyze, improve, and change their practice, and allows them to explore their work processes in order to find the most appropriate solutions.

ACKNOWLEDGEMENT

This research received no specific grant from any funding agency in the public, commercial, or not-for-profit sectors.

REFERENCES

Azizan, N., Smith, R., Cooper, V., & Abu Bakar, N. (2018). Qualitative Content Analysis: A Case Study of a Malaysian Government Education Website. *Malaysian Journal of Science Health & Technology, 1*(1). Advance online publication. DOI: 10.33102/mjosht.v1i1.14

Baan, J., Gaikhorst, L., & Volman, M. (2019). Stimulating teachers' inquiring attitude in academic and professional teacher education programmes. *European Journal of Teacher Education, 43*(3), 352–367. DOI: 10.1080/02619768.2019.1693994

Bergmark, U. (2020). Teachers' professional learning when building a research-based education: Context-specific, collaborative and teacher-driven professional development. *Professional Development in Education, 49*(2), 210–224. DOI: 10.1080/19415257.2020.1827011

Caspersen, J., & Smeby, J.-C. (2023). Research-based teacher education in Norway – a longitudinal perspective. *International Journal of Educational Research, 119*, 102177. Advance online publication. DOI: 10.1016/j.ijer.2023.102177

Čepienė, A., & Vaitiekienė, N. (2019). Studentų profesinės praktikos organizavimo proceso kolegijoje analizė. *Mokslo Taikomieji Tyrimai Lietuvos Kolegijose*, (14), 53–62. Retrieved from https://ojs.kaunokolegija.lt/index.php/mttlk/article/view/256

Cochran-Smith, M., Barnatt, J., Friedman, A., & Pine, G. (2009). Inquiry on Inquiry: Practitioner Research and Student Learning. *Action in Teacher Education, 31*(2), 17–32. DOI: 10.1080/01626620.2009.10463515

Crawford-Garrett, K., Anderson, S., Grayson, A., & Suter, C. (2015). Transformational practice: Critical teacher research in pre-service teacher education. *Educational Action Research, 23*(4), 479–496. DOI: 10.1080/09650792.2015.1019902

Creswell, J. W. (2012). *Educational Research: Planning, Conducting, and Evaluating Quantitative and Qualitative Research* (4th ed.). Pearson.

Dana, N. F., & Yendol-Hoppey, D. (2019). *The Reflective Educator's Guide to Classroom Research: Learning to Teach and Teaching to Learn Through Practitioner Inquiry*. Corwin.

Darwin, S., & Barahona, M. (2019). Can an outsider become an insider? Analysing the effect of action research in initial EFL teacher education programs. *Educational Action Research, 27*(5), 709–725. DOI: 10.1080/09650792.2018.1494616

Elliott, J. (2009). Building educational theory through action research. In *The SAGE Handbook of Educational Action Research* (pp. 28–38). SAGE Publications Ltd., DOI: 10.4135/9780857021021

Feekery, A. (2023). The 7 C's framework for participatory action research: Inducting novice participant-researchers. *Educational Action Research, 32*(3), 332–347. DOI: 10.1080/09650792.2023.2234417

Guskey, T. R. (2002). Professional Development and Teacher Change. *Teachers and Teaching, 8*(3), 381–391. DOI: 10.1080/135406002100000512

Horbačauskienė, J., Ratkevičienė, M., Kasperė, R., Staniškienė, E., Simonaitienė, B., Stankevičiūtė, Ž., & Ramanauskaitė, J. (2024). How to enhance EdTech integration in secondary education: teachers' and managers' experiences // *Navigating teaching in an era of social distancing and global conflict* / Shalin Hai-Jew (ed.). Hershey, PA: IGI Global. ISBN 9798369333822. eISBN 9798369333839. p. 127-154. DOI: DOI: 10.4018/979-8-3693-3382-2.ch006

Jakavonytė-Staškuvienė, D., Jašinauskas, L., & Kudinovienė, J. (2023). An analysis of prospective primary school teachers' action research: The case of Lithuania. *Educational Action Research*, •••, 1–16. DOI: 10.1080/09650792.2023.2298744

Kemmis, S. (2009). Action research as a practice-based practice. *Educational Action Research, 17*(3), 463–474. DOI: 10.1080/09650790903093284

Kemmis, S., McTaggart, R., & Nixon, R. (2014). *The action research planner: Doing critical participatory action research.* Springer. DOI: 10.1007/978-981-4560-67-2

Khasinah, S. (2013). Clasroom action research. *Jurnal Pionir, 1*(1), 107–114.

Koshy, V. (2010). Action research for improving educational practice: A step-by-step guide. *Sage (Atlanta, Ga.).*

Lattimer, H. (2012). Action Research in Pre-service Teacher Education: Is There Value Added*? i.e.Inquiry in Education, 3*(1), 5.

Matos, J. F., Piedade, J., Freitas, A., Pedro, N., Dorotea, N., Pedro, A., & Galego, C. (2023). Teaching and Learning Research Methodologies in Education: A Systematic Literature Review. *Education Sciences, 13*(2), 173. DOI: 10.3390/educsci13020173

Mertler, C. (2024). *Action Research: Improving Schools and Empowering Educators.* SAGE Publications, Inc.

Phillips, D. K., & Carr, K. (2010). *Becoming a Teacher through Action Research: Process, Context, and Self-Study* (2nd ed.). Routledge., DOI: 10.4324/9780203861776

Pine, G. J. (2009). *Teacher Action Research: Building Knowledge Democracies.* SAGE Publications. DOI: 10.4135/9781452275079

Puustinen, M., Säntti, J., Koski, A., & Tammi, T. (2018). Teaching: A practical or research-based profession? Teacher candidates' approaches to research-based teacher education. *Teaching and Teacher Education, 74*, 170–179. DOI: 10.1016/j.tate.2018.05.004

Reinhardt, K. S., Robertson, P. M., & Johnson, R. D. (2021). Connecting inquiry and Universal Design for Learning (UDL) to teacher candidates' emerging practice: Development of a signature pedagogy. *Educational Action Research, 31*(3), 437–454. DOI: 10.1080/09650792.2021.1978303

Schmuck, R. A. (2006). *Practical Action Research for Change*. Corwin Press, Inc.

Seale, C. (2012). Researching Society and Culture. *Sage (Atlanta, Ga.)*.

Shava, G. N., Hleza, S., Tlou, F., Shonhiwa, S., & Mathonsi, E. (2021). Qualitative Content Analysis, Utility, Usability and Processes in Educational Research. *International Journal of Research and Innovation in Social Science* [IJRISS]. *International Journal of Research and Innovation in Social Science, 5*(07), 553–558.

Smith, K., & Flores, M. A. (2019). The Janus faced teacher educator. *European Journal of Teacher Education, 42*(4), 433–446. DOI: 10.1080/02619768.2019.1646242

Somekh, B., & Zeichner, K. (2009). Action research for educational reform: Remodelling action research theories and practices in local contexts. *Educational Action Research, 17*(1), 5–21. DOI: 10.1080/09650790802667402

Spencer Clark, J., Porath, S., Thiele, J., & Jobe, M. (2020). *Action Research*. NPP eBooks. 34. Retrieved from https://newprairiepress.org/ebooks/34

Ulvik, M. (2014). Student-teachers doing action research in their practicum: Why and how? *Educational Action Research, 22*(4), 518–533. DOI: 10.1080/09650792.2014.918901

Ulvik, M., Riese, H., & Roness, D. (2017). Action research – connecting practice and theory. *Educational Action Research, 26*(2), 273–287. DOI: 10.1080/09650792.2017.1323657

Volk, K. S. (2010). Action research as a sustainable endeavor for teachers: Does initial training lead to further action? *Action Research, 8*(3), 315–332. DOI: 10.1177/1476750309351358

Yosief, A., Sulieman, M. S., & Biede, T. (2024). Improving the practices of teacher educators through collaborative action research: Challenges and hopes. *Educational Action Research, 32*(2), 204–221. DOI: 10.1080/09650792.2022.2066147

KEY TERMS AND DEFINITIONS

Action Research: Action research is a structured and contemplative exploration conducted by people or groups, such as educators, community leaders, or organizational members, to tackle practical issues, refine practices, or boost comprehension within a specific environment. It merges theoretical concepts with practical action, as participants actively partake in research to implement and evaluate significant changes in real-world situations. The aim of action research is to resolve issues, create new insights, empower those involved, and encourage a culture of ongoing learning and development.

Pre-service Teacher: Pre-service teachers generally participate in teacher education programs at universities or alternative certification paths, where they study educational theories, specific subject matter, classroom management techniques, and other vital teaching skills.

Professional Growth: It refers to the continuous process of improving a teacher's knowledge, skills, and effectiveness in the classroom. It involves ongoing learning, self-reflection, and professional development activities such as attending workshops, pursuing advanced degrees, engaging in peer collaboration, and adopting new teaching strategies. Professional growth helps teachers stay updated with educational trends, improve student learning outcomes, and enhance their career progression.

Research Competence: In teacher education it refers to a teacher's ability to systematically investigate educational issues, apply research methods, and use evidence-based findings to enhance teaching and learning. It includes skills such as critical thinking, data collection and analysis, academic writing, and the application of research outcomes to classroom practice. Developing research competence enables teachers to make informed decisions, improve instructional strategies, and contribute to the advancement of educational knowledge.

Secondary Education: Secondary education is the phase of schooling that comes after primary or elementary education and before higher or tertiary education, generally catering to students from around ages 12 to 18. Its goal is to provide students with the knowledge, skills, and qualifications required to continue to higher education or to join the workforce.

Teacher Education: It refers to the formal training and preparation that individuals undergo to become qualified teachers. It includes both theoretical knowledge and practical skills necessary for effective teaching. Teacher education programs typically cover subjects such as pedagogy, classroom management, educational psychology, subject-specific instruction, and assessment methods. These programs can be offered at colleges, universities, and teacher training institutions, leading to certifications, diplomas, or degrees required for teaching in schools.

Teaching Practice: Itrefers to the hands-on experience that aspiring teachers gain while applying pedagogical theories and methods in real classroom settings. It is a crucial component of teacher education programs, allowing trainees to develop and refine their teaching skills under the guidance of experienced mentors. Teaching practice includes lesson planning, classroom management, student engagement, assessment of learning, and adapting instructional strategies to meet diverse learner needs. This practical training helps future teachers build confidence and competence before entering the profession full-time.

Chapter 2
African Wisdom Unveiled Through Integrating African Knowledge Using Action Learning and Research

Joseph Nkosana Mnguni
https://orcid.org/0000-0001-7713-1875
Motshegofadiwa Primary School, South Africa

Tomé Awshar Mapotse
https://orcid.org/0000-0002-7234-2785
University of South Africa, South Africa

ABSTRACT

The transformative potential of integrating African Indigenous Knowledge Systems (AIKS) into contemporary research practices. It highlights the limitations of traditional text-based methods and argues for the value of Action Research and Learning (AR/AL) methodologies as a framework for fostering collaboration between researchers and African communities. The paper examines how AR/AL and technology can bridge the gap between embodied knowledge and digital representation, fostering knowledge co-creation and empowering African communities. Ethical considerations are highlighted, ensuring cultural sensitivity and respect for intellectual property rights. The paper concludes by discussing the broader impact of AR/AL on African communities, including fostering knowledge sharing, building research skills, and advocating for policy change. This approach empowers African communities to become stewards of their sustainable development.

INTRODUCTION

African communities have gained a lot of knowledge about nature and human societies over many years. This knowledge, known as the "African Indigenous Knowledge System" (AIKS), includes traditions, practices, and ways of thinking that have been shared from one generation to the next. In most cases, circumstances and resources have disappeared but it is "incompetence-related knowledge and skills"

DOI: 10.4018/979-8-3693-3542-0.ch002

(Ihejirika 2024). Limited research coverage has left a wide area uncultivated, but a broad approach has recognised that something useful can be obtained from AIKS in managing contemporary challenges.

Hence, the chapter views action research (AR) and action learning (AL) as practically blending with AIKS's presence in ongoing systematic inquiry (Mnguni & Mapotse 2023; Zuber-Skerritt, 2015: Zuber-Skerritt 2002; Watkins, Marsick, & Wasserman 2019). Through Action Research and Action Learning (AR/AL), research approaches can be developed where people would work together, and this would carry forward structured approaches into their cultural context. It will facilitate learning and building knowledge, though the most welcomed outcome would further the potential for creating synergies that can produce innovations other than that of driving ideas in a positive direction (Bezemer & Cowan 2022). Such an approach would bring together researchers and African communities in seeking solutions to complex issues and possibly an amicable means of resolving such issues (Mnguni, 2013; Shrivastava, Smith, O'Brien & Zsolnai, 2020).

Key challenges in combining AR/AL and AIKS will also be explored to see how this partnership can address important topics like ethics, sustainability, and giving African communities more power (Mnguni & Mapotse, 2024). From examples in all fields, we will show the transformative potential that AR/AL has in driving positive change and advocating social justice through diverse African contexts. Finally, the chapter seeks to pave the road for a future in which research becomes inclusive, is culturally sensitive, and able to produce meaningful change in African communities (Zuber-Skerritt, 2015 & 2021).

UNVEILING THE RICH LEGACY OF AFRICAN INDIGENOUS KNOWLEDGE SYSTEMS (AIKS)

This heritage is shown in AIKS, which are detailed systems that include a wide range of traditions, practices, and ideas collected over thousands of years and handed down through generations. Africa is a large and varied continent with a rich and important history of knowledge (Mnguni & Mapotse, 2024; Yishak & Gumbo, 2015). This history is clear in AIKS, which are complex systems made up of traditions, practices, and ideas gathered over many centuries. AIKS represents a unique body of knowledge that is often ignored by mainstream research (Ihejirika, 2024). This subheading focuses on the precise content and essence of AIKS, revealing its depth and diverse character. By demonstrating all the sectors covered by AIKS, we can begin to grasp its enormous worth and establish the framework for its fruitful integration with modern troubleshooting approaches (Mnguni, 2013). Across the vast and diverse African continent lies a treasure trove of wisdom accumulated over millennia: AIKS. These intricate systems encompass a rich tapestry of traditions, practices, and insights passed down through generations. AIKS encompasses various domains, including:

- Ecological Knowledge: Deep understanding of local flora and fauna, sustainable resource management practices, and traditional methods of agriculture and animal husbandry.
- Medicine and Healthcare: Extensive knowledge of medicinal plants, herbal remedies, and traditional healing practices.
- Social Organisation and Governance: Complex social structures, conflict resolution mechanisms, and customary laws that foster community cohesion and well-being.
- Cosmology and Spirituality: Profound belief systems that connect humanity to the natural world, offering guidance for ethical living and environmental stewardship.

The richness of AIKS lies not only in the broad context in which it is placed but also in its mechanism, which has been integrated with physical, spiritual, and social dimensions of human existence for a symbiotic coexistence of people with the environment. (Matsekoleng & Mapotse, 2023) Unfortunately, such a rich reservoir of knowledge that AIKS embodies has often been undermined or relegated to the fringes by mainstream research and development paradigms (Mnguni, 2013). Such marginalisation resulted in disconnections between AIKS and WIKS approaches toward problem-solving (Mnguni & Mapotse, 2023).

AIKS assumes greater importance in the context of the contemporary needs of African communities beyond being of historical interest (Smith, 2012). The system is complex, embodying ecological knowledge combined with an understanding of biodiversity at the local level and manageable resource use, agriculture, and animal husbandry (Bele, Nielsen, Fairclough & Herring, 2024). Similarly, the large knowledge base concerning medicinal plants and herbal remedies embedded in AIKS is a rich potential reservoir of solutions for, and supplement to, modern medicine, especially where conventional healthcare services are out of reach for communities. The social structures, mechanisms of conflict resolution, and customary laws underlying AIKS concern social organisation and governance that have contributed toward sustained community cohesion and well-being. Finally, the deep belief systems of AIKS, such as cosmology and spirituality, link humanity to the natural world and provide guidelines for ethical living and environmental care (Mnguni & Mapotse, 2024).

The rich AIKS base related to medicine and healthcare in AIKS includes some medicinal plants and herbal remedies. Such a knowledge system has immense potential in providing solutions and supplementing modern medicine, especially when conventional healthcare facilities are not adequately available. Eventually, such integration of traditional practices will contribute toward more robust and accessible healthcare systems for local communities in Africa. The fact that AIKS embodies social structures, mechanisms of conflict resolution, and customary laws has provided a corner in cohesion and well-being for the community for centuries (Matsekoleng, Mapotse & Gumbo, 2024). Western indigenous knowledge systems (WIKS) can borrow useful frameworks concerning social organisation and governance that assist in the propagation of peace, equity, and social justice within African society during modern days.

The profound belief systems of AIKS (cosmology and spirituality) connect humanity to the natural world, offering guidance for ethical living and environmental stewardship (Mnguni & Mapotse, 2024). In an era where environmental degradation is a pressing concern, these belief systems can serve as a powerful foundation for fostering a more sustainable relationship between humans and the environment. This has unveiled the rich heritage of AIKS, which could solve some of the contemporary problems. It opens the door toward a more sustainable and equitable future in Africa. The subsequent sections discuss how AR/AL methodologies can be put into place to bridge the gaps that exist between AIKS and modern scientific knowledge to ensure such wisdom effectively becomes a driver for change.

Specific uses of AR/AL methodologies in harnessing AIKS for the empowerment of African communities and addressing their challenges are discussed in the sections that follow. The quest for sustainable development in Africa requires new approaches that respond peculiarly to the challenges facing the African continent. From the top-down planning of development in the past to varied fortunes, people have now woken up to the realization that local communities should be in control and actively decide on their futures. Here, AR/AL methodologies have proven to be a strong driving force for change.

First and foremost, AR/AL go beyond traditional paradigms of research in that they foster collaboration, reflection, and continuous improvement (White & Cooper, 2022). Through these methodologies, African communities can leverage their rich cultural heritage and indigenous knowledge systems to address contemporary challenges and create a more sustainable future (Mnguni & Mapotse, 2024).

THE EVOLVING LANDSCAPE OF AR/AL

What is important in the search for sustainable development in AIKS is finding creative solutions that tackle many of the continent's challenges. Moyo (2009) observes that traditional top-down approaches to development have had mixed success and that there is a need for new ways of thinking. More focus is now being placed on local communities playing an active role in shaping their futures. This is where AR/AL methods come in as strong tools for creating change (Watkins, Marsick & Wasserman, 2019).

The AR/AL are the two collaborative ways through which change and improvement in any complex situation can be brought about. Zuber-Skerritt (2021) defined AR as a critical approach that combines theory with practice, hence allowing people to collaboratively analyse and solve their problems within their environment. Much emphasis is given to this methodology regarding experiential learning and systemic thinking; thus, AR is positioned as a dynamic process of problem-solving and innovation. With its cyclical nature of identifying issues, implementing solutions, and evaluating outcomes, AR keeps the research contextually relevant and empowering for the participants.

On the other hand, Raelin (2006) defines AL as an approach that focuses on organizational members' collective learning as they go about their activities. The approach creates an open dialogue and calls upon the participants to publicly question the underpinning assumptions of the practice, bringing about real curiosity and transparency. The AL approach empowers individuals and teams to take responsibility for their learning and development; this promotes shared efforts to solve challenging problems. Taken together, AR/AL provide robust frameworks within which transformative learning and effective problem-solving could be harnessed, most especially in contexts such as Africa, where integrating AIKS into contemporary research methodologies could yield culturally appropriate and sustainable solutions.

In the ever-changing landscape of research methodologies, AR/AL stand out for its collaborative and action-oriented approach (Watkins, Marsick & Wasserman, 2019). These dynamic tools transcend traditional research paradigms by emphasising participation, reflection, and continuous improvement. AR/AL are not static concepts with a singular origin story (Almond, 2021). They emerged organically from a confluence of ideas within critical social science throughout the 20th century. Early influences on AR include philosophers like John Dewey, who emphasised learning through experience (Simpson & Jackson, 2021), and educational reformer Kurt Lewin, who established the iterative cycles that remain central to AR today (Simpson & Jackson, 2021). Mid-20th century scholars such as Kurt Lewin and Wilfred Bion further developed AR by emphasizing the role of critical reflection and collaboration (Bion, 1961). This was also a time when Participatory Action Research, led by Paulo Freire, emerged to focus on empowering participants in research studies (Lewin, 1946: Mnguni & Mapotse, 2023; Holst, 2020). This paper shows that AR in the late 20th and the 21st century is continuous, with contributions from feminist scholars like Sandra Harding and Patricia Maguire, critical race theorists like Gloria Ladson-Billings, and Indigenous researchers like Linda Tuhiwai Smith, who have provided a variety of theoretical frameworks for understanding AR (Harding, 1987; Maguire, 1987; Ladson-Billings, 2021; Smith, 2012). Other renowned scholars in the African context include Lyotse Mbomba and Tome Mapotse, who have called for decolonising methodologies in AR as a way of ensuring that the research practices are culturally appropriate, empowering, and respond to particular needs and knowledge systems of African communities (Mbomba 2007; Mapotse 2018 & 2023; Mnguni & Mapotse 2024).

AR fosters a cyclical process where researchers and participants work together to identify problems, implement solutions, and evaluate their effectiveness. This collaborative approach ensures the research is not only relevant to the specific context but also empowers participants to take ownership of the solutions developed. AR has its roots in critical social science, with scholars like John Dewey, Kurt Lewin and Tome Mapotse laying the groundwork for its iterative and participatory nature.

AL focuses on collaborative problem-solving within a team or organizational setting. Through facilitated discussions and shared experiences, participants learn from each other and refine their approaches to address real-world challenges. Revans is credited with pioneering AL in the mid-20th century, emphasising the importance of shared experiences and collective reflection within teams (Revans, 1982). Like AR/AL has evolved beyond its initial focus on business education, finding applications in diverse fields like education and community development.

The synergy between AR/AL lies in their shared emphasis on action and reflection. Both methodologies move beyond simply identifying problems, they equip participants with the tools and knowledge to act and create positive change. This is particularly relevant in the context of Africa, where communities possess a rich legacy of AIKS. By integrating AIKS with modern research approaches facilitated by AR/AL, communities gain the power to address contemporary challenges and create a more sustainable future that is culturally appropriate and ecologically sound (Mnguni & Mapotse, 2024). The following sections will delve deeper into the specific applications of AR/AL methodologies in harnessing the power of AIKS, empowering African communities, and addressing their unique challenges.

Building upon the rich legacy of AIKS, this chapter explores the transformative potential of AR/AL methodologies for integrating AIKS wisdom into contemporary research practices. While AIKS offers a deep understanding of the natural world and human societies within the African context, it has often been marginalized by mainstream research. AR/AL approaches bridge this gap by fostering collaborative and culturally sensitive research partnerships specifically with African communities.

Moving beyond knowledge sharing, AR/AL creates a space for co-creation, where researchers and African communities work together to address complex challenges. Ethical considerations are paramount, ensuring African communities retain ownership and control over their knowledge through established protocols. This chapter showcases the practical applications and impact of AR/AL through real-world examples, such as co-managing natural resources by combining AIKS ecological knowledge with scientific methods. Furthermore, AR/AL empowers African communities by integrating their wisdom into research and equipping them with skills for active participation. Ultimately, AR/AL-guided research fosters a path towards inclusive and culturally sensitive research practices that honour Indigenous traditions and pave the way for positive change alongside these communities. This approach strengthens research by incorporating diverse perspectives and empowers African communities through collaboration and knowledge sharing.

The following sections of the chapter will delve deeper into the theoretical foundations of AR/AL, explore ethical considerations, and showcase practical applications across various fields, solidifying the transformative potential of this approach for integrating AIKS into contemporary research, specifically within the context of Africa.

UNVEILING THE TRANSFORMATIVE POTENTIAL OF AR/AL

The chapter explores the opportunities of AR/AL methodologies in integrating the AIKS into current research practice. According to Mnguni (2013), although AIKS bears high value in understanding the natural and human world, it has been excluded from mainstream research; therefore, the AR/AL approach closes the gap by bringing collaborative and culturally sensitive partnerships not only to respect Indigenous knowledge but also elevate it within the academy.

AIKS is a huge collection of knowledge that has been gathered over many years by native groups (Mnguni & Mapotse, 2024). It includes deep knowledge about how nature works, how to use resources in a way that doesn't harm the environment, and the traditions and ways of life that influence how people live together. Unfortunately, this knowledge has frequently been overlooked by traditional research methods, leading to a disconnect between valuable indigenous insights and modern research practices (Mnguni & Mapotse, 2023). It does not only deplete AIKS of this dimension but also upholds the one-dimensionality of knowledge, unable to represent African realities in all their complexities.

The gap in these critical issues could well be overcome through transformative approaches using AR/AL methodologies. More than acquiring knowledge of the communities in Africa, as indicated by Castro (2017), they go further toward collaborative and culturally sensitive research partnerships. In this framework, AR/AL, researchers, and African communities are co-creators who jointly define research questions, methodologies, and solutions to address complex challenges. A collaborative process such as this encourages mutual learning and respect for diversity in knowledge systems, therefore enriching the research landscape (Mnguni & Mapotse, 2023).

The whole process of research must be guided by ethical considerations. Informed consent and the protection of intellectual property are established conventions ensuring that African communities retain ownership and control over their knowledge (Mnguni & Mapotse, 2023). The researcher should not assume the role of facilitator or co-learner, as a matter of respect for indigenous traditions of research, hence allowing for equitable partnerships. This ethical approach is very important for creating trust and making sure that the benefits from research are shared fairly among everyone involved.

The power of AR/AL to bring change is not just an idea; it also shows real-life examples of how these methods are used and the difference they can make. AR/AL can be applied to the co-management of natural resources by integrating AIKS ecological knowledge with scientific methods. The result is more sustainable and culturally appropriate resource management strategies that prove that such a fusion between Indigenous and scientific knowledge will indeed yield innovative solutions to pressing environmental challenges (Mnguni & Mapotse, 2023).

AR/AL empowers not only the researchers but also African communities by making them integral in the research process, giving them research skills that make them active participants in knowledge sharing. This empowers research through diversity in perspective, opening ways for positive change to take place with African communities. AR/AL research begets a way leading to inclusivity, culturally sensitive research that upholds African tradition and ensures the voices of the Indigenous communities are not only heard but also valued within the wider academic discourse. According to Mnguni and Mapotse (2023), AR/AL has emerged as the silver lining for a fair and respectful research frontier since the unparalleled contribution of AIKS could only be acknowledged in such a revolutionary manner.

Transformative Power of African Indigenous Knowledge Systems (AIKS)

The transformative power of AIKS is great, and the potential they possess for resolving some of the contemporary challenges besetting African communities cannot be overemphasised. Centuries of observation, adaptation, and cultural practices inculcate into AIKS a wealth of wisdom that informs sustainable resource management, ecological stewardship, and social cohesion. A fast-changing world increasingly recognizes that the integration of diverse knowledge systems into research and development is paramount; AIKS therefore becomes an important resource to inform us about the natural environment and human interaction within it. This section reflects on how engagement with AIKS enriches academic discourses while nurturing collaborative partnerships between researchers and Indigenous communities. By showing the relevance of AIKS, we show the role this kind of knowledge could play regarding sustainable development, cultural resilience, and innovative responses fully attuned to the contexts of societies on the African continent. In these chapters, through investigation, the application of AIKS to contemporary endeavours shows that integrating it is indeed much more than an exercise for academics-it holds the power toward inclusive, equitable development on the African continent.

The quest for sustainable development in Africa necessitates a critical re-examination of how knowledge is traditionally understood and utilised. While top-down models have yielded mixed results (Moyo, 2009), a growing emphasis lies in harnessing the wisdom embedded within AIKS. This chapter embarks on a transformative journey, drawing on the scholarship of Gumbo and Williams (2023) and others, to unveil the transformative potential of AR/AL methodologies for integrating AIKS wisdom into contemporary research practices.

AIKS represents a rich tapestry of knowledge, philosophies, and practices accumulated by African communities over generations. This knowledge base offers a deep understanding of the natural world, fostering harmonious relationships between humans and their environment, as explored by Berkes (2009). Examples include traditional agricultural practices like crop rotation and intercropping, which promote biodiversity and soil health (Gliessman, 2016). African communities also possess a deep understanding of plant properties used for medicinal purposes, with the potential to inform drug discovery efforts in modern medicine (Mnguni & Mapotse, 2024; Heinrich et al., 2006). Despite this richness, AIKS has often been marginalized by mainstream research, hindering its potential contribution to solving complex challenges faced by African communities today (Mnguni, 2013; Mapotse, 2015).

A profound exploration of AIKS goes beyond simply recognizing them as a reservoir of wisdom passed down through generations. As Msila and Gumbo (2016) emphasize, AIKS is a knowledge system built upon profound observation, adaptation, and continuous refinement through experience. Take, for example, traditional ecological knowledge. African communities have developed intricate ways of classifying plants and animals, understanding weather patterns, and managing natural resources sustainably. This knowledge is not static but rather evolves through ongoing interactions with the environment (Berkes, 2009).

Unveiling the transformative potential of AIKS necessitates a paradigm shift within contemporary research landscapes. This requires acknowledging the enduring relevance of AIKS, as emphasized by Msila & Gumbo (2016). Historically, mainstream research practices have often marginalized indigenous knowledge. Bridging this gap hinges on decolonising methodologies and fostering culturally sensitive research partnerships (Mnguni & Mapotse, 2024).

This aligns perfectly with the principles championed by Linda Tuhiwai Smith (1999) in her seminal work Decolonizing Methodologies: Research and Indigenous Peoples. Smith (1999) critiques the historical power imbalances inherent in traditional research approaches and advocates for methodologies that empower African communities and respect their knowledge systems. Decolonizing methodologies emphasize dismantling these imbalances, centring the voices of indigenous communities, and fostering collaborative research.

By integrating AIKS through culturally sensitive partnerships, we unlock the immense potential to enrich and inform research practices across various fields. For instance, combining African ecological knowledge with scientific methods can lead to more sustainable and culturally appropriate strategies for natural resource management. Similarly, integrating AIKS into agricultural research can contribute to developing climate-resilient crops and improving food security. These are just a few examples of the transformative potential that lies at the intersection of AIKS wisdom and contemporary research practices.

Building upon the foundation of decolonizing methodologies and culturally sensitive research partnerships, AR/AL methodologies emerge as powerful frameworks for unlocking the transformative potential of AIKS. AR/AL approaches emphasize collaboration, reflection, and action-oriented research, fostering a space for the co-creation of knowledge between researchers and African communities (Mnguni, 2013). This inherently collaborative approach ensures that AIKS are not merely extracted from communities, but rather integrated in a way that respects ownership and control by AIKS holders. The following sections will delve deeper into the theoretical foundations of AR/AL, explore the crucial considerations for ethical research practices within this context, and showcase practical applications that demonstrate the transformative power of AR/AL in integrating AIKS wisdom for addressing Africa's development challenges.

Having established the profound value of AIKS and the critical need for culturally sensitive research integration, we now set our sights on harnessing this potential for transformative change. This chapter champions AR/AL methodologies as powerful frameworks to unlock the transformative power of AIKS. The following sections delve into the theoretical foundations of AR/AL, exploring how these approaches foster collaboration, co-creation of knowledge, and respect for Indigenous ownership of knowledge systems. We will navigate the crucial considerations for ethical research practices within this context. Most importantly, we will showcase real-world examples that demonstrate the transformative power of AR/AL in integrating AIKS wisdom. These applications will illuminate how AR/AL empowers African communities by ensuring their knowledge systems are not merely acknowledged but actively harnessed to address Africa's development challenges. Through this exploration, we will witness the profound impact of AR/AL in bridging the gap between AIKS wisdom and contemporary research practices, paving the way for a more inclusive and equitable approach to knowledge production on the African continent.

The quest for sustainable development in Africa necessitates a critical re-examination of how knowledge is traditionally understood and utilized. While top-down models have yielded mixed results (Moyo, 2009), a growing emphasis lies on harnessing the wisdom embedded within African Indigenous Knowledge Systems (AIKS). This chapter embarks on a transformative journey, drawing on the scholarship of Gumbo & Williams (2023) and others, to unveil the transformative potential of Action Research (AR) and Action Learning (AL) methodologies for integrating AIKS wisdom into contemporary research practices.

AIKS is a treasure trove of knowledge, beliefs, and practices that have been developed by African communities throughout the ages. Thus, this set of knowledge is a comprehensive one that gives intuitive knowledge of "real reality," develops closeness between people and their environment, and thus, leads us to live in a more harmonious world (Berkes, 2009). To elaborate, some of the conventional agricultural methods like crop rotation and intercropping have the effect of increasing biodiversity and enhancing soil health, which is beneficial (Gliessman, 2016). Moreover, the traditional knowledge of the Indigenous people about medicinal plants is still a significant gap in the drug discovery process (Heinrich et al., 2006). Nonetheless, despite its depth, AIKS is getting less attention and does not have the potential to cope with the complex challenges of African communities today (Mnguni, 2013).

A close analysis of AIKS discloses the fact that it is not only a collection of age-old wisdom but also something else. According to Mapotse (2015), AIKS is totally dependent on the system of knowledge that is based on the people's long-standing experience in observation, adaptation to the environmental changes, and the analysis of the structured knowledge in the continuous practice. For instance, traditional ecological knowledge shows how Indigenous communities have classified the diverse plants and animals as well as how they have categorized weather patterns and thereby utilized the natural resources in a sustainable way. The local knowledge is constantly entering a process of transformation through its entanglements with the environment (Berkes, 2009).

Having established the deep value of AIKS and the crucial requirement for research integration in culturally sensitive ways in previous chapters, this chapter develops an investigation of its potential for transformative change. In this light, we consider that AR/AL methodologies hold potent frames for realizing such potentials for the creation of transformation linked to AIKS.

The subsequent sections engage with the theoretical framing of AR/AL, in the sense that the approaches further encourage collaboration, co-creation of knowledge, and respect for Indigenous ownership of knowledge systems. We will understand some of the important considerations toward ethical research practices in this context. More importantly, we will share examples from practice that illustrate the potential of transformative power that AR/AL have brought into the integration of AIKS wisdom. These applications will illustrate how AR/AL empowers African communities through ensuring that their knowledge systems are recognized and used in the effort to address development challenges facing Africa. By doing so, we shall see how AR/AL makes profound contributions toward closing the gap between AIKS wisdom and current research practice in ways that beget an inclusive and equitable approach to knowledge production on the African continent.

INTEGRATING AIKS THROUGH AR/AL

Building on this, the following chapters detail how AIKS is implemented through AR/AL approaches. We will discuss, in-depth, what needs to be enacted practically to develop a culturally responsive research relationship between researchers and Indigenous communities. We will explore the cyclical nature of AR/AL: the way the research questions emanate from the concerns of the community, are explored collectively, and then generate actions and reflections which themselves may produce more questions. We unpack at each step how AR/AL secures protection for informed consent and intellectual property rights in Indigenous knowledge and its holders within a research process. We will witness some case studies that are great examples of the successful integration of AIKS into their specific research processes by taking the AR/AL approach, considering possibly transformative effects for those concerned. Real-life examples to be shown will demonstrate that AR/AL does not concern AIKS alone but places Indigenous peoples as active agents and co-creators of knowledge in the research landscape.

Having established both the profound value of AIKS and the limitation of traditional text-based research methods, we can now explore the dynamic approaches that exist to integrate AIKS wisdom within contemporary research practice. In such a framework, AR/AL emerge as strong tools, as identified by

Boboc et al. (2022). The research methodologies of AR/AL allow room for co-construction of knowledge through collaboration between the researcher and the Indigenous community. Such collaboration has the potential to provide some powerful intervention strategies in bridging the gap between AIKS wisdom and contemporary research paradigms. There is, however, a need to realize unequal power relations that might occur when researchers and African communities collaborate, warned by Smith (1999) in her seminal work. Ethically sensitive, culturally aware partnerships are axiomatic if AR/AL methodologies are to function, as they were designed, first, to balance out imbalances in knowledge creation, participation, and ownership across the whole research process.

In contrast, AR/AL methodologies provide a new paradigm for integrating AIKS beyond what was previously possible with traditional text-based approaches. Within this framework, Augmented Reality ET now emerges as one of the most promising tools (Almond, 2021). Imagine a highly knowledgeable elder in AIKS pointing to a specific plant (Hincapié, Díaz, Zapata-Cárdenas, Rios, Valencia & Güemes-Castorena, 2021). With Augmented Reality educational technology (ET) glasses, information is provided instantly: the traditional name of the plant, the documented medicinal uses according to AIKS, and the cultural importance the plant holds in the community (Hincapié, Díaz, Zapata-Cárdenas, Rios, Valencia & Güemes-Castorena, 2021:Boboc, Băutu, Gîrbacia, Popovici & Popovici, 2022). Therefore, this ability to visualize the "invisible"-this rich knowledge embedded within AIKS and perhaps not quite so obvious a deep realisation and appreciation of that wisdom for its members and people outside the community Mapotse (2018) has noted that this is only one example of how AR/AL, along with Augmented Reality ET, closes the gap between the knowledge of AIKS communities, represented by their embodiment, and the digital world. Coupled with Augmented Reality ET, the AR/AL methodologies will ensure an inclusive knowledge landscape that is both interactive and inclusive in terms of safeguarding AIKS through and for future generations.

AR/AL further enriches the process of integrating AIKS by leveraging its unique strengths in data analysis. The methodologies of AR/AL are especially good at processing big datasets like the wealth of African ecological knowledge passed down through generations within African communities (Almond, 2021). These could include everything from the most minute observations of animal behaviour to growth cycles in plants and weather patterns, right up to interactions within an ecosystem. Applying AR/AL techniques can help researchers carve out patterns and connections not so evident in these rich repositories.

For example, let's consider how an AR/AL system developed to analyse AIKS data on animal and plant migration might work. Members of Indigenous communities may have recorded knowledge about the time and behaviour of animal migrations, which could extend centuries (Gumbo, 2024). Putting this into a system of AR/AL may allow researchers to find patterns that have not been well-known before like the links between migratory patterns and specific environmental factors. Similarly, AR/AL may be applied in the analyses of AIKS regarding plant growth cycles and might even discover any linkage between plant phenology and minor climatic changes (Gumbo, 2024). The analysis will not only help confirm the wisdom of AIKS but also add value to scientific research by assuming a different dimension. Eventually, the integration of AIKS through AR/AL methodologies contributes to a deeper and richer understanding of the natural world, allowing more holistic approaches to environmental research and management (Mapotse, 2018).

Methodologies for AR/AL represent a novel paradigm to merge AIKS in one collaborative co-creation space between holders of AIKS and researchers. For example, Augmented Reality ET platforms bridge the chasm between embodied knowledge and its digital representation: imagine an elder, steeped in AIKS, who walks across his traditional lands with Augmented Reality ET glasses (Almond, 2021). With these glasses (Augmented Reality ET), they can identify specific locations and annotate those locations with AEK directly within the Augmented Reality ET environment. The AEK could include information on medicinal plant properties, seasonal shifts in resource availability, or cultural practices related to specific locations. This Augmented Reality ET information may then be used by the researchers to document and study it within a scientific context.

This approach transcends the constraints of conventional research methodologies, often based on extractive practices in which AIKS are collected and analysed by a researcher in isolation. On the contrary, AR/AL provides an equal and respectful research dynamic (Martin, Nicolas, Burq & Rives,2019). Elders and community members retain ownership and control of their knowledge and are active contributors to any type of research in developing and sharing new understanding as co-researchers. A second strength of AR/AL approaches is in supporting critical reflection and iterative learning cycles. It is by collectively analyzing the data produced via AR/AL with AEK that researchers and African communities can iteratively build insights into the natural world, arriving at more robust and culturally grounded scientific knowledge. This collaboration provides benefits beyond the empowerment of African communities, like enhancing the scientific quality of the research outcomes themselves (Mnguni & Mapotse, 2023).

This collaboration between the researcher and African communities does, however, have potential power imbalances, a concern raised by Smith (1999). Thoughtful use of AR/AL could minimise these concerns (Mapotse, 2023). For example, full ownership of data collected, and information dissemination should be left to the African communities. In this respect, Augmented Reality TE platforms can be designed in such a way that the way information is presented is left to the African community (Martin, Nicolas, Burq & Rives,2019). Additionally, research protocols should be developed collaboratively, ensuring informed consent and fair benefit-sharing (Bezemer & Cowan, 2022). AR/AL tools can be used to document these protocols and ensure they are followed. Finally, developing and nurturing African community members using AR/AL tools empowers them to participate actively in research and knowledge production. This fosters a more equitable partnership between researchers and African communities (Smith, 1999).

While AIKS is a treasure chest of wisdom and possible solutions for most of today's problems, there is an immense gap in this traditional wisdom from the position of scientific study (Watkins, Marsick & Wasserman, 2019). Such marginalization has led to many lost opportunities regarding advances in perception related to ecosystems, healthcare management, and societal frameworks. The AR/AL methodologies do bear immense promise in bridging gaps between these poles. These collaborative approaches create a non-hierarchical learning environment where researchers and indigenous knowledge holders enter into a continuous spiral process of inquiry, action, reflection, and further action. This kind of respect requires collaboration that ensures AIKS is integrated into scientific research with due sensitivity to the issues of cultural ownership and intellectual property rights. AR/AL gives indigenous communities the potential of active partners in the research process; therefore, owning the knowledge shared. In so doing, collaborative research can fill this knowledge gap and unlock the potential of AIKS to address some of the intractable problems facing Africa today and contribute toward an equitable and sustainable future for all (Gumbo & Williams, 2023.

As an example, a collaborative research project in Namibia (Jones & Williams, 2022) and Sekhukhune tea makers (Trapido, 2018) employed AR/AL to combine Indigenous ecological knowledge (IEK) with scientific methods for co-managing a communal wildlife conservancy. This project involved the Himba and Sekhukhune tea makers community, who possess a deep understanding of their local ecosystem. In AR/AL workshops and discussions, researchers and community members jointly identified the challenges related to wildlife poaching and habitat degradation. By integrating traditional knowledge of animal and plant behaviour and tracking techniques with scientific monitoring methods, projects have come up with a more holistic and sustainable management plan for wildlife. This co-production of knowledge enabled the Himba and Sekhukhune tea makers community to have greater agency in decision-making and resource management with a sense of ownership and responsibility for their environment. This case illustrates the potential for transformation by AR/AL to integrate Indigenous knowledge and to advance collaborative research for the benefit of researchers and communities.

By closely examining case studies and examples, we elucidate how AR/AL methodologies, when implemented with a long-term commitment, can foster collaborative and participatory approaches that not only honour but actively incorporate Indigenous perspectives (Cunningham & Barrow, 2000). A crucial aspect is ensuring fair and equitable partnerships. Establishing clear benefit-sharing agreements upfront is essential to ensure communities have a stake in the research outcomes and receive fair compensation for their knowledge (Bowrey, Cochrane, Hadley, McKeough, Pappalardo & Weatherall, 2024). AR/AL projects can further empower Indigenous communities by incorporating capacity-building initiatives. This equips them with the skills and knowledge to not only contribute their wisdom but also actively participate in the research process, from co-designing the project to analysing findings (Wallerstein, 2006)

Through practical illustrations, we highlight the transformative potential of AR/AL in seamlessly integrating traditional knowledge into research contexts across various fields (Mapotse, 2018). By illustrating how these methodologies ease collaborative and participatory approaches, we highlight their efficacy in honouring AIKS perspectives and fostering inclusive research practices. This integration not only amplifies the voices of AIKS communities but also contributes to the development of more holistic and culturally sensitive research endeavours (Mnguni, 2013). As we navigate the terrain of AR/AL, we recognise their pivotal role in bridging the gap between traditional wisdom and contemporary research methodologies, paving the way for sustainable, community-centred research endeavours that prioritize cultural preservation, social justice, and fair knowledge exchange (Mapotse, 2018; Mnguni & Mapotse, 2023).

NAVIGATING ETHICAL CONSIDERATIONS

Interactive elements within AR experiences can bring complex concepts and narratives to life, fostering deeper engagement and promoting a more holistic understanding of Indigenous cultures for both Indigenous and non-Indigenous audiences. AR technology overlays can be instrumental in visualising AEK related to plant and animal life (Bates, 2014). Merging such knowledge with AR onto the physical landscape can therefore consider sharing Indigenous community understanding with several stakeholders and land managers on practices of land management in a sustainable way. This represents an approach with considerable potential contribution toward the preservation of biodiversity and the environment.

The potential of AIKS to address contemporary challenges in Africa is undeniable. However, unlocking this potential requires careful consideration of the ethical implications. While AR/AL methodologies offer a collaborative framework, navigating ethical considerations becomes paramount (Mapotse, 2018). This section delves into the multifaceted ethical dilemmas that arise when integrating AIKS into research. Of concern to us, however, is how much knowledge from the AIKS communities will be continuously owned and controlled throughout the research process (Ladson-Billings, 2021). In fact, according to Tobin (1999), the careful handling of such ethics will enhance a respectful and empowering manner of integrating AIKS into the AR/AL framework.

The integration of AIKSK into AR/AL frameworks has great potential. This, however, calls for a rather cautious approach to ethical issues regarding the process of integration (Mapotse, 2018). This subheading therefore discusses the multi-dimensional ethical issues that arise in incorporating AIKS in research. The focus has been on ownership and control of knowledge by the AIKS communities within the research processes (Mapotse, 2023).

These range from issues of cultural sensitivity to those around creating community involvement and empowerment. For instance, the establishment of clear and specific community protocols in advance is paramount (Da Silva, 2001). Such protocols would spell out how access, use, and benefit-sharing of AIKS within AR/AL projects are to be handled (Mnguni, 2013). This would enable the community to make their decisions on the acceptable use of their knowledge, ensuring equitable benefit in contributing towards the research findings.

It is not merely a question of following protocol but rather one that involves an uncompromising commitment to integrity and respect in the research process (Mnguni & Mapotse, 2023). This involves insisting on the rights and views of African communities as equal partners in research. This section shall guide how some of these ethical dilemmas can be navigated to make certain that the integration of AIKS within AR/AL frameworks is done with due care and sensitivity to cultural contexts (Mnguni, 2013). This involves the right of communities to withhold certain knowledge. Consent agreements, therefore, ought to be customised to appreciate these cultural sensitivities (Bowrey et al.,2024).

Although AR/AL has great potential, it is necessary that one recognise that ethical consideration is an essential component of integrating AIKS into AR technology. This would mean that the African community owns what manifests as AR experiences (Smith, 2012). The collaboration and co-creation therein should be attuned to the community values and protocols for the development of AR. The cultural sensitivities of the people should also be observed throughout the process, proper transmission protocols of traditional knowledge AIKS respected, and proper attribution and ownership (Gyamerah, 2024.

This, however, offers great opportunities: AR/AL for the safeguarding and passing on of AIKS, improvements in education, and sustainable land management. Ethics do, however, play a huge role herein. Prioritising African leadership and respect for cultural protocols could prove a very powerful enabler to help bridge the gap in cultural divides toward a more sustainable future.

Augmented Reality ET presents a unique opportunity to bridge the gap between AIKS and contemporary society (Hincapié, Díaz, Zapata-Cárdenas, Rios, Valencia & Güemes-Castorena, 2021). This technology can be harnessed for a variety of purposes, with the potential to benefit both cultural preservation and environmental sustainability. With AR, immersive experiences related to the documentation and sharing of cultural practices, languages, and stories are created in a really engaging way. That can be simple: pointing the camera of a smartphone toward a certain plant and seeing its traditional use come alive in 3D, narrated by a community elder (Lui et al., 2020). This will go a long way in furthering the cultural understanding of the group and is a very modern device to transfer knowledge across generations in indigenous groups. For these reasons, an application of AR technology could be planned to be developed in schools or museums to create an utterly different form of presentation AIKS (Martin, Nicolas, Burq & Rives, 2019).

PRACTICAL APPLICATIONS AND IMPACT

These are just some of how AR/AL methodologies can be used to empower African communities (Zuber-Skerritt, 2015). Linked with their AIKS, African communities are thus able to make valuable inputs into planning their futures and solving problems that confront them. It is this kind of collaboration and sensitivity to culture that AR/AL breeds, allowing the future to be one in which African cultures thrive alongside their ecologies (Gyamerah, 2024).

In addition, such practical applications have gone further to illustrate the use of AR/AL as a means through which inclusivity and cultural sensitivity can be realised in research practices (Zuber-Skerritt, 2002 & 2015). Such integration of AIKS within research contexts further enhances not only the depth and breadth of research outputs but also empowers African communities toward self-determination through AR/AL initiatives (Khumalo & Mapotse, 2023). It is in these that training and skills development programs in African communities can be the most empowering in the AR/AL projects. They were, therefore, well placed to share their wisdom in actively participating in research processes: co-designing the project and analyzing findings (Wallerstein, 2006). Using such a participatory approach that draws upon AR/AL methodologies enables a better treading through complex puzzles posed by society in general, besides upholding African knowledge and value systems (Mnguni,2013).

Harnessing the Transformative Potential of AR/AL for Sustainable Development in Africa

This captures the broader influence of AR/AL in empowering African communities through the integration of AIKS into contemporary research approaches (Zuber-Skerritt, 2002). In allowing a participatory research environment, AR/AL has community members also become co-creators of the research along with the researcher; as such, solutions that have been developed are not only contextually relevant but also culturally appropriate. This collaborative framework enhances community ownership and greatly heightens the potential for sustainability over time.

Beyond this, the AR/AL projects are a far cry from just problem identification; they build research capacities and skills within the community (Mnguni & Mapotse, 2024). They learn how to conduct research, analyse data, and make meaningful conclusions from such findings, thus making them capable of addressing immediate challenges while building a culture of independent knowledge generation. This is important in terms of self-capacitating the communities toward different circumstances and innovating solutions that best suit their situation.

Furthermore, AR/AL has provided avenues through which important knowledge is shared across communities with challenges (Zuber-Skerritt, 2002). This is especially true of interactive AR/AL projects, where ideas, strategies, and best practices are shared to enhance problem-solving capacity within African communities. The sharing of successful approaches by the communities enables them to capitalise on the strengths of their diverse experiences in pursuit of accelerated progress toward sustainable development (Simpson & Jackson, 2021).

Documentation and validation through AR/AL research provide a voice for communities to demand policy recognition and support for their knowledge systems (Ladson-Billings, 2021: Holst,2020). Strong data and evidence from these projects shall, therefore, serve to illustrate the value and efficiency of AIKS practices, thus enabling communities to push for policies that integrate their traditional knowledge into broader development strategies (Harding, S. (1987).

Ultimately, AR/AL goes beyond solving the current challenges; it plants African communities as active agents in the construction of their futures. This means adopting a collaborative and culturally sensitive approach to foster a future in which African communities can thrive in all their unique cultural heritage and ecological knowledge (Ladson-Billings, 2021: Holst,2020). AR/AL is equipping them for the stewardship of their sustainable development; it not only secures a better future for generations to come but also calls for a need to integrate AIKS into mainstream contemporary research practices.

The Potential of Transformation in AR/AL for the Himba Community

The potential applications of AR/AL among the Himba encapsulate the transformative power these methodologies can have for wider contexts across Africa (Krishna Panicker, Nair & Oommen, 2024). By embedding African communities within the processes surrounding the rich AIKS through advanced research methodologies, AR/AL allow such communities to build competencies towards given changes and allow for far-reaching social change. This is particularly true of the varied responses of the Himba people to the commercialisation of Commiphora wildii, which reflects a complex interrelationship between economic, cultural, and environmental considerations (Galloway, Wynberg & Nott, 2016). On one hand, commercialisation does provide undeniable economic benefits since it creates new avenues for the Himba to earn an income and enhance their livelihoods.

As a result of the sale of Commiphora wildii, increased financial resources available can be put to social use such as basic services like food, education and health (Krishna Panicker, Nair & Oommen, 2024). Since women are the major beneficiaries of the economic activities regarding the plant, improvement in their status in society could result from this. These economic benefits, however, are tempered by cultural concerns in that the Himba express apprehension about the potential exploitation of their traditional knowledge and practices. Its commodification raises fears of diluting their cultural identity since the traditional uses of the plant are inalienable parts of their cultural heritage. Besides, environmental sustainability is a critical issue, as the Himba are concerned with the impacts of overharvesting and the long-term viability of Commiphora wildii. A balance must be reached between economic gain and the conservation of their natural resources. This calls for the implementation of sustainable practices to reduce ecological harm. The relationship between the Himba and the commercialisation of Commiphora wildii is thus a complex one: benefits are patent, while associated risks and challenges need management.

These methodologies have ensured that the rights and traditions of the Himba are upheld by creating a participatory environment in the wake of changing economic fortunes (Saruchera & Xaba, 2024). Ultimately, the integration of AR/AL with the Himba community will provide, not just a necessary response to immediate issues related to the commercialization of Commiphora wildii, but even to representative, transformative practices across Africa for resilience and agency building in indigenous communities within globalization and economic change.

Moreover, environmental sustainability emerges as a critical issue, with the Himba concerned about the impacts of overharvesting and the long-term viability of Commiphora wildii. Balancing economic gain with the preservation of their natural resources is essential, necessitating the implementation of sustainable practices to mitigate ecological harm. Thus, the relationship between the Himba and the commercialization of Commiphora wildii is multifaceted; while the benefits are evident, the associated risks and challenges require careful management.

Enhancing Artisanal Practices with AR/AL: The Case of Sekhukhune Tea Makers

Within the context of the Sekhukhune tea makers, AR/AL has been an essential vehicle to augment these artisan practices (Dudgeon, Bray, Walker, Wright & Sutherland, 2024). These methods let the tea makers engage in reflective practice, which provides a chance for the critical observation of their approach while incorporating input from academics and other relevant community stakeholders for continuous processing improvement of the tea and their value addition to these products (Mnguni & Mapotse, 2023). The Sekhukhune tea makers exemplify artisanal and interdisciplinary thinking, skillfully crafting handcrafted tea products from indigenous African plants. Among these innovators is Retang Phaahla, also known as Nondumiso, who brings valuable community development experience to the initiative (Trapido, 2018).

These Sekhukhune-based holders and practitioners of AIKS are a diverse category including subsistence farmers, foragers, agroecologist academics, youth community activists, traditional healers, millennial hipsters, and even ancestral spirits (Trapido, 2018). This sets a rich tapestry of contributors, underlining that the nature of the project is collaborative, hence, adding to it from the academic fraternity at the University of Limpopo (Nkambule, Ngubane & Mashiane-Nkabinde, 2024). These researchers are contributing much to the scientific validation of Indigenous healers' knowledge about the health-conferment qualities of tepane and diya, thus bridging a gap between AIKS and contemporary scientific study (Trapido, 2018: Sofowora,1996). This collaboration gave rise to the very successful craft company that has lately applied for organic certification with the European Union through the Deutsche Gesellschaft für Internationale Zusammenarbeit.

The company motto, "Bapedi. A re Boeleng Setsong ka Sepedi," which translates to "Let us return to our culture," reflects a deep commitment to cultural heritage and the promotion of AIKS practices (Trapido, 2018). In one concerted effort, Sekhukhune stakeholders have developed a line of trendy tea blends honouring not only AIKS and processing infusions but also urban, hipster, chai tea-drinking WIKS markets (Street & Prinsloo, 2013). The effort will not only preserve and promote AIKS but also demonstrate how traditional practices, when combined with WIKS market needs, could be applied to achieve sustainable economic development. Therefore, AR/AL applications within the Sekhukhune community of tea makers stand to provide a case for the described frameworks and how they could be applied to support artisan practices through cultural preservation and economic sustainability.

CONCLUSION

AR/AL provide a participatory framework that empowers African communities through the integration of AIKS into WIKS research practices. Involving community members as co-creators in the research process, these methodologies ensure that solutions are contextually relevant and culturally appropriate, hence engendering a sense of ownership and enhancing the sustainability of interventions.

Beyond the identification of a problem, the AR/AL projects develop in communities the capacity for research through their skills in conducting research, analysing data, and arriving at meaningful conclusions. This empowerment helps not only with immediate challenges but also in developing a culture of independent knowledge generation that empowers communities to adapt and innovate in response to evolving circumstances.

Besides, AR/AL enables the sharing of knowledge between communities in the same regions facing related issues and enhances problem-solving capacity in the community. The successful strategies identified through such projects provide them with evidence that can help advocate for policies incorporating their traditional knowledge systems into larger development strategies.

Ultimately, AR/AL moves beyond the immediacies of problem-solving and locates African communities at the centre of their destiny. Collaboration and cultural sensitivity will hopefully usher in a future wherein communities will be able to prosper but also actively draw on their rich cultural heritage and ecologically relevant knowledge. Integration of AIKS within mainstream research practice is not only of value but indeed forms a necessary ingredient of holistic and sustainable development throughout Africa.

Examples from the Sekhukhune tea makers and the Himba community illustrate the transformative potential of AR/AL methodologies. These examples indicate that when communities are active partners in integrating AIKS with WIKS research, sustainable economic development is achieved in conjunction with the conservation of cultural identity. AR/AL methods promote ongoing dialogue and problem-solving by the community to address complex economic changes while maintaining respect for tradition.

This is indeed the commitment that paves the way toward the integration of AIKS through AR/AL methodologies, including culturally sensitive research initiatives that foster community change in the interest of ensuring a secure future for all generations to come.

REFERENCE

Almond, A. (2021). *We Are All Related:(Re)*. Storying With Augmented Reality to Build Indigenous-Settler Relations.

Bates, D. (2014). Traditional ecological knowledge and natural resource management. *Local Environment, 19*(7), 826–841. https://www.nebraskapress.unl.edu/nebraska/9780803232464

Bele, B., Nielsen, V. K. S., Fairclough, G., & Herring, P. (2024). A Heritage Without Boundaries: Nature-Culture Synergies in and around Norwegian Protected Landscapes. *Human Ecology: an Interdisciplinary Journal, 52*(1), 1–18. DOI: 10.1007/s10745-023-00468-z

Bezemer, J., & Cowan, K. (2022). We are exploring reading in social semiotics: Theory and methods. In *Reading* (pp. 107–118). Routledge. DOI: 10.4324/9781003208648-9

Bion, W. R. (1961). *Experiences in Groups*. Tavistock Publications.

Boboc, R. G., Băutu, E., Gîrbacia, F., Popovici, N., & Popovici, D. M. (2022). Augmented reality in cultural heritage: An overview of the last decade of applications. *Applied Sciences (Basel, Switzerland), 12*(19), 9859. DOI: 10.3390/app12199859

Bowrey, K., Cochrane, T., Hadley, M., McKeough, J., Pappalardo, K., & Weatherall, K. (2024). Managing ownership of copyright in research publications to increase the public benefits from research. *Federal Law Review, 52*(1), 3–33. DOI: 10.1177/0067205X231213676

Castro, B. M. (2017). Ijiam: Augmented Reality Technologies for the Adaptation of Ecuadorian Intangible Cultural Heritage (Doctoral dissertation, OCAD University).

Dudgeon, P., Bray, A., Walker, R., Wright, M., & Sutherland, S. (2024). Caring for Country: Indigenous Well-being, Law, and Environmental Justice. In *Climate Change and Mental Health Equity* (pp. 383–407). Springer International Publishing. DOI: 10.1007/978-3-031-56736-0_15

Galloway, F. B., Wynberg, R. P., & Nott, K. (2016). Commercialising a perfume plant, Commiphora wildii: Livelihood implications for indigenous Himba in north-west Namibia. *International Forestry Review, 18*(4), 429–443. DOI: 10.1505/146554816820127541

Gumbo, M. T. (2024). In defence of African philosophy. *South African Journal of Higher Education, 38*(2), 120–140. DOI: 10.20853/38-2-6003

Gumbo, M. T., & Williams, P. J. (Eds.). (2023). *Indigenous Technology Knowledge Systems: Decolonizing the Technology Education Curriculum*. Springer Nature. DOI: 10.1007/978-981-99-1396-1

Gyamerah, K. (2024). The Role of African Indigenous Knowledge Systems and Pedagogies in Decolonizing and Transforming Mathematics and Science in Ghana (Doctoral dissertation).

Harding, S. (1987). *Feminism and Method: Incorporating Standpoint Theory*. Open University Press.

Hincapié, M., Díaz, C., Zapata-Cárdenas, M. I., Rios, H. D. J. T., Valencia, D., & Güemes-Castorena, D. (2021). Augmented reality mobile apps for cultural heritage reactivation. *Computers & Electrical Engineering, 93*, 107281. DOI: 10.1016/j.compeleceng.2021.107281

Holst, J. D. (2020). Fifty Years of Pedagogy of the Oppressed: A Book Review Essay.

Ihejirika, C. (2024). Harnessing African Indigenous Knowledge Systems for Knowledge Production: A Redefinition of a Culture-Centric Epistemology. *Journal of Contemporary Philosophical and Anthropological Studies*, *2*(1). Advance online publication. DOI: 10.59652/jcpas.v2i1.103

Khumalo, S. B., & Mapotse, T. A. (2023). Building Modern Technology Innovation on Indigenous Knowledge Systems in Technology Education. In *Indigenous Technology Knowledge Systems: Decolonizing the Technology Education Curriculum* (pp. 43–55). Springer Nature Singapore. DOI: 10.1007/978-981-99-1396-1_4

Krishna Panicker, L., Nair, A. S., & Oommen, O. V. (2024). ABS-Bioprospecting Research and Benefit Sharing. In *Biodiversity and Business* (pp. 201–215). Springer. DOI: 10.1007/978-3-031-71674-4

Ladson-Billings, G. (2021). *Critical race theory in education: A scholar's journey*. Teachers College Press.

Lewin, K. (1946). Action research and minority problems. *The Journal of Social Issues*, *2*(1), 34–46. DOI: 10.1111/j.1540-4560.1946.tb02295.x

Maguire, P. (1987). Doing Participatory Research: A Feminist Approach.

Mapotse, T. (2014). Technology Education and Action Research: Two Complementary Processes to Capacitate Teachers. *Mediterranean Journal of Social Sciences*, •••, 5. DOI: 10.5901/mjss.2014.v5n3p338

Mapotse, T. A. (2018). Development of a technology education cascading theory through community engagement site-based support. *International Journal of Technology and Design Education*, *28*(3), 685–699. DOI: 10.1007/s10798-017-9411-6

Mapotse, T. A. (2023). Ways in Which Technology Education Teachers Can Integrate Indigenous Technology Through Action Learning. In *Indigenous Technology Knowledge Systems: Decolonizing the Technology Education Curriculum* (pp. 315–327). Springer Nature Singapore. DOI: 10.1007/978-981-99-1396-1_20

Martin, S., Nicolas, C., Burq, H., & Rives, A. (2019). Enhancing cultural heritage education with augmented reality technologies. *Journal of Computer Assisted Learning*, *35*(1), 74–87. https://link.springer.com/book/10.1007/978-3-030-42156-4

Matsekoleng, T. K., & Mapotse, T. A. (2023). Technology Teachers' Use of Indigenous Knowledge to Integrate Environmental Education into Technology Education. In *Indigenous Technology Knowledge Systems: Decolonizing the Technology Education Curriculum* (pp. 283–295). Springer Nature Singapore. DOI: 10.1007/978-981-99-1396-1_18

Matsekoleng, T. K., Mapotse, T. A., & Gumbo, M. T. (2024). The role of indigenous games in education: A technology and environmental education perspective. *Diaspora, Indigenous, and Minority Education*, *18*(1), 68–82. DOI: 10.1080/15595692.2022.2160436

Mnguni, J. N. C. (2013). *Some challenges in teaching natural sciences in the context of National Curriculum Statement, MEd*. University of South Africa.

Mnguni, J. N. C., & Mapotse, T. A. (2023). Examining the Technological Divide Between Africa and the Western World: A Case of South Africa. In Gumbo, M. T., & Williams, P. J. (Eds.), *Indigenous Technology Knowledge Systems. Contemporary Issues in Technology Education*. Springer., DOI: 10.1007/978-981-99-1396-1_14

Mnguni, J. N. C., & Mapotse, T. T. (2024). Decolonizing Postgraduate Studies Embracing African Spirituality in South African Higher Education. In Global Perspectives on Decolonizing Postgraduate Education (pp. 55-81). IGI Global. DOI: 10.4018/979-8-3693-1289-6.ch005

Nkambule, B., Ngubane, S. A., & Mashiane-Nkabinde, B. (2024). Management of Context-Specific Knowledge Creation and Application through Collective Social Responsibility in School? Community Partnerships. International Journal of Learning. *Teaching and Educational Research*, *23*(7), 499–524.

Saruchera, M., & Xaba, M. B. (2024). Embracing African Traditional Knowledge Health Systems in a HIV/AIDS and COVID-19 Pandemic Era. *Journal of Black Studies*, *55*(2), 95–116. DOI: 10.1177/00219347231213373

Shrivastava, P., Smith, M. S., O'Brien, K., & Zsolnai, L. (2020). Transforming sustainability science to generate positive social and environmental change globally. *One Earth*, *2*(4), 329–340. DOI: 10.1016/j.oneear.2020.04.010 PMID: 33501419

Simpson, D. J., & Jackson, M. J. (2021). *Educational reform: A Deweyan perspective*. Routledge. DOI: 10.4324/9780203825617

Smith, T. L. (2012). *Decolonizing methodologies: Research and Indigenous peoples*. Zed Books.

Sofowora, A. (1996). Research on medicinal plants and traditional medicine in Africa. *Journal of Alternative and Complementary Medicine (New York, N.Y.)*, *2*(3), 365–372. DOI: 10.1089/acm.1996.2.365 PMID: 9395671

Street, R. A., & Prinsloo, G. (2013). Commercially important medicinal plants of South Africa: A review. *Journal of Chemistry*, *2013*(1), 205048. DOI: 10.1155/2013/205048

Trapido, A. (2018, December 21). 'Meet the tea makers of Sekhukhune who honour traditions while building a modern business. City Press, 21 December, viewed 8 February 2022. https://www.news24.com/citypress/news/meet-the-tea-makers-of-sekhukhune-who-honour-traditions-while-building-a-modern-business-20181221

Watkins, K. E., Marsick, V. J., & Wasserman, I. (2019). Action research, action learning, and appreciative inquiry: Interventions that build individual and group capacity for EBOCD. In *Evidence-based initiatives for organizational change and development* (pp. 76–92). IGI Global. DOI: 10.4018/978-1-5225-6155-2.ch004

White, R. E., & Cooper, K. (2022). Action Research. In *Qualitative Research in the Post-Modern Era: Critical Approaches and Selected Methodologies* (pp. 387–439). Springer International Publishing. DOI: 10.1007/978-3-030-85124-8_10

Yishak, D., & Gumbo, M. (2015). A stand-alone, blended or restructured indigenisation approach to curriculum? A critical perspective. International Journal of African Renaissance Studies-Multi-. *Inter- and Transdisciplinarity*, *10*(1), 60–82.

Zuber-Skerritt, O. (2002). A model for designing action learning and action research programs. *The Learning Organization*, *9*(4), 143–149. DOI: 10.1108/09696470210428868

Zuber-Skerritt, O. (2015). Participatory action learning and action research (PALAR) for community engagement: A theoretical framework. *Educational research for social change, 4*(1), 5-25.

Zuber-Skerritt, O. (Ed.). (2021). *Action research for change and development.* Routledge. DOI: 10.4324/9781003248491

Chapter 3
Classroom Teachers Constructing Their Own Knowledge and Interpretation Epistemology, Ontology of Action Research and Action Learning:
Interpretation of Classroom Teachers Indigenous Knowledge System

Zuber Peermohammed Shaikh
University of Pune, India

Satish Balkrishna Bhise
Sinhgad Institute's Smt. Kashibai Navale College of Pharmacy, India

ABSTRACT

Educational stakeholders [Indigenous knowledge System(IKS)] rely on research to make informed decisions that ultimately affect the quality of schooling for their students. Accordingly, the purpose of educational research is to engage in disciplined inquiry to generate knowledge on topics significant to the students, teachers, administrators, schools, and other educational stakeholders. Just as the topics of educational research vary, so do the approaches to conducting educational research in the classroom. Your approach to research will be shaped by your context, your professional identity, and paradigm (set of beliefs and assumptions that guide your inquiry). These will all be key factors in how you generate knowledge related to your work as an educator. This paper discusses the research process from ontological, epistemological and methodological perspectives to explore and understand the social phenomena. The goal of this review is to provide a clear understanding of different research methodologies based on ontology and epistemology.

DOI: 10.4018/979-8-3693-3542-0.ch003

INTRODUCTION

The research methodologies can generally be subdivided into quantitative and qualitative research methodology. Quantitative methodology generates knowledge by investigating things which we could measure in some way. Qualitative methodology increases understanding of why things are and the way they are in the social world, and why people act the ways they do. The differences amongst research methodologies are based on a philosophical and theoretical view of research that guides researchers in their social science research. Furthermore, researchers can select the research methodology in social science and that depends on philosophical issues, which are related to ontology (the nature of reality) and epistemology (the nature of knowledge). This paper is an attempt to make the reader aware of the research methodologies and their basis of difference from ontological, epistemological and methodological perspectives.

Review on Action Research Concept/Reflection: Action research is an approach to educational research that is commonly used by educational practitioners and professionals to examine, and ultimately improve, their pedagogy and practice. In this way, action research represents an extension of the reflection and critical self-reflection that an educator employs on a daily basis in their classroom. When students are actively engaged in learning, the classroom can be dynamic and uncertain, demanding the constant attention of the educator. Considering these demands, educators are often only able to engage in reflection that is fleeting, and for the purpose of accommodation, modification, or formative assessment. Action research offers one path to more deliberate, substantial, and critical reflection that can be documented and analyzed to improve an educator's practice.

Figure 1. Reflection process of classroom teachers

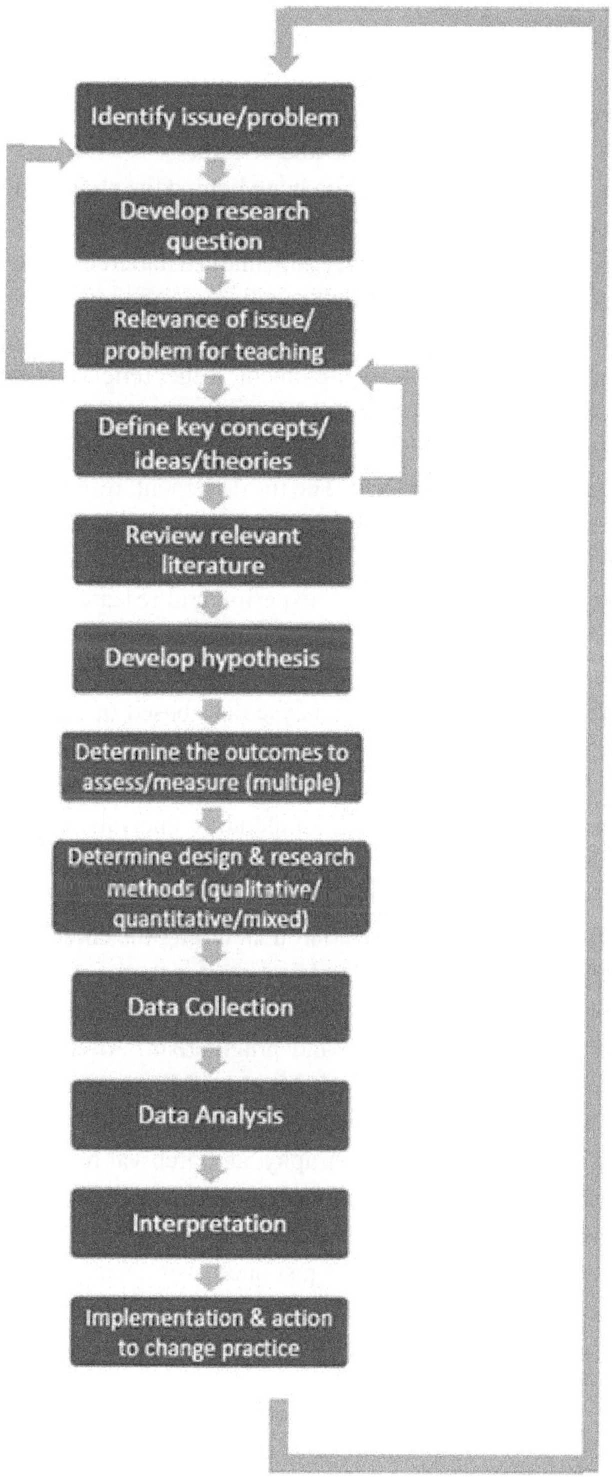

Basic Elements of Research Process

Crotty defined the as the following:

Epistemology: "The theory of knowledge embedded in the theoretical perspective and thereby in the methodology".

Theoretical Perspective: "The philosophical stance informing the methodology and thus providing a context for the process and grounding its logic and criteria".

"The strategy, plan of action, process or design lying behind the choice and use of particular methods and linking the choice and use of methods to the desired outcomes". *(Smith G.,1987)*

"The techniques or procedures used to gather and analyze data related to some research question or hypothesis". The first element, epistemology is described as inherent in the theoretical perspective as "a way of looking at the world and making sense of it". Major types of epistemology are: objectivism, constructionism, and subjectivism. The second element, a theoretical perspective describes the philosophical stance that lies behind the chosen methodology. It grounds the main assumptions of choosing methodology. Main types of theoretical perspective are: positivism (and post-positivism), interpretivism, critical inquiry, feminism, postmodernism, etc. The third element, research methodology represents the strategy and plan of action, which refers to the research design that shapes chosen research methods. Methodology provides a rationale for the choice of methods and the particular forms in which the methods are employed. Major types of methodology are: experimental research, survey research, ethnography, phenomenological research, grounded theory, heuristic inquiry, action research, discourse analysis, feminist standpoint research, etc. The last element, research methods refer to the techniques or procedures that including certain activities to collect and analyse data based on research question or hypotheses. Some of major research methods such as: sampling measurement and scaling, questionnaire, observation, interview, focus group, case study, life history, narrative, visual ethnographic methods, statistical analysis, data reduction, theme identification, comparative analysis, cognitive mapping, interpretive methods, document analysis, content analysis, conversation analysis, etc. On the other hand, Saunders, Lewis, & Thornhill explained the research process as an onion including several important layers; each layer leads to another till the centre of research onion including the choice of data collection techniques and data analysis procedures as shown. The first layer represents the research philosophy which relates to the nature and development of knowledge. Thus, the researchers can adopt different research philosophies as positivism, interpretivism, realism, and pragmatism. After selecting research philosophy, the researchers have to select research approach Mukhles M. Al-Ababneh in the second layer either deductive or inductive. In the third layer, different research strategies such as experiment, survey, case study, action research, grounded theory, ethnography, and archival research, can be applied to answer the research questions. Research method in the fourth layer can be different types to collect data, namely mono method, mixed methods, and multimethod. Time is important aspect in the research in the fifth layer, and therefore the research can collect data just once over a short period of time 'Cross-sectional' or collect data several times over a period of time 'longitudinal'. Finally, the centre of onion is the core of research that including data collection techniques and procedures of data analysis·

BACKGROUND

Literature about what is Epistemology: Epistemology involves knowledge and embodies a certain understanding of what is entailed in knowing, that represents how we know what we know. Epistemology is concerned with the acceptable of knowledge in the study field. Major types of epistemology are objectivism, constructionism, and subjectivism. Objectivism means that meaning and meaningful Linking Ontology, Epistemology and Research Methodology reality exists as such apart from the operation of any consciousness, and it represents "the position that social entities exist in reality external to social actors". This view supports that subject and object emerge as partners in the generation of meaning. Finally, subjectivism refers to the meaning comes from anything but the object to which it is ascribed, that means the object itself makes no contribution to the meaning that is imposed on the object by the subject. Saunders et al.considered subjectivist view as "social phenomena are created from the perceptions and consequent actions of social actors". On the basis that the phenomena to be investigated exist independent of consciousness and individuals are in direct contact with reality through sensory perception, and therefore this research leans more towards an objectivist epistemology.

Theorizing Action Research for the Classroom: All research, at its core, is for the purpose of generating new knowledge and contributing to the knowledge base of educational research. Action researchers in the classroom want to explore methods of improving their pedagogy and practice. The starting place of their inquiry stems from their pedagogy and practice, so by nature the knowledge created from their inquiry is often contextually specific to their classroom, school, or community. Therefore, we should examine the theoretical underpinnings of action research for the classroom. It is important to connect action research conceptually to experience; for example, Levin and Greenwood make these connections:

Action research is context bound and addresses real life problems.

Action research is inquiry where participants and researchers cogenerate knowledge through collaborative communicative processes in which all participants' contributions are taken seriously.

The meanings constructed in the inquiry process lead to social action or these reflections and action lead to the construction of new meanings. (Stenhouse L.1975)

The credibility/validity of action research knowledge is measured according to whether the actions that arise from it solve problems (workability) and increase participants' control over their own situation. Educators who engage in action research will generate new knowledge and beliefs based on their experiences in the classroom. It is these experiences, beliefs, and theories that are often discounted when more official forms of knowledge (e.g., textbooks, curriculum standards, districts standards) are prioritized. These beliefs and theories based on experiences should be valued and explored further, and this is one of the primary purposes of action research in the classroom. These beliefs and theories should be valued because they were meaningful aspects of knowledge constructed from teachers' experiences. Developing meaning and knowledge in this way forms the basis of constructivist ideology, just as teachers often try to get their students to construct their own meanings and understandings when experiencing new ideas.

Classroom Teachers Constructing their Own Knowledge: Most of you are probably at least minimally familiar with constructivism, or the process of constructing knowledge. However, what is constructivism precisely, for the purposes of action research? Many scholars have theorized constructivism and have identified two key attributes. (Koro-Ljungberg M, 2009)

Knowledge is not passively received, but actively developed through an individual's cognition;

Human cognition is adaptive and finds purpose in organizing the new experiences of the world, instead of settling for absolute or objective truth. Considering these two attributes, constructivism is distinct from conventional knowledge formation because people can develop a theory of knowledge that orders and organizes the world based on their experiences, instead of an objective or neutral reality. When individuals construct knowledge, there are interactions between an individual and their environment where communication, negotiation and meaning-making are collectively developing knowledge. For most educators, constructivism may be a natural inclination of their pedagogy. Action researchers have a similar relationship to constructivism because they are actively engaged in a process of constructing knowledge. However, their constructions may be more formal and based on the data they collect in the research process. Action researchers also are engaged in the meaning making process, making interpretations from their data. These aspects of the action research process situate them in the constructivist ideology. Just like constructivist educators, action researchers' constructions of knowledge will be affected by their individual and professional ideas and values, as well as the ecological context in which they work. The relations between constructivist inquiry and action research is important, as Lincoln states: much of the epistemological, ontological, and axiological belief systems are the same or similar, and methodologically, constructivists and action researchers work in similar ways, relying on qualitative methods in face-to-face work, while buttressing information, data and background with quantitative method work when necessary or useful. While there are many links between action research and educators in the classroom, constructivism offers the most familiar and practical threads to bind the beliefs of educators and action researchers.

FOCUS OF THE ARTICLE

Epistemology, Ontology, and Action Research: It is also important for educators to consider the philosophical stances related to action research to better situate it with their beliefs and reality. When researchers make decisions about the methodology they intend to use, they will consider their ontological and epistemological stances. It is vital that researchers clearly distinguish their philosophical stances and understand the implications of their stance in the research process, especially when collecting and analyzing their data. In what follows, we will discuss ontological and epistemological stances in relation to action research methodology. Ontology, or the theory of being, is concerned with the claims or assumptions we make about ourselves within our social reality. In relation to the discussion of constructivism, generally action researchers would consider their educational reality as socially constructed. Social construction of reality happens when individuals interact in a social system. Meaningful construction of concepts and representations of reality develop through an individual's interpretations of others' actions. These interpretations become agreed upon by members of a social system and become part of social fabric, reproduced as knowledge and beliefs to develop assumptions about reality. Researchers develop meaningful constructions based on their experiences and through communication. Educators as action researchers will be examining the socially constructed reality of schools. In the United States, many of our concepts, knowledge, and beliefs about schooling have been socially constructed over the last hundred years. Action researchers will use methods of data collection that support their ontological stance and clarify their theoretical stance throughout the research process. Koshy offers another example of addressing the ontological challenges in the classroom: A teacher who was concerned with increasing her pupils' motivation and enthusiasm for learning decided to introduce learning diaries which the children could take home. The teacher reported in her field diary that the learning diaries stimulated the children's interest in her lessons, increased their capacity to learn, and generally improved their level of participation in lessons (Fade S., 2009).

The challenge for the teacher here is in the analysis and interpretation of the multiplicity of factors accompanying the use of diaries. The diaries were taken home so the entries may have been influenced by discussions with parents. Another possibility is that children felt the need to please their teacher. Another possible influence was that their increased motivation was as a result of the difference in style of teaching which included more discussions in the classroom based on the entries in the dairies. The students' engagement and interactions with this new style of learning were all based upon their socially constructed notions of learning inside and outside of the classroom. A researcher with a positivist ontological stance would not consider these factors, and instead might simply conclude that the dairies increased motivation and interest in the topic, as a result of introducing the diaries as a learning strategy. Epistemology, or the theory of knowledge, signifies a philosophical view of what counts as knowledge – it justifies what is possible to be known and what criteria distinguishes knowledge from beliefs. Positivist researchers, for example, consider knowledge to be certain and discovered through scientific processes. Action researchers collect data that is more subjective and examine personal experience, insights, and beliefs. Having given these definitions of ontology, it is now worth identifying the ontology of the first study. The first study uses an ontology which is essentially of a social world of meanings. In this world, researchers have to assume that the world they investigate is a world populated by human beings who have their own thoughts, interpretations and meanings. The researchers' investigation of this world is clearly manifested in their use of the different research methods and techniques of the interpretive design such as interviews and reaction papers written in a pronunciation course in response to challenging readings in order to interpret the students' feelings, and inner thoughts. Moreover, using the case study as a methodology of the present study focuses the students' opinions, feelings, experiences and inner thoughts. (Freese A. R., 1999).

Figure 2. Research methodology of classroom teachers

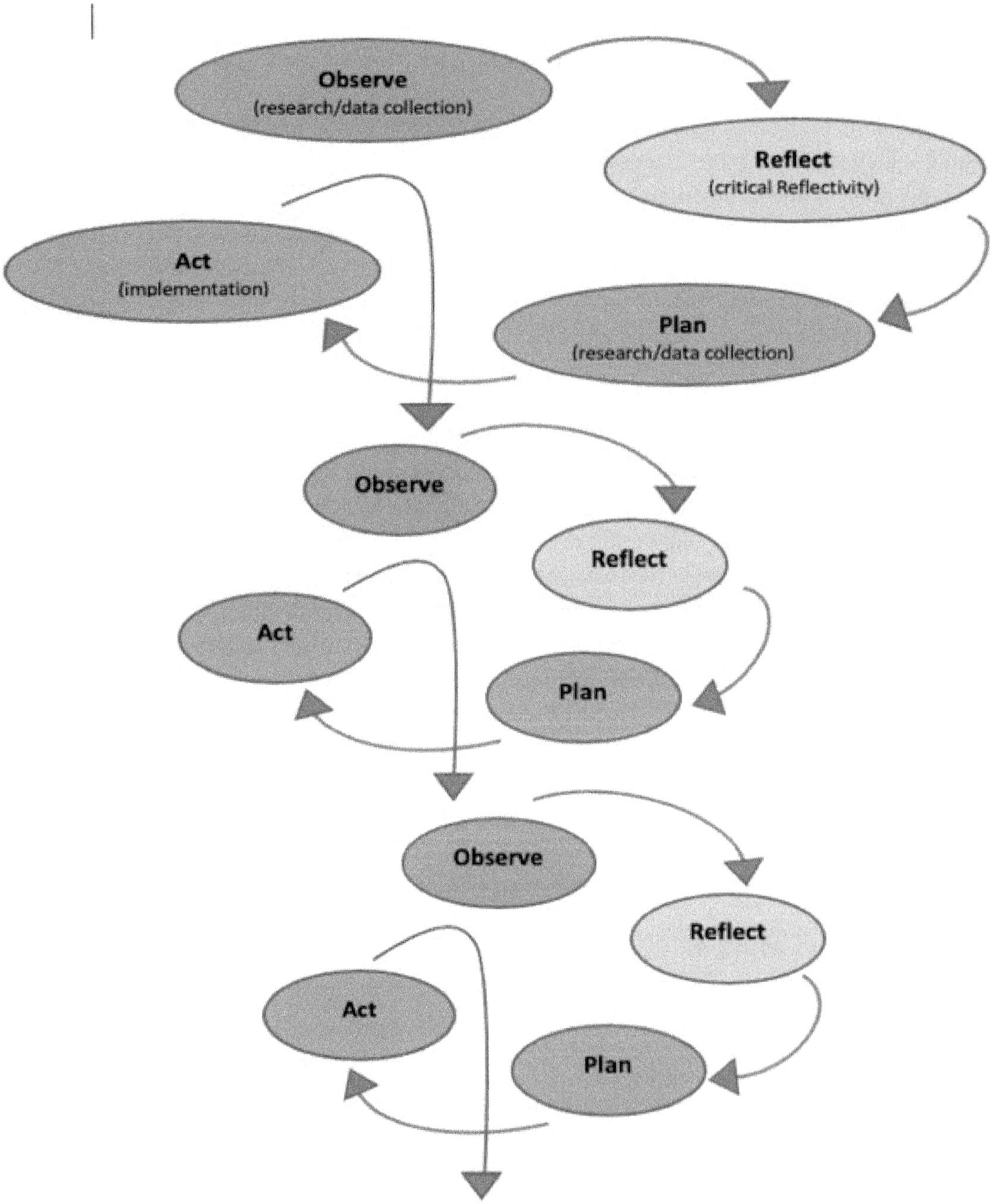

Purpose of Action Research: As one of many approaches to educational research, it is important to distinguish the potential purposes of action research in the classroom. This book focuses on action research as a method to enable and support educators in pursuing effective pedagogical practices by transforming the quality of teaching decisions and actions, to subsequently enhance student engagement and learning. Being mindful of this purpose, the following aspects of action research are important to consider as you contemplate and engage with action research methodology in your classroom:

- Action research is a process for improving educational practice. Its methods involve action, evaluation, and reflection. It is a process to gather evidence to implement change in practices.
- Action research is participative and collaborative. It is undertaken by individuals with a common purpose.
- Action research is situation and context-based.
- Action research develops reflection practices based on the interpretations made by participants.
- Knowledge is created through action and application.
- Action research can be based in problem-solving, if the solution to the problem results in the improvement of practice.
- Action research is iterative; plans are created, implemented, revised, then implemented, lending itself to an ongoing process of reflection and revision.
- In action research, findings emerge as action develops and takes place; however, they are not conclusive or absolute.

SOLUTIONS AND RECOMMENDATIONS

Objective of Current Context of Research: In thinking about the purpose of action research, it is helpful to situate action research as a distinct paradigm of educational research. I like to think about action research as part of the larger concept of living knowledge. Living knowledge has been characterized as "a quest for life, to understand life and to create… knowledge which is valid for the people with whom I work and for myself". Why should educators care about living knowledge as part of educational research? As mentioned above, action research is meant "to produce practical knowledge that is useful to people in the everyday conduct of their lives and to see that action research is about working towards practical outcomes". However, it is also about: creating new forms of understanding, since action without reflection and understanding is blind, just as theory without action is meaningless. The participatory nature of action research makes it only possible with, for and by persons and communities, ideally involving all stakeholders both in the questioning and sense making that informs the research, and in the action, which is its focus. Action research does not derive its meaning from an abstract idea, or a self-contained discovery – action research's meaning stems from the way educators negotiate the problems and successes of living and working in the classroom, school, and community. While we can debate the idea of action research, there are people who are action researchers, and they use the idea of action research to develop principles and theories to guide their practice. Action research, then, refers to an organization of principles that guide action researchers as they act on shared beliefs, commitments, and expectations in their inquiry. Many teacher educators recognize reflection as a critical component of teacher. Further, the reflective dimensions of action research have become the centerpiece of many preserves programs that view reflectivity as the central value of teacher education. Through the methodology of action research, teachers gain an understanding of self-assessment and reflection in order to raise questions about theory and practice. (Gilstrap, D. L.,2008).

Reflectivity of Research Objective: It is sometimes difficult to define, but when teaching is reflective, judgements become skilled and thoughtful, and practice is enriched and improved. Reflective practice incorporates active, persistent, and careful consideration of beliefs and practices with an understanding of the reasons behind and the subsequent consequences of specific actions. Teaching students to become reflective practitioners includes both strategy instruction as well as instruction on decision making that recognizes the interplay between beliefs, actions, and the influence of context. The concept of action research has assuming that learning the skills of action research automatically moves teachers toward reflective practice. As researchers remind us, there are many dimensions to reflectivity (e.g., making meaning through an examination of decision making in context, problem framing. However, it is highly misguided to believe that reflection will naturally emerge from the accumulation of teaching strategies and techniques. (Gustafson K.,1999).

Figure 3. Classroom teachers research analysis and data interpretation

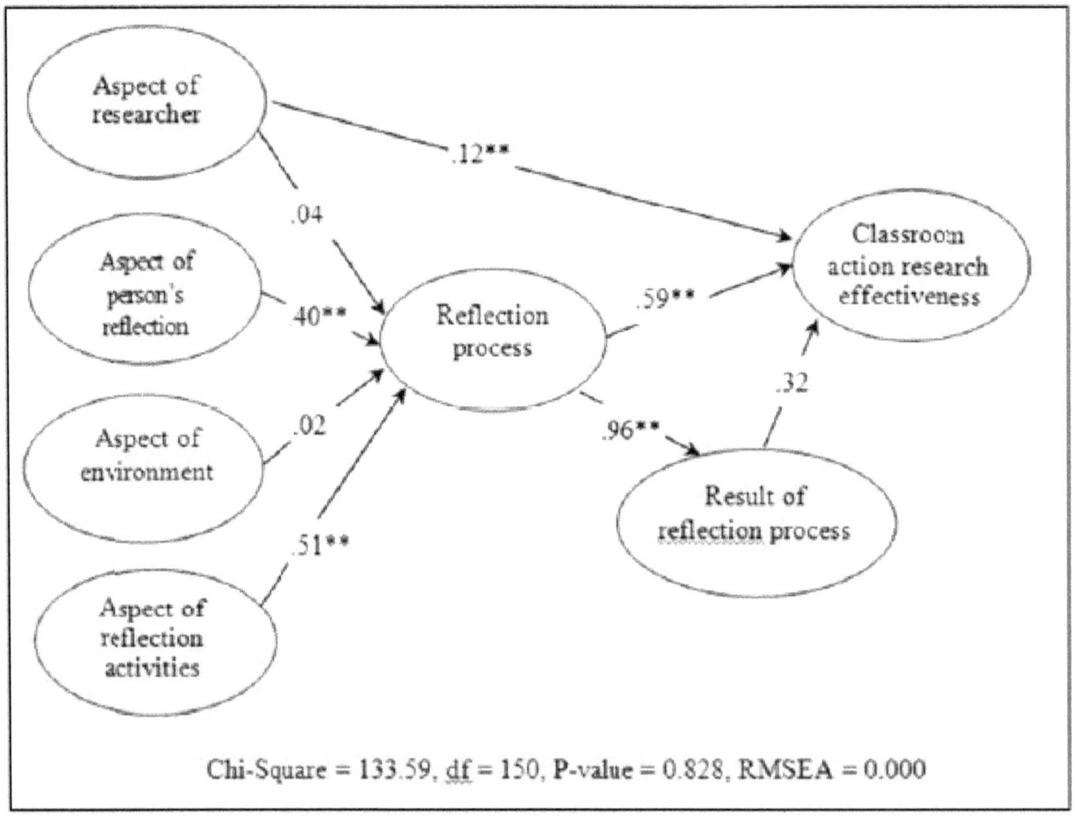

Data Sources: The study reported here used both quantitative and qualitative methods to investigate the impact of modeling action research on preserves teachers. I included an exit questionnaire with both closed-ended and open-ended questions. The quantitative portion of the questionnaire investigated preserves teachers' perceptions of the value of observing action research as a method for teaching the technical and reflective elements of an action research project. Exit questionnaires were scored and quantitatively analyzed. To capture the process of completing a selfstudy, I treated my teaching and self-reflection as case study. Utilizing my action research project as a model for preserves teachers, the case study chronicled a yearlong investigation of introducing action research to teacher education students in a fifth year, teacher licensure program. Additional analyses were conducted on the content of preserves teachers' action research projects, on transcripts from exit interviews with a representative sample of teacher candidates, and on teaching journals kept by the researcher. Open-ended questions on the exit questionnaire provided additional data for analysis. This case study was part of a larger study that investigated the impact of this teacher educator's action research project on he reading practices of preserves secondary teachers in a theories of instruction course. Qualitative data were analyzed using grounded theory. A matrix was constructed to facilitate data analysis. The researcher reviewed exit interview transcripts, student work samples, and journal entries. Through a process of constant comparison initial categories for coding were identified. A complex matrix was created that sought to reveal changes within preserves teachers as they experienced a modeled action research project. Triangulation was established through the examination of survey data, exit interviews, and through students' action research projects and the researcher's journal analyses. (Hoban, G., 2006)

Questions Arising of Research: Describe the ways in which observing an action research project helped you to understand the process required for completing a similar project on your own. Describe the ways in which the action research project conducted by the instructor in class prompted you to think about you own project. . Describe whether the steps required for completing an action research project seem clear. Discuss your understanding of the role of reflection when conducting action research. Don't worry so much about whether the project was a success per se. Describe some of the strengths of observing the instructor's action research project. Describe some of the limitations of observing the instructor's action research project. How could the process of modeling action research be improved? That is, when the instructor teaches her students about conducting action research in the future, what should she do differently? Is it important for an instructor to model an action research project? Why or why not? (Kreber C., 2005).

Procedure and Data Analysis: Participants: Twenty-eight preserves teachers were enrolled in a year-long, secondary licensure program. The preserves teachers were post-baccalaureate students across multiple content areas. All sought secondary licensure and were participants in a program in which cohort students completed course work in curriculum and instruction, action research, and student teaching. As the student teaching experience proceeds, candidates identify areas in their teaching in need of improvement for the focal points of their action research projects. For example, discrepancies between teacher expectations and student performance are identified through reviews of personal journals and data gathered from peer and/or site teacher interviews and teaching self-analyses. Following this baseline data collection, teacher candidates complete a four-stage process which includes the identification of a problem statement, the development of a plan of action, further data collection, and finally, an evaluation of outcomes in their teaching. During the final stage of the project, candidates gather and evaluate data stemming from their problem focus while evaluating the relationship between instruction, curriculum, and students. That is, candidates are asked to evaluate the merit and outcomes of systematically evaluating their teaching over time. This new perspective, known as social constructivism, considers the act of learning to be more than the construction of new meanings. In fact, it results from a dynamic interaction between personal change and social circumstances, and it involves cognition as well as feelings. An adequate view of teacher learning, therefore, should recognize this connection between internal development and the person's social landscape. To take a concrete example, having to explain our ideas so that other teachers can understand them and can interact with us and with each other forces team teachers to clarify meanings and find words for thoughts that might otherwise have been realized only through action. The collaboration required to do this provides teachers with rich opportunities to understand their tacit knowledge and support changing views of themselves as teachers (Roberts 1998). Research shows a link between a favourable school climate—one where teachers are encouraged to talk to each other—and teacher development. (Malkani J., 2005)

Result and Interpretation: There are many other models that supplement the basic process of action research with other aspects of the research process to consider. For example, illustrates a spiral model of action research proposed by Kemmis and McTaggart (2004). The spiral model emphasizes the cyclical process that moves beyond the initial plan for change. The spiral model also emphasizes revisiting the initial plan and revising based on the initial cycle of research:

Interpretation of action research spiral: Other models of action research reorganize the process to emphasize the distinct ways knowledge takes shape in the reflection process. O'Leary's model, for example, recognizes that the research may take shape in the classroom as knowledge emerges from the teacher's observations. O'Leary highlights the need for action research to be focused on situational understanding and implementation of action, initiated organically from real-time issues:

Interpretation of O'Leary's cycles of research: Lastly, Macintyre's model, offers a different characterization of the action research process. Macintyre emphasizes a messier process of research with the initial reflections and conclusions as the benchmarks for guiding the research process. Macintyre emphasizes the flexibility in planning, acting, and observing stages to allow the process to be naturalistic. Our interpretation of Macintyre process is below:

Interpretation of the action research cycle, Macintyre: We believe it is important to prioritize the flexibility of the process, and encourage you to only use these models as basic guides for your process. Your process may look similar, or you may diverge from these models as you better understand your students, context, and data. (Moran M. J., 2007).

FUTURE RESEARCH DIRECTIONS

Observation of Current Research: Each student was given a series of questions related to the project and told to write as many answers as possible. They then shared their responses with each other and I wrote them on the board. Most of the students said that they liked the project work, which is fundamental in keeping motivation high. They also said they felt comfortable in their groups and, more important, experienced a sense of belonging. Apparently, student-to-student interaction fosters confidence because it allows students to talk freely without any authority judging their ideas. The data gathered through the Students' rating scale, particularly Characteristics A, B, and I, supports this, as most of my students enjoyed working in groups and appeared to have been active participants who listened to each other. However, two groups admitted they had experienced discomfort due to the lack of affinity between the members of the group. This could explain the result for Characteristic C of the student scale, which shows that 60% of the students rarely encouraged or praised members of their groups. Their learning styles might have influenced their behaviour in groups, since those who prefer working individually and are fond of teacher-centred instruction might not have felt at ease, and therefore might have played a less cooperative role. As regards the use of the target language, the majority of students acknowledged that their interactions with other students were in Spanish, which is consistent with my colleague's observation in relation to Characteristic D (on the student scale) and my diary. According to the students, they sometimes needed help from me, and I guided their work. One student commented: "She did not impose opinions. She helped us and she was always looking at what we were doing." From this comment I infer that, although some students had a degree of dependence on me, most perceived me as monitoring their work and serving as an adviser. This conforms with my colleague's observation of my behaviour, specifically the results for Characteristics A, B, D, and E. Other students affirmed that they had shared information to help each other or solved problems independently. They also made comments suggesting that they were able to manage their own learning and could rely on themselves. Thus, there seems to be a correlation between the students' opinions about their own cooperative and independent behaviour and what my colleague found as she observed them, particularly with respect to Characteristic E (on the student scale). The students expressed satisfaction with certain aspects of cooperative teaching, such as assessing each other's work, which earlier—before I learned to relinquish control— might have seemed foreign to them. They indicated their satisfaction by such comments as "I liked looking for information!" and "I enjoyed correcting my partners' work!" Such comments support the results for both the students' behaviour (Characteristic F) and the teacher's behaviour (Characteristic C). The students also found dividing work into different roles very successful, as evidenced by such comments as "I was the editor and I learnt how to control what was going on in the group!" and "We could organise ourselves in a better way!" These statements are consistent with the information collected by my colleague (student Characteristic H), who detected a high degree of effective work. The most interesting task for most of the students was creating their own comic strips. Other students were pleased about getting to know the history of comic strips. Some groups said they were most frustrated by being unable to edit videos. Still others claimed that they did not work cooperatively and in a responsible manner. However, this last claim does not seem to correspond with the data gathered by the observer, which indicated that the students worked cooperatively to a high degree (Characteristic G). To improve the project, the students suggested getting started earlier to allow more time to organise themselves and carry out their tasks. (Randle H., 2007)

Discussion: The basic principle underpinning action research: It is that this research involves "identifying a problematic area, imagining a possible solution, trying it out, evaluating it (did it work?), and changing practice in the light of the evaluation". This is a basic problem-solving process. However, to turn it into an action research process, researchers need to state why they want to examine or explore a particular issue and collect information or data to show the process. Such information or data act as evidence in terms of whether the researchers believed they were moving in the direction they were anticipating reaching in the first place. Essentially, the methodology of action research is that researchers need to evaluate what they are researching, and continually ensure that what they are researching is actually working and reaching the desired objective/s. Action research cannot be conducted on a once-off basis, but is rather a continuous process – hence its cyclical or spiral nature. Instead of a linear model, action research advances through cycles, 'starting' with reflection on action, and proceeding round to new action which is then further researched. (Redmond B., 2004).

Participatory action research: Although there does not always seem to be a clear distinction between action research and participatory action research (PAR), and in fact it is at times used interchangeably, we have felt it necessary to differentiate between the two. Reason talks about action research, participatory research, action learning, action science, action enquiry and co-operative enquiry, which in fact are "all contemporary forms of action oriented research which place emphasis on a full integration of action and reflection, so that the knowledge developed in the inquiry process is directly relevant to the issues being studies". Huizer points out that in grassroots work in Third World countries a form of action research has emerged which tried to utilize the research itself as well as knowledge acquired through it, to enhance the grip of the local people, the participants on their own communities. Greenwood et al., continue by reasoning that participatory action research "enhances problem formulation, hypothesis formulation, data acquisition, data analysis, synthesis, and application". Thus, conducting participatory action research is a very collaborative process between researchers and members of a group or organization. Participatory action research has as its main aim the finding of solutions to tangible and concrete problems and disagreements. The results of such research, though, if performed systematically and consistently, also contribute to a greater knowledge of conflict-solving methods as such, which can be applied to a range of concrete situations. This is essential for the replication elsewhere of general theoretical knowledge in this area, from which grassroots groups in different circumstances can benefit (Huizer, 1997). The distinctive features of participatory research are, first, the group or community's participation in the whole research activity which, second, is a process in which research is directly related to transformative actions. (*Sambrook, S., 2008)*

Selener provides participatory action research approaches which have been developed and applied in four broad areas: Participatory research in community development has its roots in Latin America where it developed into its present form and where its main ideology evolved in the early 1960s. Liberation theology and the sociology of liberation were two participator research approaches that showed the commitment of Latin American social scientists to become active participants in the liberation of the poor and oppressed. Participatory research has been inspired by Paulo Freire, critical consciousness, conscientization, and empowerment. Freire introduced the concept of conscientization, meaning "the identification and critical analysis of social, political, and economic contradictions, leading to organized action to solve immediate problems and to counter the oppressive aspects of society. Dresser tells us that a "common element among schools reporting academic success is that they all have a systematic process for promoting students' social-emotional skills." The caring and empathy exuded by students in these schools permeates the school culture and climate. Dresser demonstrates the effect of embedding social-emotional standards within the language arts classroom. Dresser has found that most of her preserves teacher candidates did not have understanding of the importance of emotions in student learning and achievement (*Seng T. W., 2008).*

The Relationship between Different Orientations of Action Research and Reflective Practice: Examination of the action research literature reveals that its goals like those of reflective practice are change and improvement. The ways in which the reflective inquiry processes within action research are framed and interpreted, however, relate to the underpinning epistemology of the action research model being employed – whether, for example, the goal is for teachers to become more effective or efficient or empowered. The relationship between particular models of action research and types of reflective processes is graphically demonstrated by Grundy's typology, that distinguishes between technical, practical and emancipator models of action research. Technical action research seeks to deliver more efficient effective practice 'through the practical skill of the participants'. With its emphasis on orderly sequencing of research and the use of methods associated with the social science paradigm, it seems to require technical research expertise which may be externally provided, and thus this model presents less opportunity for the development of teachers' capacities for reflection in or on practice. Altrichter & Posch warn that this can also restrict practitioners to a subordinate, technical-rational approach to teaching without 'developing reflective features of professional action'. Reflection in this approach is likely to be related to solving immediate pressing problems of efficient and effective delivery of curricula. It is likely to be orientated towards the development of pedagogical strategies, skills and tactics In contrast, practical action research aims to improve practice through the application of practical judgement and the accumulated personal wisdom of the teacher. Models of action research falling within this frame are as concerned with process as the end product of inquiry. They rely not only on the exercise of moral and practical judgement by teachers, but also their capacities to identify issues or problems that are salient to this in these professional contexts. Reflective processes within this action research approach are likely to focus upon building teachers' capacities to self. Research within cognitive science demonstrates also that the art of problem finding (a highlevel meta-cognitive skill) is a much rarer commodity than that of problem solving. Teachers themselves have also been shown to hold a tendency to 'technologies' action – reflection cycles and, in so doing, trivialize the opportunities for real engagement with and ownership of a meaningful, selfreflexive inquiry. This results not only from personal history and disposition, but also from school culture and the busyness of school life. So for personal, social, contextual and psychological reasons, there are limitations in both the conceptualization and conduct of action research and reflection *(Taylor.,2010).*

Discussion Drawned Definitions of Action Research and Examples: At this point, it may be helpful for readers to have a working definition of action research and some examples to illustrate the methodology in the classroom. Bassey offers a very practical definition and describes "action research as an inquiry which is carried out in order to understand, to evaluate and then to change, in order to improve educational practice." Cohen and Manion situate action research differently, and describe action research as emergent, writing: essentially an on-the-spot procedure designed to deal with a concrete problem located in an immediate situation. This means that ideally, the step-by-step process is constantly monitored over varying periods of time and by a variety of mechanisms (questionnaires, diaries, interviews and case studies, for example) so that the ensuing feedback may be translated into modifications, adjustment, directional changes, redefinitions, as necessary, so as to bring about lasting benefit to the ongoing process itself rather than to some future occasion. Lastly, Koshy describes action research as: a constructive inquiry, during which the researcher constructs his or her knowledge of specific issues through planning, acting, evaluating, refining and learning from the experience. It is a continuous learning process in which the researcher learns and also shares the newly generated knowledge with those who may benefit from it. These definitions highlight the distinct features of action research and emphasize the purposeful intent of action researchers to improve, refine, reform, and problem-solve issues in their educational context (*Watts M., 2009).*

To better understand the distinctness of action research, these are some examples of action research topics:

Examples of Action Research Topics

i. Flexible seating in 4th grade classroom to increase effective collaborative learning.
ii. Structured homework protocols for increasing student achievement.
iii. Developing a system of formative feedback for 8th grade writing.
iv. Using music to stimulate creative writing.
v. Weekly brown bag lunch sessions to improve responses to PD from staff.

vi. Using exercise balls as chairs for better classroom management.

Theorizing Action Research for the Classroom: All research, at its core, is for the purpose of generating new knowledge and contributing to the knowledge base of educational research. Action researchers in the classroom want to explore methods of improving their pedagogy and practice. The starting place of their inquiry stems from their pedagogy and practice, so by nature the knowledge created from their inquiry is often contextually specific to their classroom, school, or community. Therefore, we should examine the theoretical underpinnings of action research for the classroom. It is important to connect action research conceptually to experience; for example, Levin and Greenwood make these connections:

i. Action research is context bound and addresses real life problems.
ii. Action research is inquiry where participants and researchers cogenerate knowledge through collaborative communicative processes in which all participants' contributions are taken seriously.
iii. The meanings constructed in the inquiry process lead to social action or these reflections and action lead to the construction of new meanings.
iv. The credibility/validity of action research knowledge is measured according to whether the actions that arise from it solve problems (workability) and increase participants' control over their own situation.

Educators who engage in action research will generate new knowledge and beliefs based on their experiences in the classroom. Let us emphasize that these are all important to you and your work, as both an educator and researcher. It is these experiences, beliefs, and theories that are often discounted when more official forms of knowledge (e.g., textbooks, curriculum standards, districts standards) are prioritized. These beliefs and theories based on experiences should be valued and explored further, and this is one of the primary purposes of action research in the classroom. These beliefs and theories should be valued because they were meaningful aspects of knowledge constructed from teachers' experiences. Developing meaning and knowledge in this way forms the basis of constructivist ideology, just as teachers often try to get their students to construct their own meanings and understandings when experiencing new ideas. *(Wiratchai N., 2009).*

Application of Action Research in Theory: Action research-based inquiry in educational contexts and classrooms involves distinct participants – students, teachers, and other educational stakeholders within the system. All of these participants are engaged in activities to benefit the students, and subsequently society as a whole. Action research contributes to these activities and potentially enhances the participants' roles in the education system. Participants' roles are enhanced based on two underlying principles:

communities, schools, and classrooms are sites of *socially mediated actions,*and action research provides a greater understanding of self and new knowledge of how to negotiate these socially mediated environments; Social systems are based on beliefs, and generally, beliefs about what will benefit the greatest amount of people in that society. Beliefs, and more specifically the rationale or support for beliefs, are not always easy to demonstrate or observe as part of our reality. While we can observe the students writing, and we can assess their ability to develop a written argument, it is difficult to observe the students' understanding of argumentative writing and its purpose in their future.

communities, schools, and classrooms are part of social systems in which humans interact with many cultural tools, and action research provides a basis to construct and analyze these interactions. In our quest for knowledge and understanding, we have consistently analyzed human experience over time and have distinguished between types of reality. Humans have constantly sought "facts" and "truth" about reality that can be empirically demonstrated or observed. The knowledge that is conveyed in a classroom is bound to, and justified by, a social system. A postmodernist approach to understanding our world seeks knowledge within a social system, which is directly opposed to the empirical or positivist approach which demands evidence based on logic or science as rationale for beliefs. Action research does not rely on a positivist viewpoint to develop evidence and conclusions as part of the research process. Action research offers a postmodernist stance to epistemology (theory of knowledge) and supports developing questions and new inquiries during the research process. In this way action research is an emergent process that allows beliefs and decisions to be negotiated as reality and meaning are being constructed in the socially mediated space of the classroom. (Sekaran U. 2003)

The Role of Critically Reflective Practice in Teachers' Professional Development and its Relationship with Action Research: 'Reflection' and 'critically reflective practice' are considered central to the professional development of teachers as teaching and learning involve complex processes that occur in diverse multidimensional environments and contexts. There is a general assumption that there is not necessarily one right approach to teaching and learning. Evaluating teaching through deliberation, articulation and reconstructing past understandings, experiences and practices can lead to improvement and avoid perpetuating ineffective practice (Silverman D., 2010)

Critically reflective practice provides a means from which to challenge personal assumptions and values about teaching and the theories that are used to drive the approaches employed. In this respect, its relationship to 'action research' becomes even more pronounced as being a critically reflective practitioner is considered essential to the professional development and growth of teachers who are regarded as inquirers engaging in both self-study and collaborative research within or outside of their educational contexts, schools and classrooms. Driven by the Aristotelian concept of praxis, Grundy's participatory action research sought to understand how external societal influences may underpin individual practices. Central to this work is the concept of 'criticality', the turning points that provide insight as to where the boundaries of knowledge and understanding begin and end, and the degree of their permeability on developing practice. (Tashakkori A., 2003)

Critical praxis for educators is intended to "move beyond the constraints of formal teaching, knowledge and curriculum and instead encourages communities, teachers and students to work together in producing new understandings and practices". Grundy advocated the need for deliberate and deep reflection that required of practitioners that they make their understandings of their work explicit, which required an examination of how those understandings emerged and were shaped by the conditions and contexts of their work. Accountability and standardization continue to dictate the culture of positivism, which in turn exerts a profound impact on the nature of educational research. If teachers are to be researchers, provide evidence-based practice and create knowledge about teaching, then they need to develop critical perspectives and understanding of the ontology of positivism. "Only then will teachers understand positivistic research and be empowered to act in opposition to the policy implications which come from it". This paradigm shift in the role of the teacher being one that implements existing theory in practice to generating knowledge of practice is consistent with contemporary expectations of the profession. It not only ensures that teaching and learning is progressive, innovative and responsive to the needs and requirements of society, but also that teachers can provide evidence for what they do, how and why. Factors that influence the extent to which this may occur can be attributed to teacher experience and career stage, the effects of organizational and cultural contexts that may impact on opportunities for professional growth, and because teachers may have varying ways to evaluate practice through reflection, if at all. Crawford's study that explored teachers' implementation of high impact teaching strategies exemplified the importance of establishing a consistent and rigorous approach for engaging with critically reflective practice. The principles and techniques that are embedded within action research, provide both the systematic rigour and flexibility necessary to develop an evidence-base for practice in a range of diverse and complex education contexts and settings such as that in the study mentioned. (Zikmund W. G., 2003)

CONCLUSION AND FUTURE SCOPE

Epistemology, Ontology, Reflection and the Process of Action Research for Classroom Teachers for Indigenous Knowledge System(IKS): When an individual engages in reflection on their actions or experiences, it is typically for the purpose of better understanding those experiences, or the consequences of those actions to improve related action and experiences in the future. Reflection in this way develops knowledge around these actions and experiences to help us better regulate those actions in the future. The reflective process generates new knowledge regularly for classroom teachers and informs their classroom actions. Unfortunately, the knowledge generated by educators through the reflective process is not always prioritized among the other sources of knowledge educators are expected to utilize in the classroom. Educators are expected to draw upon formal types of knowledge, such as textbooks, content standards, teaching standards, district curriculum and behavioral programs, etc., to gain new knowledge and make decisions in the classroom. While these forms of knowledge are important, the reflective knowledge that educators generate through their pedagogy is the amalgamation of these types of knowledge enacted in the classroom. Therefore, reflective knowledge is uniquely developed based on the action and implementation of an educator's pedagogy in the classroom. Action research offers a way to formalize the knowledge generated by educators so that it can be utilized and disseminated throughout the teaching profession. Research is concerned with the generation of knowledge, and typically creating knowledge related to a concept, idea, phenomenon, or topic. Action research generates knowledge around inquiry in practical educational contexts. Action research allows educators to learn through their actions with the purpose of developing personally or professionally. Due to its participatory nature, the process of action research is also distinct in educational research. There are many models for how the action research process takes shape. I will share a few of those here. Each model utilizes the following processes to some extent: Plan a change; Take action to enact the change; Observe the process and consequences of the change; Reflect on the process and consequences; Act, observe, & reflect again and so on.

Classroom Teachers Constructing their Own Knowledge Most of you are probably at least minimally familiar with constructivism, or the process of constructing knowledge. However, what is constructivism precisely, for the purposes of action research? Many scholars have theorized constructivism and have identified two key attributes:

- Knowledge is not passively received, but actively developed through an individual's cognition;
- Human cognition is adaptive and finds purpose in organizing the new experiences of the world, instead of settling for absolute or objective truth.

Managerial Implications and Challenges: Considering these two attributes, constructivism is distinct from conventional knowledge formation because people can develop a theory of knowledge that orders and organizes the world based on their experiences, instead of an objective or neutral reality. When individuals construct knowledge, there are interactions between an individual and their environment where communication, negotiation and meaning-making are collectively developing knowledge. For most educators, constructivism may be a natural inclination of their pedagogy. Action researchers have a similar relationship to constructivism because they are actively engaged in a process of constructing knowledge. However, their constructions may be more formal and based on the data they collect in the research process. Action researchers also are engaged in the meaning making process, making interpretations from their data. These aspects of the action research process situate them in the constructivist ideology. Just like constructivist educators, action researchers' constructions of knowledge will be affected by their individual and professional ideas and values, as well as the ecological context in which they work.

The relations between constructivist inquiry and action research is important, as Lincoln states: much of the epistemological, ontological, and axiological belief systems are the same or similar, and methodologically, constructivists and action researchers work in similar ways, relying on qualitative methods in face-to-face work, while buttressing information, data and background with quantitative method work when necessary or useful. While there are many links between action research and educators in the classroom, constructivism offers the most familiar and practical threads to bind the beliefs of educators and action researchers. It is also important for educators to consider the philosophical stances related to action research to better situate it with their beliefs and reality. When researchers make decisions about the methodology they intend to use, they will consider their ontological and epistemological stances. It is vital that researchers clearly distinguish their philosophical stances and understand the implications of their stance in the research process, especially when collecting and analyzing their data. In what follows, we will discuss ontological and epistemological stances in relation to action research methodology. Ontology, or the *theory of being,* is concerned with the claims or assumptions we make about ourselves within our social reality – what do we think exists, what does it look like, what entities are involved and how do these entities interact with each other. In relation to the discussion of constructivism, generally action researchers would consider their educational reality as socially constructed. Social construction of reality happens when individuals interact in a social system. Meaningful construction of concepts and representations of reality develop through an individual's interpretations of others' actions. These interpretations become agreed upon by members of a social system and become part of social fabric, reproduced as knowledge and beliefs to develop assumptions about reality. Researchers develop meaningful constructions based on their experiences and through communication. Educators as action researchers will be examining the socially constructed reality of schools. In the United States, many of our concepts, knowledge, and beliefs about schooling have been socially constructed over the last hundred years. For example, a group of teachers may look at why fewer female students enroll in upper-level science courses at their school. This question deals directly with the social construction of gender and specifically what careers females have been conditioned to pursue. We know this is a social construction in some school social systems because in other parts of the world, or even the United States, there are schools that have more females enrolled in upper level science courses than male students. Therefore, the educators conducting the research have to recognize the socially constructed reality of their school and consider this reality throughout the research process. Action researchers will use methods of data collection that support their ontological stance and clarify their theoretical stance throughout the research process.

Koshy offers another example of addressing the ontological challenges in the classroom: A teacher who was concerned with increasing her pupils' motivation and enthusiasm for learning decided to introduce learning diaries which the children could take home. They were invited to record their reactions to the day's lessons and what they had learnt. The teacher reported in her field diary that the learning diaries stimulated the children's interest in her lessons, increased their capacity to learn, and generally improved their level of participation in lessons. The challenge for the teacher here is in the analysis and interpretation of the multiplicity of factors accompanying the use of diaries. The diaries were taken home so the entries may have been influenced by discussions with parents. Another possibility is that children felt the need to please their teacher. Another possible influence was that their increased motivation was as a result of the difference in style of teaching which included more discussions in the classroom based on the entries in the dairies. Here you can see the challenge for the action researcher is working in a social context with multiple factors, values, and experiences that were outside of the teacher's control. The teacher was only responsible for introducing the diaries as a new style of learning. The students' engagement and interactions with this new style of learning were all based upon their socially constructed notions of learning inside and outside of the classroom. A researcher with a positivist ontological stance would not consider these factors, and instead might simply conclude that the dairies increased motivation and interest in the topic, as a result of introducing the diaries as a learning strategy. Epistemology, or the *theory of knowledge,* signifies a philosophical view of what counts as knowledge – it justifies what is possible to be known and what criteria distinguishes knowledge from beliefs. Positivist researchers, for example, consider knowledge to be certain and discovered through scientific processes. Action researchers collect data that is more subjective and examine personal experience, insights, and beliefs. Action researchers utilize interpretation as a means for knowledge creation. Action researchers have many epistemologies to choose from as means of situating the types of knowledge they will generate by interpreting the data from their research. For example, Koro-Ljungberg et al., identified several common epistemologies in their article that examined epistemological awareness in qualitative educational research, such as: objectivism, subjectivism, constructionism, contextualism, social epistemology, feminist epistemology, idealism, naturalized epistemology, externalism, relativism, skepticism, and pluralism. All of these epistemological stances have implications for the research process, especially data collection and analysis.

Researcher Positionality in Action Research: In this first chapter, we have discussed a lot about the role of experiences in sparking the research process in the classroom. Your experiences as an educator will shape how you approach action research in your classroom. Your experiences as a person in general will also shape how you create knowledge from your research process. In particular, your experiences will shape how you make meaning from your findings. It is important to be clear about your experiences when developing your methodology too. This is referred to as researcher positionality. Maher and Tetreault define positionality as: Gender, race, class, and other aspects of our identities are markers of relational positions rather than essential qualities. Knowledge is valid when it includes an acknowledgment of the knower's specific position in any context, because changing contextual and relational factors are crucial for defining identities and our knowledge in any given situation. By presenting your positionality in the research process, you are signifying the type of socially constructed, and other types of, knowledge you will be using to make sense of the data. As Maher and Tetreault explain, this increases the trustworthiness of your conclusions about the data. This would not be possible with a positivist ontology. We will discuss positionality more in chapter 6, but we wanted to connect it to the overall theoretical underpinnings of action research.

REFERENCES

Fade, S. (2009). *Learning and assessing through reflection: a practical guide.* [Online]. Available from: from www.practicebasedlearning.org [2009, June 12]

Freese, A. R. (1999). The role of reflection on preservice teachers' development in the context of a professional development school. *Teaching and Teacher Education, 15*(8), 895–909. DOI: 10.1016/S0742-051X(99)00029-3

Gilstrap, D. L., & Dupree, J. (2008). A regression model of predictor variables on critical reflection in the classroom: Integration of the critical incident questionnaire and the framework for reflective thinking. *Journal of Academic Librarianship, 34*(6), 469–481. DOI: 10.1016/j.acalib.2008.09.019

Gustafson, K., (1999). Issue and difficulties in promoting learner reflection: results from a three – year study. *Critical reflection.*

Hoban, G., & Hastings, G. (2006). Developing different forms of student feedback to promote teacher reflection: A 10-year collaboration. *Teaching and Teacher Education, 22*(8), 1006–1019. DOI: 10.1016/j.tate.2006.04.006

Koro-Ljungberg, M., Yendol-Hoppey, D., Smith, J. J., & Hayes, S. (2009). (E)pistemological awareness, instantiation of methods, and uninformed methodological ambiguity in qualitative research projects. *Educational Researcher, 38*(9), 687–699. DOI: 10.3102/0013189X09351980

Kreber, C. (2005). Reflection on teaching and the scholarship of teaching: Focus on science instructors. *Higher Education, 50*(2), 323–359. DOI: 10.1007/s10734-004-6360-2

Malkani, J., & Allen, J. (2005). *Cases in teacher education: beyond reflection into practice.*

Moran, M. J. (2007). Collaborative action research and project work: Promising practices for developing collaborative inquiry among early childhood preservice teachers. *Teaching and Teacher Education, 23*(4), 418–431. DOI: 10.1016/j.tate.2006.12.008

Randle, H., & Tilander, K. (2007). Reflection: A method for organizational and individual development. *US-China Education Review., 4*(9), 1–10.

Redmond, B. (2004). *Reflection in action: developing reflective practice in health and social services.* Ashgate.

Sambrook, S., & Stewart, J. (2008). Developing critical reflection in professional focused doctorates: A facilitator's perspective. *Journal of European Industrial Training, 32*(5), 359–373. DOI: 10.1108/03090590810877085

Sekaran, U. (2003). *Research methods for business: A skill building approach* (4th ed.). John Wiley & Sons, Inc.

Seng, T. W. (2008). *Measuring practicum student teachers' reflectivity: the reflective pedagogical thinking scale.*

Silverman, D. (2010). *Doing qualitative research* (3rd ed.). Sage publications, Inc.

Smith, G. (1987). Whatever happened to educational priority areas? *Oxford Review of Education*, *13*(1), 23–39. DOI: 10.1080/0305498870130103

Stenhouse, L. (Ed.). (1975). *An Introduction to Curriculum Research and Development London*. Heinemann.

Tashakkori, A., & Teddlie, C. *Handbook of mixed methods in social & behavioral research*. Califorina: Sage publications, Inc. 2003.

Taylor. (2010). *Transformative learning theory-An overview.*

Watts, M., & Lawson, M. (2009). Using a meta-analysis activity to make critical reflection explicit in teacher education. *Teaching and Teacher Education*, *25*(5), 609–616. DOI: 10.1016/j.tate.2008.11.019

Wiratchai, N. (2009). *Research and statistics: questions and answers.* Bangkok: Icon printing.

Zikmund, W. G. (2003). *Business research methods* (7th ed.). Thomson/South-Western.

KEY TERMS AND CONDITIONS

Educational Research: It provides a vast landscape of Indigenous knowledge System(IKS) on topics related to teaching and learning, curriculum and assessment, students' cognitive and affective needs, cultural and socio-economic factors of schools, and many other factors considered viable to improving schools.

Reflection: It is 'the deliberate and purposeful act of thinking which centers on ways of responding to problem situations'.

Approaches of Action Research: Action research is often referred to as practitioner-based research as it involves practitioners, and is also known as self-reflective practice as it involves individuals or practitioners reflecting on their own work.

Participatory Action Research: Dick states that "participation, by building shared understanding and shared commitment, increases the motivation for collective and collaborative action". Wadsworth stresses that such research "is not participation followed by research and then hopefully action. Instead there are countless tiny cycles of participatory reflection on action, learning about action and then new informed action which is in turn the subject of further reflection".

Epistemology: It deals with 'the nature of knowledge, its possibility, scope and general basis' as cited by Crotty. Furthermore, Maynard as cited by Crotty explains that 'epistemology is concerned with providing a philosophical grounding for what kinds of knowledge are possible and how we can ensure that they are both adequate and legitimate'.

The Social Element of Action Research: It implies that the social constructivist perception recognizes collaborative dialogue and talk to be central to "the process of continual intellectual, experiential and attitudinal growth of teachers".

Constructionism: It refers to the meaning that comes into existence in and out of human engagement with the realities in the world due to there is no truth waiting to discover it as well no meaning without a mind.

Chapter 4
Community-Driven Holistic Food Systems for Sustainable Food Security:
Lessons From Dayalbagh, India

Jyoti Gogia
https://orcid.org/0000-0002-4968-5915
Dayalbagh Educational Institute, India

Neha Sikarwar
https://orcid.org/0000-0003-0587-4308
Dayalbagh Educational Institute, India

ABSTRACT

An inclusive and participative food system is essential to ensure sufficient nutritious food for all while considering sustainability. Agroecological practices require scientific expertise and farmers' familiarity with the local situation, across food systems for sustainable transformations. The co-creation of two knowledge systems—practice and research—rather than a top-down approach, can significantly contribute to comprehensive agroecological evolution while preserving community traditions. The United Nations Food and Agriculture Organization (FAO) has stressed the importance of integrating spiritual elements into the food system. Many valuable practices rooted in indigenous and local knowledge can be observed in diverse communities. Consequently, embracing and learning from these traditions can greatly support the sustainable advancement of humanity. The Dayalbagh community has inventively ensured food security by building a robust and sustainable food system that is both holistic and resilient. Additionally, it upholds social, economic, environmental, and cultural values.

INTRODUCTION

> "The individual effort has led to selfishness and inhumanity, whereas the collective effort can abate both the evils if it does not remove them altogether". – Mohandas Karamchand Gandhi

DOI: 10.4018/979-8-3693-3542-0.ch004

The United Nations established the Sustainable Development Goals (SDGs) as a comprehensive framework to foster sustainable development for all. These goals highlight the significance of inclusivity and collective advancement in creating a brighter future for everyone. They specifically aim to eliminate hunger, food insecurity, and all forms of malnutrition by the year 2030. The Global Hunger Index (2023) indicated that the global hunger score had dropped less than one point from 19.1 in 2015 to 18.3 in 2023. This suggests that progress in reducing hunger worldwide has been stationary. Additionally, an estimated 28.9 percent (2.33 billion) of the world's population experienced moderate or severe food insecurity in 2023 (FAO et al., 2024). Despite significant economic growth, many people continuously experience poverty, food insecurity, hunger, and malnutrition (Narayanan, 2015; World Bank, 2018). Globally in 2022, 149 million children under 5 were estimated to be stunted (too short for age), 45 million were estimated to be wasted (too thin for height), and 37 million were overweight or living with obesity (WHO, 2024). The pursuit of economic progress often leads to social and environmental challenges. The global food systems need transformation to become more sustainable, resilient, and inclusive to tackle these issues, as emphasized in the 2030 Agenda of Sustainable Development goals.

Sustainable Development Goal 2, one of the 17 radical goals, strives to eradicate hunger, ensure food security, enhance nutrition, and foster sustainable agriculture, aiming for Zero Hunger by 2030. The Food and Agriculture Organization (2018) also imagines a future in which nourishing food is easily accessible to everyone and where natural resources are managed to maintain ecosystem integrity, guaranteeing the satisfaction of present and future human needs. An inclusive, robust, well-coordinated and participatory food system has the potential to provide an adequate supply of nutritious food while adhering to principles of sustainability. It is essential to focus on the intricate connections within food systems and how they intersect with various dimensions of food security (Sikarwar and Gogia, 2023). Recognizing these interrelationships allows us to effectively address the challenges of feeding the global population in a manner that is both equitable and environmentally responsible.

The food system is a complex network of processes that incorporates interconnected activities from production to consumption (farm to fork). Agroecological practices require scientific expertise and farmers' familiarity with the local situation, across food systems for sustainable transformations. Indigenous knowledge is gradually acknowledged as an invaluable asset in addressing global issues like climate change, biodiversity loss, and food insecurity. It provides distinctive perspectives and solutions that can supplement modern scientific knowledge. The essential contribution of Indigenous peoples was acknowledged in the 2007 UN Declaration on the Rights of Indigenous Peoples (UNDRIP). The Sustainable Development Goals (SDGs) reference Indigenous peoples and affirm that sustainable development cannot be achieved without safeguarding their traditional knowledge and territories (Campbell, J.Y., 2019).

Indigenous peoples play a key role in advancing sustainable development goals, particularly in addressing hunger, poverty, well-being, inequality, justice, and the impacts of climate change and unsustainable global food supplies (FAO, 2021; Kuhnlein and Chotiboriboon, 2022). They have profound relationships with food systems, reflecting vital knowledge for climate resilience. Indigenous communities, as custodians of a vast portion of the world's remaining biodiversity, possess a deep understanding of sustainable agriculture that can provide invaluable guidance for transforming global food systems (Nicoline, 2024; Vijayan et al., 2022).

Although traditional techniques have demonstrated their effectiveness through farmer-to-farmer transfers or local technology transfers, they have not been sufficiently documented and are often replaced by more modern methods (Melash et al., 2023; Mishra et al., 2020; Sharma et al., 2020). Unfortunately, the valuable insights from Indigenous practices are frequently neglected in policy and practical applications.

Many Indigenous peoples embrace a holistic view of health, encompassing physical, social, mental, and spiritual dimensions. This perspective aligns with the Food and Agriculture Organization's emphasis on the importance of spiritual elements within food systems. These are closely connected to cultural food practices and the biodiversity that supports food system resilience (Kuhnlein and Chotiboriboon, 2022). Indigenous knowledge systems are becoming vital with an institutional resilience approach i.e. learning local actions to have a global impact (Lugo-Morin, 2020). Therefore, it is essential to adopt a holistic approach, effectively utilized by many Indigenous communities, who have drawn upon centuries of wisdom, knowledge, and a deep connection to the land. Their practices can serve as valuable guidance in our efforts toward sustainability (Nicoline 2024).

By combining the knowledge gained from practical experience and formal research, instead of a top-down approach, a significant contribution can be made to the comprehensive development of agro-ecology while maintaining community traditions. This method offers a dynamic and adaptable approach that can positively address the increasingly intricate challenges encountered within the agriculture and food system (Utter et al., 2021).

METHODS AND MATERIALS

The study has been carefully crafted with captivating and interactive approaches to gather the most in-depth and precise information possible. This includes conducting key informant interviews, engaging in focus-group discussions, embarking on field visits, administering stimulating tests, and thorough validation follow-ups.

The Dayalbagh community has been chosen as a case for in-depth study. There are overall six *mohallas* (colonies) in Dayalbagh. To ensure a representative sample, 10 households were randomly selected from each *Mohalla*, thus, 60 households from the entire community. A pre-tested self-structured questionnaire was employed to investigate the residents' consumption patterns, food security, dietary diversity, and health and hygiene awareness. Additionally, interviews were conducted with community officials to assess the status of the food systems. Quantitative and qualitative data analysis has been incorporated into the study. The following statistical tools and techniques have been incorporated-

The Household Food Insecurity Access Scale (HFIAS) measures household-level food security by focusing on access. It includes nine occurrence questions and nine follow-up questions, referring to experiences of the past 30 days. The method encapsulates household experiences and responses related to food insecurity, summarizing them on a scale.

The Household Food Security Survey Module (HFSSM) is an 18-item questionnaire developed by the USDA to assess household food security over 12 months. It uses a linear scale and provides a single numeric value between 0 and 18 to indicate the level of food security within a household.

The Household Dietary Diversity Score (HDDS) assesses the variety in households' food consumption. It includes twelve different food groups and is used to gauge the overall nutritional quality of the diet. This scoring system assesses the diversity of food intake over the last 24 hours, providing valuable insight into the adequacy of nutrients and the range of foods consumed by individuals within the household.

The Food Consumption Score (FCS) measures household dietary diversity based on nine food groups and their consumption frequencies over the past seven days. The score is calculated by weighing the food groups based on their nutritional significance and multiplying them by their consumption frequencies. The resulting composite score classifies households into three categories: poor (0-21), borderline (21.5-35), or acceptable (>35). This score serves as an indicator of a household's caloric availability.

FAO'S GUIDING PRINCIPLES FOR SUSTAINABLE FOOD SYSTEMS

Addressing the challenges of food security and nutrition requires a collaborative effort, as they are complex and interconnected issues. By engaging multiple disciplines and institutions, we can foster productive discussions that lead to innovative solutions and improvements in these critical areas. Global food systems face difficulties arising from interactions across various scales and levels. As a result, effectively tackling the complexities of food systems necessitates a more holistic, comprehensive, and coordinated approach.

A food systems approach includes all its elements, relationships, and effects, rather than focusing on individual sectors or disciplines. This approach broadens the analysis of food-related issues and considers the interconnected activities and feedback within the system. It considers all relevant factors contributing to a problem and the social, environmental, and economic impacts of potential solutions (Calicioglu et al., 2019) to bring about transformative systemic changes.

A sustainable food system ensures food and nutrition security while emphasizing the economic, social, and environmental needs of future generations. A food system is considered economically sustainable if the activities conducted by each food system actor or support service provider are commercially or fiscally viable. The activities should generate benefits, or economic value-added, for all categories of stakeholders: wages for workers, taxes for governments, profits for enterprises, and food supply improvements for consumers. On the social dimension, a food system is considered sustainable when there is equity in the distribution of the economic value added, considering vulnerable groups categorized by gender, age, race, and so on. Food system activities must contribute to advancing socio-cultural outcomes, such as nutrition and health, traditions, labour conditions, and animal welfare. On the environmental dimension, sustainability is determined by ensuring that the impacts of food system activities on the surrounding natural environment are neutral or positive, taking into consideration biodiversity, water, soil, animal and plant health, the carbon footprint, the water footprint, food loss and waste, and toxicity.

The sustainable food system approach is based on five principles that outline 20 actions that balance sustainability's social, economic, and environmental dimensions to achieve the Sustainable Development Goals presented in Table 1.

Table 1. FAO's guiding principles and actions for achieving food sustainability

Principles	Actions
1. Increase productivity, employment, and value addition in food systems	1. Facilitate access to productive resources, finance, and services (SDGs- 1,2,5,7,8,9,14,15)
	2. Connect smallholders to markets (SDGs- 1,2,5,8,9,11,12,14,15)
	3. Encourage diversification of production and income (SDGs- 1,2,8,11,15)
	4. Build producers' knowledge and develop their capacities (SDGs- 1,2,4,5,8,9,13,15)
2. Protect and enhance natural resources	1. Enhance soil health and restore the land (SDGs- 1,2,11,12,15)
	2. Protect water and manage scarcity (SDGs- 1,2,6,7,8,9,11,12,14,15)
	3. Mainstream biodiversity conservation and protect ecosystem functions (SDGs- 1,2,6,11,12,14,15)
	4. Reduce losses, encourage reuse and recycle, and promote sustainable consumption (SDGs- 2,7,12)
3. Improve livelihoods and foster inclusive economic growth	1. Empower people and fight inequalities (SDGs- 1,2,5,8,10,13,14,16)
	2. Promote secure tenure rights for men and women (SDGs- 1,2,5,8,9,13,14)
	3. Use social protection tools to enhance productivity and income (SDGs- 1,2,5,8,10,13,14)
	4. Improve nutrition and promote balanced diets (SDGs- 1,2,3,5,12)
4. Enhance the resilience of people, communities and ecosystems	1. Prevent and protect against shocks: enhance resilience (SDGs- 1,2,9,11,13,14)
	2. Prepare for and respond to shocks (SDGs- 1,2,9,13,14)
	3. Address and adapt to climate change (SDGs- 1,2,11,13,14)
	4. Strengthen ecosystem resilience (SDGs- 1,2,13,14,15)
5. Adapt governance to new challenges	1. Enhance policy dialogue and coordination (SDGs- 1,2,5,6,7,11,12,13,14,15,16,17)
	2. Strengthen innovation systems (SDGs- 1,2,5,9,12,14)
	3. Adapt and improve investment and finance (SDGs- 1,2,9,12,13,14,15,16,17)
	4. Strengthen the enabling environment and reform the institutional framework (SDGs- 1,2,5,6,7,10,11,12,13,14,15,16,17)

Note. Adapted from FAO, 2018

FOOD SYSTEMS AT DAYALBAGH: INSIGHTS ON FOOD SECURITY AND SUSTAINABILITY

Geographic Context

Dayalbagh, the 'Garden of the Merciful,' is a remarkable example of community-based holistic sustainability (Dua et al., 2023). It was established on January 20, 1915, by Huzur Sahabji Maharaj, the Fifth Revered Leader of the Radhasoami Faith. The foundation of this serene and harmonious community was laid with a simple ceremony of planting a mulberry tree on the auspicious day of Basant Panchami. Situated one and a half kilometers north of the enchanting city of Agra, Uttar Pradesh, India, Dayalbagh covers an area of 6.51 square kilometers. It is a private colony consisting of six *mohallas* (colonies) and has evolved into a healthcare habitat and an eco-village while still offering modern amenities to its residents. Dayalbagh is also the headquarters of the Radhasoami Satsang Community, maintaining its peaceful coexistence with nature.

Local Demographics and Social Organization

The survey findings demonstrate that the Dayalbagh community exemplifies a strong commitment to education, as evidenced by the absence of illiteracy among its members. This dedication is reflected in the community's development of formal and informal educational pathways aligning with national educational goals of universal foundational literacy and numeracy. Furthermore, the community is notable for its inclusivity, with individuals from various socio-economic backgrounds coexisting without discrimination. Additionally, it is remarkable that every member of the working-age group is engaged in productive activities (absence of unemployment), highlighting the community's rejection of laziness and idleness.

The research also revealed that only 6.66 percent of the respondents had loan liabilities. These loans were obtained from institutional sources, which can be attributed to the sound educational background of the people and the presence of three cooperative banks, including one *Dayalbagh Mahila Cooperative Bank Ltd.*, in the area. It reflects an effort to empower women and ensure their economic and financial stability. Further, the study found that none of the respondents had ever taken out a loan to purchase or provide food for themselves or their family members, likely due to the unique food systems in place.

The essence of Dayalbagh revolves around selfless service (Seva) and aims to establish a new world order where men and women work harmoniously for the service of humanity. Citizens of all ages follow a lifestyle prioritizing physical, mental, emotional and spiritual well-being, incorporating physical exercises, yoga, a vibrant routine, and a nutritious Lacto-vegetarian diet. Dayalbagh is gaining global recognition for promoting the eternal messages of *Better Worldliness* and the *Fatherhood of God and Brotherhood of Man*.

The State of Food Security at Dayalbagh

Household food insecurity has several health implications including a greater prevalence of diabetes, heart disease, asthma, musculoskeletal diseases, intestinal disorders, and mental health disorders (Kirkpatrick et al., 2010; Vozoris and Tarasuk, 2003). The food security analysis conducted in Dayalbagh employed two assessment scales. The Household Food Security Survey Module (HFSSM) evaluated that all households in the area are entirely food secure. Complementing this, the Household Food Insecurity Access Scale (HFIAS) showed a complete absence of food insecurity. Furthermore, the food consumption score for residents was recorded at 69.6, which is consistent with an acceptable level of food security.

The community in question is characterized by its Lacto-vegetarian dietary practices, rooted in its religious beliefs. To evaluate the dietary diversity among households, researchers identified twelve food categories: cereals, roots and tubers, vegetables, fruits, meat/poultry/offal, eggs, fish and seafood, pulses/legumes/nuts, dairy products, oils/fats, sugar/honey, and miscellaneous items. Notably, all respondents from the community consumed foods from each of the nine categories that align with their Lacto-vegetarian diet. Households are classified into three dietary diversity categories: low (scores of 0-3), medium (4-5), and high (6 and above). Despite all respondents having eliminated non-vegetarian food options, they continue to exhibit a high level of household dietary diversity. The average dietary diversity score among these respondents stands at an impressive 8.62. This situation highlights a stark contrast to the Indian government's ongoing concerns about achieving sufficient dietary diversity and nutrition for the targeted population (Basole and Basu, 2015; Deaton and Eze, 2009; Khera and Somanchi, 2020; Swaminathan, 2001).

Ensuring food security involves more than just making food available; it necessitates guaranteeing that all individuals have access to an adequate and nutritious diet for a healthy and active life (FAO, 2002). This encompasses not only the consistent availability of food but also the stability of the food supply and the means for individuals to obtain food in a dignified manner. Maintaining a healthy lifestyle requires consuming nutritious food and following hygienic procedures in food production and consumption (Gillespie and Kadiyala, 2012). It is imperative for physical and mental development, reproductive health, and disease prevention. According to WHO (2020), one in ten individuals becomes ill as a result of consuming contaminated food worldwide.

Low education is closely linked to poor sanitation and hygiene practices (BP, 2021). Food literacy encompasses a comprehensive understanding of three main areas: agroecological knowledge, which pertains to the origins and types of food produced; cultivation and production knowledge, focusing on how food is grown; and processing and consumption knowledge, detailing how food is prepared and consumed. This knowledge can be acquired informally, through everyday practices at home and in the community, and formally, through structured education in academic institutions (Gartaula et al., 2020).

Dayalbagh residents demonstrate a high level of awareness and conscientiousness regarding food safety, particularly regarding hygiene and sanitation practices. This awareness is largely attributed to their strong educational foundations and supportive community environment. Respondents were assessed on a four-point Likert scale (never=1, sometimes=2, often=3, and ever=4). It was found that residents were significantly aware of food nutrition (3.93), food ingredients (3.77), food hygiene, and sanitation practices (4.0).

The analysis of respondents' attitudes towards food wastage reveals a strong sense of responsibility among individuals for minimizing waste. Community members prioritize mindful eating and actively plan their meals to match their needs, ensuring food is savoured and utilized. In the uncommon instances of leftover food, they efficiently collect and redistribute it to dairy cattle, supported by a well-established food collection system within the community.

Holistic Food Systems Approach of Dayalbagh Community

Gandhiji was a staunch advocate of 'Gram Swaraj' (village self-rule) and viewed it as the cornerstone of a decentralized economy. He ardently supported the local village ecosystem, particularly agriculture and agro-based industries, which aimed to be self-sustaining by providing food to all and generating supplementary income for rural communities. He strongly believed in prioritizing cooperation over competition in the lives of farmers, emphasizing the significance of their livelihoods. Gandhiji also expressed skepticism about market-oriented economies, asserting that they often fall short in addressing the socio-economic challenges of communities.

A well-planned, community-driven, holistic food system offers a steady supply of wholesome, nourishing food. Dayalbagh serves as a notable example of the successful implementation of agroecology-cum-precision farming to fulfill the food requirements of its residents, as well as those of neighbouring colonies and villages (Satsangi and Dhir, 2024). The Dayalbagh community's food system considers all interconnected food subsystems to achieve food security while upholding several social, economic, environmental, and cultural principles.

Food Production and Land-Based Productive Activities

In 1943, agricultural activities began in Dayalbagh, with women participating in agriculture from as early as 1955-56, demonstrating the importance given to gender equality (Satsangi and Dhir, 2024). Dayalbagh holds a specialized agricultural administration and management department committed to advancing agricultural research and development. The allocation of land for various crops, cropping sequences, and diversification of crop production are key decisions made by agricultural experts at the Scientific Committee on Agriculture. They ensure a consistent supply of wholesome and nutritious food for everyone and maintain self-sufficiency in food production. Table 2 depicts the diverse crops produced at Dayalbagh-

Table 2. Crops at Dayalbagh

Categories	Crops
Cereal crops	Paddy (Rice), Wheat, Maize, Bajra, Jowar (Sorghum), Barley
Pulse crops	Moong, Urad, Lobia, Arhar, Chana (Gram), Kulth, Masoor
Edible oilseeds	Groundnut, Mustard, Taramira/Jamba (Arugula)
Sugar crops	Sugarcane
Fruits and Vegetables	Banana, Strawberry, Grapes, Mandarin (Kinnow), Guava, Java Plum, Aegle Marmelos (Bael), Tomatoes, Lady Finger, Spinach, Cucumber, Bottle gourd, Bitter gourd, Ridge Gourd, Brinjal, Carrot, Radish, Potato, Onion, Lemon, Mushroom, Cauliflower, Cabbage, Beetroot, Jackfruit
Spices	Green Chilli, Curry leaves (Sweet neem leaves), Garlic, Sesame seed
Medicinal crops	Ashwagandha, Akarkara, Arand, Bhumi Amla, Chitrak, Giloy, Katsaria, Karanjwa, Mulethi, Pippali, Sahajan, Vasaka, Amla (Indian Gooseberry)
Aromatic crops	Mentha (Mint), Tulsi, Lemongrass

continued on following page

Table 2. Continued

Categories	Crops
Forage crops	Green fodder, Maize, Barseem, Sorghum, Bajra, Napier grass
Other crops in association with DEI	Mango, Ginger, Tamarind, Olives, Tissue-cultured date palms, Rose, Jatropha, Aloe vera, Watermelon

Note. Adapted from the Ph.D. thesis of Neha Sikarwar

Figure 1. Agricultural practices at Dayalbagh (Seva by community members and students)

Note. Adapted from the Ph.D. thesis of Neha Sikarwar

The Dayalbagh community embodies the principle of selfless service, with no individual ownership of land to maximize the benefits of large-scale production. Farm sites benefit from community members' dedicated efforts, including students, who generously volunteer their time twice daily, all year round (Figure 1). It promotes respect for the dignity of labour and a spirit of self-reliance. In Dayalbagh, various ecosystems thrive harmoniously, encompassing crop fields, vegetable gardens, fruit orchards, fodder fields, medicinal herb gardens, and biodiverse parks. In Dayalbagh, the principle of diversity is effectively applied by practicing vertical, spatial, and temporal diversity (Satsangi and Dhir, 2024). Various crops are being cultivated including Kharif, Rabi, and Zaid, as well as medicinal and aromatic crops, aiming for optimal land resource utilization. They grow various forage crops for their livestock to ensure that milk and dairy production remains cost-effective.

Organic farming has always remained the hallmark of the Dayalbagh agriculture system. Instead of chemical fertilizers, the emphasis is being given to bio-fertilizers, organic manure (Jivamrit liquid form), and bio-pesticide (Trichoderma). The community prioritizes judicious use of chemicals to control pests, diseases, and weed pressure, and minimizes the intensive use of chemical fertilizers.

The harmonious integration of a diverse livestock herd and successful agricultural production ensures the community's nutritional needs are met year-round (Gogia and Sikarwar, 2019). The Dayalbagh Dairy, an indigenous dairy, was established in 1926 and was regarded as the best in Asia (Dua et al. 2023 from Radhasoami Satsang Sabha, 2019). The animal community encompasses 864 cows, 206 buffaloes, 43 goats, and 13 camels, encompassing males, females, and calves. A classic blend of agriculture and dairy, a circular economic approach, is an important feature of the Dayalbagh food production. Animal waste serves as a valuable organic fertilizer that improves soil fertility and promotes crop growth. This enhanced crop yield, in turn, helps meet the nutritional needs of diverse livestock. It emphasizes the interconnectedness of livestock and agriculture, underscoring a sustainable framework that respects heritage while embracing inventive solutions for the betterment of humans and animals. The Dayalbagh *Gaushala* is a shining example of a holistic approach, placing equal importance on the well-being of both humans and animals. approach integrates modern practices and scientific techniques to effectively manage the nutrition, feeding, health, and breeding of cattle. This is complemented by traditional wisdom in rearing the herd, creating a comprehensive method for livestock management (Satsangi and Dhir, 2024).

Considering the health issues of different age groups, full cream, and toned milk are available to keep people healthy and safe. Additionally, several milk products, like cream, ghee, butter, curd, milk-made sweets (srikhand, kalakand), chocolates, and confectionery items (fudges, truffles), etc. are also available. The *Khandsari* unit was established at the dairy compound for the sake of manufacturing Desi Sugar so that it could further supply *Ilaichidana* Units and Molasses to Gaushala for further use.

The water sourced from the dairy farm is used for irrigation in vegetable farming as part of the solar agriculture farming project and greenhouse cultivation. Figure 2 portrays an advanced 200-kilowatt peak solar power system that has been set up on raised platforms to support solar energy generation and agricultural activities.

Figure 2. Cultivation under solar plant

Note. Adapted from the Ph.D. thesis of Neha Sikarwar

To conserve the groundwater and to augment water availability for irrigation, the community has given land to the government of Uttar Pradesh, India for a Sewage Treatment Plant (STP), with an understanding that the treated water within the defined parameters will be supplied for irrigation of agricultural fields. They have also incorporated rainwater harvesting systems to ensure sustainable irrigation practices. It reflects the eco-sensitive approach of the community.

Precision farming techniques, including site-specific sampling and input management using sensors, as well as supply chain monitoring technologies like smart truck tracking, play a critical role in reducing waste, improving efficiency, and lowering economic and environmental costs (Satsangi and Dhir, 2024).

The collaboration with the Dayalbagh Educational Institute (Deemed to be University) integrates traditional/indigenous knowledge with modern science. This partnership has resulted in the establishment of two international research institutes: the International Center for Agroecology (ICA) in New Jersey and the International Center for Applied Systems Science and Sustainable Development (ICASSSD) in Toronto, which focus on agroecology, applied systems science, and sustainable development. These institutes contribute to research, policy development, and decision-making processes, intending to promote a culture of sustainability (Satsangi and Dhir2024).

Figure 3 highlights the unique concept of the community's common kitchen (Dayal Bhandar). It supplies cooked food to its residents and acts as a *langar* providing free meals to pilgrims during special occasions.

Figure 3. Common kitchen (Dayal Bhandar) at Dayalbagh

Figure 3.1: Women making chapattis

Figure 3.2: Women chopping vegetables

Figure 3.3: Food distribution

Figure 3.4: Common dining hall

Note. Adapted from the Ph.D. thesis of Neha Sikarwar

There is a well-defined menu of meals that fulfills the dietary diversity needs of people. There is a provision of a monthly card or food can be taken by depositing *Bhent* (Rs.1 to Rs.10) as deemed appropriate. The dining options include on-site meals in the shared kitchen or takeout for convenient consumption at home. The food at the community kitchen is prepared and served in a highly safe and hygienic environment.

The emphasis is on using sustainable energy sources, particularly solar energy. The kitchen utilizes a biomass gasifier and a solar thermal cooking system, along with other energy sources, to cook food on a large scale, effectively reducing costs and pollution. The communal kitchen provides food to all and saves women's time for other activities and community service.

Mahila (women) Associations in Dayalbagh are actively involved in organizing and regulating social activities for women. These associations foster community collaboration and empower women by providing a structured platform for engagement. Among the diverse range of initiatives they undertake, one notable focus is on sourcing and distributing high-quality, unadulterated spices and pickles. By ensuring that these products meet rigorous standards of authenticity and quality, the associations contribute to preserving traditional culinary practices, thereby enriching the dietary experiences of families while prioritizing health and safety.

The entire production process is carried out with a spirit of service for the greater good. As a result, the cost of labour, which makes up a significant portion of the overall food cost, is not factored into the production expenses. This approach helps to keep food prices reasonable and affordable for everyone.

Food Distribution

Enhancing food production is important but not sufficient to achieve food security. While increased production ensures availability, proper distribution is essential to augment food accessibility. Chronic hunger in certain regions starkly illustrates the profound disparities in resource distribution. Unlike India's public distribution system which has been ineffective in ensuring that food reaches the intended recipients (George and McKay, 2019; Gupta and Mishra, 2018; Gupta and Singh, 2016; Narayanan, 2015; Tanksale and Jha, 2015; Varadharajan et al., 2013), the food distribution system in Dayalbagh is effective, transparent, and well-monitored. There is a significant amount of waste in the distribution chain in India, leading to high costs and limited benefits (Dreze and Khera, 2015; Khera, 2011; Mahamallik and Sahu, 2011; Rao, 2000). However, the Dayalbagh system does not experience any loss of benefits.

To ensure a consistent supply of food, various establishments such as fruit and vegetable shops (Figure 4), kiosks, and canteens are located throughout the area, offering food at affordable prices. Ration cards are issued to households to prevent duplicate or excessive access to food items. Every resident has equitable access to inexpensive raw and prepared food based on their family size, rather than their purchasing power. A fair and equal distribution system of Dayalbagh promotes long-term economic efficiency and sustainable community growth.

Figure 4. Vegetable shop at Dayalbagh

Note. Adapted from the Ph.D. thesis of Neha Sikarwar

The government's mid-day meal program is effectively executed in the primary schools of Dayalbagh. Prioritizing students' health, the school offers hot milk, roasted gram, bananas, and biscuits. This initiative promotes the awareness and practice of consuming nutritious food from an early age, discouraging the consumption of junk food. Additionally, during fieldwork, individuals are provided with flavoured milk, tea with rusk, hot milk, and seasonal fruits such as banana, kinnow, or mango.

Food Consumption

"Food to a large extent is what holds a society together, and eating is closely linked to deep spiritual experiences." - *Peter Farb and George Armelagos*

The basis of food consumption should be the nutritional needs and overall health of an individual. People must make informed choices that support their well-being. However, when consumption becomes intertwined with social status, it can result in extravagant displays and unnecessary waste. The unconscious food choices result in health disturbances, economic instability, social inequality, environmental degradation, and food insecurity. This highlights the importance of promoting mindful and sustainable food practices.

The food choices we make have a big impact on the economy, society, and the environment. We can help to feed others or drive-up prices by increasing the demand. The concept of socially responsible food consumption is becoming increasingly important, especially considering that one-third of the food produced for human consumption is wasted annually worldwide, while many people in other regions face hunger. There is a growing consensus that sustainability issues are not solely the responsibility of supply chains; consumers must also take accountability for their actions. Individual consumer choices play a pivotal role in food waste, alongside the production and distribution processes (Rasool, et al., 2024).

Based on the doctrine that *eating is necessary for spiritual growth*, the people in Dayalbagh adopt conscious food choices (voluntary simplicity) that help establish a balance between an individual's satisfaction and food sustainability. In the case of Dayalbagh, food demand is driven by necessity. People don't stockpile food unnecessarily, regardless of their wealth, as affordability doesn't determine food access. Preference is given to locally produced food items that help in saving transportation costs and help in reducing the negative impact on the environment. Thus, contributing to a more sustainable and resilient food system.

People in the community have a sense of social and environmental responsibility. Intake of non-vegetarian food, alcohol, or other intoxicants like drugs is strictly prohibited thereby saving them from a 'crowding-out effect' (John, 2008). Conspicuous food consumption is discouraged. A vegetarian meal's carbon footprint is lower than a non-vegetarian meal (Scarborough et al., 2014). In this way, the Dayalbagh people are helping to nourish themselves and the planet. They willingly avoid consuming junk food and processed oil leading to obesity and non-communicable disease (Arnalte-Mur et al., 2020; GBD, 2015; Sannigrahi, 2016). Under the motto 'Waste Nothing,' the Dayalbagh community ensures that every resource, including food, is utilized to its fullest potential. People are committed to maximizing efficiency and minimizing waste in all aspects of their lives.

Dayalbagh Food Systems Approach to Food Security and Sustainability

Abundant food stock is not a sufficient condition to attain food security. The food system requires an uninterrupted supply of nutritious food for all. Unexpected price surges and influences of other factors push millions of people into poverty, aggravating income inequalities and threatening food security. FAO has been advocating for the recognition and preservation of Globally Important Agricultural Heritage Systems (GIAHS), which are living ecosystems where human communities have a deep connection with their cultural and agricultural landscape and the broader social environment. Dayalbagh serves as a remarkable illustration of a GIAHS, which, with its various features, sets an example to accomplish food security and sustainable development. (Figure 5).

Figure 5. The sustainable food systems approach to food security at Dayalbagh

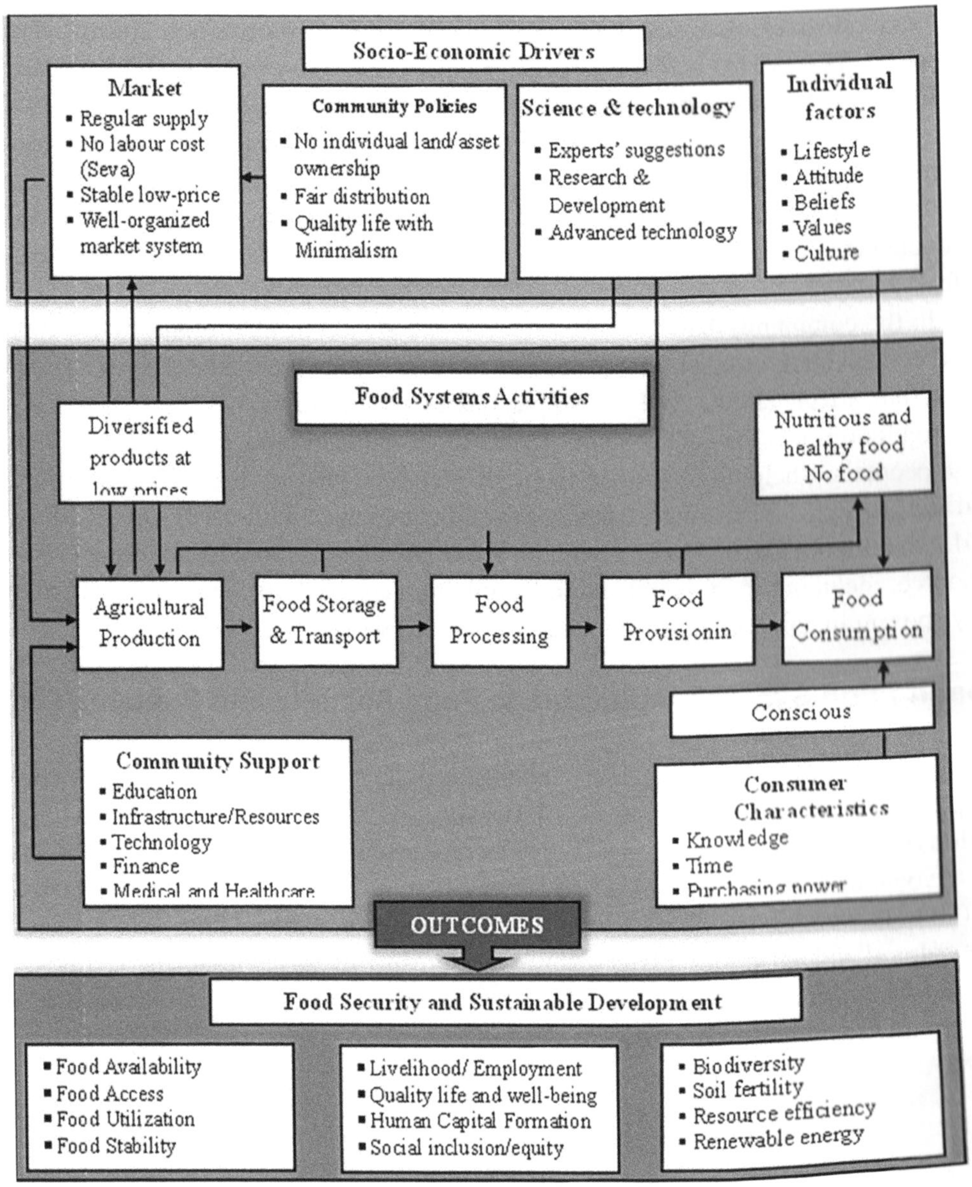

Note. Adapted from the Ph.D. thesis of Neha Sikarwar

Food Practices as Resilience Strategies in Dayalbagh Amidst COVID-19 Pandemic

The global Coronavirus (COVID-19) outbreak has disrupted food systems worldwide, leading to significant changes in dietary practices and people's nutritional and health statuses. It has led to a rise in hunger and food insecurity, which has been worsened by the economic impact of the pandemic and escalating global inflation (Kanter et al., 2023). The pandemic has exposed vulnerabilities in the global food systems, prompting communities to turn to local food systems to address the crisis.

During the challenging times of the COVID-19 pandemic, the Dayalbagh community demonstrated remarkable adaptability in their food practices. This exploration delves into the innovative ways they sustained their food systems, showcasing resilience, creativity, and a commitment to sustainability. From novel recipes that utilized local ingredients to community-driven initiatives that ensured food security, the story of Dayalbagh is one of hope and transformation in the face of adversity. Dayalbagh's food practices have successfully enabled its residents to overcome food-related challenges during the pandemic, ensuring an ample food supply to support their well-being.

The lockdown imposed in India due to COVID-19 resulted in a significant economic shock. It brought the economy to a standstill as all sectors were forced to close their enterprises completely. Although agricultural activities were exempt from some restrictions, there were considerable disruptions to the agriculture value chain. Throughout the pandemic, there was a reverse migration of farm labour from urban areas to rural regions, and social distancing measures adversely affected agricultural production in various parts of the country (NABARD, 2020; Singh, B. et al., 2020). Dayalbagh successfully sustained its food production levels amidst the challenges of the pandemic. This was largely due to the community's commitment to selfless service, which guaranteed a continuous food supply. As a result, residents could uphold a healthy lifestyle by staying physically active during these difficult times.

During the pandemic, wearing headgear and masks and using sanitizers was mandatory for everyone. All agricultural vehicles and tools have been thoroughly sanitized, ensuring the safety and health of field workers during transportation. All items were made available at nominal prices and free of cost for those who could not afford them. Additionally, potassium permanganate ($KMNO_4$) has been utilized to disinfect fruits and vegetables before their consumption. The government and WHO prescribed norms of social distancing and safe practices, which were strictly followed during the production, distribution, and consumption processes.

The offering of free medical services and expert guidance from specialized doctors was a key initiative during the pandemic. Moreover, Ayurvedic Kalmegh Kadha and preventive homeopathic medicines, as endorsed by the AYUSH Ministry, were made available. Ultimately, the innovative integrated systems at Dayalbagh have significantly contributed to empowering its residents to overcome the challenges posed by the pandemic.

DAYALBAGH FOOD SYSTEMS FRAMEWORK AND FAO'S FIVE GUIDING PRINCIPLES

The Food and Agriculture Organization has outlined guiding principles for a sustainable and equitable food system (Table 1). These principles underscore the importance of promoting food security, nutrition, and biodiversity while minimizing negative environmental impacts. The FAO has emphasized that by upholding these principles, a resilient and inclusive food system capable of meeting the needs of future generations can be established. The Dayalbagh community's food system effectively aligns with all the guiding principles, contributing significantly to achieving Sustainable Development Goals (SDGs) through its outcomes as illustrated in Table 3.

Table 3. Food systems of Dayalbagh and FAO's five guiding principles

Principles	Actions	Outcomes
Increase productivity, employment, and value addition in the food system	Access to communal resources (resource and knowledge sharing), Free voluntary services (Seva), Natural resource management practices (close connectivity among people and with nature), Use of advanced technologies (youth participation), Crop and animal diversity, Integrated crop-livestock system, Arrangements of production to consumption (farm to fork), Dedicated research and policies, Free expert advisory services, Multistakeholder participatory processes	Equity, social cohesion, and well-being (supporting culture, social and spiritual manifestations), Circular solidarity and reciprocity, Technology-oriented system (youth retention), Sustained community, Food and nutritional security, Dietary diversity, Food price stability
Protect and enhance natural resources	Renewable energy, Organic and seasonal farming, Sustainable land use practices (optimum utilization, circular economy, waste management through reusing and recycling), Conscious decisions, Nutrition, hygiene, and sanitation education	Preservation of the environment and biodiversity, Sustainable lifestyle (Nutritionally rich diverse food, safe drinking water, hygiene, and sanitation)
Improve livelihoods and foster inclusive economic growth	Equal distribution, No individual ownership of resources (fair access, no conflicts), Community participation, Women and Youth Empowerment, Social protection schemes for security (mid-day meal, healthcare habitat, Rapid Action Force), Responsible governance	Inclusive growth, A gender-free community, Social protection, Equality
Enhance the resilience of people, communities, and ecosystem	Prevention against shocks, Adaptability to climate change, Generational transmission of knowledge and attitude, Core values of solidarity, reciprocity, and communal work, Community engagement, collective decision-making, co-responsibility, Strengthened governance, Use of ICT, Agriculture Research & Development, Safe storage facilities, Ease of finance	Strong social organization, Sustainable production, Resilient to climate change, Reduced greenhouse gas emissions, Globally Important Agricultural Heritage Systems (GIAHS)
Adapt governance to new challenges	Coordinated and integrated mechanisms, Innovative and flexible system, Inclusive and incremental institutional development, Mobilization of resources, Shared objectives, Agriculture Research & Development, Ease of finance, Regular monitoring, Alignment with SDGs	Empowered stakeholders, Intense agroecology-cum-precision farming model, Absence of food insecurity

Note. Authors' Creation

CONCLUSIONS

"A community which wishes to go ahead with its work must remain united in purpose, thought, and action; it must also conserve all its resources of men, money, and material."

- Param Guru Huzur Mehtaji Maharaj, the Sixth Revered Leader of the Radhasoami Faith

Food is an essential requirement for every individual's survival. But the utility of food should not just be limited to the mere existence of mankind; rather it should also ensure a healthier and qualitative life for all. Adequacy of the food supply means that (a) the overall supply potentially covers the overall nutritional needs in terms of quantity and quality; (b) the food must be safe (free of toxic factors and contaminants) and of good quality (taste and texture); and (c) the types of foodstuffs commonly available should be culturally acceptable; (d) lastly, the complete food process must be stable and sustainable.

The values, norms, customs, and beliefs that define a society are fundamentals of its culture and often remain consistent across generations. However, a deep-seated commitment to these nonfunctional or traditional values is essential; it fortifies the social fabric and ensures enduring sustainability for human communities and the surrounding ecology (Gupta, 2011). Indigenous knowledge is a valuable repository of sustainable farming practices that can significantly contribute to global efforts in achieving food security, enhancing biodiversity, and combating climate change. Case studies from various regions illustrate the success of Indigenous wisdom in fostering sustainable agriculture. These examples highlight how, when acknowledged and supported, Indigenous knowledge can be a potent resource in tackling contemporary agricultural challenges (Dhivya et al., 2024).

Indigenous communities are rich in traditional knowledge and practices related to food that significantly enhance food production, safety, and quality. Their deep understanding of local ecosystems and agricultural techniques allows them to cultivate crops and manage resources. Combining established Indigenous knowledge and scientific advancements is essential for tackling the critical issue of food security. This approach may help ensure consistent access to nutritious food. By leveraging such food practices, one can gain valuable insights into enhancing food systems for everyone (Kumbhar and Kharwar, 2023; Melash et al., 2023).

Dayalbagh has created a robust and sustainable agriculture and food system model that delivers healthy and nutritious food while upholding social, economic, environmental, and cultural values. The system emphasizes healthy diets, sustainable food production, equal access to food and nutrition, hygiene and sanitation awareness, efficient governance, and minimal waste. Ultimately, Dayalbagh aims to foster a hunger-free community, prioritizing healthy living and promoting overall well-being.

The ideology embraced by the Dayalbagh community nurtures a spirit of cooperation and encourages an organized, structured approach to living. This philosophy emphasizes the harmonious interconnectedness of production, distribution, and consumption, creating a seamless flow between these vital processes. This way of life is deeply rooted in strong philosophical principles, which promote a sense of unity among community members while also improving their overall effectiveness. By prioritizing collaboration and mutual support, the Dayalbagh community cultivates an environment where each member can contribute to and benefit from the collective well-being.

ACKNOWLEDGMENT

The authors sincerely thank Prof. Prem Kumar Kalra, Senior Director at Dayalbagh Educational Institute (Deemed to be University), for his invaluable support in conceptualizing the study. They are thankful to the administrative and agricultural management staff and the residents of Dayalbagh for their unwavering guidance and support. Special thanks are also extended to Ms. Ayushi Kukreja and Mr. Deepak Gogia for their contributions to finalizing the chapter.

CONFLICT OF INTEREST

The authors declare no conflict of interest.

REFERENCES

Arnalte-Mur, L., Ortiz-Miranda, D., Cerrada-Serra, P., Martinez-Gómez, V., Moreno-Pérez, O., Barbu, R., & Šūmane, S. (2020). The drivers of change for the contribution of small farms to regional food security in Europe. *Global Food Security*, *26*, 100395. Advance online publication. DOI: 10.1016/j.gfs.2020.100395

Basole, A., & Basu, D. (2015). Fueling calorie intake decline: Household-level evidence from rural India. *World Development*, *68*, 82–95. DOI: 10.1016/j.worlddev.2014.11.020

Borgen Project. (2021). Education and Improving Sanitation in Developing Countries.

Calicioglu, O., Alessandro, F., Bracco, S., Lorenzo, B., & Ralph, S. (2019). The Future Challenges of Food and Agriculture: An Integrated Analysis of Trends and Solutions. *Sustainability (Basel)*, *11*(1), 222. DOI: 10.3390/su11010222

Campbell, J. Y. (2019). No Sustainable Development Without Indigenous Peoples https://sdg.iisd.org/commentary/guest-articles/no-sustainable-development-without-indigenous-peoples/

Deaton, A., & Eze, J. (2009). Food and nutrition in India: Facts and interpretations. *Economic and Political Weekly*, *44*(7).

Dhivya, C, Monika, A., Jayashree, V. and Kamali, S P. (2024). The Role of Indigenous Knowledge in Sustainable Farming Practices. GreenariA- An Agricultural Monthly e-Magazine. 2 (10), 2584-153X

Dreze, J., & Khera, R. (2015). Understanding leakages in the Public Distribution System. *Economic and Political Weekly*, *50*(7), 39–42.

Dua, P., Dhir, A., Naranyan, A., & Allamraju, A. (2023). A Holistic Approach to Achieving SDGs: A Case Study of Dayalbagh, India. https://uwspace.uwaterloo.ca/ handle/10012/19181

FAO. (2002). State of Food Insecurity in the World. Retrieved from: www.fao.org

FAO. (2018). *Transforming Food and agriculture to achieve the SDGs*. Food and Agriculture Organization of the United Nations.

FAO. (2021). Indigenous Peoples' Food Systems: Insights on Sustainability and resilience from the front line of climate change.

FAO, IFAD, UNICEF, WFP and WHO. (2024). The State of Food Security and Nutrition in the World 2024 – Financing to end hunger, food insecurity, and malnutrition in all its forms. Rome.DOI: 10.4060/cd1254en

Gartaula, H., Patel, K., Shukla, S., & Devkota, R. (2020). Indigenous knowledge of traditional foods and food literacy among youth: Insights from rural Nepal, Journal of Rural Studies, Volume 73, Pages 77-86, ISSN 0743-0167, DOI: 10.1016/j.jrurstud.2019.12.001

George, N.A. and McKay, F.H. (2019). The Public Distribution System and Food Security in India. International Journal of Environmental Research and Public Health, 3;16(17):3221. . PMID: 31484374; PMCID: PMC6747310.DOI: 10.3390/ijerph16173221

Gillespie, S., & Kadiyala, S. (2012). *Exploring the agriculture nutrition disconnect in India*. International Food Policy Research Institute.

Global Hunger Index. (2023). *The power of youth in shaping food systems*. Welt Hunger Hilfe.

Gogia, J. and Sikarwar, N. (2019). Agriculture and Dairy Sectors contribution to Food Security in India, International Journal of Basic and Applied Research, Vol. 9, No. 4

Gupta, A., & Mishra, D. K. (2018). Public Distribution System in Uttar Pradesh: Access, Utilization and Impact. *Indian Journal of Human Development.*, *12*(1), 20–36. DOI: 10.1177/0973703018774668

Gupta, A. D. (2011). Does Indigenous Knowledge have anything to deal with Sustainable Development? Antrocom Online Journal of Anthropology. 7 (1). https://antrocom.net/wp/wp-content/uploads/2024/05/gupta-docs-indigenous-knowledge-deal-sustainable-development.pdf

Gupta, P., & Singh, U. (2016). Efficiency of Public Distribution System. *Business Review*, *17*, 79–95.

John, R. M. (2008). Crowding out effect of tobacco expenditure and its implications on household resource allocation in India. *Social Science & Medicine*, *66*(6), 1356–1367. DOI: 10.1016/j.socscimed.2007.11.020 PMID: 18187245

Kanter, R., Kennedy, G., & Boza, S. (2023). Editorial: Local, traditional, and indigenous food systems in the 21st century to combat obesity, undernutrition, and climate change. *Frontiers in Sustainable Food Systems*, *7*, 1195741. DOI: 10.3389/fsufs.2023.1195741

Khera, R., & Somanchi, A. (2020). A review of the coverage of PDS. Ideas for India: https://www.ideasforindia.in/topics/poverty-inequality/a-review-of-the-coverage-of-pds.html

Kirkpatrick, S. I., McIntyre, L., & Potestio, M. L. (2010). Child hunger and long-term adverse consequences for health. *Archives of Pediatrics & Adolescent Medicine*, *164*(8), 754–762. DOI: 10.1001/archpediatrics.2010.117 PMID: 20679167

Kuhnlein, H. V., & Chotiboriboon, S. (2022). Why and How to Strengthen Indigenous Peoples' Food Systems with examples from two unique Indigenous communities. *Frontiers in Sustainable Food Systems*, *6*, 808670. DOI: 10.3389/fsufs.2022.808670

Kumbhar, S., & Kharwar, P. (2023). Indigenous Knowledge System and Food Security. *Contemporary Social Science*, *32*(4).

Lugo-Morin, D. (2020). Indigenous communities and their food systems: A contribution to the current debate. *Journal of Ethnic Foods*, *7*(1), 6. DOI: 10.1186/s42779-019-0043-1

Mahamallik, M., & Sahu, G. B. (2011). Identification of the poor: Errors of exclusion and inclusion. *Economic and Political Weekly*, *46*(9), 71–77.

Melash, A. A., Bogale, A. A., Migbaru, A. T., Chakilu, G. G., Percze, A., Ábrahám, É. B., & Mengistu, D. K. (2023) Indigenous agricultural knowledge: A neglected human based resource for sustainable crop protection and production. Heliyon 9(1): e12978 9 p. ISSN: 2405-8440

Mishra, A., Singh, S. R. K., & Raut, A. A. (2020). Traditional Knowledge in Agriculture. Division of Agricultural Extension, ICAR-Agricultural Technology Application Research Institute, New Delhi. https://icar.org.in/sites/default/files/2022-06/IITKA_Book_Traditional-Knowledge-in-Agriculture-English_0_0.pdf

NABARD. (2020). Impact Assessment of COVID-19 on Indian Agriculture and Rural Economy. Department of Economic Analysis and Research. National Bank for Agriculture and Rural Development. https://www.nabard.org/auth/writereaddata/tender/1211203145Impact%20Assessment%20of%20COVID.pdf

Narayanan, S. (2015). Food Security in India: The Imperative and Its Challenges. *Asia & the Pacific Policy Studies*, *2*(1), 197–209. DOI: 10.1002/app5.62

Nicoline, D. H. (2024). Transforming Food Systems Must Include Indigenous Peoples' Rights, Voices and Knowledge. Food Security and Nutrition. Farming First. https://farmingfirst.org/2024/08/transforming-food-systems-must-include-indigenous-peoples-rights-voices-and-knowledge/

Rao, C. H. H. (2000). Declining demand for food grains in rural India causes and implications. *Economic and Political Weekly*, *35*(4), 112–123.

Rasool, S., Cerchione, R., Centobelli, P., Oropallo, E., & Salo, J. (2024). Sustainability and socially responsible food consumption: An empirical investigation based on self-awareness and self-congruity theories. *Asia Pacific Journal of Marketing and Logistics*, *36*(4), 993–1023. DOI: 10.1108/APJML-07-2022-0559

Sannigrahi, K. (2016). Dietary Transition from Traditional to Modernization increase Non-Communicable Diseases. *Indian Journal of Applied Research*, *6*(8).

Satsangi, P. S., & Dhir, A. (2024). *Role of Communities in Achieving Sustainable Development*. Academic Foundation India.

Scarborough, P., Appleby, P. N., Mizdrak, A., Briggs, A. D., Travis, R. C., Bradbury, K. E., & Key, T. J. (2014). Dietary greenhouse gas emissions of meat-eaters, fish-eaters, vegetarians, and vegans in the UK. *Climatic Change*, *125*(2), 179–192. DOI: 10.1007/s10584-014-1169-1 PMID: 25834298

Sharma, I. P., Kanta, C., Dwivedi, T., & Rani, R. (2020). *Indigenous Agricultural Practices: ASupreme Key to Maintaining Biodiversity* (Goel, R., Eds.). Microbiological Advancements for Higher Altitude Agro-Ecosystems & Sustainability, Rhizosphere Biology, Springer Nature Singapore Pte Ltd., DOI: 10.1007/978-981-15-1902-4_691

Sikarwar, N., & Gogia, J. (2023). Linkages between food systems and food security. *Research Journal of Humanities and Social Sciences.*, *14*(4), 179–186.

Singh, B., Shirsath, P.B., Jat, M.L., McDonald, A.J., Srivastava, A.K., Craufurd, P., Rana, D.S., Singh, A.K., Chaudhari, S.K., Sharma, P.C., Singh, R., Jat, H.S., Sidhu, H.S., Gerard, B. and Braun, H. (2020). Agricultural labour, COVID-19, and potential implications for food security and air quality in the bread-basket of India. Agricultural Systems, 185, 0308-521X, .DOI: 10.1016/j.agsy.2020.102954

Swaminathan, M.S. (2001). A future attack on the PDS, Frontline. 18 (2).

Tanksale, A., & Jha, A. K. (2015). Implementing National Food Security Act in India: Issues and challenges. *British Food Journal*, *117*(4), 1315–1335. Advance online publication. DOI: 10.1108/BFJ-07-2014-0239

Utter, A., White, A., Mendez, V. E., & Morris, K. (2021). Co-creation of knowledge in agroecology. *Elementa*, *9*(1), 1. DOI: 10.1525/elementa.2021.00026

Varadharajan, K. S., Thomas, T., and Kurpad, A. V. (2013). Poverty and the state of nutrition in India. Asia Pacific Journal of Clinical Nutrition, 22(3), 326–339. https://doi.org/. 3.19.DOI: 10.6133/apjcn.2013.22

Vijayan, D., Ludwig, D., Rybak, C., Kaechele, H., Hoffmann, H., Schönfeldt, H. C., Mbwana, H. A., Rivero, C. V., & Löhr, K. (2022). Indigenous knowledge in food system transformations. *Communications Earth & Environment*, *3*(1), 213. DOI: 10.1038/s43247-022-00543-1

Vozoris, N. T., & Tarasuk, V. S. (2003). Household food insufficiency is associated with poorer health. *The Journal of Nutrition*, *133*(1), 120–126. DOI: 10.1093/jn/133.1.120 PMID: 12514278

WHO. (2020). Food safety. https://www.who.int/news-room/fact-sheets/detail/food-safety

WHO. (2024). Malnutrition. https://www.who.int/news-room/fact-sheets/detail/malnutrition

World Bank. (2018). Food-borne Illnesses Cost US$ 110 Billion Per Year in Low- and Middle-Income Countries. https://www.worldbank.org/en/news/press-release/2018/10/23/food-borne-illnesses-cost-us-110-billion-per-year-in-low-and-middle-income-countries

Chapter 5
Decolonising Education Systems Through Action Research and Action Learning Methodologies

Muhammad Usman Tariq
https://orcid.org/0000-0002-7605-3040
Abu Dhabi University, UAE & University College Cork, Ireland

ABSTRACT

This chapter examines how Action Learning (AL) and Action Research (AR) approaches are being used to decolonize educational systems. In order to create more inclusive and culturally relevant learning environments it emphasizes the significance of tearing down colonial legacies in education and incorporating Indigenous Knowledge Systems (IKS) into curricula. With the help of AR and AL educators and students can co-create knowledge solve real-world problems and advance educational equity. These are reflective participatory methods. The benefits of IKS for diverse learning communities are highlighted as the chapter covers methods for implementing it in educational settings. It also looks at how professional development affects teachers highlighting the necessity of preparing them to act as co-researchers in the decolonization process.

INTRODUCTION

With the current global emphasis on education decolonization and action research/action learning as approaches to education system decolonization is gaining popularity. This approach aims at discrediting the colonial frameworks that are found in education systems and bringing into practice those that embrace IKS, community involvement and cultural appropriate practices. In chapter, the author's intention is to explain how the aim of decolonizing education can be met through the implementation of the participatory action frameworks such as action research and action learning. In doing so, it solves the problem, which has been prevalent in the past, that education is offered and constructed within the context of Western paradigms and paradigms that do not respect the Indigenous and marginalized population's heritage and knowledge systems. The aim of this chapter is to prescribe coverage showing why decolonizing education matters; how action research and action learning could be instrumental in the process; and how these approaches can indeed be transformative – based on the lived experience. More recently, the

DOI: 10.4018/979-8-3693-3542-0.ch005

realization of coloniality in education systems that reproduces colonialism has been more evident leading to calls for transformation of education systems (Cox et al., 2021). This discussion broadens to the ways that action research and action learning can enhance the formation of education practice at a local level through involving the local people, teachers and learners to play an active role in constructing educational systems that might embody their cultures, values and experiences. The process of decolonization is significant to liberate education systems to achieve equity and competency. It has also been discovered that a large influx of education systems across the globe continues to function within frameworks that obscures non-Western epistemologies. This marginalization does not only restrict the contents that can be taught but also excludes students of Indigenous and non-Western origins. Decolonizing education is about creating room for these voices, and it is about the presence of curricula and methods which are accepted in different cultures. Thus, by promoting Indigenous knowledges and other non-metropolitan epistemologies, one can establish learning paradigms allowing including, contextual, and empowering peculiar for every learner (Wilson et al., 2020).

The call to decolonize the education system is very crucial as it is evident in the education systems across the world. Since the 1970s, education has been applied as a mechanism of colonial domination, by forcing Indigenous people and minorities to adapt to such Western ideas as history and languages. This has resulted in a negative way where students have been pulled away from their culture, this leading to culture shock and inferiority complex. A decolonization of education aims at reversing such a process in a bid to recognize indigenous cultures, language and knowledge practices. According to Ajaps (2021), the education systems in the developing nations must be changed or aligned to the current developments and dreams of the people and not to continue with the colonial mentality that was imported from the developed countries. I therefore conclude that, Action research is an important insight in the process of decolonizing education. It is a cyclical process of defining and solving problems and reflecting through participation where the researchers address issues with the communities. In the context of education, action research lets those teachers, students and the communities themselves speak and be heard. This is used to avert the top-down process that is typical of the educational reform initiatives where policies are laid down without considering of the conditions of various groups of people in society (Cornish et al., 2023). As these stakeholders are engaged in the process of developing and implementing changes in education through the action research, this framework reconstructs colonialism and offers solutions to foster fair and relevant education. For example, the participatory action research has been effectively applied to the Indigenous communities to transform curricular as well as teaching practices. For instance, Wilson et al. (2020) employed action research in a nursing programme to integrates an Indigenous health curriculum framework. Originally Presented by the University of Alberta, this project entails working with Indigenous people to develop an education curriculum that comprises both the traditional Indigenous people's knowledge and the modern conventional health practices. Through decade long planning, deliberation, sensitization and consultation, Frost devised an adaptable and culturally sensitive programme, thus improving preparedness for the students in Indigenous contexts, therefore, enhancing the respect towards Indigenous health practices.

Likewise, action learning has been useful in the process of promoting decolonization of education. Action learning in contrast with other models of learning that involve the process of transferring knowledge from the instructors to the students, action learning involves first-hand experiences where learners are engaged in learning activities as they carry out their tasks. Unlike the traditional method of note taking where learners do most of the writing, this method makes the students to be proactive in their learning by asking questions, and self-reflecting on what they have learned. In the current context of decolonizing education, the Action Learning can be an effective way to question the imposed paradigms and stimulate students to reflect on relations and power structures underlying the education process. For example, Manathunga (2020) explains how action learning can enable the development of spaces for Southern knowledge systems in the higher learning institutions. As result of the incorporation of multiple forms of learning and knowledge into the action learning paradigm, it is more effective in the creation of a climate that is conducive for the peaceful coexistence of all students. Traditional/indigenous systems of knowledge are vital in the attainment of the process of decolonization of education. Traditional knowledge or indigenous knowledge systems which have evolved over a long period are elaborate systems of acquiring knowledge about the world, its resources and the universe. Although they have tried to elaborate theoretical frameworks and empirical studies of how diverse societies conceptualize knowledge and engage with it, they have frequently been marginalized or even devalued in the official curricula that represent the universal model of scientific rationality drawn from the Western tradition. According to Mhembe decolonizing education implies acknowledging the legitimacy of IKS and trying to incorporate them into education in acceptable ways. It also has positive effects on indigenous students by providing for their cultural and knowledge validation as well as supporting the expansion and enhancement of knowledge acquisition by the other students by embracing different and unique ways of learning and perceiving reality (Dudgeon et al., 2020; Tariq, 2024).

For instance, in a participation realities paper by Lepore et al. (2021), capacity-building for community-based participatory research that involved; international partnership was done under a decolonizing lens. The project aimed at bringing in researchers from multiculturalism so they could design research methodologies that would embrace Indigenous epistemologies and ken. It was not only a response to the problematic of colonialism in revealing research methodologies of the Western tradition but a piece of evidence of how IKS is capable of confronting global issues. Education using IKS appears relevant for not only cultural accreditation, but also for the current problems like climate change, sustainable development, and social equity. There is knowledge about the indigenous people or local communities of how to live with the environment and act sustainably. Implementation of this knowledge in the system enhances the tourism education systems to have a clue on how to mitigate the current environmental challenges. According to Bourgeois et al. (2024) decolonization of educational systems to integrate IKS is not only the right thing to do for Indigenous formations but also the only way to support the future. Furthermore, its integration into the education system empowers Indigenous people and make them proud to be the way they are. Students are more likely to learn when their cultures and histories are incorporated in the curriculum as this makes them focus on their work. This is true especially given that in some systems, Indigenous students have been denied education and hence when they join schools they are regarded as subpar. The promotion of IKS in education therefore enhances achievement gaps in equal ways for all children hence aimed at decolonising education (Martinez-Vargas, 2020; Tariq, 2025).

Therefore, decolonizing education systems through action research and action learning methodologies is an important starting point towards the much needed inclusive, equitable and culturally relevant education. It is therefore important that communities are involved in the process of reforming education and valuing the Indigenous knowledge systems and practices by assimilating them into our education systems we start to work towards weaning off the colonial education regime. In this chapter, the concepts such decolonization of education, action research and action learning and indigenization of education in the contemporary society have been explained. In adopting these methodologies, the educators and the policy makers will be in a better position to develop education systems that will be in tandem with the cultures and the knowledge systems of the world and thus improve the results of the learners.

BACKGROUND

About-Decolonization in education can best be described and understood as a critical process, which seeks to question and eradicate colonialism influences that have shaped education systems. Indigenization as a concept involves the endeavor to decolonize and regain the Indigenous knowledge systems, languages and cultures that were subdued by colonialist regimes. This process is pursued to ensure that the process of education is rational, sensible, and consciously developed to take form encompassing the indigenous structures of knowing and learning. Martinez-Vargas (2020) is right, decolonization means understanding that the colonial apparatus has not gone away and working diligently to get rid of it. The critical importance of decolonization in education is tethered in the fact that it caters for previous injustices and imbalance. Historically during colonization each colonial power used to introduce their model of education in all the colonized countries and denigrated Indigenous knowledge systems. This has led to generations of student's Especially Indigenous and marginalized students being stripped off their cultural identities and forced into assimilation of the Anglo-American model (Wilson et al., 2020). Decolonizing education on the other hand tries to exactly undo this by ensuring that students are placed in learning environments where such voices and ways of knowing are incorporated into the curriculum so that learners have a representation of themselves in what they are learning. Another important reason for decolonial project in education is to equip the learners with the knowledge and skills to enable them to question power relation and thus transform the social reality. In their recent analysis, Cox et al., (2021) argues that decolonization of curriculum does not only concern what is taught but how it is taught. This entails critiquing of existing epistemological frameworks like the western epistemology that has been touted to be superior to other epistemologies. In other words, decolonization contributes to the achievement of a fairer context for global learning, by giving a plural place to similar knowledge (Tariq, 2025).

This chapter argues that historical background is critical in explaining the need to decolonize education systems, especially concerning colonialism. Pre-colonial education systems can be seen in the territories that were colonized by the European powers in which education was offered mainly with the purpose of imparting European culture and values. These education systems were established to socialize Indigenous people into colonial society, that is, to impart to them the cultures, language and history of the colonizers and denying them their own cultural belief systems and practices. This process was particularly evident in countries such as Canada, Australia and South Africa where, in the operation of residential schools' Indigenous children were taken away from families and communities, to assimilate the societies into western conventions (Ajaps, 2021). Effects of colonial education system remain rather significant and persistent ones. A big number of educational systems in use today retain colonial qualities and many of the qualities of colonialism: Eurocentric and often trivializing of any other form of knowledge. It has therefore caused a high level of inequality in education where indigenous and marginalized students perform dismally as compared to their counterparts in western education systems (Dudgeon et al., 2020). Decolonizing education entails reparation for such injustices to have a fair education system for all. Decolonization of education has been described by several scholarships as having several guiding principles, one of which is considering and incorporating IKS. Indigenous Knowledge Systems (IKS) entails the understanding systems and models that have been developed over time by Indigenous groups. There is always a strong correlation of such systems with earth environments and the community as a whole and therefore these systems comprise of information critical to aspects like, sustainability, health among others. Nevertheless, they have been excluded or marginalized in Western school education that tends to focus on science and personal success. The process of de-colonizing education requires an understanding of the significance of IKS and searching for techniques by which they could be included in the teaching and learning activity (Lepore et al., 2021).

For instance, the Māori people of New Zealand experience the process of the integration of the indigenous knowledge and language into the education system as the key element in the decolonization process. Kura Kaupapa Māori or Māori language immersion schools is education that is Māori-centric grounded in Māori epistemology and appears as a departure from the colonized education system that has relinquished Māori epistemology. This approach has proved effective in increasing educational achievements of the Māori students as well as revival of Māori language and customs (Bourgeois et al., 2024). The second of the main points when implementing the decolonization of education is the diversification of the material to be studied. This means changing the traditional perspective in history, science and cultural systems and adopting a multiculturalism outlook of knowledge in the world. According to Manathunga (2020), decolonizing education is the process of fashioning curricular with an objective of embracing the diverseness of human experiences and acknowledging the non-western civilizations' input of gaining knowledge. For instance, approaching colonization from the point of view of the colonized instead of the colonizers enables students to understand global history in a different approach as it goes against the conventional history lessons that students must learn in school. Decolonizing education also means reconsidering the relations of power within the classroom itself. Conventional classroom paradigms limit the flow of information and reduce the teacher to the single role of information provider with learners expected to simply accept and receive this information. This hierarchical system entails the Western way of perceiving the world in accordance with the colonial masters' assertion that they are superior to the colonized. By decolonizing education, what is meant is increasing the interaction in the classrooms and using the participatory method where students are presented with the knowledge and experiences and asked to take an active role in the learning process. This design of learning allows students to have control or reign over everything that they are learning and has functions of debunking the notion that learning is an activity that the teacher always initiates and the learner always receiving (Cornish et al., 2023).

Moreover, action research and action learning as methodologies to support the decolonization of education is impossible. These methods focus on people's involvement and self-reflection, which means that educators, students and other community members are engaged in the process of defining and designing more effective and equal education systems. Action research for instance is a process that uses cycles of findings, working on a plan of action and reviewing the effects of the plan. This approach is especially relevant during the decolonization process as it can be implemented as a long process of feedback and learning where the marginalized communities' views will be incorporated and implemented accordingly (Ajaps, 2021). Decolonization in education is the process that aims to revolutionize the education system to eliminate colonial legacy and belief system. This includes valuing Indigenous knowledge systems, accepting differences and encouraging increased representation and equality in the learning process. Through exposing the wrongs of the colonial past and advocating for a better and more diverse learning environment, there lies capabilities of a better future education of learners. Owing to such activities and approaches such as Action research, Education systems can thus be developed to fit the diverse human knowledge and experience (Tariq, 2025).

ROLE OF ACTION RESEARCH (AR) AND ACTION LEARNING (AL) IN DECOLONIZATION

AR and AL are participative with the capacity to drive a process of decolonization of education systems. They are both underpinned by collaborative, reflective and improvement orientated practices which lend themselves well to the radical process of decolonization. AR is a cyclical process where individuals or organizations try to assess issues, decide on ways of solving problems, put the identified solutions into practice and then evaluate the outcomes. AL, on the other hand, is centered on learning derived from the problem-solving process and working in teams, individuals in groups practice and analyze the problems and change their behavior in the process (McNiff, 2020). In combined, the AR and the AL offer the usability to negotiate the oppressive educational paradigms, ensure the incorporation of the marginalized and encourage the embrace of Indigenous and local knowledge systems. Thus, possibilities of decolonization offer students, educators and communities the role of stakeholders who can actively shape education with the help of AR and AL. To establish one's postcolonial epistemology, it is therefore important to note that the process of decolonization involves a rejection of structural and hierarchical dispensation of knowledge. AR, and AL promote partnership and children's learning, and this principle is achieved as well. These methodologies develop forums through which the participants are afforded an opportunity to speak about the matters of concern, the experiences and come up with collectively acceptable approaches for dealing with the emerging issues. For instance, in AR, the search for knowledge is not enclosed within academicians but encompasses ordinary people affected by the problem at hand, they are also involved in decision making on results (Smith, 2021). This makes the solutions that are developed to be most suitable to the context needed, hence decolonizing the educational systems by also taking into consideration the ideas of the normally sidelined groups (Tariq, 2025).

In the same process, it can be also important to point out that the use of AR and of AL can be helpful in the process of the critical unsettling of hegemonic epistemologies of education by inviting critical self-reflection on the current practices and by the introduction of multivocal knowledges. This of course is very true in post-colonial societies where education systems are structured with western orientations at the detriment of the indigenous knowledge systems. AR gives a chance to consider these problems systematically because the educators as well as the communities can discuss how the colonial managerial impact appeared and whether it is still present in curriculum, instructional approach, and organizational policies. Everyone agreed on doing a cycle of action reflection so that participants were able to note which areas are still colonized and how to act in such regards. Like peer assessment, AL enables participants to look at the practical problems faced by their peers while using their experiences to address such problems which leads to a more pluralistic approach in learning (Kember & McKay, 2022). The first example is using AR to help decolonize the curriculum is through the incorporation of Indigenous Knowledge Systems. It also will help educators and community members in the process of reviewing the existing curriculum, understanding the gaps where Indigenous perspectives are missing or misrepresented in it and creating new content that reflects Indigenous knowledge and values. For instance, a school in Australia may engage Indigenous elders and cultural experts to incorporate Aboriginal history, focus on the issue of environmental sustainability, language among other aspects. AR makes constant appraisal, and modifications are possible to ensure the curriculum grows as a responsive organ of the community, and as individual students' cultural mirrors. Likewise, it can be instrumental in decolonization whereby the use of AL enables the educators and students to collectively solve problems and reflect critically. In AL, the participants are grouped into small and sophistically known as 'learning sets' where they are supposed to work on real-life problems or issues, and this entices them to use their experiences and knowledge in making solutions. It can be very helpful in the decoloniality of education processes as the use of reflection practices how colonial power structures operate and fosters the elaboration of measures for making the education context more open and just (Revans, 2021).In a university setting, for example, an AL group might concentrate on addressing the underrepresentation of Indigenous knowledge in the course offerings of the institution. To ensure that Indigenous perspectives are given the same respect as Western knowledge, they could collaborate with faculty members to create new courses that are centered on these perspectives through reflection and action.

Several decolonization projects that used AR and AL in diverse educational contexts have been successful. The use of AR by Māori educators in New Zealand to decolonize their schools through the inclusion of Māori language and cultural practices in the curriculum is one noteworthy example. The impetus behind this initiative stemmed from the realization that Māori students, who frequently felt cut off from their cultural identity, were not being adequately catered to by traditional Western education systems. Teachers were able to work with Māori communities to rethink the curriculum, implement innovative teaching strategies, and establish a more culturally sensitive learning environment by participating in AR. Improved educational outcomes for Māori students, who now have greater access to culturally relevant content and feel more connected to their education, reflect this initiative's success (Ritchie & Rau, 2018). In the South African context, where universities have employed these approaches to address the colonial legacy in higher education, AR and AL are also assisting decolonization efforts. In response to calls for the decolonization of South African universities made by the #RhodesMustFall and #FeesMustFall movements, numerous academic institutions have embraced AR and AL as tools for involving staff, students, and communities in conversations about curriculum reform. Through these processes, participants have collaborated to pinpoint the ways in which colonialism has influenced academic programs and policies at universities. They have also created action plans to address issues of racial and social inequality on campus and integrate African knowledge systems (Cornish et al., 2023). The way that education is provided in South African universities has changed significantly because of this participatory approach, which has given staff and students the confidence to take charge of the decolonization process.

Because AR and AL prioritize the perspectives and experiences of people who have historically been marginalized in educational decision-making, they are highly effective tools for decolonization efforts. These approaches offer a framework for challenging prevailing narratives and encouraging cooperation, both of which are necessary to develop more inclusive and equitable educational systems. Educators, students, and communities can collaborate to confront colonial legacies and establish learning environments that are sensitive to the various needs of every learner by participating in cycles of action and reflection (Wilson et al., 2020). To sum up, augmented reality and artificial intelligence (AI) are potent approaches that can help decolonize educational systems. Through the facilitation of cooperation, introspection, and ongoing enhancement, these methodologies enable communities to contest prevailing knowledge structures and establish more comprehensive and fair learning environments. Case studies from South Africa and New Zealand demonstrate how AR and AL, by elevating the voices of marginalized people and incorporating Indigenous knowledge into the curriculum, can bring about significant change in education. By offering a dynamic and adaptable framework for continuous decolonization initiatives, these approaches guarantee that educational systems will keep changing to meet the demands of various communities (Tariq, 2024).

INTEGRATING INDIGENOUS KNOWLEDGE SYSTEMS (IKS) INTO THE CURRICULUM

One of the most important stages in the decolonization of education is the inclusion of Indigenous Knowledge Systems (IKS) in curriculum. Indigenous knowledge, which includes the ancient customs, ways of life, and wisdom of Indigenous peoples, offers a wealth of information that has traditionally been disregarded by Western educational institutions. Integrating Indigenous knowledge systems (IKS) into contemporary education can help close the knowledge gap between Indigenous and Western cultures, promoting a more accepting and validating learning environment. The goal of incorporating IKS into curricula is to create a more equitable and culturally responsive educational system through a variety of strategies. Working with Indigenous communities is a crucial strategy for integrating IKS into educational curricula. Teachers, policymakers, and Indigenous elders in the area can collaborate to guarantee that the curriculum accurately represents the customs, languages, and cultural values of the community. For instance, schools and Aboriginal communities in Australia have collaborated to incorporate traditional ecological knowledge into science curricula. Through this partnership, students have had the opportunity to learn about Indigenous land management techniques that provide an alternative to conventional methods of environmental sustainability, such as fire stick farming and water conservation (Smith & Ballard, 2020). Incorporating Indigenous communities into the process of developing curricula transforms education into a collaborative effort that honors and elevates indigenous knowledge. An additional tactic is to incorporate Indigenous viewpoints into a variety of subjects as opposed to just one or two courses. This all-encompassing strategy guarantees that IKS is viewed as a crucial component of the learning process rather than as a tokenistic inclusion or "add-on." Indigenous mathematical systems, for example, can be incorporated into math classes where students study customary methods of measurement, counting, and pattern recognition. As part of their studies in language arts, students can examine oral histories, storytelling customs, and Indigenous literature. Teachers foster a multidisciplinary understanding of knowledge that dissolves boundaries between Western and Indigenous epistemologies by integrating IKS into a variety of subject areas (Meyer, 2021).

Professional development and teacher training are also essential for properly incorporating IKS into the curriculum. Indigenous knowledge is unfamiliar to many educators, and they may feel unprepared to teach it. Giving educators the tools and training they need can help them become more competent and confident when integrating IKS into their courses. For instance, teacher professional development programs in New Zealand aim to raise understanding of Māori language and cultural traditions. These courses equip teachers with useful resources to include Māori viewpoints in their instruction, fostering a learning environment that is inclusive of all student cultures (Ritchie & Rau, 2018). Teachers who get ongoing professional development are guaranteed to be able to not only introduce IKS but also to interact with it in a courteous and relevant way. For different learning contexts, the inclusion of IKS in curriculum offers various advantages. First, it helps Indigenous kids, who frequently feel alienated by standard Western curriculum that do not represent their experiences or beliefs, feel proud of their culture and sense of identity. Respecting and incorporating Indigenous knowledge into the classroom helps students feel more connected to their culture and more motivated to study. It also helps to make education more relevant to their daily life (Smith, 2021). For instance, the incorporation of Indigenous languages and cultural practices into South African schools has increased academic performance for Indigenous children, who now perceive that their education reflects their history (Bourgeois et al., 2024). Second, not just Indigenous kids but all students benefit from the incorporation of IKS in their education. Education becomes a more dynamic and diverse experience when pupils are exposed to a variety of worldviews. IKS frequently places a strong emphasis on connections, links, and holistic thinking, all of which can help students develop their critical thinking abilities and widen their viewpoints. Students can get an appreciation for the value of sustainability and the interdependence of nature and human systems, for example, by studying about Indigenous environmental stewardship methods (Wilson et al., 2020). Students who have a wider perspective on the world are better equipped to be more adaptive, internationally conscious, and sympathetic in a world that is becoming more linked.

Additionally, IKS offers realistic, regional answers to today's problems. Knowledge systems have their roots in the environmental, social, and cultural contexts in which they have developed in many Indigenous communities. Because of this, IKS is extremely pertinent for tackling regional problems including resource sustainability, health, and land management. For instance, indigenous people's knowledge of medicinal plants has been exploited to create remedies for a range of illnesses, advancing contemporary medicine. Students learn about the scientific method as well as the importance of conventional wisdom and locally available remedies when this information is incorporated into science courses (Kimmerer, 2013). IKS can help address real-world issues and increase the contextual relevance of education in this way.

There are several instances of IKS being successfully integrated into educational settings. Kura Kaupapa Māori, also known as Māori immersion schools, offer an educational approach in New Zealand where the curriculum is fully centered around Māori language, culture, and values. In addition to offering pupils a comprehensive, culturally appropriate education, these schools provide an alternative to the mainstream educational system and have been extremely effective in conserving and reviving Māori language and culture (Bishop & Glynn, 1999). These schools' achievements show how important it is to provide Indigenous knowledge systems a place in official educational institutions and emphasize the advantages of giving kids an education that is based on their cultural identity. Indigenous viewpoints are now part of the curriculum in many Canadian schools, part of a larger attempt to address the historical marginalization of First Nations, Métis, and Inuit peoples. For example, the social studies curriculum in several provinces now requires the study of Indigenous history and residential schools. Alongside this, initiatives have been made to include Indigenous knowledge into science and geography, among other academic areas. These programs teach non-Indigenous students about the rich histories and cultures of Canada's Indigenous peoples in addition to assisting Indigenous students in seeing themselves represented in the curriculum (Battiste, 2013). The curriculum's integration of IKS is not without its difficulties. The continuance of colonial mindsets in educational institutions, where Western knowledge is frequently viewed as superior to Indigenous knowledge, is one of the primary challenges. To overcome this, educators and legislators must adopt a new perspective, acknowledging the importance of IKS and pledging to develop a more inclusive and fairer curriculum. Involving Indigenous communities in decision-making processes and providing resources like financing for curriculum development and teacher training are also necessary (Smith & Ballard, 2020). Despite these obstacles, incorporating IKS into the classroom is a vital step in decolonizing educational systems and developing inclusive, egalitarian, and varied knowledge system-reflective learning environments.

PROFESSIONAL DEVELOPMENT FOR EDUCATORS

To successfully apply Action Research (AR) and Action Learning (AL) in the decolonization of education, educators must engage in professional development. The goal of this professional development is to provide educators with the information and abilities they need to question conventional, frequently colonial pedagogies and include Indigenous views and knowledge into their instruction. In addition to being learning facilitators, educators are co-researchers and active participants in the creation and transformation of knowledge, and they are therefore essential to the advancement of decolonization activities. The first step in preparing educators to use AR and AL techniques for decolonization is to assist them in comprehending the theoretical foundations of these strategies. In the reflective process of action research, teachers and students collaborate to discover issues, create solutions, and assess the results. Because it is participative and centers on real-world issues that are pertinent to the experiences of the learners, it is especially helpful for decolonization (Koshy, 2010). In contrast, action learning entails using knowledge to enhance practice, reflecting on actions, and tackling real-world problems in groups. Both approaches make it possible for learning to be collaborative and iterative, which makes them perfect for tackling the challenges of decolonization in education.

In a Canadian professional development program that emphasized decolonization, for instance, teachers received training on using augmented reality (AR) to investigate how their curriculum and methods of instruction reinforced colonial ideas. To include Indigenous Knowledge Systems (IKS) into topics like science, history, and language arts, the instructors collaborated with Indigenous communities to co-develop lesson plans. These IKS included storytelling, land-based education, and traditional ecological knowledge (Battiste, 2013). Teachers who participated in this reflective process acquired practical experience in developing more inclusive and culturally relevant teachings, in addition to developing their awareness of Indigenous viewpoints. To guarantee that educators can confidently use AR and AL in the classroom to facilitate decolonization, this type of training is crucial. Acknowledging teachers' contributions as co-researchers in the decolonization process is another essential component of professional development. Teachers actively participate in the development and distribution of information rather than only being passive consumers of it. Within the framework of decolonization, this implies that teachers collaborate with students, Indigenous communities, and other relevant parties to co-create curriculum and instructional strategies that represent a range of viewpoints. To make sure that their teaching methods are culturally sensitive and consistent with Māori values, for example, educators at New Zealand's Māori immersion schools do continuous research in collaboration with the region's Māori elders (Ritchie & Rau, 2018). This cooperative approach guarantees that the educational system represents the needs and values of the Indigenous population while also improving the professional growth of the instructors.

Case studies of decolonization-focused professional development programs show the transformational power of this kind of training for both teachers and students. A noteworthy instance originates from South Africa, where a professional development course was created to assist educators in incorporating IKS into science and math courses. Teachers were able to gain information from both academic and community-based sources through the program's seminars, group research projects, and mentoring from Indigenous knowledge bearers (Smith & Ballard, 2020). Students in this curriculum showed better engagement and comprehension of the subject, especially in reference to local environmental challenges. Teachers reported feeling more confident in their ability to teach Indigenous knowledge alongside Western scientific principles. In a similar vein, an Australian professional development initiative aimed to incorporate traditional ecological knowledge into environmental science courses for teachers working in isolated Indigenous communities. Teachers worked in groups to identify local environmental concerns, such as water management or land conservation, and then developed lesson plans that integrated both Indigenous and Western knowledge to solve these challenges. The program employed Action Learning as its framework (Kimmerer, 2013). Teachers involved in the program indicated that when students could relate the subject to their own lives and communities, it increased their engagement level in the classroom. Since Indigenous community members were viewed as co-educators in the educational process, the program also assisted educators in forging closer bonds with them.

The significance of consistent and cooperative professional growth for decolonization is emphasized by these case studies. Decolonizing education is a continual process that calls on teachers to interact with a variety of viewpoints, constantly evaluate their own methods, and modify their curricula to suit the requirements of their pupils. Teachers who participate in professional development programs that emphasize AR and AL approaches will be equipped with the necessary skills to participate in this process in a meaningful and efficient manner. Professional development is essential to equipping teachers with the skills necessary to apply action research and action learning for decolonization. Professional development programs may contribute to the creation of more inclusive and culturally responsive educational systems by giving teachers the tools they need to reflect on their work, work with Indigenous communities, and incorporate Indigenous knowledge into their curricula. By co-creating knowledge with their students and communities, teachers, as co-researchers in the decolonization process, contribute to the change of education. Case studies from Australia, New Zealand, South Africa, Canada, and New Zealand show how professional development initiatives centered on decolonization may have a profound impact, emphasizing the value of continuous learning and cooperation in establishing fair and pertinent educational opportunities for all students. Teachers may question colonial narratives in education and create curriculum that represent the many knowledge systems of Indigenous peoples throughout the world by utilizing AR and AL. In addition to helping instructors, this professional development also gives students deeper, more meaningful learning experiences, which promotes an inclusive and equitable educational system.

COMMUNITY ENGAGEMENT AND PARTICIPATION

Those indigenous people should be involved in the process of decolonization to foster the culturally sensitive education system that embraces and extends recognition to IKS. Considering these definitions, indigenous communities have a wealth of knowledge of their past, culture and the environment hence are in a vantage position for decolonization of education. Thus, such change processes fail to transform education systems adequately and may lack the legitimacy of democratization of education effort if the people targeted by decolonization are not involved in the process. Successful engagement with the stakeholders in indigenous communities mean that the education systems and institutions benefit from resources such as knowledge and wisdom that enhance the curricula, teaching methodologies as well as the learning processes. Furthermore, it encourages ownership and relativity to make sure that it is not forced by outside sources but the ideas of the natives of the territories in question to decolonize themselves (Mbah & Bailey, 2022). As a result of the above, the following strategies can be adopted by education systems that seeks to meaningfully engage the indigenous role players: One of these tactics is the long-term cooperation between the non-indigenous scholars and indigenous people, in which the voices are valued and taken into consideration. It can be attained through consultation on the policies, involvement of indigenous people in decision making processes as well as ensuring that they have proximity to policy making and formulation. On the same note, including indigenous leaders or elders, or having indigenous people sit as advisors or teachers in the classroom aids in eradicating the difference between traditional and formal schooling (Ray, 2021). Schools should also afford indigenous people an opportunity to participate in the development of other curriculum where they shall be able to find representation of their values, language and perception of the world. All these efforts help toward the attainment of a just society that respects cultural differences and embraces multiculturalism between people of different origins (Cummings & Others, 2021).

Reflections from actual community-engaged Action Research (AR) and Action Learning (AL) projects show how collaboration fosters decolonization. For example, an AR project implemented in Canada is called Indigenous Quality Assurance project which engaged indigenous communities into re-imagining higher education quality frameworks. Thus, the success of the project is based on the use of Two-Eyed Seeing – the combination of indigenous and Western approaches to learning and knowing – that the academic and the indigenous community co-designed a curriculum that meets both the universities and the indigenous people's expectations. Another example is PALAR project implemented in South Africa in which the communities collaborated with the schools to create context solutions. It was also very important because this partnership meant that the community's goals like indigenous languages and cultures were assimilated into the school system so that the learning process was sensitive to students' needs (Rangana, Geduld, & Sathorar, 2023). Explicit consideration of communities and indigenous people can help enhance decolonization's efficiency and longevity. Such collaborations do more than make education more relevant, but it allows indigenous people to have a say in education's direction. By employing AR and AL methodologies, the communities and the educational institutions may find ways to support construction of systems that embrace a wider range of cultural knowledge and education as a social change agency and justice activist (Omodan & Dastile, 2023).

ADDRESSING CHALLENGES IN DECOLONIZATION

Putting education systems back into the black is an elaborate process everyone embarks on when decentralizing education to face several challenges. Another major issue is the issue of resistance to change which considers both the context of the educational institutions and other stakeholders. Most of the teachers, principals, and policy makers are still trapped in the paradigm that has been used systematically for many years now which has a bias towards the European systems. Such models tend to exclude Indigenous Knowledge Systems (IKS) and other eclectic forms of learning hence infringing the current attempts to implement decultured curricula and pedagogy (Ray, 2021). Besides, there can be misunderstandings about what decolonization means and the changes it should imply and therefore the dynamics can be only symbolic, without eradicating the mechanisms of coloniality in education (Omodan & Dastile, 2023). A third factor is the poor resource and infrastructure that are available to aid the eradication of colonialism. A lot of education systems, especially in developing countries, face challenges such as low capita funding, lack of teachers and antiquated infrastructure. These are some of the challenges that hinder the process of disseminating new curricula, training teachers or teaching indigenous languages and culture in school (Mbah & Bailey, 2022). Further, there are difficulties in developing Culturally Relevant and Representative (CRR) historical education content and context to explain the indigenous people's history and experiences, primarily because educational organizations have limited cooperation or no access to indigenous peoples (Cummings et al., 2021). Possible strategies for addressing these challenges are stressing on developing meaningful relationships with the indigenous peoples and promoting effective cooperation among all the system's levels. For example, the teacher training can add to the decolonization principles in teacher training by providing solutions of how teachers can practice culturally relevant pedagogy. Such programs may also stress other practices of using Action Research (AR) and Action Learning (AL) approaches that help educators reflect on their practices, cooperate with indigenous communities in improving their knowledge (Rangana, Geduld, & Sathorar, 2023).

This is further compounded by engaging the indigenous leaders, elders, and cultural experts in the development and delivery of the curricula to ensure that education is meaningful and culturally appropriate. Through the presented case studies, it is possible to identify extremely innovative and complex approaches to the challenges of decolonization efforts. One such decolonization agenda at the University of Queensland in Australia was the introduction of an Indigenous Studies major prepared with an input of the indigenous people. It not only integrates the indigenous knowledges to the curriculum but also gives opportunities for the students for engaging in the community learning activities that enhance the understanding of the indigenous cultures and reconciliation (Ray, 2021). Since the Māori language has been included as one of the languages to be used in schools and since the education system has been embracing bilingualism, New Zealand has gone a long way in directing its policies in form of education system to be more culturally sensitive (Omodan & Dastile, 2023). Therefore, it can be stated that despite the problems mentioned above the process of decolonizing education is possible and should be continued. Working together, challenging status quo thinking and being steadfast in the quest for true transformative change, it is possible to design education contexts that embrace people's wisdom as well as their history. Effective decolonization processes are therefore good reference points of how education can be reformed to be more emancipative, inclusive and pluralist (Mbah & Bailey, 2022).

EVALUATING THE IMPACT OF DECOLONIZATION INITIATIVES

It is therefore important for the effects of decolonization measures towards education to be assessed to determine the extent to which objectives of espousing IKS and altering curricula are being achieved. It is noteworthy that the efficiency of these measures can be evaluated by using several methods. One typical approach is the quantitative research where the views of educators, students and other members of a particular community can be explored by conducting interviews and focus groups and by using case studies. In this way, receiving such feedback, institutions would be aware of the extent to which decolonization activities are effective and which aspects may require additional development (Ray, 2021). It is also possible to use qualitative measures and quantitative data analysis in surveys to monitor shifts in learner comprehensiveness and performance, dropout rates and, more specifically, indigenous learners' degree of decolonization efforts (Omodan & Dastile, 2023). It is crucial to measure the effectiveness of decolonization efforts using key performance indicators and related metrics. Another usage of IKS is in the curriculum and its integration is another significant KPI. Incorporation of indigenous views may be measured using the number of schools and universities that have implemented indigenous views and knowledge, and languages by the number of courses that have adopted indigenous view and extent or proportion of indigenous perspectives in the practices and assessment materials used in the schools and universities (Mbah & Bailey, 2022). Another important indicator is the level of participation of the indigenous peoples in the decision-making processes that concern them, including curriculum design, teachers training processes as well as governance structures. The extent of the indigenous organizations' involvement and the number of community-based partnerships can also help to gauge the effectiveness of decolonizing approach in educational activities (Cummings et al., 2021). Also, the graduation rates as well as the academic performance and other achievements of the indigenous students as the learners are some of the essential benchmarks that can show if the decolonization procedures enhance education fairness (Rangana, Geduld, & Sathorar, 2023).

The best evaluation frameworks have been established as dominant featuring both qualitative and quantitative methodologies in efforts to capture the decolonization efforts impact. For instance, a framework that has been developed to evaluate decolonization in Canadian universities consists of an emphasis on the integration of indigenous knowledge into curricula and research as well as indigenous faculty and employees and services for indigenous students. These perspectives help to put all areas of the educational process into consideration ranging from the subject that is taught to the environment of an institution, (Ray, 2021). In New Zealand, there is another kind of evaluation model regarding the Bilingual Education Success Checklists which monitors the progress of bilingual programs by using students' Maori language learning, as well as the views and interviews with students, parents and teachers to assess the cultural values and equity of the curriculum (Omodan & Dastile, 2023). Another example is from Australia only universities that adopted some type of decolonisation measures conduct annual assessments of their curricula where representatives consider how indigenous knowledge is embedded in a curriculum and to what extent it meets the Australian government reconciliation targets. That is why using feedback from indigenous students and faculty is an important consideration in the process because evaluation cannot only or should not be led by institutional priorities but with reference to the needs and demands of indigenous peoples (Mbah & Bailey, 2022). Thus, the assessment of the effectiveness of decolonization measures is highly liberal in the combination of approaches that should be taken and methods that should be used. Using the above-described KPIs and involved with the communities these concerns try to address, the educational institutions can monitor positive changes and tweak their actions to guarantee that decolonization initiatives create sustainable shifts for the better (Rangana, Geduld, & Sathorar, 2023).

FUTURE DIRECTIONS AND INNOVATIONS

Decolonizing education through Action Research (AR) and Action Learning (AL) – is an ongoing process and several emerging trends indicate the new course. Another trend of the decolonization process is the engagement of digital tools and technologies. Web-based teaching and learning space, learning management systems, and digital narratology are as well applied to engage Indigenous voices and epistemologies in a learning environment (Omodan & Dastile, 2023). These technologies provide chances to disseminate the IKS cross culturally and therefore catalyze the infusion of cultural perspectives into the conventional curricula. Furthermore, with the advancement of digital technology it becomes easier for indigenous people to engage with academic institutions; where they share local and other forms of knowledge, which is important in the global community. Future trends concerning the status of IKS in education show that as more awareness is created on the need to develop education systems that can solve current social problems, IKS is likely to receive an increased endorsement. When the decolonization process unfolds, the learning entities are expected to pay more attention to co-construct curriculums with the indigenous people that can be both relevant and scholarly (Mbah & Bailey, 2022). This may foster the reorientation of equally ranked IKS and western knowledge systems, which is the ethos of this paper. Also, the use of IKS is still predicted to extend its scopes to the humanities, social sciences, and innovation processes as well as staking its foothold in STEM disciplines. For instance, indigenous principles in matters of environmentalism and natural resources conservation hold a great potential in influencing the future crop of scientists and engineers in their thinking and action. Current scholarship in decolonization includes ongoing efforts toward identifying more appropriate methodologies and epistemologies for the inclusion of IKS. Main interest is given to the idea of educators as active participant in the decolonial journey. The engagement of teachers and indigenous knowledge bearers in action research means that schools can develop the curriculum based on the learners' life experiences and the social realities of the community. Also, the investigation of decolonializing of Teacher training programs is another topic of interest since it aims at training teachers to embrace culturally sensitive education (Cummings et al., 2021). Furthermore, through the contemporary literature, one witnesses an increasing interest in policy and practice by which segregationists seek to respond to decolonization imperatives, be this in curriculum-making, in assessment, or in governance (Rangana, Geduld, & Sathorar, 2023).

Another area for growth that Decolonize found can be explored in the future is that of international cooperation in the process of decolonizing education. It will be expected that more collaborations between indigenous people together with educational facilities and researchers from various countries will occur, thus enhancing the flow of knowledge and successful approaches (Omodan & Dastile, 2023). For instance, it is possible to note that various joint research initiatives of academic institutions of Africa, Latin America, and Australia have already shown the potential of learning collaboration in supporting decolonialization process. As these partnerships grow, they're going to help produce a world knowledge of the decolonization in addition to IKS that'll enable larger and permanent educational revolution. Thus, it can be noticed that the future of decolonizing education through AR and AL is rather bright, rich in potential types of development. The integration, the practice of IKS, the use of digital technologies and partnerships between indigenous peoples and academic institution will always have an impact in changing education systems as observed by Ray, (2021). During the advancement in research and development within this learning area, it could be expected that more culturally sensitive and competent practices will be developed that will enhance the students' learning experience.

CONCLUSION

In this chapter, the movement for transforming the education systems using AR and AL frameworks has been discussed in this chapter. We started with decoloniality as the process of undoing coloniality that persists in conventional education systems. European colonization dominates curricular, pedagogy, and the epistemology, and it disregards IKS and other cultures. Decolonizing education is thus a process of moving towards a culturally sensitive, and fairly education environment in learning institutions for all learners. This process implies recognition of the colonial history of education and the promotion of the strategies which would help indigenous people to contribute to the modern curricula. From this work, it can be stated that the inclusion of AR and AL methodologies into decolonization processes is a helpful tool. These participatory approaches enable the educators and students to engage with the indigenous people hence entering the communities' point of view into the process. By constructing teachers as researchers and engaging students as knowledge producers, AR and AL support the rejection of colonialist epistemologies in education that can manifest in form of vertical power relations. This chapter also included the exploration on how AR and AL had been effectively applied and applied in decolonialization practices in which proved the advantages of these approaches to disrupt conventional pedagogy. This shows that the process of decolonizing education entails the use of IKS in the curriculum. We talked about the possible approaches to implement IKs such as partnership with indigenous communities to co-design curricula, as well as using culturally appropriate content, and Indigenous narratives as well as knowledge in education.

The rationale for embedding IKS is as follows; It promotes meaningful learning among students and improvement of contextual knowledge and understanding of issues affecting society. Therefore, integration of IKS into schools' mainstream curricula provides nations with an opportunity of constructing diverse learning contexts that capture the multi-faceted nature of knowledge in the modern society. Further, the place of professional development in preparing educators to decolonize was also explored. The process of decolonization is not an easy one, and teachers are the major players in this process, and it is important to empower educators with knowledge and skills on how to implement AR, AL and IKS in their classrooms. Professional development programs that address decolonization not only benefit the teachers, but it also fosters a continuation of decolonization process in schools and institutions as the process becomes institutionalized. Several examples of PD programs, which adopted AR and AL to decolonization were demonstrated to exemplify the potential of such initiatives in changing the current education system.

Moreover, community engagement appeared to be an important factor that ought to be taken into consideration when trying to facilitate the process of decolonization. Indigenous participation in the education process enhances the process of partnership with the curriculum and ensures it caters for the wants and needs of the society. In AR and AL initiatives implemented by our research communities, we have observed how schooling frameworks can transform and become more sensitive to context, which means that First Peoples' voices are not only listened to but that they are the core and focus of learning. Nonetheless, difficulties persist in the process of decolonizing education as foreseen in its decolonized future. Challenges like institutionalization, lack of resources, and ignoring indigenous knowledge are some of the challenges that persist. However, all these challenges can be overcome by using new methods and by committing educators, researchers and policymakers. Examples from the previous and current global endorse the strategies that have been deployed to ensure that some of the barriers highlighted herein are eliminated to pave the way for the adoption of Height of Prejudice and Discrimination in the Somali Community to course CHWs to adapt or die 'solutions. All in all, the decolonization of education through the deployment of AR and AL methodologies is a very valuable perspective that can bring significant changes in traditional statistical systems in education. IKS must be integrated and meaningful community engagement, and training of educators on how to provoke change, we can improve on the opportunity of providing for inclusion and equity of learners. So, the decolonization process is not only about the question and themes of the past but about the present and future of education too. It is therefore a wakeup call to educators, researchers and policy makers to further engage in the collective search and search for new and more effective strategies to earnestly advance indigenous knowledge and the voices of indigenous peoples into education. This is the time to wake up and change and the value of decolonization can only be achieved if everyone takes an action, and this should be a long-term change. Another feature that creates further potential for using AR and AL in the future vision of education's nature is that every knowledge system may be valued and respected as a part of a learning process infrastructure.

REFERENCES

Abu Moghli, M., & Kadiwal, L. (2021). Decolonising the curriculum beyond the surge: Conceptualisation, positionality and conduct. *London Review of Education*, *19*(1), 1–16. DOI: 10.14324/LRE.19.1.23

Agherdien, N., Pillay, R., Dube, N., & Masinga, P. (2022). What does decolonising education mean to us? Educator reflections. *Scholarship of Teaching and Learning in the South*, *6*(1), 55–78. DOI: 10.36615/sotls.v6i1.204

Ajaps, S. O. (2021). Decolonising education for environmental conservation: A participatory action research (Doctoral dissertation, New York University).

Bilal, A., & Chand, A. (2023). Decolonising education and research: An Indianised approach to pedagogy. *The Arab Journal of the Social Sciences*, *22*(2), 23–40.

Bourgeois, R., Karuri-Sebina, G., & Feukeu, K. E. (2024). The future as a public good: Decolonising the future through anticipatory participatory action research. *foresight, 26*(4), 533-549.

Cornish, F., Breton, N., Moreno-Tabarez, U., Delgado, J., Rua, M., de-Graft Aikins, A., & Hodgetts, D. (2023). Participatory action research. *Nature Reviews. Methods Primers*, *3*(1), 34. DOI: 10.1038/s43586-023-00214-1

Cox, R., Heykoop, C., Fletcher, S., Hill, T., Scannell, L., Wright, L., & Plush, T. (2021). Creative action research. *Educational Action Research*, *29*(4), 569–587. DOI: 10.1080/09650792.2021.1925569

Cummings, S., Munthali, N., & Shapland, P. (2021). A systemic approach to the decolonisation of knowledge. In *The politics of knowledge in inclusive development and innovation* (pp. 65-79). DOI: 10.4324/9781003112525-7

Dudgeon, P., Bray, A., Darlaston-Jones, D., & Walker, R. (2020). Aboriginal participatory action research: An indigenous research methodology strengthening decolonisation and social and emotional wellbeing.

Fahlberg, A. (2023, March). Decolonizing sociology through collaboration, co-learning and action: A case for participatory action research. *Sociological Forum*, *38*(1), 95–120. DOI: 10.1111/socf.12867

Glowach, T., Mitchell, R., Bennett, T., Donaldson, L., Jefferson, J., Panford, L., & Hemmings, E. (2023). Making spaces for collaborative action and learning: Reflections on teacher-led decolonising initiatives from a professional learning network in England. *Curriculum Journal*, *34*(1), 100–117. DOI: 10.1002/curj.186

Gough, K. V., Veléz-Torres, I., Ruette-Orihuela, K., Fayad, J., Bueno, B., Corredor, G., & van Grootheest, S. (2023). Engaged pedagogic research: Transforming societies through co-learning and social action. *Environment and Planning C. Politics and Space*, *41*(1), 109–129.

Gullion, J. S., & Tilton, A. (2020). *Researching with: A decolonizing approach to community-based action research* (Vol. 6). Brill. DOI: 10.1163/9789004424852

Harrison, N., & Clarke, I. (2022). Decolonising curriculum practice: Developing the indigenous cultural capability of university graduates. *Higher Education*, *83*(1), 183–197. DOI: 10.1007/s10734-020-00648-6

Hayes, A., Luckett, K., & Misiaszek, G. (2021). Possibilities and complexities of decolonising higher education: Critical perspectives on praxis. *Teaching in Higher Education, 26*(7-8), 887–901. DOI: 10.1080/13562517.2021.1971384

Igwe, P. A., Madichie, N. O., & Rugara, D. G. (2022). Decolonising research approaches towards non-extractive research. *Qualitative Market Research, 25*(4), 453–468. DOI: 10.1108/QMR-11-2021-0135

Itwaru, S. (2024). Decolonizing higher education pedagogy: From theory to practice. *Collected Essays on Learning and Teaching, 15*(1).

Kirshner, J., & Kamberelis, G. (2021). *Decolonizing transcultural teacher education through participatory action research: Dialogue, culture, and identity*. Routledge. DOI: 10.4324/9781003111573

Lepore, W., Hall, B. L., & Tandon, R. (2021). The Knowledge for Change Consortium: A decolonising approach to international collaboration in capacity-building in community-based participatory research. *Canadian Journal of Development Studies. Canadian Journal of Development Studies, 42*(3), 347–370. DOI: 10.1080/02255189.2020.1838887

Maine, M. K. (2022). Towards a decolonised research curriculum in psychology: A participatory action research exploration of students' perspectives (Doctoral dissertation, University of Pretoria).

Manathunga, C. (2020). Decolonising higher education: Creating space for Southern knowledge systems. *Scholarship of Teaching and Learning in the South, 4*(1), 4–25. DOI: 10.36615/sotls.v4i1.138

Martinez-Vargas, C. (2020). Decolonising higher education research: From a uni-versity to a pluri-versity of approaches. *South African Journal of Higher Education, 34*(2), 112–128. DOI: 10.20853/34-2-3530

Mbah, M. F., & Bailey, M. (2022). Decolonisation of research methodologies for sustainable development in Indigenous settings. In *Indigenous Methodologies, Research and Practices for Sustainable Development* (pp. 21–48). Springer International Publishing. DOI: 10.1007/978-3-031-12326-9_2

Omodan, B. I., & Dastile, N. P. (2023). Analysis of participatory action research as a decolonial research methodology. *Social Sciences (Basel, Switzerland), 12*(9), 507. DOI: 10.3390/socsci12090507

Rangana, N., Geduld, D., & Sathorar, H. (2023). Multiple stakeholder collaborations to co-develop school improvement plans in Community schools: A Participatory Action Learning and Action Research (PALAR) study. *South African Journal of Educational Leadership and Management, 3*(1), 5–25.

Rasool, S., & Harms-Smith, L. (2021). Towards decoloniality in a social work programme: A process of dialogue, reflexivity, action and change. *Critical African Studies, 13*(1), 56–72. DOI: 10.1080/21681392.2021.1886136

Ray, L. (2021). Decolonizing action research through Two-Eyed Seeing: The Indigenous quality assurance project. *The Canadian Journal of Action Research, 21*(3), 95–114. DOI: 10.33524/cjar.v21i3.510

Tariq, M. U. (2024). Generative AI in Curriculum Development in Higher Education. In Fields, Z. (Ed.), *Impacts of Generative AI on Creativity in Higher Education* (pp. 227–258). IGI Global., DOI: 10.4018/979-8-3693-2418-9.ch009

Tariq, M. U. (2024). Harnessing Persuasive Technologies for Enhanced Learner Engagement and Motivation. In M. Sanmugam, D. Lim, N. Mohd Barkhaya, W. Wan Yahaya, & Z. Khlaif (Eds.), *Power of Persuasive Educational Technologies in Enhancing Learning* (pp. 30-62). IGI Global. https://doi.org/ DOI: 10.4018/979-8-3693-6397-3.ch002

Tariq, M. U. (2024). Navigating the Personalization Pathway: Implementing Adaptive Learning Technologies in Higher Education. In Minh Tung, T. (Ed.), *Adaptive Learning Technologies for Higher Education* (pp. 265–291). IGI Global., DOI: 10.4018/979-8-3693-3641-0.ch012

Tariq, M. U. (2025). Gamification and Virtual Reality: Transforming Early Childhood Language Learning. In Manditereza, B. (Ed.), *Digital Pedagogy in Early Childhood Language Development* (pp. 25–52). IGI Global Scientific Publishing., DOI: 10.4018/979-8-3693-7437-5.ch002

Tariq, M. U. (2025). Integrating Theory and Practice: Transformative Approaches in Practice-Led Research for Professional Development. In Kwok, S., Yu, P., & Le Corre, J. (Eds.), *Exploring Practice-Led Research for Professional Development* (pp. 419–436). IGI Global Scientific Publishing., DOI: 10.4018/979-8-3693-6376-8.ch016

Tariq, M. U. (2025). Beyond the Classroom: The Sustained Influence of Educators on Student Success. In West, E. (Ed.), *Former Educators' Roles in Supporting Student Growth and Learning* (pp. 321–346). IGI Global Scientific Publishing., DOI: 10.4018/979-8-3693-6427-7.ch013

Wilson, C., Heinrich, L., Heidari, P., & Adams, K. (2020). Action research to implement an Indigenous health curriculum framework. *Nurse Education Today*, *91*, 104464. DOI: 10.1016/j.nedt.2020.104464 PMID: 32526618

Ziai, A., Bendix, D., & Müller, F. (2020). Decolonizing knowledge orders, research methodology and the academia: An introduction. In Bendix, D., Müller, F., & Ziai, A. (Eds.), *Beyond the Master's Tool* (pp. 1–16).

KEY TERMS AND DEFINITIONS

Decolonization of Education: The process of dismantling colonial frameworks in education to embrace Indigenous Knowledge Systems (IKS), cultural relevance, and local community involvement.

Action Research: A participatory, cyclical research approach where educators and communities collaborate to identify, address, and reflect on issues in education.

Action Learning: A problem-solving approach in education where individuals or groups learn through active engagement, reflection, and collaboration within real-world contexts.

Indigenous Knowledge Systems (IKS): The holistic, localised knowledge, skills, and cultural practices developed and sustained by Indigenous communities over generations.

Coloniality in Education: The continued influence of colonial ideologies, structures, and epistemologies in modern education systems, marginalising non-Western perspectives.

Chapter 6
Decolonizing Education Systems Through Action Research and Action Learning

Muhammad Usman Tariq
https://orcid.org/0000-0002-7605-3040
Abu Dhabi University, UAE & University College Cork, Ireland

ABSTRACT

The chapter investigates the transformative potential of coordination of Inborn Information Frameworks (IKS) into instructive systems utilizing techniques such as Activity investigation (AR) and Activity Learning (AL). This approach challenges the bequest of colonialism in instruction by centering inborn points of view, information, and hones within the educational modules' improvement preparation. Activity Investigation and Activity Learning provide collaborative and iterative forms through which teachers, policymakers, and innate communities can co-create instructive activities. These strategies prioritize the dynamic interests of all partners, guaranteeing that activities are responsive to the differing needs and points of view of innate people groups. Teachers can refine their hones to better join IKS in instructing and learning encounters by locking in persistent reflection and adjustment. The chapter analyzes case considerations and best hones from different settings, illustrating the viability of AR and AL in decolonizing instruction frameworks.

INTRODUCTION

The chronicled effect of colonisation on instruction frameworks is significant and persevering. Colonisation reshaped the instructive scenes of numerous nations, frequently forcing remote structures, dialects, and educational modules while modules marginalising innate information and hones. This chapter digs into the colonial legacy in instruction, looking at the strong impacts and the coming detachment between ordinary instruction and inborn information. Colonial powers, essentially European, presented their instruction frameworks to the locales they colonised. These frameworks were outlined to serve the interface of the colonisers, centring on making a lesson of nearby directors and labourers who would help keep up colonial run the show (Ashcroft et al., 2000). Instruction was an instrument for social digestion, upholding the predominance of the coloniser's culture and dialect. For case, in British

DOI: 10.4018/979-8-3693-3542-0.ch006

colonies like India, the Macaulay Diminutive of 1835 emphasised English instruction over conventional Indian learning frameworks (Viswanathan, 1989). The objective was to make "a course of people, Indian in blood and colour, but English in taste, conclusions, ethics, and judgment skills" (Macaulay, 1835). This approach is driven by the foundation of English-medium schools and the marginalisation of innate dialects and information frameworks. The instruction framework was additionally planned in African colonies to serve colonial needs. The French approach of absorption pointed to teaching a small first class who would receive French culture and dialect, whereas the lion's share of the populace remained uneducated or received negligible professional training (Hargreaves, 1967). This created a critical hole between the taught first class and the masses, a partition that endures in numerous post-colonial social orders nowadays.

Disengagement Between Ordinary Instruction and Inborn Information

One of the foremost critical legacies of colonial instruction frameworks is their detachment between ordinary instruction and innate information. Colonisers frequently see innate information frameworks as second-rate or unimportant, driving their orderly avoidance of formal instruction (Semali & Kincheloe, 1999). This has had enduring suggestions for the esteem set on inborn information and social hones.

Marginalisation of Inborn Dialects

Colonial instruction frameworks prioritised the coloniser's dialect, frequently driven to the marginalisation and decay of inborn dialects. For occurrence, in numerous African nations, the medium of instruction in schools is still a colonial dialect, such as English, French, or Portuguese. This has noteworthy suggestions for cognitive improvement and social personality. Children learn best in their mother tongue, and instruction in an outside dialect can prevent comprehension and scholarly execution (Pinnock, 2009). The burden of remote dialects in instruction, moreover, contributes to the disintegration of inborn dialects. Numerous dialects are imperilled since they are not utilised in formal instruction or official communication (Skutnabb-Kangas, 2000). This misfortune of dialect compares to a misfortune of social legacy, as dialect could be essential for transmitting social information and conventions.

Concealment of Innate Information Frameworks

Colonial instruction frameworks regularly neglected inborn information frameworks, regarding them as primitive or informal. For example, conventional African instruction, which was all-encompassing and community-based, emphasising ethical values, viable aptitudes, and social duty, was supplanted by Western-style instruction that centred on personal accomplishment and hypothetical information (Fafunwa, 1974). In numerous inborn societies, information is transmitted orally and through hone, frequently inserted in ceremonies, stories, and communal exercises (Battiste, 2002). Colonial instruction frameworks, with their accentuation on composed writings and formal classroom settings, frequently expelled these strategies as second-rate. This has led to the cheapening of innate ways of knowing and a detachment between taught people and their social roots. Different countries examples are as below:

India

The colonial instruction framework in India could be a prime case of colonisation disturbing inborn instruction. Sometime recently, the British ran the show; India had a differing cluster of instructive education, counting conventional gurukuls and madrasas, which gave instruction in nearby dialects and consolidated devout and philosophical lessons (Kumar, 2005). The British presented an instruction framework based on the English language and Western educational programs, marginalising these conventional instruction shapes. The presentation of English-medium instruction made a lesson for English-speaking elites who frequently separated themselves from innate information and social hones. This disengagement is still apparent nowadays, as English-medium instruction is frequently seen as prevalent, driving the disregard of innate dialects and information frameworks (Batra, 2013; Fahlberg, 2023).

Africa

In Africa, the colonial bequest in instruction is characterised by a double framework of prior prioritised knowledge over innate information. The formal instruction framework, presented by the colonisers centred on scholastic subjects instructed in colonial dialects, whereas inborn instruction, which was down-to-earth and community-oriented, was generally overlooked (Ocitti, 1994). In instance, an instruction framework was set up by British in Kenya that served the party and a couple of African elites (Sheffield, 1973). Innate knowledge and practices, such as conventional medication, rural procedures, and social ceremonies, were prohibited from the educational programs. This has driven a detachment between taught people and their communities, as formal instruction frequently did not prepare them with the information and aptitudes important to their neighbourhood setting.

Australia

In Australia, the colonial instruction system had devastating impacts on Indigenous communities. The British colonised their instruction framework, which pointed to absorbing Innate children into Western culture. Inborn children were frequently expelled from their families and set in private schools, where they could not speak their dialects or hone their societies (McConaghy, 2000). The misfortune of Innate dialects and social hones drives this arrangement of absorption. Numerous Innate Australians nowadays battle with the bequest of these approaches, as the instruction system still needs to completely coordinate Inborn information and points of view (Nakata, 2007). Endeavours are being made to join Innate information into the educational programs, but critical challenges remain in bridging the crevice made by colonial instruction approaches.

BACKGROUND

For a long time, there has been a developing recognition of the esteem of innate information and its coordination in formal instruction frameworks. Different nations and communities are taking steps to decode coloniser instruction frameworks and reconnect with inborn information. One approach to tending to the disengagement between customary instruction and inborn information is through change in educational programs. In Unused Zealand, for illustration, the instruction system has undergone critical changes to incorporate Māori information and points of view. The presentation of the Te Whāriki early childhood curriculum, which is based on Māori standards and values, could be a step towards making a comprehensive instruction framework that regards and values innate information (Ritchie & Rau, 2006). Essentially, in Canada, there have been endeavours to join, to begin with, countries', Métis', and Inuit's points of view in the educational programs. The Truth and Compromise Commission of Canada has called for instructive changes that incorporate educating Inborn history, culture, and information in schools (Truth and Compromise Commission of Canada, 2015). These endeavours point to making a more comprehensive instruction framework that recognises the commitments of innate people groups.

Community-Based Instruction

Community-based instruction is another approach to reconnecting routine instruction with innate information. This model emphasises the inclusion of the community in instruction preparation and integrates local information and hones into the educational modules. For instance, in The Frozen North, The Frozen North Local Information Organise advances community-based instruction that consolidates inborn information and social hones into the formal instruction framework (Barnhardt & Kawagley, 2005). This approach not only protects innate information but also makes instruction more pertinent and important for indigenous students. By including the community in the instruction preparation, understudies can learn from older folks and community members, gaining a deeper understanding of their social legacy and its significance to modern life.

Dialect Revitalisation Programs

Dialect revitalization grams are vital for reconnecting instruction with indigenous knowledge, as dialect may be a key vehicle for transmitting social information and values. Different nations have executed programs to revitalise or consolidate dialects into the instruction framework. In Hawaii, the Pūnana Leo preschools offer immersion programs within the Hawaiian dialect, pointing to resuscitating the dialect and protecting Hawaiian culture (Wilson & Kamana, 2001). These programs have effectively expanded the number of Hawaiian speakers and coordinated Hawaiian social hones into the instruction framework. In Ridges, the Welsh government implemented arrangements to promote the Welsh language in schools, including Welsh-medium education and bilingual programs. These endeavours increased the number of speakers who combined social information in the modules (Jones 2013). The colonial bequest in instruction frameworks has had significant and enduring impacts, causing a critical disengagement between ordinary instruction and indigenous knowledge. Colo colonisation has colonised foreign education structures, marginalised dialects, and debased inborn knowledge systems. This bequest is clear in numerous post-colonial social orders nowadays, where formal instruction frameworks frequently fall flat to coordinated innate points of view and information. Endeavours to address this disconnect are underway, with different nations and communities executing educational module changes, community-based education models, and dialect revitalisation programs. These activities aim to form more comprehensive instruction frameworks that respect and esteem innate knowledge, helping to bridge the crevice made by colonial education arrangements. By reconnecting instruction with inborn information, social orders can advance social differences, protect important conventions, and make more significant instructive encounters for all students (Solva et al., 2022).

Foundations of Action Research and Action Learning

Activity Investigate (AR), and Activity Learning (AL) are strategies that have significantly affected instructive change. They emphasize intelligent hone, collaborative request, and iterative problem-solving, making them especially successful within the energetic and complex instruction field. This part investigates the establishments of AR and AL, their significance to instructive change, and the standards and forms included in applying these strategies inside instructive settings. Action Research (AR) could be a participatory and intelligent investigative strategy that centres on tackling real-world issues through a cycle of arranging, acting, watching, and reflecting. Kurt Lewin, a social clinician, is frequently credited with spearheading AR within the mid-20th century. Lewin's work emphasized the significance of including practitioners in the inquiry about preparing to simultaneously create down-to-earth arrangements and modern information (Lewin, 1946). Within the instructive setting, AR permits teachers to methodically explore their hones, actualize changes, and assess the results to progress instructing and learning (McNiff & Whitehead, 2010). Activity Learning (AL), created by Reg Revans, could be a handle where a little gather of people, known as an "activity learning set," work on genuine issues, reflect on their encounters and learn from each other. Revans defined AL in the 1940s based on the rule that learning is best accomplished through activity and reflection (Revans, 1982). AL promotes collaborative problem-solving and proficiency in instructive settings as members handle bona fide challenges and reflect on their methodologies and results (Pedler et al., 2005).

Pertinence to Instructive Change

The pertinence of AR and AL to instructive change lies in their capacity to cultivate a culture of persistent change, intelligent honing, and collaboration among teachers. Conventional instructive models regularly emphasize top-down approaches and inactive educational programs, which can be detached from the advancing needs of understudies and teachers. In contrast, AR and AL advance a bottom-up approach, where teachers identify issues, test arrangements, and refine hones based on proof and reflection (Somckh & Zeichner, 2009). Joining AR into an instructive hone energises instructors to end up analysts of their claim classrooms, driving them to more pertinent and compelling educational procedures. For occurrence, an educator might utilize AR to investigate new methods of expanding understudy engagement in science classes. By efficiently executing and assessing diverse approaches, the teacher can recognize the foremost successful methodologies and share these discoveries with colleagues, contributing to broader school improvement (Kemmis et al., 2014; Tariq, 2024a). Additionally, AL improves instructive change by cultivating collaborative learning communities. When instructors, chairpersons, and other partners work together in activity learning sets, they share assorted points of view and create inventive arrangements for common challenges. For illustration, a group of school pioneers might utilize AL to address educational module usage issues. Collectively reflecting on their encounters and testing unused approaches can create more effective strategies and support each other's proficient development (Pedler et al., 2005; Bourgeois et al., 2024).

Standards and Forms of AR and AL within the Setting of Instruction

AR and AL standards and processes are grounded in participatory, intelligent, and iterative approaches to problem-solving and learning. The activity inquiry's essential standards incorporate participation, reflection, and relevant significance. Support implies including all stakeholders—teachers, understudies, and administrators—in the investigative handle to guarantee that the arrangements are common and broadly acknowledged (Stringer, 2007). Reflection is vital because it allows educators to analyze their hones fundamentally, get the results of their activities, and make educated decisions about future steps (Schon, 1983). Relevant pertinence guarantees that the inquiry addresses the needs and conditions of the instructive setting, making the discoveries straightforwardly pertinent and useful (Herr & Anderson, 2014).

THE AR PREPARATION REGULARLY TAKES AFTER A PATTERNED DESIGN

Distinguishing a problem, planning an intercession, taking activity, collecting information, reflecting on the results, and changing the arrangement based on what was learned. If a bunch of instructors note that understudies are battling with perusing comprehension, they might arrange an intercession including modern guidelines methodologies. They would implement these strategies, collect information on understudy execution, reflect on the viability of the intercession, and alter their approach in like manner (McNiff & Whitehead, 2010). Activity Learning is built on collaborative learning, reflection, and real-world problem-solving standards. Collaborative learning involves members working together in action learning sets to use their collective information and involvement. Reflection could be a ceaseless process where members fundamentally examine their activities and results in picking up more profound bits of knowledge and moving forward their hones (Revans, 1982). Real-world problem-solving guarantees that the centre remains on tending to actual issues faced by the members, making the learning handle straightforwardly significant and impactful (Zuber-Skerritt, 2002; Tariq, 2025b). The AL handle includes shaping an activity learning set, distinguishing and prioritizing issues, creating and actualizing activity plans, reflecting on what comes about, and sharing knowledge with the gather. Occasionally, a set of instructors might recognize a challenge related to inspiration from the understudy. They would collaboratively create procedures to improve inspiration, actualize these techniques in their classrooms, reflect on the results, and share their encounters with the gather to memorize from each other and refine their approaches (Pedler et al., 2005).

Practical Examples in Instruction

A practical example of Activity inquiry in instruction includes a group of basic school instructors pointing to move forward understudy engagement during math lessons. They might begin by recognizing engagement issues, such as low participation or off-task behaviour. After arranging and executing exercises and technology-based tools intelligently, they collect information through perceptions, understudy criticism, and execution assessments. Reflecting on this data, they may discover that certain exercises boost engagement and join more of these exercises into their lessons. They share their discoveries with colleagues, cultivating a collaborative environment of nonstop change (Kemmis et al., 2014). An illustration of Activity Learning seems to include school directors tending to the challenge of joining modern innovation into the educational modules. Shaping an activity learning set, they talk about the obstacles they face, such as the need for instructor preparation or understudy to get gadgets. Through standard gatherings, they create and test arrangements like proficient advancement workshops for teachers and activities to extend gadget openness. Reflecting on their advance, they learn which procedures are most viable and support each other in executing them more broadly over their schools (Pedler et al., 2005). Activity Inquire and Action Learning are foundational techniques for cultivating instructive change. They emphasize participatory, intelligent, and iterative approaches, enabling teachers to address real-world challenges collaboratively and efficiently. By implanting these strategies into instructive hone, schools can cultivate a culture of ceaseless advancement, development, and proficiency development. The standards and processes of AR and AL guarantee that arrangements are relevantly relevant, practically effective, and collaboratively created, eventually improving the quality of instruction and profiting all partners included (Kayira et al., 2022).

INDIGENOUS KNOWLEDGE SYSTEMS AND EDUCATION

Inborn Information Frameworks (IKS) include the social, biological, and social understandings created by inborn communities over eras. These frameworks are profoundly established within the settings and situations in which inborn people groups live and are fundamental to their social characters and ways of life. Integrating IKS into standard instruction offers critical openings and imposing challenges. This talk gives a diagram of IKS and its significance in instruction and looks at the complexities of IKS coordination in modern instructive systems.

Inborn Information Frameworks (IKS)

Innate Information Frameworks (IKS) are all-encompassing systems that include a wide range of information, including biological administration, conventional medication, otherworldly hones, and social administration. These frameworks differ from storehouses of viable data but exemplify innate communities' values, convictions, and worldviews. IKS is ordinarily transmitted through verbal conventions, ceremonies, and hands-on hones, guaranteeing that information is relevant and custom-made to the community's requirements (Battiste, 2002). The importance of IKS in instruction is complex. To begin with, it approves the social legacy and character of innate understudies, cultivating a sense of pride and having a place. When understudies see their social hones and information reflected in their instruction, it upgrades their engagement and inspiration. Moment, IKS offers interesting viewpoints and arrangements that can enhance the broader instructive educational programs. For instance, inborn environmental information can provide profitable experiences for maintainable natural hones, contributing to worldwide endeavours to address climate alteration (Barnhardt & Kawagley, 2005; Tariq, 2025c).

Challenges in Joining IKS into Standard Instruction

The coordination into standard instruction presents a few challenges, published in chronicled, social, and systemic variables. One of the essential challenges is the bequest of colonialism, which has regularly cheapened and marginalised innate information. Colonial instruction frameworks ordinarily forced Western information ideal models, rejecting innate ways of knowing as second-rate or insignificant. This authentic setting led to an underestimation of IKS in instructive teaching (Smith, 2012). Another challenge is the epistemological distinction between IKS and Western information frameworks. Whereas Western instruction frequently emphasises explanatory and theoretical consideration, IKS is experiential, all-encompassing, and context specific. Bridging these epistemological partitions requires teachers to create socially responsive pedagogies that regard and coordinate assorted ways of knowing. This errand is complicated because numerous teachers have been prepared to follow Western standards and may need recognition or understanding of IKS (McKinley & Smith, 2019). Commonsense challenges moreover emerge within the integration handle. For illustration, the transcendently verbal nature of IKS poses troubles in an instruction framework that depends intensely on composed writings and standardised evaluations. Moreover, there's regularly a need for instructional assets and materials that precisely reflect inborn information and points of view. Creating such assets requires collaboration with innate communities and a commitment to guaranteeing their information is spoken to truly and deferentially (Battiste & Henderson, 2000; Mahadew and Hlalele, 2022).

Openings in Coordination IKS into Standard Instruction

Despite these challenges, there are critical openings for coordinating IKS into standard instruction that can lead to more comprehensive, differing, and successful learning situations. One promising approach is the advancement of socially responsive educational modules that consolidate IKS nearby Western information. For case, in Modern Zealand, the Te Whāriki early childhood educational programs system coordinates Māori concepts and values, creating a bicultural instructive approach that benefits all students (Ritchie & Rau, 2006). Instructor instruction programs play a vital part in this integration handle. By giving teachers preparation in socially responsive instructional methods and information about IKS, educator planning programs can prepare instructors with the aptitudes and understanding vital to joining innate points of view into their instruction. In Canada, the College of Saskatchewan offers a Lone Ranger of Instruction program that incorporates courses on Inborn information and pedagogies, planning future instructors to form more comprehensive classrooms (Battiste, 2013). Community inclusion is basic for the fruitful integration of IKS into instruction. Collaborating with inborn older folks, pioneers, and information attendants guarantees that educational activities are grounded within innate communities' lived encounters and skills. Such organisations can also help in creating educational module materials and instructing methodologies that are socially fitting and educationally viable. For instance, in Gold Country, the Gold Country Local Information Arrange works with nearby communities to coordinate inborn information into the state's instruction framework, advancing socially responsive instruction (Barnhardt & Kawagley, 2005). Moreover, computerised advances offer unused conceivable outcomes for protecting and dispersing IKS. Computerised narration, online stages, and multimedia assets can capture the lavishness of innate information and make it open to a larger audience. These apparatuses can help bridge the crevice between verbal conventions and modern educational practices, giving energetic and locked-in ways for understudies to memorise almost IKS (McConaghy, 2000; Tariq, 2025d).

Illustrations of Effective Integration

A few illustrations illustrate the fruitful integration of IKS into standard education. In Australia, the consolidation of Native viewpoints into the national educational modules has been a key centre, with activities like the "Two-Way Science" program in Western Australia. This program mixes Western science with innate biological information, giving understudies a comprehensive understanding of the characteristic world (Oliver, 2013). Within the joined-together States, the Navajo Country works its claim schools, such as the Navajo Preliminary School in Modern Mexico, which coordinates Navajo dialect and culture into the educational modules. This approach not only preserves the Navajo legacy but also enhances students' scholastic execution by creating a significant social learning environment (Johnson, 2010). Inborn Information Frameworks hold critical esteem for instruction, advertising interesting points of view, cultivating social character, and contributing to more feasible and comprehensive hones. The integration of IKS into standard instruction presents both challenges and openings. This preparation is basic to overcoming chronicled legacies of marginalisation, tending to epistemological contrasts, and creating suitable assets. The potential benefits, such as upgraded understudy engagement, enriched educational programs, and more comprehensive instructive situations, make this integration beneficial. By grasping IKS, teachers can make learning encounters that are not as it were scholastically vigorous but too socially confirming and pertinent for all students (Tariq, 2024e).

STRATEGIES FOR DECOLONIZING EDUCATION

Decolonizing instruction includes changing instructive hones to recognize, coordinate, and regard Innate Information Frameworks (IKS). Instructive education can make more comprehensive and socially significant learning situations by implanting IKS in education about program improvement, instruction, and appraisal through strategies like Activity Inquire about (AR) and Activity Learning (AL). Furthermore, locking in and collaborating with innate communities within the co-creation of instructive substance guarantees that these endeavours are grounded in genuineness and regard for innate points of view. This dialogue investigates inventive approaches for decolonizing instruction and highlights the significance of collaboration with innate communities.

Imaginative Approaches to Insert IKS in Educational Modules Advancement

Coordination of IKS into educational module advancement requires a principal move from conventional Western-centric models to more comprehensive and assorted systems. Activity Inquire about (AR) offers a participatory approach to educational program advancement, where teachers collaborate with inborn communities to recognize instructive needs and co-create substance that reflects inborn information and values. This preparation includes persistent cycles of arranging, acting, watching, and reflecting, permitting for iterative enhancements based on criticism and perceptions (McNiff & Whitehead, 2010). For case, in Modern Zealand, integrating Māori information into the national educational programs is guided by standards of association and cooperation. The Te Whāriki educational programs for early childhood instruction consolidate Māori concepts and values, advancing a bicultural approach that benefits all understudies (Ritchie & Rau, 2006). This show illustrates how AR can encourage the co-development of educational programs honouring inborn conventions and assembly instructive guidelines. Activity Learning (AL) bolsters educational program advancement by cultivating collaborative problem-solving and intelligent honing among teachers. In an AL system, teachers shape activity learning sets to investigate and address challenges related to integrating IKS into their instruction. These sets give a steady environment for sharing encounters, testing with modern approaches, and reflecting on results (Pedler et al., 2005; Sridhar et al., 2023).

Inserting IKS in Instructing and Appraisal

To successfully insert IKS into education, teachers must embrace socially responsive pedagogies that regard and consolidate innate ways of knowing. This includes moving past the tokenistic incorporation of inborn substance to make learning encounters profoundly established from innate points of view. AR can assist teachers who investigate and refine these pedagogies through persistent cycles of request and reflection. For occurrence, an educator might utilize AR to create and test instructing techniques that consolidate inborn narrating strategies to upgrade understudy engagement and understanding of social stories (Kemmis et al., 2014). The evaluation must also be decolonized to reflect innate values and learning styles. Conventional appraisals, which frequently prioritize composed and standardized testing, may not satisfactorily capture the information and aptitudes of inborn understudies. AR can encourage the improvement of alternative assessment strategies that are more adjusted with IKS. For case, performance-based evaluations, verbal introductions, and community-based ventures can give more all-encompassing and socially significant measures of understudy learning (Battiste, 2013). AL can advance and bolster the integration of IKS in instruction and evaluation by empowering teachers to create and refine these hones collaboratively. Through activity learning sets, teachers can share fruitful techniques, receive criticism, and ceaselessly progress in their approaches. This collaborative preparation guarantees that instruction and evaluation hones are socially responsive and academically compelling (Zuber-Skerritt, 2002).

Engagement and Organisation with Inborn Communities

Locking in and collaborating with innate communities is significant for the true integration of IKS into instruction. Such partnerships ensure that instructive substance and hones are grounded within the lived encounters and abilities of inborn individuals. This collaborative approach regards the office of inborn communities and makes a difference in constructing belief and common regard. One viable association technique is the establishment of admonitory chambers or committees composed of inborn senior citizens, pioneers, and information guardians. These chambers can guide educational module improvement, education hones, and appraisal strategies, guaranteeing they are socially suitable and conscious. For example, in The Frozen North, the Frozen North Local Information Arrange collaborates with local communities to coordinate innate information into the instruction framework, advancing socially responsive instruction (Barnhardt & Kawagley, 2005). Co-creation of instructive substance with inborn communities is another crucial viewpoint of this organisation. This handle includes working with community individuals to create educational module materials, lesson plans, and appraisal devices that reflect their information and values. For occasion, the College of Saskatchewan's College of Instruction has created programs incorporating courses on Innate information and pedagogies, planned in collaboration with innate communities (Battiste, 2013). Different instructive activities illustrate fruitful engagement and organisation. In Australia, the "Two-Way Science" program mixes Western logical information with Native ecological knowledge, creating educational modules that are both experimentally thorough and socially significant (Oliver, 2013). This program was created through broad meetings and collaboration with Native communities, guaranteeing that the substance precisely reflects their information and points of view. Decolonizing instruction through integrating Inborn Information Frameworks requires inventive approaches to implementing education and education. Activity Inquire about and Activity Learning gives profitable systems for this integration, advancing collaborative, intelligent, and iterative hones. Locking in and collaborating with inborn communities is essential for guaranteeing the genuineness and social pertinence of instructive substance and hones. By grasping these methodologies, instructive teachers can make more comprehensive, different, and compelling learning situations that honour and regard inborn information and points of view (Tariq, 2025f).

CASE STUDIES AND SUCCESS STORIES

The effective integration of Innate Information Frameworks (IKS) into instructive systems through Activity Inquiry (AR) and Activity Learning (AL) has been illustrated in different topographical and social settings. This case thinks about giving profitable bits of knowledge into successful methodologies and best-hones, highlighting the transformative potential of IKS in instruction. This discourse analyzes a few striking cases, emphasizing the lessons learned and best hones that can be connected broadly.

Unused Zealand's Te Whāriki Educational Modules

Unused Zealand's Te Whāriki early childhood educational modules could be a spearheading illustration of inserting IKS into instruction through a bicultural approach that consolidates Māori information and values. The improvement of Te Whāriki included broad collaboration with Māori communities, reflecting standards of association and support central to AR (Ritchie & Rau, 2006). The educational modules emphasize all-encompassing improvement, familial connections, and the interconnecting of all learning perspectives, central components of Māori epistemology. Using Te Whāriki illustrates the importance of honest-to-goodness engagement with inborn communities within the curriculum improvement preparation. By including Māori teachers, senior citizens, and families in the creation and continuous refinement of the educational modules, Te Whāriki guarantees that the instructive substance is socially significant and conscious. This collaboration has driven more prominent social pride among Māori understudies and made strides in instructive results (Service of Instruction, 2017). Goodness organisations with inborn communities are significant for making socially important educational modules. Inserting IKS into instructive systems can upgrade understudy engagement and results.

Gold Country Local Information Organise

The Gold Country Local Information Organise (ANKN) is an activity that integrates innate information into Alaska's instruction framework. This program uses AR to collaboratively create instructive content and educate procedures that integrate the frozen north's local ways of knowing. ANKN's approach incorporates the documentation and spread of inborn information through computerized files, educational program guides, and proficient advancement for teachers (Barnhardt & Kawagley, 2005). ANKN has effectively coordinated IKS into different subjects, including science and social thinking, by adjusting conventional environmental information with Western logical concepts. For illustration, the "Gold Country Local Science Educational programs" combine inborn knowledge of neighbourhood biological systems with standard science curricula, giving understudies a more comprehensive understanding of their environment. Making open assets that archive inborn information is crucial for its integration into standard instruction. Preparing teachers in socially responsive pedagogies upgrades their capacity to consolidate IKS viably.

Navajo Preliminary School, Unused Mexico

The Navajo Preliminary School (NPS) in Modern Mexico shows how indigenous-run schools can effectively integrate IKS into their instructional systems. NPS employs AL to ceaselessly refine its educational programs and educate hones, guaranteeing that they adapt to Navajo values and traditions. The school's programs incorporate courses on Navajo dialect, history, and logic, which are near standard scholarly subjects (Johnson, 2010). NPS's approach emphasizes experiential learning and community association. For occurrence, understudies are interested in ventures that apply Navajo standards to real-world issues, such as natural preservation endeavours established in conventional environmental information. This hands-on approach strengthens scholarly concepts and fortifies students' social character and community ties. Hands-on ventures that apply IKS to real-world issues and upgrade learning and social association. Involving the community in instructive exercises cultivates a sense of proprietorship and pertinence.

Two-Way Science in Western Australia

The "Two-Way Science" program in Western Australia coordinates Native environmental information with Western logical standards. This program was created through AR, including collaboration between Native communities, researchers, and teachers. The educational modules centre on nearby natural issues, mixing innate and logical information to create an all-encompassing learning encounter (Oliver, 2013). Two-way science utilizes place-based education, where learning happens within the local environment and integrates indigenous perspectives on arrival administration and biology. This approach not only gives understudies down-to-earth logical information but also cultivates regard for Native culture and conventions. Learning within the nearby environment and joining innate points of view makes the instruction more significant and locks in. Combining IKS with logical standards improves the educational modules and broadens students' understanding.

Best Hones from Assorted Settings

The taking after best hones emerges from this case ponders and can be connected to other instructive settings. Honest-to-goodness associations with innate communities are basic for the effective integration of IKS. This includes counselling innate pioneers and information attendants and including them in the co-creation and progressing advancement of instructive substance (Smith, 2012).

Socially Responsive Pedagogies

Teachers must be prepared to use socially responsive educating strategies that regard and join innate ways of knowing. This includes understanding the epistemological establishments of IKS and adjusting education procedures to reflect these standards (McKinley & Smith, 2019).

All-encompassing and Experiential Learning

Emphasizing all-encompassing and experiential learning approaches adjusts well with IKS. Exercises that include hands-on learning, narrating, and community ventures can make instruction more significant and socially pertinent for innate understudies (Battiste, 2013).

Intrigue Curriculum Development

Coordination of IKS over distinctive subjects, such as science, history, and social considerations, gives a more comprehensive and interconnected understanding of information. This intriguing approach can improve students' learning encounters and advance a more profound appreciation of inborn viewpoints (Barnhardt & Kawagley, 2005).

Persistent Reflection and Adjustment

Utilizing AR and AL methodologies permits nonstop reflection, adjustment, and enhancement of instructive hones. This iterative preparation guarantees that the integration of IKS remains responsive to the wants and criticism of innate communities (Kemmis et al., 2014).

Integrating Indigenous Information Frameworks into instruction through Activity Investigate and Activity Learning has been effective in different settings, from Modern Zealand to the Frozen North and Australia. These case studies highlight the significance of collaborative improvement, socially responsive pedagogies, and community involvement. By receiving these best hones, instructive teachers can make more comprehensive, pertinent, and compelling learning situations that honour and regard inborn information and points of view.

IMPLICATIONS AND FUTURE DIRECTIONS OF DECOLONIZED EDUCATION SYSTEMS

Decolonising instruction frameworks by joining Innate Information Frameworks (IKS) has significant societal and instructive suggestions. These changes can cultivate more noteworthy social inclusivity, advance social equity, and enhance the instructive encounters of all understudies. One noteworthy effect of decolonised instruction is the approval and revitalisation of inborn societies. Generally, colonial instruction frameworks have marginalised innate information, contributing to social disintegration and personality misfortune. By joining IKS as an educational program, instructive teachers can help in inverting these patterns and advancing social pride and progression among innate communities (Battiste, 2013). For occasion, incorporating inborn dialects and social hones within the classroom fortifies students' social characters, cultivating a sense of place and self-worth (McKinley & Smith, 2019). Instructively, decolonised frameworks can improve students engagement and scholastic victory. When understudies see their societies and information frameworks reflected in their instruction, they are more likely to feel persuaded and associated with their learning. This social pertinence can decrease dropout rates and improve scholastic results for innate understudies. In addition, non-indigenous understudies have an advantage from the introduction to different points of view, which broadens their understanding and cultivates intercultural competence (Barnhardt & Kawagley, 2005). Societally, decolonised instruction can contribute to more noteworthy social value and cohesion. By recognising and vilifying IKS, social orders can address chronicled treacheries and advance a more even-handed control flow. This move can assist in disassembling systemic bigotry and separation, cultivating more comprehensive communities. Moreover, coordination IKS can back economic improvement hones, as inborn information regularly incorporates profitable experiences into natural stewardship and asset administration (Smith, 2012).

Future Patterns and Challenges within the Integration of IKS

A few patterns and challenges will shape the integration of IKS inside ideal models of worldwide instruction. Mechanical headways, advancing educational hones, and expanding acknowledgement of the value of innate information all suggest promising directions for the long run.

Mechanical Integration

Progressions in innovation display critical openings for the dispersal and integration of IKS. Computerised stages can report and share innate information broadly, making it open to broader gatherings of people. For illustration, advanced narrating and online chronicles can protect verbal conventions and social hones, giving profitable instructive assets (McConaghy, 2000). Besides, virtual reality and increased reality innovations can make immersive learning encounters that bring innate situations and hones into the classroom. Innovative integration poses challenges, guaranteeing that computerised representations of IKS are exact and conscious. When inborn information is shared in its unique setting, social allotment or distortion can occur. It is pivotal to involve indigenous communities in creating and administrating advanced assets to preserve realness and regard (Battiste & Henderson, 2000).

Advancing Academic Hones

Future academic hones will likely advance towards more holistic and experiential learning models that align with IKS. Project-based learning, open-air instruction, and community-based ventures can all support the integration of IKS by emphasising hands-on and context-specific learning encounters. These approaches encourage understudies to apply their information in real-world settings, cultivating more profound understanding and maintenance (Oliver, 2013). However, scaling these academic hones over different instructional frameworks can be challenging. Many teachers are trained in conventional Western ideal models and may need more abilities or certainty to implement unused strategies successfully. Therefore, proficient improvement and instructor instruction programs must prioritise preparing in socially responsive pedagogies and IKS (McKinley & Smith, 2019).

Acknowledgment and Regulation Back

Increasing recognition of the esteem of IKS inside worldwide instruction ideal models may be a positive drift. Universal organisations and instructional teaching are starting to recognise the significance of indigenous information about worldwide challenges such as climate change and biodiversity disasters. Arrangements that support the consideration of IKS in national educational modules and educational standards can assist in institutionalising these hones (Smith, 2012). Institutionalising IKS inside formal instruction frameworks requires considerable changes in arrangement, educational program improvement, and asset assignment. Securing long-term funding and bolstering regulations to maintain these endeavours is basic. Additionally, it is imperative to explore the potential pressures between standardised instructive systems and the differing, localised nature of IKS (Barnhardt & Kawagley, 2005).

Collaborative Research and Activity

Action Research (AR) and Activity Learning (AL) will significantly coordinate IKS in instruction. These strategies cultivate collaborative, iterative forms that lock in inborn communities as dynamic members in instructive inquiry about and advancement. By ceaselessly refining hones based on input and perceptions, AR and AL guarantee that instructive activities stay responsive to innate people groups' wants and points of view (Kemmis et al., 2014). Be that as it may, compelling collaboration requires building belief and mutual respect between analysts, teachers, and inborn communities. This preparation can be time-consuming and complex but ensuring that instructive activities are socially fitting and maintainable is basic. Furthermore, analysts and teachers must be willing to challenge their suspicions and predispositions and grasp a truly decolonised approach to instruction (Smith, 2012). Integrating Innate Information Frameworks into instruction holds critical potential for making more comprehensive, impartial, and socially pertinent learning situations. By approving indigenous cultures, upgrading understudy engagement, and advancing social cohesion, decolonised instruction can contribute to broader societal change. Looking forward, mechanical progressions, advancing educational hones, and expanding regulation back offer promising bearings for the longer term. These endeavours must be supported by veritable collaboration with inborn communities, progressing proficient improvement, and a commitment to challenging and changing existing instructive ideal models. By grasping these standards, instructive frameworks can move towards a more comprehensive and impartial future.

CONCLUSION

Decolonizing instruction frameworks through the integration of Innate Information Frameworks (IKS) and utilizing strategies such as Activity investigation (AR) and Activity Learning (AL) presents a transformative opportunity for teachers, policymakers, and analysts. This conclusion summarizes key experiences and gives noteworthy proposals to advance broader appropriation and successfully execute these approaches.

Social Approval and Revitalization

Decolonizing instruction approves and revitalizes innate societies, neutralizing the verifiable marginalization of these information frameworks. Joining IKS into instructive systems cultivates social pride and progression, which is pivotal for the character and self-worth of innate understudies (Battiste, 2013). This social integration benefits innate understudies and enhances the instructive encounter for all learners by uncovering them to differing points of view.

Improved Instructive Results

Integrating IKS into an educational module can improve understudy engagement and scholastic execution. When understudies see their social information reflected in their instruction, it improves their inspiration and association with learning (McKinley & Smith, 2019). For illustration, the Te Whāriki educational modules in Modern Zealand, which coordinate Māori information, have been related to made strides in instructive results and social pride among Māori understudies (Ritchie & Rau, 2006).

Social Value and Cohesion

Decolonized instruction contributes to more prominent social value by tending to authentic shameful acts and advancing more impartial control flow. By recognizing and valuing IKS, instructional frameworks can help disassemble systemic prejudice and cultivate comprehensive communities. The Gold Country Local Information Arrange represents this by joining inborn biological information with standard instruction, advancing social cohesion and natural stewardship (Barnhardt & Kawagley, 2005).

All-encompassing and Experiential Learning

IKS adjusts well with all-encompassing and experiential learning models, emphasizing hands-on and context-specific learning encounters. These approaches empower understudies to apply their information in real-world settings, cultivating more profound understanding and maintenance (Oliver, 2013). Programs like the Navajo Preliminary School in Modern Mexico highlight the viability of experiential learning in improving instructive results and social association (Johnson, 2010).

Collaborative and Iterative Forms

Utilizing AR and AL techniques advances collaborative, iterative forms that lock in inborn communities as dynamic members in instructive investigation and improvement. These techniques guarantee that educational initiatives stay responsive to innate people groups' wants and points of view (Kemmis et al., 2014).

Create Collaborative Associations

Building honest-to-good organisations with innate communities is fundamental to the fruitful integration of IKS. Teachers and policymakers ought to lock in inborn pioneers, older folks, and information guardians within the co-creation of educational modules and instructive arrangements. This collaborative approach guarantees that instructive substance is socially pertinent and aware (Smith, 2012).

Actualize Socially Responsive Pedagogies

Instructor instruction and proficient advancement programs must prioritize preparing in socially responsive pedagogies. Teachers' ought to be prepared with the aptitudes and information to integrate IKS into their teaching hones viably. This incorporates understanding inborn epistemologies and adjusting teaching techniques to reflect these standards (McKinley & Smith, 2019).

Advanced All-encompassing and Experiential Learning

Schools should receive all-encompassing and experiential learning models that align with IKS. These will include project-based learning, open-air instruction, and community-based ventures that provide hands-on learning encounters. These approaches not only upgrade understudy engagement but also advance the viable application of information (Battiste, 2013).

Use Mechanical Progressions

Utilize computerized stages and technologies to report and spread IKS. Online files, computerized narrating, and virtual reality can make innate information available to broader gatherings of people while protecting social hones. In any case, it is vital to include inborn communities in overseeing these assets to guarantee exactness and regard (McConaghy, 2000).

Institutionalise Bolster for IKS

Policymakers ought to create and execute approaches that support the integration of IKS into national educational modules and instructional benchmarks. This incorporates securing long-term financing and organisation to maintain these endeavours. Guaranteeing that instructional systems suit the differing, localized nature of IKS is additionally basic (Barnhardt & Kawagley, 2005).

Energise Persistent Reflection and Adjustment

AR and AL strategies should be broadly embraced to encourage persistent reflection, adjustment, and enhancement of instructive hones. These iterative forms permit teachers to refine their approaches based on criticism and perceptions, guaranteeing that activities remain responsive to the advancing needs of inborn communities (Kemmis et al., 2014).

REFERENCES

Barnhardt, R., & Kawagley, A. O. (2005). Indigenous Knowledge Systems and Alaska Native Ways of Knowing. *Anthropology & Education Quarterly*, *36*(1), 8–23. DOI: 10.1525/aeq.2005.36.1.008

Battiste, M. (2013). *Decolonizing Education: Nourishing the Learning Spirit*. Purich Publishing.

Battiste, M., & Henderson, J. Y. (2000). *Protecting Indigenous Knowledge and Heritage: A Global Challenge*. Purich Publishing. DOI: 10.59962/9781895830439

Berger, P. L., & Luckmann, T. (1966). *The Social Construction of Reality: A Treatise in the Sociology of Knowledge*. Anchor Books.

Bourgeois, R., Karuri-Sebina, G., & Feukeu, K. E. (2024). The future as a public good: decolonising the future through anticipatory participatory action research. *foresight, 26*(4), 533-549.

Brayboy, B. M. J. (2005). Toward a Tribal Critical Race Theory in Education. *The Urban Review*, *37*(5), 425–446. DOI: 10.1007/s11256-005-0018-y

Cajete, G. A. (1994). *Look to the Mountain: An Ecology of Indigenous Education*. Kivaki Press.

Cajete, G. A. (1999). *Ignite the Sparkle: An Indigenous Science Education Model*. Kivaki Press.

Dei, G. J. S. (1996). *Anti-Racism Education: Theory and Practice*. Fernwood Publishing.

Deloria, V., & Wildcat, D. (2001). *Power and Place: Indian Education in America*. Fulcrum Publishing.

Denzin, N. K., Lincoln, Y. S., & Smith, L. T. (2008). *Handbook of Critical and Indigenous Methodologies*. Sage Publications. DOI: 10.4135/9781483385686

Emeagwali, G., & Dei, G. J. S. (2014). *African Indigenous Knowledge and the Disciplines*. Sense Publishers. DOI: 10.1007/978-94-6209-770-4

Fahlberg, A. (2023, March). Decolonizing Sociology Through Collaboration, Co-Learning and Action: A Case for Participatory Action Research 1. *Sociological Forum*, *38*(1), 95–120. DOI: 10.1111/socf.12867

Freire, P. (1970). *Pedagogy of the Oppressed*. Herder and Herder.

Gay, G. (2010). *Culturally Responsive Teaching: Theory, Research, and Practice*. Teachers College Press.

Grande, S. (2004). *Red Pedagogy: Native American Social and Political Thought*. Rowman & Littlefield Publishers.

Johnson, L. R. (2010). The Cultural Implications of Navajo Language Shift. *International Journal of the Sociology of Language*, *2010*(201), 149–158.

Kawagley, A. O. (1995). *A Yupiaq Worldview: A Pathway to Ecology and Spirit*. Waveland Press.

Kayira, J., Lobdell, S., Gagnon, N., Healy, J., Hertz, S., McHone, E., & Schuttenberg, E. (2022). Responsibilities to decolonize environmental education: A co-learning journey for graduate students and instructors. *Societies (Basel, Switzerland)*, *12*(4), 96. DOI: 10.3390/soc12040096

Kemmis, S., McTaggart, R., & Nixon, R. (2014). *The Action Research Planner: Doing Critical Participatory Action Research*. Springer. DOI: 10.1007/978-981-4560-67-2

King, T. (2003). *The Truth About Stories: A Native Narrative*. House of Anansi Press.

Kovach, M. (2009). *Indigenous Methodologies: Characteristics, Conversations, and Contexts*. University of Toronto Press.

Loughran, J. J. (2002). Effective Reflective Practice: In Search of Meaning in Learning about Teaching. *Journal of Teacher Education*, *53*(1), 33–43. DOI: 10.1177/0022487102053001004

Mahadew, A., & Hlalele, D. J. (2022). Understanding inclusion in early childhood care and education: A participatory action learning and action research study. *South African Journal of Childhood Education*, *12*(1), 1073. DOI: 10.4102/sajce.v12i1.1073

McConaghy, C. (2000). *Rethinking Indigenous Education: Culturalism, Colonialism and the Politics of Knowing*. Post Pressed.

McKinley, E., & Smith, L. T. (2019). *Handbook of Indigenous Education*. Springer. DOI: 10.1007/978-981-10-3899-0

Meyer, M. A. (2008). *Indigenous and Authentic: Hawaiian Epistemology and the Triangulation of Meaning*. AltaMira Press.

Mihesuah, D. A. (2004). *Indigenous American Women: Decolonization, Empowerment, Activism*. University of Nebraska Press.

Nakata, M. (2007). *Disciplining the Savages: Savaging the Disciplines*. Aboriginal Studies Press.

Oliver, R. (2013). The Role of Indigenous Knowledges in Science Education: Learning from Two-Way Science. *Australian Journal of Indigenous Education*, *42*(2), 179–189.

Paris, D., & Alim, H. S. (2017). *Culturally Sustaining Pedagogies: Teaching and Learning for Justice in a Changing World*. Teachers College Press.

Pihama, L., Smith, L. T., Taki, M., & Lee, J. (2004). A Literature Review on Kaupapa Maori and Maori Education Pedagogy. *International Journal of Learning*, *11*, 181–188.

Ritchie, J., & Rau, C. (2006). Enacting a Te Whāriki Curriculum: The Complexities of Bicultural Involvement. *Early Childhood Folio*, *10*, 20–24.

Semali, L. M., & Kincheloe, J. L. (1999). *What is Indigenous Knowledge? Voices from the Academy*. Falmer Press.

Silva, C. M. D., Sauerbronn, F. F., & Thiollent, M. (2022). Decolonial studies, non-extractive methods, and participatory action research in Accounting. *RAC. Revista de Administração Contemporânea*, *26*(04), e210014. DOI: 10.1590/1982-7849rac2022210014.en

Smith, L. T. (1999). *Decolonizing Methodologies: Research and Indigenous Peoples*. Zed Books.

Smith, L. T. (2012). *Decolonizing Methodologies: Research and Indigenous Peoples* (2nd ed.). Zed Books.

Snively, G., & Corsiglia, J. (2001). Discovering Indigenous Science: Implications for Science Education. *Science Education*, *85*(1), 6–34. DOI: 10.1002/1098-237X(200101)85:1<6::AID-SCE3>3.0.CO;2-R

Sobel, D. (2004). *Place-Based Education: Connecting Classrooms & Communities*. The Orion Society.

Sridhar, S., Alizadeh, F., Ratner, L., Russ, C. M., Sun, S. W., Sundberg, M. A., & Rosman, S. L. (2023). Learning to walk the walk: Incorporating praxis for decolonization in global health education. *Global Public Health: An International Journal for Research, Policy and Practice*, *18*(1), 2193834. DOI: 10.1080/17441692.2023.2193834 PMID: 36989128

Sterenberg, G. (2013). Considering Indigenous Knowledges and Mathematics Curriculum. *Canadian Journal of Science, Mathematics and Technology Education = Revue Canadienne de l'Enseignement des Sciences, des Mathématiques et de la Technologie*, *13*(1), 18–32. DOI: 10.1080/14926156.2013.758325

Stewart, S. (2002). *Native American Education: A Reference Handbook*. ABC-CLIO.

Swisher, K., & Tippeconnic, J. (1999). *Next Steps: Research and Practice to Advance Indian Education*. ERIC Clearinghouse on Rural Education and Small Schools.

Tariq, M. U. (2025a). Transforming Pedagogy: Best Practices for SoTL in Online Learning Environments. In Rahimi, R., & Soares, L. (Eds.), *Enhancing the Scholarship of Teaching and Learning in Online Learning Environments* (pp. 21–46). IGI Global Scientific Publishing., DOI: 10.4018/979-8-3693-7595-2.ch002

Tariq, M. U. (2025b). Gamification and Virtual Reality: Transforming Early Childhood Language Learning. In Manditereza, B. (Ed.), *Digital Pedagogy in Early Childhood Language Development* (pp. 25–52). IGI Global Scientific Publishing., DOI: 10.4018/979-8-3693-7437-5.ch002

Tariq, M. U. (2025c). Integrating Theory and Practice: Transformative Approaches in Practice-Led Research for Professional Development. In Kwok, S., Yu, P., & Le Corre, J. (Eds.), *Exploring Practice-Led Research for Professional Development* (pp. 419–436). IGI Global Scientific Publishing., DOI: 10.4018/979-8-3693-6376-8.ch016

Tariq, M. U. (2025d). Empowering Learning Through Networked and Connected Education: Case Studies in Digital Engagement. In Gierhart, A. (Ed.), *Cases on Enhancing P-16 Student Engagement With Digital Technologies* (pp. 169–198). IGI Global Scientific Publishing., DOI: 10.4018/979-8-3693-5633-3.ch007

Tariq, M. U. (2025e). Beyond the Classroom: The Sustained Influence of Educators on Student Success. In West, E. (Ed.), *Former Educators' Roles in Supporting Student Growth and Learning* (pp. 321–346). IGI Global Scientific Publishing., DOI: 10.4018/979-8-3693-6427-7.ch013

Tariq, M. U. (2025f). The Belmont Report: Guiding Ethical Principles in Human Research. In Throne, R. (Ed.), *IRB, Human Research Protections, and Data Ethics for Researchers* (pp. 245–268). IGI Global Scientific Publishing., DOI: 10.4018/979-8-3693-3848-3.ch010

Thaman, K. H. (2003). Decolonizing Pacific Studies: Indigenous Perspectives, Knowledge, and Wisdom in Higher Education. *The Contemporary Pacific*, *15*(1), 1–17. DOI: 10.1353/cp.2003.0032

Wilson, S. (2003). Progressing Towards an Indigenous Research Paradigm in Canada and Australia. *Canadian Journal of Native Education*, *27*(2), 161–178.

Wilson, S. (2008). *Research is Ceremony: Indigenous Research Methods*. Fernwood Publishing.

KEY TERMS AND DEFINITIONS

Decolonizing Education: The process of challenging and dismantling colonial structures, narratives, and practices in education to create inclusive, culturally relevant, and equitable learning environments.

Indigenous Knowledge Systems (IKS): The complex set of knowledge, practices, and worldviews developed by indigenous communities

Action Research (AR): A participatory research methodology that involves iterative cycles of planning, action, observation, and reflection

Action Learning (AL): A collaborative learning process where individuals or groups work on real-world problems

Colonial Education Legacy: The enduring influence of colonial-era education systems that marginalised indigenous knowledge

Epistemic Justice: The recognition and equitable treatment of diverse knowledge systems, ensuring that indigenous perspectives, experiences, and ways of knowing are valued and integrated into educational and institutional frameworks.

Participatory Pedagogy: An educational approach that actively involves learners and communities in shaping curricula, teaching methodologies, and learning experiences

Chapter 7
Enabling Indigenous Knowledge Systems in Action Research and Action Learning:
How IKS Co-Exist in the Field of Technology Education

Princess Blose
https://orcid.org/0000-0001-9825-8090
University of South Africa, South Africa

Elliot C. Ndlovu
https://orcid.org/0000-0002-6774-0463
University of South Africa, South Africa

ABSTRACT

This book chapter explores the co-existence of Indigenous Knowledge Systems (IKS) with Technology Education, focusing on their potential to enhance learners' understanding of technology and appreciation for diverse cultural perspectives. Grounded in Constructivism theory, the study involved interviews with three experienced Senior Phase Technology teachers selected for their expertise. Thematic analysis revealed valuable insights into the benefits of incorporating IKS for teachers and learners. We acknowledge challenges such as the dominance of Western viewpoints and the need for better curriculum support. To overcome these challenges, we propose practical strategies like hands-on learning experiences and collaboration with community experts. Ultimately, the chapter emphasizes the importance of making design activities culturally relevant and promoting creative and critical thinking to effectively integrate IKS into Technology Education.

INTRODUCTION

The fusion of IKS with Technology Education redefines the boundaries of learning and innovation. It offers a transformative approach that enriches learners' understanding of technology and fosters a profound appreciation for diverse cultural perspectives. The concepts of 'incorporation' and 'integra-

DOI: 10.4018/979-8-3693-3542-0.ch007

tion' are intertwined in this chapter. The South African Technology Education policy document, for instance, does draw attention to the co-existence of IKS. For example, it stipulates that as learning takes place in a classroom, learners must be made aware of the interrelationship between technology, society, and the environment, which draws attention to the co-existence of indigenous knowledge in the curriculum [Department of Basic Education (DBE), 2011]. Nonetheless, as much as Technology Education emphasizes the co-existence of IKS in its teaching and learning, the Curriculum Assessment Policy Document Statement (CAPS) document does not give a clear integration in its teaching and learning. A study by Yeseraw (2023) showed a low level of indigenous knowledge incorporation in the contents, including exercises, project work, and graphic representations. Correspondingly, several studies have made attempts to incorporate indigenous knowledge into Western science, specifically in environmental assessment and resource management and, lately, in Technology Education (Abah et al., 2015; Blose, 2023; Ndlovu, 2023; Blose & Gumbo, 2024; Ndlovu & Gumbo, 2024). Although attempts have been made, the study's critics contend that Western science continues to be valued over Indigenous knowledge Abah et al., (2015) to the disadvantage of Indigenous peoples (Brook & McLachlan, (2005); Gilchrist et al., (2005); Nadasdy, (2003). Therefore, this chapter aims to explore the co-existence of IKS in Technology Education. Thus, the objective was to explore the benefits of integrating indigenous knowledge into Technology Education, practical strategies, challenges, and real-world applications that can inspire teachers and learners. Hence, the research question for this chapter is: How does integrating indigenous knowledge into Technology Education enhance practical strategies for co-existence for teaching and learning in the Senior Phase classroom?

BACKGROUND

In this chapter, we found it relevant to give an overview of the IKS. IKS has been found to have firm foundations within Indigenous people (Gumbo, 2015; Mpofu et al.,2014). Thus, IKS encapsulate centuries of wisdom, passed down through generations within communities worldwide. These systems offer unique insights into sustainable practices, harmonious relationships with the environment, and innovative technological solutions. According to Brondízio, et al., (2021), Indigenous peoples' and local communities' knowledge, values, and practices offer ways to understand and better address social-environmental problems. Therefore, by embracing indigenous perspectives, we can expand our understanding of technology beyond its conventional boundaries and appreciate the interconnectedness of all living beings.

Drawing from the perspectives above, IKS mostly embrace the social, economic, political, and cultural systems of how Indigenous people interconnect with their surroundings for survival which then informs the generation and maintenance of their structural knowledge. For example, Kosoe et al., (2020) contended that many African societies have long been doing things that reflect their culture, location, experience, and environment. This can further be incorporated into the Technology Education curriculum as the co-existence of the culture, location, experience, and environment with the Indigenous communities has more influence on decision-making that recognizes the Indigenous aspects. De Sousa Santos et. al., (2007), for instance, claimed that knowledge production is a social practice, and as such, all knowledge is situated (historically, politically, socially) and partial. These scholars further argue against the monoculture of knowledge, an approach that is based on positivist notions of science, legacies of colonial and postcolonial relations, and global capitalism. Furthermore, they contend that Indigenous peoples everywhere know a great deal about the environments in which they have lived for generations; and that this kind of knowledge must be valued and considered in the planning and implementation of educational content and the development of policies. Currently, in the South African context, the curriculum policies have drawn the co-existence of IKS in the planning and teaching of different subjects i.e., valuing IKS (DBE, 2011). The National Curriculum Statement (NCS) Grades R-12 aimed to produce learners who can communicate effectively using visual, symbolic, and/or language skills in various modes. Hence, NCS helps channel the curriculum to the flexibility towards the realisation of the co-existence of Indigenous knowledge in educational settings.

The literature on the concept of IKS has been gained and has been defined using a variety of definitions. Some scholars have related it to local knowledge or traditional knowledge (Muchenge et al., 2016). In this chapter, to show the relativeness or the co-existence of the IKS in the field of Technology Education, we adopted the definition of the scholars Abah et al. (2015). Thus, IKS encompasses the technology, social, economic, and philosophical learning, and governance systems of a community. This resonates well with the scope of the curriculum for Technology Education. For instance, the CAPS document for Technology Education states that while investigating, learners should be provided with opportunities to explore values and attitudes to develop informed opinions that can help them to make compromises and value judgments. Drawing further from Abah et al. (2015), IKS is better understood as practical, personal, and contextual units that cannot be detached from an individual, their community, or the environment (both physical and spiritual). Therefore, within Indigenous communities or societies lies rich different knowledge that can be very influential in the curriculum of Technology Education this is because the curriculum emphasizes the importance of knowledge as a critical aspect for the design and creation of a solution. Similarly, Abah et al., (2015) argued that in various tribes of the world, there exists indigenous knowledge that can be meaningfully integrated into Western school curricula for improved learning outcomes. Furthermore, Abah et al. (2015) claimed a notification that the artifacts that are available in traditional environments are important tools that can be used to bridge the gap between what is usually taught in the classroom and what exists outside the classroom, that is, in society. In the same view, Technology Education offers learners the opportunity to learn to use authentic contexts rooted in real situations outside the classroom with knowledge and application of knowledge as one of the key issues to teach (Ndlovu & Gumbo, 2018). From this point of view, the Technology Education curriculum has its roots in the traditional knowledge of the communities. Subsequently, Abah et al. (2015) argued that education cannot exclude cultural knowledge since the content of education has value underpinning it and is associated with a particular culture.

Therefore, IKS provides a holistic understanding of technology, culture, and the environment that transcends conventional Western paradigms. These systems are deeply embedded in indigenous communities' social, economic, political, and cultural contexts, offering valuable insights to enhance Technology Education (Yishak & Gumbo, 2017). By recognising the existence of IKS in the Technology Education curriculum, teachers can bridge the gap between traditional knowledge and modern educational practices, fostering a more inclusive and culturally responsive learning environment. This existence would not only acknowledge the significance of Indigenous wisdom but also enrich learners' learning experiences, equipping them with diverse perspectives and the ability to think critically about the interconnectedness of knowledge, culture, and the environment.

THE MAIN FOCUS OF THE CHAPTER

This chapter is grounded in the Constructivism theory of learning. According to Rannikmäe et al. (2020), Constructivism has been strongly advocated for over 50 years, when rapid changes in society and rapid educational reforms occurred. Constructivism is a theory associated with a way of teaching where learners play an active role in their learning. Rannikmäe et al. (2020) further indicate that learners' interpretations of their experiences depend on the social environment surrounding them as members of society. The choice of constructivism theory in this chapter was found more relevant in deeply understanding the co-existence of indigenous knowledge and Technology education in the Senior Phase.

LITERATURE REVIEW

The literature review focuses on the benefits of IKS integration in Technology Education, challenges experienced by Technology Education teachers, practical strategies, and real-world applications for IKS integration in Technology Education. This literature review gives an overview of the opportunities provided for the co-existence of IKS in the Technology Education Curriculum.

Benefits of Integration of IKS

IKS are relatively linked to the cultural living of indigenous communities and the environment (Triyanto & Handayan, 2020). The integration of IKS into Technology Education brings a multitude of benefits. It promotes cultural diversity, fosters creativity and innovation, and instils a sense of responsibility towards the environment. Incorporating indigenous knowledge, learners are encouraged to think critically, collaborate with diverse communities, and develop holistic and sustainable solutions (Moalosi et al., 2017). This enriching experience enhances academic learning and nurtures empathy and respect for different ways of knowing. It would further teach them to manage time and material resources effectively, provide opportunities for collaborative learning, and nurture teamwork. This can be supported through practical projects using various technological skills (investigating, designing, making, evaluating, and communicating) that suit different learning styles (DBE, 2014). While some learners can appreciate the importance of indigenous knowledge, others find it too complicated for them to understand because they lack the appropriate background and cultural sensitivity (Kruger & Vass, 2025). Under these circumstances, very little actual indigenous information is accessible to them. In order to accommodate that, in this chapter, we saw it fit to detail the benefits of integrating IKS into the Technology Education curriculum. Authors believe that detailing the benefits will expand the opportunities and co-existence of IKS presented in the Technology Education curriculum.

Enhancing Cultural Relevance and Inclusivity

This study believes that integrating IKS into Technology Education would make the curriculum more culturally relevant for learners from Indigenous backgrounds. Preserving cultural industries through science and technology can have huge benefits to society (Chahine, 2020). It would allow them to see their own knowledge systems and cultural heritage reflected in their education, which can increase engagement and motivation. This connection helps learners develop a stronger sense of identity and pride in their cultural roots (Mpuangnan & Ntombela, 2024). Moreover, incorporating IKS acknowledges and validates the diverse ways of knowing and doing, beyond Western scientific methods. It promotes inclusivity by ensuring that learners from all backgrounds see their cultural knowledge as valuable and relevant in the educational context (Thelma & Madoda, 2024). This can help reduce feelings of alienation among Indigenous learners and foster a more inclusive learning environment for all.

Bridging Traditional and Modern Technologies

Integrating IKS into Technology Education provides opportunities for learners to explore how traditional knowledge and modern technology can complement each other. Similarly, Ankiewicz (2016) attested that Technology is embedded differently in different historical-cultural contexts. Therefore, coexistence recognition of IKS in Technology Education can lead to innovative solutions that blend the best of both worlds. For example, traditional water management techniques can be combined with modern engineering to create more efficient and sustainable systems. Learning about traditional and modern technologies gives learners a broader perspective on how technologies can be adapted to meet local needs and conditions. This contextual understanding is crucial in developing technologies that are not only effective but also culturally appropriate and sustainable. In the current South African Technology school curriculum, such can be realised when dealing with topics like processing and structures. For instance, the Grade 7 curriculum on structures in term 2 emphasises discussions on how designers consider the needs of society while considering the impact on society and the environment. There lies the opportunity for learners to reflect on culturally and sustainable solutions that will reduce the impact caused by technology.

Fostering Critical Thinking and Problem-Solving Skills

IKS often involves holistic, systems-based thinking considering the interconnections between people, the environment, and technology. Integrating this approach into Technology Education encourages learners to think critically about the broader impacts of technology on society and the environment. This can enhance their problem-solving skills by prompting them to consider a wider range of factors when developing solutions. IKS are typically practical and applied, focusing on solving real-world problems faced by communities. Pophiwa and Saidi (2022), connotes that Africans have always been able to identify their own problems or issues that affected them during the pre-colonial period. In this case, for example, Pophiwa and Saidi (2022) referred to the refusal of King Shaka to be assisted by the Western army general to school him on his innovative military techniques; instead, he innovatively formed a sustainable weapon. This clearly shows the power of the IKS, specifically fostering critical thinking and problem-solving skills. This further shows that by incorporating IKS into the curriculum, learners can be exposed to hands-on, experiential learning opportunities that build their capacity to address real-life challenges in activities involving the design process or problem-solving in a Technology subject. In Grade 7, for instance, learners are required to design a cell phone tower. One of the design skills encourages learners to disguise the tower so that it blends in with the environment, avoiding visual pollution. This practical focus helps learners develop skills directly applicable to their lives i.e., thinking critically on possible Indigenous ways that blend well with the environment to avoid visual pollution by using more natural resources.

Supporting Lifelong Learning and Adaptability

IKS has evolved to adapt to changing environments and circumstances. This adaptability is a key lesson for learners in Technology Education, as it teaches them the importance of flexibility and innovation in the face of new challenges. Learning from IKS encourages learners to be lifelong learners who can adapt their knowledge and skills to a rapidly changing world. Ankiewicz (2016) Indigenous knowledge is social, collaborative, and cultural, and may therefore serve as a vehicle to advance some of the 21st-century skills. Integrating IKS into Technology Education benefits learners and empowers Indigenous communities by recognizing and respecting their knowledge. This can lead to greater collaboration between educational institutions and Indigenous communities, creating opportunities for community members to contribute to the curriculum and share their expertise.

Preserving and Revitalizing Indigenous Knowledge

Incorporating IKS into the curriculum helps preserve traditional knowledge that might otherwise be lost. This is especially important as many Indigenous communities face the erosion of their cultural practices and languages. For instance, Cindi (2021) indicated that the impacts of colonisation and globalization have meant that over time, this knowledge has been lost and not imparted to younger generations. In this chapter, we affirm that teaching IKS in the Technology Education curriculum would play a role in safeguarding and revitalising the coexistence and value of this knowledge for future generations. IKS represents the accumulated wisdom of generations. Integrating this knowledge into the education system ensures that these lessons are passed on to younger generations, allowing them to benefit from the insights and experiences of their ancestors.

Aligning Education with Global Sustainability Goals

Integrating IKS aligns with global sustainability goals, particularly those related to education, climate action, and sustainable communities. Therefore, teaching learners about IKS, schools contribute to broader efforts to achieve sustainable development, as outlined in the United Nations Sustainable Development Goals (SDGs). In their study, Blose and Gumbo (2024), claimed that IKS present opportunities to teach Technology Education towards education for sustainable development. Therefore, integrating IKS is crucial in the curriculum's teaching as it also supports equity and justice by challenging the dominance of Western-centric education and promoting a more balanced approach that respects all forms of knowledge. This contributes to a more equitable education system that serves the needs of all learners. Despite the benefits of integrating IKS teachers experience challenges with its teaching in the Technology curriculum. Therefore, we found it very important for us to discuss those challenges as this would allow for the establishment of practical strategies to overcome those challenges. Hence, the authors have outlined those challenges in the below section.

Challenges Experienced by Technology Education Teachers

While integrating IKS brings immense value to Technology Education, it also poses challenges that must be addressed with sensitivity and respect. For instance, Blose (2023), showed that there is little attention given to the integration of indigenous knowledge in the curriculum. This, therefore, makes its value less important in the curriculum. The Western paradigms continue to overpower the importance of indigenous knowledge in the teaching and learning materials, especially in most African curricula. Similarly, even though IKS are the primary sources of ecosystems and the best examples of sustainability in natural resource management, the legacies of colonialism in Africa endanger the practical aspects of life (IK and ways of learning) by replacing them with the values of Western education systems (Brondízio, et al., 2021). Therefore, teachers must navigate issues of cultural appropriation, ensure proper attribution of knowledge, and seek permission from Indigenous communities before incorporating their wisdom into the curriculum as some of the knowledge systems are classified under Chiefs or Kings. Burgess et al., (2022), cultural mentoring presents a pathway for schools and teachers to genuinely engage with local Indigenous knowledge and demonstrates the benefits to all. We are, therefore, of the view that collaboration with Indigenous communities will enable the success of the integration of the concept of IKS. This will also strengthen the confidence of teachers with their integration as some, despite showing interest in the integration, still do not have confidence in integrating it into the existing curriculum. Most of the time the integration of the concept comes as additional work from teachers. Hence, we believe that this chapter would show the co-existence of the concept within the curriculum. Therefore, to lessen the challenges, authors believe that collaboration with Indigenous partners should be based on mutual trust and ongoing dialogue, with a commitment to upholding ethical standards and honouring traditional protocols.

Nonetheless, there are challenges that the curriculum of Technology Education has which hinder the success of the integration of IKS. For instance, scholars such as Govender and Mutendera (2020) showed that Indigenous knowledge is largely neglected in the primary school curriculum, yet it espouses the history, art, nature, and traditions of the community from which learners come. The biggest threat to integrating the IKS remains in the lesser value given to it. The educational systems and policies also make it difficult for teachers to clearly understand how exactly this concept should add value to their teaching. Hence, teachers' adaption to its existence remains a challenge. As much the policy, for instance, for the Technology Education curriculum stipulates its recognition of co-existence, it is still not clear on the main integration in the curriculum in particular various topics. The currency of the topics from the Technology Education curriculum is situated in Western parts of the context, leaving room for how and when this knowledge should be featured. For example, mechanical and electrical systems and control topics are more inclined to engineering fields. The content could be integrated with the Indigenous knowledge merely through day-to-day experience when they carry loads, in their farms etc. Blose and Gumbo (2024), indicated that teachers confessed that Westernization shapes their understanding of technology, and their approaches are attributed to this situation.

In the South African context, the policy for instance states that:

> Technology, society, and the environment must be addressed throughout the syllabus wherever applicable. The activities prescribed provide ample opportunities for dealing with indigenous technologies, the impact of technology, and bias in technology.

This is contrary to the reality of the application of the concept, as teachers still wonder when and how this must be done. In this instance, the policy indicates that the technology, society, and environment aspects must be addressed throughout the syllabus wherever applicable which still leaves room for how (Ndlovu & Gumbo, 2018). Authors, therefore, argue that as much as other content areas have been well outlined with how and when its teaching should take place, authors say this should be the same with the integration of IKS as with the other areas there is no exception wherever applicable. This does leave teachers with wonders of where is applicable, but they know that the other contents and topics must be fully taught with no exception of applicability. Similarly, Swift (1992) realized that there is a potential role for Indigenous knowledge in Science and Technology Education that has received less emphasis than it might. Swift further adds that status has the potential to provide both a problem and an opportunity regarding Indigenous knowledge in Science and Technology Education. What is presented as a problem is the normally lower status of indigenous knowledge compared with imported knowledge which might lead to resistance to education courses incorporating indigenous knowledge. Nonetheless, the curriculum for South Africa has made some provision for the indigenous systems to be integrated; however, some gaps make it difficult for it to be fully taken into consideration. Where the document for the Technology Education curriculum speaks of IKS it emphasizes the terms when applicable. This is another instance mentioned in the document: learners should be made aware of different coexisting knowledge systems wherever applicable. This has led teachers to void on whether Indigenous knowledge should be integrated into the Technology Education curriculum for instance one teacher said:

> I think it is there and done, but there are still gaps. There are still gaps because when I look at the syllabus that we have or the curriculum that we are not focusing on older generation is focusing on the generation of today.

This shows that as much as the curriculum speaks to indigenous knowledge, mostly to the syllabus, it focuses on Western paradigms. Supportably, a study by Blose and Gumbo (2024), indicated that although teachers appreciate the role of Indigenous Knowledge, they still have difficulty teaching in a way that incorporates it.

Integrating IKS into Technology Education presents both an opportunity and a challenge. While there is recognition of the value that Indigenous perspectives bring, particularly in promoting sustainable practices and preserving cultural heritage, the actual implementation within the curriculum remains fraught with obstacles. These include the dominance of Western paradigms, insufficient curriculum guidance, and the perceived lower status of Indigenous knowledge (Gumbo, 2015). Teachers often find themselves navigating ambiguous policies, such as the recurring phrase "wherever applicable," which leaves room for (mis)interpretation and uncertainty about how to effectively incorporate Indigenous knowledge into their lessons. This vagueness complicates the teaching process and undermines the potential impact of IKS on learners' learning experiences.

Moreover, the legacy of colonialism and ongoing structural challenges within educational systems exacerbate the marginalisation of Indigenous knowledge. Although some provisions exist within the curriculum, the lack of clear, actionable guidelines hampers teachers' confidence and ability to integrate these systems meaningfully. Collaboration with Indigenous communities, rooted in mutual respect and ongoing dialogue, is a crucial pathway for overcoming these barriers. In fostering partnerships and engaging in cultural mentoring, schools and teachers can better navigate the complexities of inclusion, ensuring that IKS are not merely an afterthought but a vital, respected component of Technology Education. Therefore, addressing these challenges with a concerted effort to redefine the curriculum's approach to Indigenous knowledge will be key to achieving a more inclusive, culturally responsive educational landscape. Hence, the following section delved into the practical strategies for integrating IKS in the field of Technology Education.

Practical Strategies for Integration

Integrating IKS into Technology Education requires a thoughtful and collaborative approach. Teachers can begin by engaging with local indigenous communities, listening to their stories, and understanding their traditional practices. Therefore, by incorporating indigenous perspectives into lesson plans and activities, teachers can create meaningful experiences that resonate with learners of all backgrounds. Hands-on projects that tap into indigenous knowledge guides can immerse learners in a rich tapestry of knowledge and wisdom. For instance, the Technology Education curriculum emphasizes problem-solving that taps into the use of authentic contexts rooted in real situations outside the classroom (DBE, 2011). Authors, therefore, argue that integrating IKS into Technology Education involves developing practical strategies that align with specific skills, knowledge areas, and values. Table 1 below encompasses how authors believe the co-existence of IKS can align with the focus areas of investigating, designing, making, evaluating, and communicating, as well as with structures, processing, mechanical systems and control (S/C), and electrical systems and control (S/C). This will include addressing values related to the environment, impact, and bias.

Table 1. Practical strategies for consideration of Indigenous Knowledge Systems in Technology Education

Skills, Knowledge and Values	Focus area	Indigenous knowledge integration
Skills	Investigate	Learners can investigate Indigenous technologies related to their local context or study global Indigenous innovations. For example, they can research traditional water purification methods, sustainable agricultural practices, or Indigenous renewable energy solutions.
	Design	Encourage learners to use Indigenous principles of sustainability, minimal waste, and environmental harmony in their design projects. For instance, when designing a shelter, learners could incorporate materials and techniques inspired by Indigenous architecture.
	Make	Facilitate projects where learners create prototypes or models using Indigenous crafting techniques. For example, they could build solar cookers inspired by Indigenous principles of solar energy use or construct water filtration systems based on traditional methods.
	Evaluate	Develop evaluation rubrics that include criteria based on Indigenous values, such as sustainability, environmental impact, community benefit, and cultural appropriateness. Learners can evaluate their own projects or peer projects against these criteria.
	Communicate (including – reports and presentation)	Assign learners to communicate their findings and projects through reports, presentations, or storytelling formats that are culturally relevant to Indigenous practices, such as oral presentations or visual storytelling.

continued on following page

Table 1. Continued

Skills, Knowledge and Values	Focus area	Indigenous knowledge integration
Knowledge	Structures	Explore Indigenous construction methods (e.g., adobe structures, bamboo frameworks) and integrate these into lessons on structures. Compare these methods with modern techniques, highlighting the sustainable aspects of traditional practices.
	Processing	Include Indigenous food processing techniques, traditional crafting, and material processing in lessons. For instance, exploring how Indigenous methods of tanning leather or processing natural dyes can be incorporated into projects.
	Mechanical Systems and Control	Introduce Indigenous mechanical innovations, such as traditional irrigation systems, animal-powered devices, or wind-catching structures. Compare these with modern mechanical systems, analysing efficiency, sustainability, and cultural context.
	Electrical Systems and Control	Discuss Indigenous methods of energy generation and management, like windmills or solar heat harnessing, and relate these to modern electrical systems and renewable energy technologies.
Values	Indigenous Impact (Environment: social, natural, economic)	Incorporate discussions on how IKS emphasizes a balanced relationship with the environment. Use case studies to show how traditional practices can inform modern sustainable technologies. Explore how Indigenous technologies support community well-being and economic sustainability. Discuss how modern adaptations of these technologies can benefit communities economically while preserving cultural heritage.
	Bias	Educate learners on recognizing and addressing biases that may undervalue Indigenous knowledge in technology. Use comparative studies to highlight the effectiveness of Indigenous approaches alongside modern technologies.

Implementation of Indigenous Practices in Technology Education Curriculum

Real-world examples of the successful integration of IKS into Technology Education would serve as inspirational models for teachers to embark on this transformative journey. We, therefore, believe that case studies showcasing innovative projects, learner collaborations with Indigenous communities, and partnerships with tribal organizations demonstrate the power of embracing Indigenous wisdom in the classroom. Therefore, this chapter highlights the best indigenous practices and lessons teachers can incorporate in their teaching. From this perspective, teachers can gain valuable insights into implementing meaningful and ethical integrations of indigenous knowledge in their teaching practices which will reflect the co-existing of IKS in the curriculum. Examples of Indigenous practices for the co-existence of Indigenous knowledge in Technology Education are elaborated on in the following paragraphs:

Since our early ages, authors have realised that Indigenous communities have used techniques like rainwater harvesting, qanats (underground tunnels), and terraced farming to manage water resources efficiently. These systems are not only sustainable but also tailored to local environmental conditions. So, the authors highly believe that by learning from this view, learners can explore these water management techniques by designing and building models or simulations. They can compare the efficiency of these traditional methods with modern water management systems and discuss their applicability in today's context, focusing on sustainability and low environmental impact.

Another exciting example explored in this chapter is based on the natural materials that indigenous people used for survival. Indigenous peoples have long used natural materials such as hemp, bamboo, cotton, and natural dyes from plants, minerals, and insects for making textiles and other items. For example, the Zulus and Swati Indigenous people used beads to make jewellery and clothing items and made a lot of items using special grasses. Regarding Swati indigenous people, Crush's (2001) view, functional utilization of grass was primarily in the form of building shelters and developed simultaneously into creating objects fulfilling a range of domestic requirements. Therefore, in a Technology Education context, learners can learn about processing these materials and exploring how traditional methods minimize environmental harm. Using these techniques, they can be tasked with making their own small items, such as cloth or rope. This connects to lessons on material properties, processing techniques, and sustainable production methods.

Many Indigenous communities have utilized natural elements such as wind and sunlight for energy generation long before modern technologies became available. For example, Pueblo people in the Southwestern United States designed their homes to maximize solar gain in winter and minimize it in summer, demonstrating an early form of passive solar energy use. Learners can explore these passive solar designs or small-scale wind energy solutions. They might design and build models that reflect these traditional practices, such as solar ovens or wind catchers, learning about energy efficiency and sustainable design in the process.

Indigenous peoples in Africa have used materials like grass, mud, wood, and stones primarily for construction, creating buildings that were not only sustainable but also resilient to local environmental challenges. For example, grass houses are known for their energy efficiency. The grass acts as an insulator, trapping heat during the winter and keeping things cool in the summer. In Technology Education, learners can explore these building techniques by constructing small models or studying the thermal properties of these materials compared to conventional building materials. This practice connects to structures, energy efficiency, and sustainable design lessons.

Indigenous communities have developed a wide variety of tools, from fishing hooks made of bone to bows and arrows tailored to specific environments. These tools reflect a deep understanding of material properties, mechanics, and the environment. Learners can study the design and function of traditional tools, and then create their own using both traditional and modern materials. This hands-on approach helps learners understand the evolution of technology and the principles of design, force, and motion.

Moreover, IKS often emphasizes community-centered approaches to technology, where tools and technologies are developed with direct input from the community to meet specific needs, such as communal cooking areas or shared water access points. Learners can engage in projects involving community-based design processes, identifying local challenges and developing solutions inspired by Indigenous approaches. This might include designing a community garden, a shared solar charging station, or rainwater collection systems.

Finally, we believe that highlights how Indigenous practices prioritize balance and respect for natural ecosystems, learners can be encouraged to consider these values in all technology projects. Like showing the importance of technology that benefits the community and emphasizing collaborative and inclusive approaches that are characteristic of many Indigenous practices. Moreover, biases by positioning Indigenous knowledge as a valuable and integral part of the broader technological landscape would challenge the misconception that modern technology is inherently superior. Therefore, integrating Indigenous practices into the Technology Education curriculum not only enhances learners' technical skills but also instills values of sustainability, respect for diversity, and community-centered design. Therefore, we believe that these practical indigenous strategies would provide learners with a rich, culturally relevant learning experience that honours Indigenous knowledge and prepares them for a future where technology and tradition can coexist harmoniously.

METHODOLOGY

The qualitative approach and a descriptive case study design were used to explore the co-existence of IKS and Technology education in the Senior Phase classroom in depth. A sample of seven Senior Phase teachers comprised the inquiry. Seven Senior Phase Technology teachers were purposively selected based on their at least B.Ed. qualifications in Technology education and their experience in teaching the subject. The selection of the seven teachers was motivated by the fact that their knowledge and experience in teaching Technology would assist in gaining a deep understanding of their experience with integrating IKS into their teaching. Semi-structured interviews were used. The guide questions were developed to enable the participants to elaborate freely, and for authors to make follow-up questions. The data was collected and collated from the district of Ehlanzeni Mpumalanga Province. This district has indigenous sites surrounded by indigenous people who still value and appreciate their diverse cultures. Thematic analysis was used to analyse data and descriptions and narratives were the core roots of the data analysis and themes for the co-existence were identified as detailed in the results section below (Clarke & Braun, 2017). According to Cedervall and Åberg (2010), thematic analysis can be used to analyse large and small data sets – from case study research with 1–2 participants to large interview studies with 60 or more participants (Mooney-Somers et al., 2008) – and homogenous and heterogeneous samples. Hence, in this chapter, the focus was only on seven Senior Phase Technology teachers teaching the subject in Grades 7-9. In this instance, we followed Braun & Clarke's (2006) six steps for doing a thematic analysis i.e., 1: Become familiar with the data, 2: Generate initial codes, 3: Search for themes, 4: Review themes, 5: Define themes, 6: Write-up.

Table 2 below represents some of the identified sub-themes from the main theme that presented the opportunities and challenges in Technology Education. The research question(s) guided us in the process of identifying the sub-themes. For ethics considerations, pseudonyms were used i.e., Teachers A, B, C, etc.

RESULTS AND DISCUSSION

This chapter was aimed at exploring the co-existence of IKS in the field of Technology Education. Table 2 below outlines the themes and sub-themes that guide the discussions of findings in this section. Each sub-theme's description is outlined in the table.

Table 2. Identified sub-themes

Bridging Cultural Contexts in Educational Design and Problem-Solving Activities: Opportunities in Integrating Indigenous Knowledge and Challenges		
Questions	**Sub-themes**	**Description**
How can design and problem-solving activities be modified to incorporate traditional knowledge and make them more culturally relevant to learners?	Cultural Relevance in Design Activities	Design and problem-solving tasks can be culturally relevant by incorporating traditional knowledge, such as designing shelters or using indigenous methods for processing materials.
What types of scenarios can be used in the curriculum to better reflect learners' everyday lives and integrate IKS in a meaningful way?	Integration of Indigenous Scenarios	Using scenarios that resonate with learners' everyday lives, such as designing a kraal, allows learners to apply IKS in practical contexts, bridging traditional and modern knowledge.
How can teachers encourage learners to bring their cultural knowledge into the design process to enhance creativity and critical thinking skills?	Fostering Creative and Critical Thinking	Allowing learners to integrate their cultural knowledge into design processes fosters creativity and critical thinking, bridging indigenous and modern technological knowledge.
What practical activities can be incorporated into the curriculum to help learners appreciate and apply indigenous methods alongside modern technological skills?	Emphasis on Practical Application	Hands-on activities involving traditional practices, like farming techniques or material processing, help learners appreciate indigenous methods alongside modern technologies.
How can Indigenous knowledge holders be involved in the educational process to provide learners with authentic insights into traditional practices and technologies?	Involvement of Indigenous Experts	Engaging Indigenous knowledge holders, such as community elders, in the classroom provides authentic insights into traditional practices, enhancing learning relevance and context.
What should a structured framework for integrating IKS into Technology Education include to ensure consistent and culturally relevant teaching practices?	Need for a Structured Framework	A structured framework is needed to guide teachers in incorporating IKS into Technology Education, ensuring consistent and culturally relevant integration of traditional knowledge.

Bridging Cultural Contexts in Educational Design and Problem-Solving Activities: Opportunities in Integrating Indigenous Knowledge

The integration of IKS into Technology Education presents a unique opportunity to bridge cultural contexts within the learning environment. Teachers recognize that IKS and modern technologies often coexist, though not always consciously integrated. This coexistence is seen in everyday practices where traditional methods, such as grinding food, persist alongside modern technological solutions. However, there is also a sense of loss or replacement, where modern advancements overshadow traditional practices. In their study, Adom et al., (2024) found that despite IKS being rooted in community-based development activities is often unclear and questioned. This highlights the need for deliberate efforts to balance these systems, valuing indigenous knowledge while embracing technological innovations.

Cultural Relevance in Design and Problem-Solving Activities is a key opportunity identified by teachers. Design tasks and problem-solving activities can be made culturally relevant by incorporating elements of traditional knowledge, allowing learners to connect with their heritage while developing technological skills. For instance, activities like designing shelters or using indigenous methods for processing materials (e.g., grinding mealies or peanuts) enable learners to engage with cultural practices. Correspondingly, Adom et al., (2024), in their study argued that Indigenous knowledge is easily accessible when addressing local problems. These tasks not only make learning more relatable but also provide a platform for learners to apply traditional knowledge within a modern educational context. For instance, Teacher D said:

> *"I think in processing that is where we can relate to the cultural concepts because most of the processing procedures Indigenous ways of processing materials, they were practiced, and they're still being practiced in most homes and families".*

The use of Indigenous scenarios in educational activities further supports the integration of IKS into Technology Education. Teachers have highlighted the effectiveness of using scenarios that resonate with learners' everyday lives or cultural experiences. Teacher E said: *"They are things that they are familiar with things that are part of their lives"*. An example is the design of a kraal—a traditional enclosure for livestock—which engages learners in practical problem-solving that draws on indigenous knowledge of rural life and animal management. By embedding such culturally familiar scenarios into the curriculum, teachers can foster a deeper connection between learners' cultural contexts and the technological skills they are acquiring. In the same view, a study by Muigua (2021), highlighted Indigenous knowledge as the best alternative to modern techniques in most areas.

Another significant opportunity lies in fostering creative and critical thinking through culturally relevant activities. Teacher C said: *"Yes, mam the activities make them relate to their cultural context"*. When learners are encouraged to incorporate their cultural knowledge into design processes, it enhances their creativity and critical thinking skills. This bridging of indigenous and modern technological knowledge not only enriches the educational experience but also promotes inclusivity. For instance, Adom (2016) showed that IKS, including cosmological beliefs, taboos, and totems, have prevented resource exploitation in the Ashanti region. Therefore, we argue that in Technology Education open-ended projects that allow learners to draw upon both indigenous and contemporary knowledge systems can stimulate innovative solutions, making education more engaging and culturally responsive.

Therefore, the authors argue that emphasising practical application over theory is another important aspect of integrating IKS into Technology Education. In the same put, teachers stress the importance of hands-on learning, where learners actively participate in traditional practices alongside modern technological applications. Examples include incorporating traditional farming techniques and material processing methods as part of technology lessons. Such practical experiences help learners appreciate indigenous methods and see their relevance in today's technological world, making learning more impactful and meaningful. Teacher B said: *"must be well structured, well designed, and we must ensure that it fits in the fourth industrial revolution that we are in actually".*

The involvement of Indigenous Knowledge holders, such as community elders or traditional practitioners, is also seen as a valuable opportunity to enrich the learning experience. By bringing indigenous experts into the classroom, learners would gain firsthand insights into traditional practices, making the learning process more authentic and contextually relevant. This is evident in their study by Adom et al., (2024), that the transfer of Indigenous knowledge from generation to generation requires the involvement of more knowledgeable and experienced elderly people. Teacher G said: *"Therefore, I can say no for now, but it will be a great thing if I can be able to invite one of them in the future".* This approach validates IKS and bridges the gap between traditional and modern knowledge, fostering a sense of respect and continuity for cultural heritage within the educational setting.

Challenges Encountered to Realize the Co-Existence of IKS in Technology Education Curriculum

Integrating IKS into Technology Education presents various challenges that underscore the need for a structured framework to facilitate this process. The coexistence of IKS and modern technology in educational settings is not straightforward, and several obstacles must be addressed to create an inclusive and balanced learning environment. In the below sections, the authors discuss the challenges that were identified in the field of Technology Education.

Challenges in relating to cultural contexts are evident when design tasks and problem-solving activities are based on scenarios that do not align with learners' cultural experiences. For example, rural learners may find it difficult to relate to tasks set in urban contexts, which can make the learning process less engaging and relevant. Teacher A said:

> "Some, some they don't, okay. It's not all the tasks that are relating to their context, because I'm teaching in most in sort of a rural area, but you will find the scenario talking about a municipality around town".

While some teachers try to adapt these tasks to be more culturally relevant, this requires additional effort and customization that may not always be feasible. The disconnect between the task context and learners' cultural backgrounds highlights the need for more thoughtful integration of IKS in educational design activities. In their study, Firdaussy et al. (2024) argued that the successful integration of IKS lies in the strong curriculum foundations. Thus, the curriculum should foreground the importance of maintaining strong connections to ancestral knowledge and values, allowing Indigenous students to learn their heritage, history, cultural customs, and standard academic subjects.

Another significant challenge is the lack of alignment with Indigenous knowledge within the current educational curriculum and resources. Teacher F said: *"We hardly do because the teaching resources that we have are not in line with the Indigenous technology."* Many participants pointed out that existing teaching materials, including the CAPS document, do not sufficiently incorporate Indigenous technologies. This omission makes it difficult for teachers to align classroom content with learners' cultural contexts, often leaving IKS underrepresented in Technology Education. Similarly, Firdaussy et al. (2024) presented that, at times the national education systems in Indonesia do not consider the special needs and characteristics of indigenous peoples. As a result, learners miss opportunities to connect their learning with their cultural heritage, and IKS remain marginal in the broader educational landscape.

The findings strongly indicate the need for a framework to guide the effective incorporation of IKS into teaching practices. Teacher A said: *"There should be a framework because if you check the CAPS document, it's a national police document. And it does not speak to the context of each school"*. Such a framework would provide teachers with a structured approach to integrating Indigenous knowledge into Technology Education, ensuring that content is relevant to local community contexts. By establishing clear guidelines and strategies for incorporating IKS, this framework would help bridge the gap between traditional and modern technological knowledge, fostering a more inclusive and culturally responsive educational experience.

Challenges and resource limitations further complicate the integration of IKS in Technology Education. Practical issues such as a lack of resources, inadequate support, and limited access to culturally relevant materials make it difficult for teachers to effectively include Indigenous knowledge in their lessons. Teacher F said: *"But the resources that you have the likes of the textbook, the teaching aids that we have most of them, they don't have a relationship with indigenous knowledge"*. Teachers express a need for more professional development opportunities, including training and workshops, to deepen their understanding of the relationship between indigenous and modern technologies. Without these resources and support, the integration of IKS remains a challenging and often inconsistent endeavour. In the same view, Firdaussy et al., (2024), Indigenous children confront significant challenges in accessing quality education due to economic factors, minimal facilities and infrastructure, and a shortage of teachers.

Overall, the findings suggest that IKS can coexist and be effectively integrated into Technology Education using culturally relevant design and problem-solving activities. However, this integration requires thoughtful adaptation and deliberate efforts to create inclusive learning environments that honour and value Indigenous knowledge alongside modern technological advancements. Developing educational frameworks that consciously embed IKS into the Technology curriculum will be crucial in achieving this balance, ultimately enhancing learners' learning experiences and fostering an appreciation for both traditional and contemporary knowledge systems.

Implications

Integrating IKS into Technology Education can make the curriculum more culturally relevant for learners from Indigenous backgrounds. This can increase engagement, motivation, and a sense of identity and pride in their cultural heritage. It promotes inclusivity by validating diverse ways of knowing and doing, beyond Western scientific methods, which can help reduce feelings of alienation among Indigenous learners. The coexistence would provide opportunities for learners to explore how traditional knowledge and modern technology can complement each other, leading to innovative solutions that blend the best of both worlds. This approach can help learners develop a broader perspective on technology, making it more adaptable to local needs and conditions. IKS involve holistic, systems-based thinking that considers the interconnections between people, the environment, and technology. This can enhance learners' problem-solving skills by prompting them to consider a wider range of factors. Practical, hands-on learning opportunities rooted in IKS can build learners' capacity to address real-life challenges. IKS teach the importance of flexibility and innovation in the face of new challenges, encouraging learners to be lifelong learners. Recognising and respecting Indigenous knowledge can lead to greater collaboration between educational institutions and Indigenous communities. incorporating IKS into the

Technology Education curriculum would help preserve traditional knowledge that might otherwise be lost, safeguarding it for future generations. This integration would ensure that the accumulated wisdom of generations is passed on, benefiting younger generations. The integration of IKS aligns with global sustainability goals, particularly those related to education, climate action, and sustainable communities. It supports equity and justice by challenging the dominance of Western-centric education and promoting a more balanced approach that respects all forms of knowledge.

RECOMMENDATIONS

The implementation of the co-existence of IKS in the Technology Education curriculum would provide a broad aspect of its value in the teaching and learning materials of the curriculum as well as its incorporation in the teaching practices by the teachers. This could also provide curriculum developers with an overview of the concepts and help them come up with ways to contribute to the actual actions to decolonize the Technology Education curriculum.

CONCLUSION

Integrating IKS into Technology Education offers a transformative approach that enriches learners' understanding of technology and fosters a profound appreciation for diverse cultural perspectives. This integration not only acknowledges the significance of Indigenous wisdom but also enriches the learning experiences of learners, equipping them with diverse perspectives and the ability to think critically about the interconnectedness of knowledge, culture, and the environment.

The chapter concludes that while there are challenges, such as the dominance of Western paradigms and insufficient curriculum guidance, these can be addressed through deliberate efforts, including creating a guiding framework, incorporating practical learning experiences, and involving community experts. By embracing diverse perspectives, fostering cross-cultural collaboration, and honouring traditional wisdom, teachers can empower learners to become global citizens who value interconnectedness and respect for all forms of knowledge.

This book chapter serves as a testament to the profound impact of integrating Indigenous knowledge into Technology Education, paving the way for a more vibrant and harmonious future for all learners.

REFERENCES

Abah, J., Mashebe, P., & Denuga, D. D. (2015). Prospect of integrating African Indigenous knowledge Systems into the teaching of sciences in Africa. *American Journal of Educational Research, 3*(6), 668–673.

Adom, D. (2016). *Asante IKS: Repositories of Conservation Ethics for Ghana's Biodiversity. Sub-Sahara indigenous knowledge and new discoveries in the new millennium interdisciplinary approach: Uthman Danfodio University, Sokoto, PTF I Hall, Sokoto, Nigeria. 7. Sokoto State*. Uthman Danfodio University.

Adom, D., Asare-Aboagye, I., Thompson, J., & Kquofi, S. (2024). Integration of Indigenous Knowledge Systems in African Higher Education in Ghana: A Post-CESA (2016-2025) Review. What Next for Africa's Higher Education?: What Should Follow the Continental Education Strategy for Africa (CESA 2016-2025), 161.

Ankiewicz, P. (2016). The relevance of indigenous technology knowledge systems (ITKS) for the 21st century classroom. In PATT 32 *conference proceedings* (pp. 22-34).

Blose, P. (2023). *The incorporation of education for sustainable development in Technology Education for an indigenous context* (Doctoral Thesis). https://uir.unisa.ac.za/handle/10500/30606

Blose, P., & Gumbo, M. T (2024). Developing a framework for integrating Indigenous knowledge Systems in Technology Education with sustainable development principles. *African Journal of Research in Mathematics, Science and Technology Education*, 1-13. https://doi.org/DOI: 10.1080/18117295.2024.2356416

Braun, V., & Clarke, V. (2006). Using thematic analysis in psychology. *Qualitative Research in Psychology, 3*(2), 77–101. DOI: 10.1191/1478088706qp063oa

Brondízio, E. S., Aumeeruddy-Thomas, Y., Bates, P., Carino, J., Fernández-Llamazares, Á., Ferrari, M. F., Galvin, K., Reyes García, V., McElwee, P., Molnár, Z., Samakov, A., & Shrestha, U. B. (2021). Locally based, regionally manifested, and globally relevant: Indigenous and local knowledge, values, and practices for nature. *Annual Review of Environment and Resources, 46*(1), 481–509. DOI: 10.1146/annurev-environ-012220-012127

Brook, R. K., & McLachlan, S. M. (2005). On using expert-based science to "test" local ecological knowledge. *Ecology and Society 10*(2): r3. [online] URL: https://www.ecologyandsociety.org/vol10/iss2/resp3/

Burgess, C., Bishop, M., & Lowe, K. (2022). Decolonising Indigenous education: The case for cultural mentoring in supporting indigenous knowledge reproduction. *Discourse (Abingdon), 43*(1), 1–14. DOI: 10.1080/01596306.2020.1774513

Cedervall, Y., & Åberg, A. C. (2010). Physical activity and implications on well-being in mild Alzheimer's disease: A qualitative case study on two men with dementia and their spouses. *Physiotherapy Theory and Practice, 26*(4), 226–239. DOI: 10.3109/09593980903423012 PMID: 20397857

Chahine, I. C. (2020). Towards African Humanicity: Re-mythogolising ubuntu through reflections on the ethnomathematics of African cultures. [CriSTal]. *Critical Studies in Teaching and Learning, 8*(2), 95–111. DOI: 10.14426/cristal.v8i2.251

Cindi, L. (2021). *Incorporating African IKS Indigenous knowledge into systems into the Basic Education curriculum: Experiences from two schools in the Gauteng and KwaZulu-Natal Provinces, South Africa* (Dissertation).

Clarke, V., & Braun, V. (2017). Thematic analysis. *The Journal of Positive Psychology*, *12*(3), 297–298. DOI: 10.1080/17439760.2016.1262613

Crush, J. (2001). Twentieth-century Swaziland When the Sleeping Grass Awakens: Land and Power in Swaziland. By Richard Levin. Johannesburg: University of Witwatersrand Press, 1997. Pp. 290. SA Rand 98.94 (ISBN 1-86814-301-5). *Journal of African History*, *42*(2), 307–352. DOI: 10.1017/S0021853701457938

De Sousa Santos, B., Nunes, J., & Meneses, M. (2007). *Another knowledge is possible: Beyond Northern epistemologies*. Verso.

Department of Basic Education. (2011). *National Curriculum Statement: Curriculum and Assessment Policy Statement (CAPS) Technology Grades 7 – 9*. Government Printers.

Firdaussy, U. F., Ningsih, S., & Asrawijaya, E. (2024). Basic education for indigenous peoples in Indonesia: Limiting children's cultural alienation and loss of identity. *Issues in Educational Research*, *34*(3), 995–1015.

Gilchrist, G., Mallory, M., & Merkel, F. (2005). Can local ecological knowledge contribute to wildlife management? Case studies of migratory birds. *Ecology and Society*, *10*(1).

Govender, N., & Mutendera, G. (2020). Teachers' and custodians' views and dilemmas arising thereof regarding the integration of indigenous knowledge in the primary school. *Alternative*, *16*(4), 356–368. DOI: 10.1177/1177180120970935

Gumbo, M. T. (2015). Indigenous technology in Technology Education curricula and teaching. In Williams, P. J., Jones, A., & Bunting, C. (Eds.), *Contemporary issues in Technology Education: The future of Technology Education* (pp. 57–74). Springer.

Kosoe, E. A., Adjei, P. O. W., & Diawuo, F. (2020). From sacrilege to sustainability: The role of IKS indigenous knowledge insystems in biodiversity conservation in the Upper West Region of Ghana. *GeoJournal*, *85*(4), 1057–1074. DOI: 10.1007/s10708-019-10010-8

Kruger, C., & Vass, G. (2025). Indigenous knowledges and learners. In *Teaching Middle Years* (4th ed., pp. 36–49). Routledge., https://www.taylorfrancis.com/chapters/edit/10.4324/9781003458586-4/indigenous-knowledges-learners-candace-kruger-greg-vass

Mooney-Somers, J., Perz, J., & Ussher, J. M. (2008). A complex negotiation: Women's experiences of naming and not naming premenstrual distress in couple relationships. *Women & Health*, *47*(3), 57–77. DOI: 10.1080/03630240802134134 PMID: 18714712

Mpofu, V., Otulaja, F. S., & Mushayikwa, E. (2014). Towards culturally relevant classroom Science: A theoretical framework focusing on traditional plant healing. *Cultural Studies of Science Education*, *9*(9), 221–242. DOI: 10.1007/s11422-013-9508-5

Mpuangnan, K. N., & Ntombela, S. (2024). Community voices in curriculum development. *Curriculum Perspectives*, *44*(1), 49–60. DOI: 10.1007/s41297-023-00223-w

Muchenje, F., Gora, R. B., & Makhuvaza, N. (2016). Interrogating the concept of time among the Shona: a post-colonial discourse. In Emeagwali, G., & Shizha, E. (Eds.), *African Indegenous Knowledge and the Sciences: Journeys into the past and present* (pp. 79–92). Sense. DOI: 10.1007/978-94-6300-515-9_6

Muigua, P. D. (2021). Revisiting the Place of Indigenous Knowledge in the Sustainable Development Agenda in Kenya. Journal of cmsd, 7(2).

Nadasdy, P. (2003). *Hunters and Bureaucrats: Power, Knowledge, and Aboriginal-State Relations in the Southwest Yukon*. UBC Press.

Ndlovu, E. C. (2023). *Senior phase technology teachers' integration of indigenous creative design processes in systems and control* (Doctoral Thesis). https://hdl.handle.net/10500/30901

Ndlovu, E. C., & Gumbo, M. T. (2018). Technology teachers' integration of technology-society-environment in teaching-learning activities. *Paper presented at the9th Annual UNISA ISTE Conference on Mathematics, Science and Technology Education.* Skhukhuza, South Africa

Ndlovu, E. C., & Gumbo, M. T. (2024). Decolonising technology education: Integrating indigenous knowledge for sustainable development in electrical and mechanical systems and control. African Journal of Research in Mathematics. *Science and Technology Education*, *28*(3), 362–372.

Pophiwa, N., & Saidi, U. (2022). Approaches to embedding IKS in Made in Africa Evaluations. *African Evaluation Journal*, *10*(1), 623. DOI: 10.4102/aej.v10i1.623

Rannikmäe, M., Holbrook, J., & Soobard, R. (2020). In Akpan, B., & Kennedy, T. J. (Eds.), *Science education in theory and practice: An introductory guide to learning theory* (pp. 259–275). Springer. DOI: 10.1007/978-3-030-43620-9_18

Swift, D. (1992). Indigenous Knowledge in the Service of Science and Technology in Developing Countries. *Studies in Science Education*, *20*(1), 1–28. DOI: 10.1080/03057269208560002

Thelma, C. C., & Madoda, D. (2024). Exploring the relationship between cultural identity and educational equity: A Zambian analysis. *International Journal of Research Publication and Reviews*, *5*(5), 11440–11448.

Triyanto & Handayani, R.D. (2020). Prospect of integrating indigenous knowledge in the teacher learning community. *Diaspora, Indigenous, and Minority Education*, •••, 1–14.

Yeseraw, A. (2023). *Indigenous Knowledge Inclusion in the New Curriculum Materials and Teachers 'Classroom Practices in the Amhara Region Primary and Middle Schools* (Doctoral dissertation).

Yishak, D., & Gumbo, M. T. (2017). The blending of indigenisation and standardisation in the Ethopian Education and Training Policy: A critical inquiry. In Gumbo, M.T. & Msila, V. (Eds.). African voices on indigenisation of the curriculum: Insights from practice, pp. 246-284. Wandsbeck: Reach Publishers.

Chapter 8
Factors Contributing to Action Research Learning and Formulation:
Strategies, Reflection, Perspective, and Ethics

Zuber Peermohammed Shaikh
University of Pune, India

ABSTRACT

Action Research/Learning means that ideally, the step-by-step process is constantly monitored over varying periods of time and by a variety of mechanisms (questionnaires, diaries, interviews and case studies) so that ensuing feedback may be translated into modifications, adjustment, directional changes, redefinitions, as necessary, so as to bring about lasting benefit to ongoing process itself rather than to some future occasion. These Chapter aiming to describes factors contributing to action research as: a constructive inquiry, during which researcher constructs his or her knowledge of specific issues through planning, acting, evaluating, refining and learning from the experience. It is a continuous learning process in which the researcher learns and also shares the newly generated knowledge with those who may benefit from it. These chapter highlight the distinct formulation features of action research and emphasize the purposeful intent of Ethics of researchers to improve, refine, reform, and problem-solve issues in their educational context.

INTRODUCTION

Factors Contributing to Action Research: At this point, it may be helpful for readers to have a working definition of action research and some examples to illustrate the methodology in the classroom. Bassey offers a very practical definition and describes "action research as an inquiry which is carried out in order to understand, to evaluate and then to change, in order to improve educational practice." Cohen and Manion situate action research differently, and describe action research as emergent, writing: essentially an on-the-spot procedure designed to deal with a concrete problem located in an immediate

DOI: 10.4018/979-8-3693-3542-0.ch008

situation. Similar to your research question, it is important to have a clear and realistic understanding of what is possible to research in your specific educational context. For example, would you be able to address unsatisfactory structures (policies and systems) within your educational context? Probably not immediately, but over time you potentially could. It is much more feasible to think of change happening in smaller increments, from within your own classroom or context, with you as one change agent. For example, you might find it particularly problematic that your school or district places a heavy emphasis on traditional grades, believing that these grades are often not reflective of the skills students have or have not mastered. Instead of attempting to research grading practices across your school or district, your research might instead focus on determining how to provide more meaningful feedback to students and parents about progress in your course. While this project identifies and addresses a structural issue that is part of your school and district context, to keep things manageable, your research project would focus outcomes on your classroom. The more research you do related to the structure of your educational context the more likely modifications will emerge. The more you understand these modifications in relation to the structural issues you identify within your own context, more you can influence others by sharing your work and enabling others to understand modification and address structural issues within their contexts. Throughout your project, you might determine that modifying your grades to be standards-based is more effective than traditional grades, and in turn, that sharing your research outcomes with colleagues at an in-service presentation prompts many to adopt a similar model in their own classrooms. It can be defeating to expect the world to change immediately, but you can provide the spark that ignites coordinated changes. In this way, action research is a powerful methodology for enacting social change. Action research enables individuals to change their own lives, while linking communities of like-minded practitioners who work towards action. (Lewin K.2021)

Educational research provides a vast landscape of knowledge on topics related to teaching and learning, curriculum and assessment, students' cognitive and affective needs, cultural and socioeconomic factors of schools, and many other factors considered viable to improving schools. Educational stakeholders rely on research to make informed decisions that ultimately affect the quality of schooling for their students. Accordingly, the purpose of educational research is to engage in disciplined inquiry to generate knowledge on topics significant to the students, teachers, administrators, schools, and other educational stakeholders. Just as the topics of educational research vary, so do the approaches to conducting educational research in the classroom. Your approach to research will be shaped by your context, your professional identity, and paradigm (set of beliefs and assumptions that guide your inquiry). These will all be key factors in how you generate knowledge related to your work as an educator.

Action research is an approach to educational research that is commonly used by educational practitioners and professionals to examine, and ultimately improve, their pedagogy and practice. In this way, action research represents an extension of the reflection and critical self-reflection that an educator employs on a daily basis in their classroom. When students are actively engaged in learning, the classroom can be dynamic and uncertain, demanding the constant attention of the educator. Considering these demands, educators are often only able to engage in reflection that is fleeting, and for the purpose of accommodation, modification, or formative assessment. Action research offers one path to more deliberate, substantial, and critical reflection that can be documented and analyzed to improve an educator's practice. (MacDonald B., 1978)

Objectives of Action Research

Purpose of Action Research: As one of many approaches to educational research, it is important to distinguish the potential purposes of action research in the classroom. This book focuses on action research as a method to enable and support educators in pursuing effective pedagogical practices by transforming the quality of teaching decisions and actions, to subsequently enhance student engagement and learning. Being mindful of this purpose, the following aspects of action research are important to consider as you contemplate and engage with action research methodology in your classroom: In thinking about the purpose of action research, it is helpful to situate action research as a distinct paradigm of educational research. I like to think about action research as part of the larger concept of living knowledge. Living knowledge has been characterized as "a quest for life, to understand life and to create… knowledge which is valid for the people with whom I work and for myself". Why should educators care about living knowledge as part of educational research? As mentioned above, action research is meant "to produce practical knowledge that is useful to people in the everyday conduct of their lives and to see that action research is about working towards practical outcomes" However, it is also about: creating new forms of understanding, since action without reflection and understanding is blind, just as theory without action is meaningless. The participatory nature of action research makes it only possible with, for and by people and communities, ideally involving all stakeholders both in the questioning and sense making that informs the research, and in the action, which is its focus. (Madaus G.F.,2022)

Figure 1. Demonstrates the 4-phase representation of the basic action inquiry cycle.

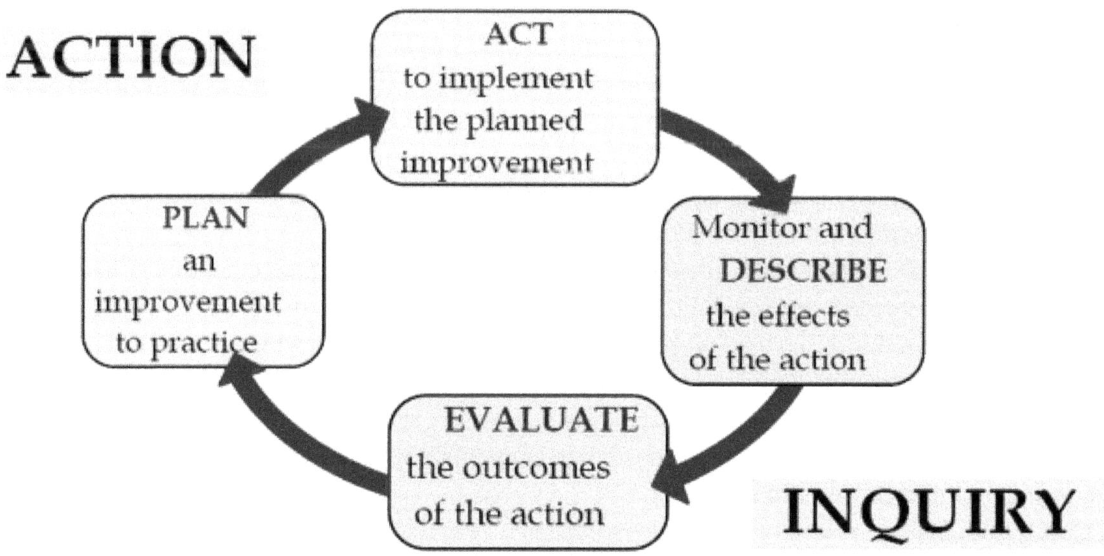

Literature Review of Current Context of Research: Although teachers entered the project voluntarily and were supposed to be aspiring to enquiry/discovery pedagogies, four teachers withdrew very soon after the commencement of the project. Having received the first documents from the central team they realized that the enquiry/discovery pedagogies were not possible to implement in their schools.

One withdrew because the school was oriented to preparing children to pass formal examinations, another thought that what he was doing was not enquiry/discovery pedagogy but some form of guided instruction. A few teachers voiced strong skepticism of the aims of the project in its initial formulation. However, these skeptical teachers remained within the project and were extremely valuable as critics and contributed a considerable number of documents about their researchers. They became committed to the project's aims, whilst reserving a detachment from drawing any firm conclusions. The teachers who had most difficulty in facing feedback from pupils, from documents written by the project teachers or from discussions, were those whose personal identity was inextricably bound up with particular views of the professional role of a teacher. These teachers underwent considerable stress, reported nightmares and insomnia and required extra support from the central team. They stayed within the project, did write documents, but developed much more slowly than the teachers who could reflect on their own practices more readily. This was the beginning of the Classroom Action Research Network (CARN), which now has international membership. Other regional initiatives in the UK include the Teacher Research Network of Northern Ireland based at the University of Ulster and the Avon Curriculum Review and Evaluation Programme. Like most programmes these are small-scale and brief, with transitory funding. In an attempt to further establish action research Jack Whitehead at the University of Bath, Pamela Lomax at Kingston University and Richard Winter at Anglia Polytechnic University supervise Master's courses which can include dissertations based upon teachers' research into their own practice. These studies are within the constraints of academic time and do not allow for the risk associated with group participatory research. However, Colin Fletcher at Cranfield Institute of Technology has developed alternative ways of approaching the supervision of participatory research since the 1970s. The tendency to individual reflexivity using selective work of Schon as the exemplar rather than group research will not promote democratic participation. Nor will the explicit yet convoluted distrust in teachers' accounts as ideologically distorted misrepresentations of reality. As Elliott (1991) argues, if claims are made to a distinction between 'practical' and 'emancipatory' action research, as is done by Carr & Kemmis (1986), they should not deny the possibility of critical reflexive practices arising out of the struggle by practitioners with their action research self-understandings. The problem of participation is in the main who is to define the issue under investigation, theorising about it and relationships in the process and in whose name is the research publicised, if at all. (Gates P., 2019)

Unfortunately it is not possible at the outset to say how people participate in a project, because this will depend on their interests and capabilities, in regard to different aspects of a project, at different times and places. In view of this, I find it helpful to refer to four different ways in which people can participate in an action research project:

a. **Compulsion:** When a participant has no choice in the matter, usually because it is some kind of a constraint or a directive from a superior.
b. **Co-Option:** When a researcher persuades someone (to choose) to help them with their research, the co-opted person in effect agreeing to provide a service to the researcher.
c. **Co-Operation:** When a researcher gets someone to agree to participate in their project, the co-operating person working as a partner in many respects (in that they are regularly consulted), but on a project that always "belongs" to the researcher (project "owner"). Most dissertation research is of this kind.
d. **Collaboration**: when people work together as co-researchers on a project in which they have equal shares.

BACKGROUND

Problem of Action Research/Learning Plan of Current Context of Research: Planning thoughtfully involves having a path in mind, but not necessarily having specific objectives. Due to your experience with students and your educational context, the research process will often develop in ways as you expected, but at times it may develop a little differently, which may require you to shift the research focus and change your research question. I will suggest a couple methods to help facilitate this potential shift. First, you may want to develop criteria for gauging the effectiveness of your research process. You may need to refine and modify your criteria and your thinking as you go. For example, we often ask ourselves if action research is encouraging depth of analysis beyond my typical daily pedagogical reflection. You can think about this as you are developing data collection methods and even when you are collecting data. The key distinction is whether the data you will be collecting allows for nuance among the participants or variables. This does not mean that you will have nuance, but it should allow for the possibility. Second, criteria are shaped by our values and develop into standards of judgement. If we identify criteria such as teacher empowerment, then we will use that standard to think about the action contained in our research process. Our values inform our work; therefore, our work should be judged in relation to the relevance of our values in our pedagogy and practice. (Halsey A., 2020)

Theoretical Framework of Current Context of Research

Researchers often work with teachers on action research for professional development, for example, and it is surprising how few consider enabling their students to participate in their work, so participation is initially seen as a matter of compulsion: the teacher decides what will happen and their students are not even consulted. Yet students can be involved at least at the co-optional level. For instance, they can assist as informants, they can help to produce data by observing and interviewing other participants, and assist in planning and implementing changes to practice. Participation has serious ethical ramifications, because how one defines participation can result in some participants being adversely affected by their participation. The World Bank, for instance, takes participation seriously, but it includes a 1-way flow of information from the project to those affected as a form of participation, whereas the recipients of the information may in fact be merely told they will lose their homes and livelihoods through being relocated to make way for an industrial development. Similarly, 'consultation' as the 2-way flow of information is also seen as a form of participation, yet that is often used as a form of espionage, the aim being to discuss a project with those affected in order to discover from them how to overcome their objections to a project's predetermined plan of action For such ethical reasons it is necessary to check on participation, not just at the proposal stage, but also throughout its duration. As a preventative, one should frequently check on what steps the researcher has taken to ensure that those affected are not being deceived, manipulated, or exploited. As a positive promotion at the whole project level, one should aim for an action research project that:

a. Addresses topics of mutual concern
b. Is based on a shared commitment to performing the research
c. Enables all those involved to actively participate in any way they desire
d. Shares control over the research processes evenly as possible
e. Produces an *input cost: outcomes value* ratio that is similarly beneficial for all participants

f. Establishes inclusive procedures for deciding matters of justice amongst participants

Focus of the Article

Action Research Benefits From Knowledge Management: First, in terms of professional and organisational development, action research is most effective when it is networked (vertically and horizontally) throughout the organization, though in my experience this is seldom achieved. Second, action research produces much practice-based knowledge that ought to be incorporated into the academic content of 'vocational' disciplines such as teaching, business, and journalism, but very little of the knowledge generated by action research is actually theorised and published in refereed academic journals. Action research should be able to bridge both the theory into practice transition, and the practice into theory transformation, though it shows little sign of doing so, perhaps because it is so orientated to improving practice. (House E., 1992)

Ethics in Action Research: Ethics has already been raised in regard to participation because there are always issues when one is making changes that affect others. A major problem for university action researchers is that they can seldom meet the medical research standards that are usually applied. My university, for instance, construed a teacher's students as the 'research subjects', and then insisted that they be given the right of withdrawal which is simply impossible in a school setting. A way was found around such Difficulties because ethical principles must underpin (and thereby legitimise) the procedures and ground rules of all research. My position is that the general ethical guideline that should be built into any action research project right from the start is that — No researcher or other participant ever engages in an activity that disadvantages another participant without their knowledge and consent. This usually rules out control group experiments, for instance, because it is disadvantageous to the control group not to have the benefit of changes that the action researcher expects to improve their practice. It also rules out the action researcher continuing with a change in practice that they can see is not improving the situation. That can be a problem for dissertation students who cannot then complete a pre- and post- research design although they have painstakingly produced base-line (or pre-change) data. The reason for allowing disadvantage with knowledge and consent is that participants often wish to make personal sacrifices in terms of their time and effort in order to improve their practice. Linked to this qualification, however, is the principle that the value of the outcomes for all participants should be tied to their input. This does not mean that outcomes will be similar in kind for all participants: when university academics work with school teachers, for instance, the valuable outcomes for the teacher tend to be in terms of improved practice, for the academic in terms of consultation fees and publications. One of the major problems with more traditional forms of research in education is that far too often teachers' professional experience and knowledge of their practice has been appropriated by the researcher with no corresponding benefit to the teachers involved. As it is hard to estimate such values, action researchers often co-publish with participants who have made a major contribution to the project. (Kohl S., 1990)

Solutions and Recommendations

Methodology of Current Context of Research: Does Your Timeline Work?: While action research is situated in the temporal span that is your life, your research project is short-term, bounded, and related to the socially mediated practices within your educational context. The timeline is important for bounding, or setting limits to your research project, while also making sure you provide the right amount of time for the data to emerge from the process. For example, if you are thinking about examining the use of math diaries in your classroom, you probably do not want to look at a whole semester of entries because that would be a lot of data, with entries related to a wide range of topics. This would create a huge data analysis endeavor. Therefore, you may want to look at entries from one chapter or unit of study. Also, in terms of timelines, you want to make sure participants have enough time to develop the data you collect. Using the same math example, you would probably want students to have plenty of time to write in the journals, and also space out the entries over the span of the chapter or unit. In relation to the examples, we think it is an important mind shift to not think of research timelines in terms of deadlines. It is vitally important to provide time and space for the data to emerge from the participants. Therefore, it would be potentially counterproductive to rush a 50-minute data collection into 20 minutes – like all good educators, be flexible in the research process.

Planning of Action Research: After the thinking about the considerations above, you are now at the stage of having selected a topic and reflected on different aspects of that topic. You have undertaken a literature review and have done some reading which has enriched your understanding of your topic. As a result of your reading and further thinking, you may have changed or fine-tuned the topic you are exploring. Now it is time for action. In the last section of this chapter, we will address some practical issues of carrying out action research, drawing on both personal experiences of supervising educator-researchers in different settings and from reading and hearing about action research projects carried out by other researchers. Engaging in an action research can be a rewarding experience, but a beneficial action research project does not happen by accident – it requires careful planning, a flexible approach, and continuous educator-researcher reflection. Although action research does not have to go through a pre-determined set of steps, it is useful here for you to be aware of the progression which we presented in Chapter 2. The sequence of activities we suggested then could be looked on as a checklist for you to consider before planning the practical aspects of your project. We also want to provide some questions for you to think about as you are about to begin.(Landsberger H.A., 2021)

- Have you identified a topic for study?
- What is the specific context for the study? (It may be a personal project for you or for a group of researchers of which you are a member.)
- Have you read a sufficient amount of the relevant literature?
- Have you developed your research question(s)?
- Have you assessed the resource needed to complete the research?

As you start your project, it is worth writing down:

- A ***working title*** for your project, which you may need to refine later
- The ***background of the study***, both in terms of your professional context and personal motivation
- The ***aims*** of the project
- The ***specific outcomes*** you are hoping for

Although most of the models of action research presented in Chapter suggest action taking place in some pre-defined order, they also allow us the possibility of refining our ideas and action in the light of our experiences and reflections. Changes may need to be made in response to your evaluation and your reflections on how the project is progressing. For example, you might have to make adjustments, taking into account the students' responses, your observations and any observations of your colleagues. All this is very useful and, in fact, it is one of the features that makes action research suitable for educational research. (Miller, 2007)

Action Research Planning Sheet: In the past, we have provided action researchers with the following planning list that incorporates all of these considerations. Again, like we have said many times, this is in no way definitive, or lock-in-step procedure you need to follow, but instead guidance based on our perspective to help you engage in the action research process. The left column is the simplified version, and the right column offers more specific advice if need.

Educators and Researchers Involved in Research Process: It is important to not isolate yourself when doing research. Many educators are already isolated when it comes to practice in their classroom. The research process should be an opportunity to engage with colleagues and open up your classroom to discuss issues that are potentially impacting your entire educational context. Think about the following relationships:

Research Participants: You may invite a variety of individuals in your educational context, many with whom you are in a shared situation (e.g. colleagues, administrators). These participants may be part of a collaborative study, they may simply help you develop data collection instruments or intervention items, or they may help to analyze and make sense of the data. While the primary research focus will be you and your learning, you will also appreciate how your learning is potentially influencing the quality of others' learning. (Calhoun, 1994)

Observers: We always tell educators to be public about your research, or anything exciting that is happening in your educational context, for that matter. In terms of research, you do not want it to seem mysterious to any stakeholder in the educational context. Invite others to visit your setting and observe your research process, and then ask for their formal feedback. Inviting others to your classroom will engage and connect you with other stakeholders, while also showing that your research was established in an ethic of respect for multiple perspectives.

Critical Friends or Validators: Using critical friends is one way to involve colleagues and also validate your findings and conclusions. While your positionality will shape the research process and subsequently your interpretations of the data, it is important to make sure that others see similar logic in your process and conclusions. Critical friends or validators provide some level of certification that the frameworks you use to develop your research project and make sense of your data are appropriate for your educational context. Your critical friends and validators' suggestions will be useful if you develop a report or share your findings, but most importantly will provide you confidence moving forward.

Potential Researchers: As an educational researcher, you are involved in ongoing improvement plans and district or systemic change. The flexibility of action research allows it to be used in a variety of ways, and your initial research can spark others in your context to engage in research either individually for their own purposes, or collaboratively as a grade level, team, or school. Collaborative inquiry with other educators is an emerging form of professional learning and development for schools with school improvement plans. While they call it collaborative inquiry, these schools are often using an action research model. It is good to think of all of your colleagues as potential research collaborators in the future. (Parsons, 2002)

Prioritize Ethical Practice: Try to always be cognizant of your own positionality during the action research process, its relation to your educational context, and any associated power relation to your positionality. Furthermore, you want to make sure that you are not coercing or engaging participants into harmful practices. While this may seem obvious, you may not even realize you are harming your participants because you believe the action is necessary for the research process. For example, commonly teachers want to try out an intervention that will potentially positively impact their students. When the teacher sets up the action research study, they may have a control group and an experimental group. There is potential to impair the learning of one of these groups if the intervention is either highly impactful or exceedingly worse than the typical instruction. Therefore, teachers can sometimes overlook the potential harm to students in pursuing an experimental method of exploring an intervention. If you are working with a university researcher, ethical concerns will be covered by the Institutional Review Board (IRB). If not, your school or district may have a process or form that you would need to complete, so it would beneficial to check your district policies before starting. Other widely accepted aspects of doing ethically informed research, include:

- What sort of considerations are necessary to take action in your educational context?
- How do you facilitate an action plan without disrupting your teaching?
- How do you respond when the unplanned happens during data collection?

An action research project is a practical endeavor that will ultimately be shaped by your educational context and practice. Now that you have developed a literature review, you are ready to revise your initial plans and begin to plan your project. This chapter will provide some advice about your considerations when undertaking an action research project in your classroom. (Altrichter H., 1993)

Identify a Clear Research Question: Your literature review should have exposed you to an array of research questions related to your topic. More importantly, *your review should have helped identify which research questions we have addressed as a field, and which ones still need to be addressed.* More than likely your research questions will resemble ones from your literature review, while also being distinguishable based upon your own educational context and the unexplored areas of research on your topic. Regardless of how your research question took shape, it is important to be clear about what you are researching in your educational context. Action research questions typically begin in ways related to "How does … ?" or "How do I/we … ?", for example:

Research Question Examples:

- How does a semi-structured morning meeting improve my classroom community?
- How does historical fiction help students think about people's agency in the past?
- How do I improve student punctuation use through acting out sentences?
- How do we increase student responsibility for their own learning as a team of teachers?

I particularly favor questions with I or we, because they emphasize that you, the actor and researcher, will be clearly taking action to improve your practice. While this may seem rather easy, you need to be aware of asking the right kind of question. One issue is asking a too pointed and closed question that limits the possibility for analysis. These questions tend to rely on quantitative answers, or yes/no answers. For example, "How many students got a 90% or higher on the exam, after reviewing the material three times? Another issue is asking a question that is too broad, or that considers too many variables. For example, "How does room temperature affect students' time-on-task?" These are obviously researchable questions, but the aim is a cause-and-effect relationship between variables that has little or no value to your daily practice. I also want to point out that your research question will potentially change as the research develops. If you consider the question:

- ***How do I improve student punctuation use through acting out sentences?:*** As you do an activity, you may find that students are more comfortable and engaged by acting sentences out in small groups, instead of the whole class. Therefore, your question may shift to:
- ***How do I improve student punctuation use through acting out sentences, in small groups?:*** By simply engaging in the research process and asking questions, you will open your thinking to new possibilities and you will develop new understandings about yourself and the problematic aspects of your educational context.*(Argryis C., 2019)*

Results of Current Context of Research: Hopefully, you found a lot a research on your topic. If so, you will now have a better understanding of how it fits into your area and field of educational research. Even though the topic and area you are researching may not be small, your study itself should clearly focus on one aspect of the topic in your classroom. It is important to maintain clarity about what you are investigating because a lot will be going on simultaneously during the research process and you do not want to spend precious time on erroneous aspects that are irrelevant to your research. Even though you may view your practice as research, and vice versa, you might want to consider your research project as a projection or megaphone for your work that will bring attention to the small decisions that make a difference in your educational context. From experience, our concern is that you will find that researching one aspect of your practice will reveal other interconnected aspects that you may find interesting, and you will disorient yourself researching in a confluence of interests, commitments, and purposes. We simply want to emphasize – don't try to research everything at once. Stay focused on your topic, and focus on exploring it in depth, instead of its many related aspects. Once you feel you have made progress in one aspect, you can then progress to other related areas, as new research projects that continue the research cycle.

Understand Your Capabilities and Know that Change Happens Slowly?: Similar to your research question, it is important to have a clear and realistic understanding of what is possible to research in your specific educational context. For example, would you be able to address unsatisfactory structures (policies and systems) within your educational context? Probably not immediately, but over time you potentially could. It is much more feasible to think of change happening in smaller increments, from within your own classroom or context, with you as one change agent. For example, you might find it particularly problematic that your school or district places a heavy emphasis on traditional grades, believing that these grades are often not reflective of the skills students have or have not mastered. Instead of attempting to research grading practices across your school or district, your research might instead focus on determining how to provide more meaningful feedback to students and parents about progress in your course. While this project identifies and addresses a structural issue that is part of your school and district context, to keep things manageable, your research project would focus the outcomes on your classroom. The more research you do related to the structure of your educational context the more likely modifications will emerge. The more you understand these modifications in relation to the structural issues you identify within your own context, the more you can influence others by sharing your work and enabling others to understand the modification and address structural issues within their contexts. Throughout your project, you might determine that modifying your grades to be standards-based is more effective than traditional grades, and in turn, that sharing your research outcomes with colleagues at an in-service presentation prompts many to adopt a similar model in their own classrooms. It can be defeating to expect the world to change immediately, but you can provide the spark that ignites coordinated changes. In this way, action research is a powerful methodology for enacting social change. Action research enables individuals to change their own lives, while linking communities of like-minded practitioners who work towards action. (Argryis C., 2019)

Planning of Research Process: Planning thoughtfully involves having a path in mind, but not necessarily having specific objectives. Due to your experience with students and your educational context, the research process will often develop in ways as you expected, but at times it may develop a little differently, which may require you to shift the research focus and change your research question. I will suggest a couple methods to help facilitate this potential shift. First, you may want to develop criteria for gauging the effectiveness of your research process. You may need to refine and modify your criteria and your thinking as you go. For example, we often ask ourselves if action research is encouraging depth of analysis beyond my typical daily pedagogical reflection. You can think about this as you are developing data collection methods and even when you are collecting data. The key distinction is whether the data you will be collecting allows for nuance among the participants or variables. This does not mean that you will have nuance, but it should allow for the possibility. Second, criteria are shaped by our values and develop into standards of judgment. If we identify criteria such as teacher empowerment, then we will use that standard to think about the action contained in our research process. Our values inform our work; therefore, our work should be judged in relation to the relevance of our values in our pedagogy and practice. (Brown A. 2001)

Does Your Timeline Work?: While action research is situated in the temporal span that is your life, your research project is short-term, bounded, and related to the socially mediated practices within your educational context. The timeline is important for bounding, or setting limits to your research project, while also making sure you provide the right amount of time for the data to emerge from the process. For example, if you are thinking about examining the use of math diaries in your classroom, you probably do not want to look at a whole semester of entries because that would be a lot of data, with entries related to a wide range of topics. This would create a huge data analysis endeavor. Therefore, you may want to look at entries from one chapter or unit of study. Also, in terms of timelines, you want to make sure participants have enough time to develop the data you collect. Using the same math example, you would probably want students to have plenty of time to write in the journals, and also space out the entries over the span of the chapter or unit. In relation to the examples, we think it is an important mind shift to not think of research timelines in terms of deadlines. It is vitally important to provide time and space for the data to emerge from the participants. Therefore, it would be potentially counterproductive to rush a 50-minute data collection into 20 minutes – like all good educators, be flexible in the research process.(Buckingham B. R., 2018)

Prioritize Ethical Practice: Try to always be cognizant of your own positionality during the action research process, its relation to your educational context, and any associated power relation to your positionality. Furthermore, you want to make sure that you are not coercing or engaging participants into harmful practices. While this may seem obvious, you may not even realize you are harming your participants because you believe the action is necessary for the research process. For example, commonly teachers want to try out an intervention that will potentially positively impact their students. When the teacher sets up the action research study, they may have a control group and an experimental group. There is potential to impair the learning of one of these groups if the intervention is either highly impactful or exceedingly worse than the typical instruction. Therefore, teachers can sometimes overlook the potential harm to students in pursuing an experimental method of exploring an intervention. If you are working with a university researcher, ethical concerns will be covered by the Institutional Review Board (IRB). If not, your school or district may have a process or form that you would need to complete, so it would beneficial to check your district policies before starting. Other widely accepted aspects of doing ethically informed research, include:

Confirm Awareness of Study **and** ***Negotiate Access*** – with authorities, participants and parents, guardians, caregivers and supervisors (with IRB this is done with Informed Consent).

- ***Promise to Uphold Confidentiality*** – Uphold confidentiality, to your fullest ability, to protect information, identity and data. You can identify people if they indicate they want to be recognized for their contributions.
- ***Ensure participants' rights to withdraw from the study at any point.***
- ***Make sure data is secured, either on password protected computer or lock drawer.****(Carr, W. 2018)*

Future Research Directions

Discussion of Current Context of Research: Prepare to Problematize your Thinking: Educational researchers who are more philosophically-natured emphasize that research is not about finding solutions, but instead is about creating and asking new and more precise questions. This is represented in the action research process shown in the diagrams in Chapter 1, as Collingwood (1939) notes the aim in human interaction is always to keep the conversation open, while Edward Said (1997) emphasized that there is no end because whatever we consider an end is actually the beginning of something entirely new. These reflections have perspective in evaluating the quality in research and signifying what is "good" in "good pedagogy" and "good research". If we consider that action research is about studying and reflecting on one's learning and how that learning influences practice to improve it, there is nothing to stop your line of inquiry as long as you relate it to improving practice. This is why it is necessary to problematize and scrutinize our practices.

Ethical Dilemmas for Educator-Researchers: Classroom teachers are increasingly expected to demonstrate a disposition of reflection and inquiry into their own practice. Many advocate for schools to become research centers, and to produce their own research studies, which is an important advancement in acknowledging and addressing the complexity in today's schools. When schools conduct their own research studies without outside involvement, they bypass outside controls over their studies. Schools shift power away from the oversight of outside experts and ethical research responsibilities are shifted to those conducting the formal research within their educational context. Ethics firmly grounded and established in school policies and procedures for teaching, becomes multifaceted when teaching practice and research occur simultaneously. When educators conduct research in their classrooms, are they doing so as teachers or as researchers, and if they are researchers, at what point does the teaching role change to research? Although the notion of objectivity is a key element in traditional research paradigms, educator-based research acknowledges a subjective perspective as the educator-researcher is not viewed separately from the research. In action research, unlike traditional research, the educator as researcher gains access to the research site by the nature of the work they are paid and expected to perform. The educator is never detached from the research and remains at the research site both before and after the study. Because studying one's practice comprises working with other people, ethical deliberations are inevitable. Educator-researchers confront role conflict and ambiguity regarding ethical issues such as informed consent from participants, protecting subjects (students) from harm, and ensuring confidentiality. They must demonstrate a commitment toward fully understanding ethical dilemmas that present themselves within the unique set of circumstances of the educational context. Questions about research ethics can feel exceedingly complex and in specific situations, educator- researchers require guidance from others. (Chambers R., 1983)

Illustration of Current Context of Research: Think about it this way. As a part-time historian and former history teacher I often problematized who we regard as good and bad people in history. I (Clark) grew up minutes from Jesse James' childhood farm. Jesse James is a well-documented thief, and possibly by today's standards, a terrorist. He is famous for daylight bank robberies, as well as the sheer number of successful robberies. When Jesse James was assassinated, by a trusted associate none-the-less, his body travelled the country for people to see, while his assailant and assailant's brother reenacted the assassination over 1,200 times in theaters across the country. Still today in my hometown, they reenact Jesse James' daylight bank robbery each year at the Fall Festival, immortalizing this thief and terrorist from our past. This demonstrates how some people saw him as somewhat of hero, or champion of some sort of resistance, both historically and in the present. I find this curious and ripe for further inquiry, but primarily it is problematic for how we think about people as good or bad in the past. Whatever we may individually or collectively think about Jesse James as a "good" or "bad" person in history, it is vitally important to problematize our thinking about him. Talking about Jesse James may seem strange, but it is relevant to the field of action research. If we tell people that we are engaging in important and "good" actions, we should be prepared to justify why it is "good" and provide a theoretical, epistemological, or ontological rationale if possible. Experience is never enough, you need to justify why you act in certain ways and not others, and this includes thinking critically about your own thinking. Educators who view inquiry and research as a facet of their professional identity must think critically about how to design and conduct research in educational settings to address respect, justice, and beneficence to minimize harm to participants. This chapter emphasized the due diligence involved in ethically planning the collection of data, and in considering the challenges faced by educator-researchers in educational contexts. (Checkland P., 1998)

How is Action Research Defined?: Action research is a form of investigation designed for use by teachers to attempt to solve problems and improve professional practices in their own classrooms. It involves systematic observations and data collection which can be then used by the practitioner-researcher in reflection, decision-making and the development of more effective classroom strategies. - Parsons and Brown (2002)

Action Research is a fancy way of saying let's study what's happening at our school and decide how to make it a better place. - Emily Calhoun (1994)

Action research is a natural part of teaching. Teachers are continually observing students, collecting data and changing practices to improve student learning and the classroom and school environment. Action research provides a framework that guides the energies of teachers toward a better understanding of why, when, and how students become better learners. - A. Christine Miller (2007)

Strategies and Factors contributing Five Phases of Action Research

Selecting an area or focus:

- Identifying an area of interest
- Focus on students
- Look at both immediate and cumulative effects

Collecting data:

- Collect existing archival data
- Use additional multiple data sources
- Collect data regularly
- Promote collective ownership of data
- Monitor data collection

Organizing data:

- Count instances, events, and artifacts
- Display data in tables and charts
- Arrange data by classroom, grade level, and school
- Organize for analysis

Analyzing and interpreting data:

- Analyze and question the data as a professional collective
- Decide what can be celebrated and what needs attention
- Determine priority area(s) for action

Studying the professional literature:

- Identify professional literature that relates to or matches the interest
- Gather research reports, research syntheses, articles, videotapes, etc.
- Analyze and interpret these materials for understanding and action
- Determine the most promising actions

Factorial Questions ask for the Design of Context of Action Research

Why am I collecting this data?

- How is the data related to the study question?
- What will the data tell us about students' learning and teaching strategies?

What exactly am I collecting?

- What kind of data will give me the best information about students learning and teaching strategies?
- How can I gather data on the same question in different ways, from different sources, and at different times (triangulation).

Where am I going to collect it?

- What kind of a sample is needed?

- Do I need to identify the student for long-term tracking?

When am I going to collect it and for how long?

- How much data is needed?
- How periodic should the collection be?

Who is going to collect it?

- Is data being collected by myself or will others be involved? (Chein S., 1948)

Figure 2. Factors contributing to action research process.

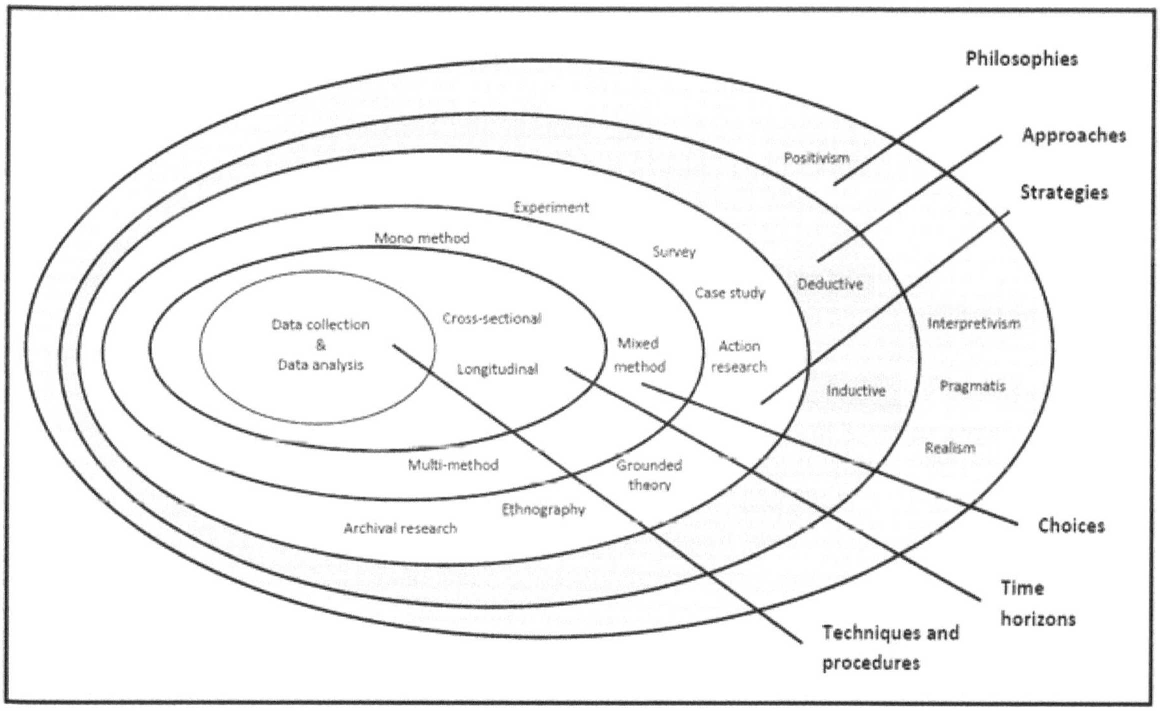

Factors Contributing of Formulation Strategies of Twelve Characteristics of Action Research

Action research should be continual rather than either continuous or occasional, because one cannot continuously action research one's practice, but one should regularly work to improve an aspect of it, so it should be more frequent than occasional.

Practice tends to be a matter of responding effectively and immediately to events as they arise, and academic research tends to operate according to set methodological protocols. Action research comes between the two because it is pro-active with regard to change, and its change is strategic in the sense that it is action based upon understanding achieved through the analysis of research information. Strategic action or 'tactical action' stands in contrast to action which is instant, a result of routine or habit, though it is informed by the wisdom of experience applied to good information which can only be produced by sound research processes. It also stands in contrast to action that is constrained by research protocols: methodology is always paramount in academic research, but in action research, research methodology should always be subservient to practice, so that one does not decide not to try to evaluate change, for instance, because one does not have a good measure or adequate baseline data, rather one seeks to make judgements on the best evidence that one can produce.(Cooke W., 2002)

Whereas routine practice tends to be the sole responsibility of the practitioner, and most research is carried out in teams these days, action research is participatory in that it includes all those involved in various ways, and it is collaborative in its ways of working.

Routine practice is naturalistic in that it is not researched, so there is no manipulation of the situation; both action research and academic research are experimental in the sense of making things happen to see what actually happens, but because action research occurs in non-manipulated social settings, it does not follow the cannons of controlled variables common to academic research, so it can be termed more generally interventionist rather than more narrowly experimental.(Blaikie N., 2002)

Routine practice does not normally allow for much examination of its procedures, values and effectiveness, but as an improvement process, action research always starts from some kind of a problem, and the term 'problematise' is often applied because action research, in common with Argrys and Schon's (1974) idea of 'double-loop learning' in reflective practice, treats 'the problem' as a problem itself. In fact, socially critical action research often starts with an examination of who owns the problem, which is one form of problematisation. Academic research, according to Kuhn, is generally a matter of proceeding with a given agenda, and that, coupled with the need for funding, means that it is generally commissioned either by government or commercial interests, or by peer review. Action research is sometimes commissioned too, of course, but even then it is far less constrained by this than purely academic research.

On-going routine practice is generally simply experienced by the participants, though when significant professional judgement is required, deliberation occurs, and the process moves more towards action inquiry, as the practitioner will usually follow up the results of the judgement in order to learn from it. Action research is always deliberative because when one intervenes in routine practice one is venturing into the unknown, so one has to make expert judgements about what, for instance, is most likely to improve the situation most effectively. Academic research is more often argued in the formal sense of in- and de-ductive theorisation. Those processes are employed in action research, of course, but not for the production of positivistic conclusions and predictions, which are very different from good professional judgements. (Collis J., 2003)

Again, action research stands somewhere between the non-recording of much that happens in routine practice and the rigorous peer review of method, data and conclusions in academic research. Action research tends to document its progress, often through compiling a portfolio of the kind of information that is regularly produced by routine practice, such as test scores or in education, client satisfaction indices in service organisations. or the minutes of production team meetings in business.

The main criterion for routine practice is that it works well, and concerns with how and why it works only arise when there are problems or improvements could be made, under which conditions the practitioner will move into an action inquiry, if not an action research mode where understanding the problem and knowing why it occurs are essential to designing changes to improve the situation. Theories are conceptual systems constructed to explain other knowledge, and they are a major concern of academic research. Although one does need to explain phenomena in action research, it is not its main purpose to construct the kind of web of explanations that comprise academic theory, and indeed, over the years the huge volume of action research has produced very little theoretical knowledge. (Creswell J. W., 1994)

All research generates knowledge, and whereas science investigates a very broad range of phenomena, the knowledge action research generates is almost exclusively about specific practices in specific social situations. On the other hand, action research is eclectic in that it generates, integrates and applies knowledge in and of a wide range of social disciplines, such as sociology, psychology, cultural studies, semiotics, information technology and administration. Furthermore, the knowledge applied to routine practice is normally knowledge gained from experience in practice, whereas the knowledge applied to practice in action research tends to be new knowledge, either that gained from academic learning or research, or that produced from the research of practice. (Crotty M. 1998)

This needs no explanation: the context, processes and results of routine practice are limited to those of the practitioner concerned, where as academic research aims for as wide generalisation as possible.

This has to do with knowledge management: knowledge gained in routine practice tends to remain with the individual practitioner, and knowledge gained in action research is more often be shared with known others in the same organisation or profession; it tends to be disseminated through networking and teaching rather than through publication as in the case of academic research. (Easterby-Smith, 1999) The fact that action research tends towards the practitioner end of these last two continua is something that needs addressing if it is to make much contribution to practitioner knowledge in the wider sphere of, for instance, the strategies of qualified practitioners across a whole occupation. To return to the point that these characteristics are a tension between action in the fields of practice and research, it is essential not to lose sight of action research as a process in which practitioners "gather evidence about their practices and critique assumptions, beliefs and values embedded in them".(Hamlyn D. W., 1995) Similarly, McNiff (2002:7) writes that action research involves becoming aware of the principles that drive us in our work: we need to be clear about both what we are doing and why we are doing it. Whilst most would agree that such an orientation is essential to action research, it is also central to other kinds of action inquiry, especially reflective practice, and without the distinction of the role of research methods in the process, the two would appear to be identical. (Remenyi D., 2005) Separating the two, however, is more a matter of emphasis than kind. For instance, an early childhood educator student of mine demonstrated this approach when reflecting on what she wanted to achieve in her action research project: "*For me that means I will not only become play-based in my approach to education, but I will also come to understand why I have become so.*"

As the supervisor of her action research, in contrast to reflective practice, for instance, I would see it as my job to ensure that she came to reorientate her practice and deepen her understanding of herself in as methodologically sound a way (rather then merely pragmatically effective) as possible. Another characteristic of the reciprocal relationship between research and improved practice is that one does not just understand practice in order to improve it in action research, one also gains an improved understanding of routine practice through improving it, so improvement is the context, means and main end of understanding. (Robson C., 2002) Context: As action research is an improvement process, one cannot action research routine practice: action research creates a moving research target by disrupting routine practice, and it leaves many loose ends in its wake (see for instance, the example of "action theorising" below). (Saunders, M., 2003) Means: As changes are reactive, monitoring what changes and how, leads not only to understanding one's own practice, but also to further understanding aspects of the situation, people, and one's practices that one has not set out to change. For example, many teachers learn a great deal about their students' perceptions of good teaching when they shift from teacher transmission to collaborative construction of knowledge. End: Dissemination and publication of the understanding of practice gained from improving it can also be made into an important spin-off of action research.(Saunders M., 2009)

CONCLUSION AND FUTURE PROSPECTS

Educational researchers who are more philosophically-natured emphasize that research is not about finding solutions, but instead is about creating and asking new and more precise questions. This is represented in the action research process shown in the diagrams, as Collingwood (1939) notes the aim in human interaction is always to keep the conversation open, while Edward Said (1997) emphasized that there is no end because whatever we consider an end is actually the beginning of something entirely new. These reflections have perspective in evaluating the quality in research and signifying what is "good" in "good pedagogy" and "good research". If we consider that action research is about studying and reflecting on one's learning and how that learning influences practice to improve it, there is nothing to stop your line of inquiry as long as you relate it to improving practice. This is why it is necessary to problematize and scrutinize our practices.

REFERENCES

Altrichter, H., Posch, P., & Somekh, B. (1993). Professors investigate their work: An introduction to methods of action research. Research With Children: Perspectives and Practices, 241-257.

Argyris, C., Putnam, R., & Smith, D. M. (1985). *Argyris, Chris, Robert Putnam, and Diane Maclain Smith, Action Science: Concepts, Methods, and Skills for Research and Intervention*. Jossey-Bass.

Argryis, C., Putnam, R., & McLain Smith, D. (2019) *Action Science: Concepts, Methods, and Skills for Research and Intervention*. New York: Jossey Bass (Out of print: http://www.actiondesign.com/action_science/)

Boomer, G. (1985). Fair Dinkum Teaching and Learning: Reflections on Literacy and Power. Boynton/Cook Publishers, Inc., 52 Upper Montclair Plaza, PO Box 860, Upper Montclair, NJ 07043.

Blaikie, N. (2000). *Designing social research: The logic of anticipation* (1st ed.). Polity Press.

Brown, A., & Dowling, P. (2001). *Doing research/reading research: A mode of interrogation for teaching*. Routledge Falmer.

Buckingham, B. R. (2018). *Research for teachers*. New York: Silver, Burdett and Co Berge, B-M. with H. Ve (2000) *Action research for gender equity*. Buckingham: Open University Press.

Calhoun, E. F. (1994). *How to Use Action Research in the Self-Renewing School*. Association for Supervision and Curriculum Development.

Carr, W., & Kemmis, S. (2018). *Becoming critical: knowing through action research*. Falmer Press.

Chambers, R. (1983). *Rural Development: Putting the Last First*. Longman Press.

Checkland, P., & Holwell, S. (1998). *Information, Systems, and Information Systems: Making Sense of the Field*. Wiley.

Chein, S., Cook, S. W., & Harding, J. (1948). The field of action research. *The American Psychologist, 3*(2), 33–44. DOI: 10.1037/h0053515 PMID: 18911932

Collis, J., & Hussey, R. (2003). *Business research: A practical guide for undergraduate and postgraduate students* (2nd ed.). Palgrave Macmillan.

Cooke, W. (Bill) (undated) *A Foundation correspondence on action research: Ronald Lippitt and John Collier*. Produced by Working Paper No.6/2002 of the Institute for Development Policy and Management.

Creswell, J. W. (1994). *Research design: Quantitative and qualitative approaches*. California: Sage publication, Inc.

Crotty, M. (1998). *The foundations of social research: Meaning and perspective in the research process* (1st ed.). Sage Publication, Inc.

Easterby-Smith, M., Thorpe, R., & Lowe, A. (1999). *Management research: An introduction* (1st ed.). SAGE Publications.

Gates, P. (2019). Developing consciousness and pedagogical knowledge through mutual observation. In Woods, P. (Ed.), *Working for Teacher Development.*

Hamlyn, D. W. (1995). History of epistemology. In Honderich, T. (Ed.), *The Oxford companion to philosophy* (pp. 242–245). Oxford University Press.

House, E. (1992). Response to 'Notes on pragmatism and scientific realism'. *Educational Researcher, 21*(6), 18–19. DOI: 10.3102/0013189X021006018

Kohl, H., & Kohl, S. (1990). *The Long Haul - an autobiography of Myles Horton*. Doubleday.

Landsberger, H. A. (2021). *Hawthorne Revisited: management and the worker, its critics and developments in human relations in industry*. Cornell University Press.

Lewin, K.(2021) Die Sozialisierung des Taylorsystems, Praktischer Sozialismus, No.4.

MacDonald, B. (1978). *The Experience of Innovation*. CARE,University of East Anglia.

Madaus, G. F. (2022). *Educational Evaluation: classic works of Ralph W. Tyler*. Kluwer.

Miller, C. A. (2007). "Action Research: Making Sense of Data." On-line article recovered 11/05/2007, www.coe.fau.edu/sfcel/sensdata.htm

Parsons, R. D., & Brown, K. S. (2002). *Teacher as Reflective Practitioner and Action Researcher*. Wadsworth/Thomson Learning.

Remenyi, D., Williams, B., Money, A., & Swartz, E. (2005). *Doing research in business and management: An introduction to process and method* (1st ed.). SAGE Publication.

Robson, C. (2002). *Real world research* (2nd ed.). Blackwell Publishing Ltd.

Saunders, M., Lewis, P., & Thornhill, A. (2003). *Research methods for business students* (3rd ed.). Pearson Education Limited.

Saunders, M., Lewis, P., & Thornhill, A. (2009). *Research methods for business students* (5th ed.). Pearson Education Limited.

Chapter 9
Fundamentals and Illustration of Action Research Modelling, Reflection, Application:
Basics in Modelling Action Research – Theory and Types

Zuber Peermohammed Shaikh
University of Pune, India

Satish Balkrishna Bhise
Sinhgad Institute's Smt. Kashibai Navale College of Pharmacy, India

ABSTRACT

Action Research develop conceptual structures that took explicit account of the defining social roles and strongly influence the process of any change in the modes of production. Action research was exemplified by the discussion of problems followed by group decisions on how to proceed. Action research must include the active participation by those who have to carry out the work in the exploration of problems that they identify and anticipate. After investigation of these problems the group makes decisions, monitoring and keeping note of the consequences. The group would decide on when a particular plan or strategy had been exhausted and fulfilled, come to nothing, and would bring to these discussions newly perceived problems. These Research elaborates are action research is an emergent process that allows beliefs and decisions to be negotiated as reality and meaning are being. Indeed fair observation demonstrated the efficacy of action research for improving productivity.

INTRODUCTION OF FUNDAMENTALS OF ACTION RESEARCH: ACTION RESEARCH

is a research method that aims to simultaneously investigate and solve an issue. In other words, as its name suggests, action research conducts research and takes action at the same time. It was first coined as a term in highly interactive method, action research is often used in the social sciences, particularly in educational settings. Particularly popular with educators as a form of systematic inquiry, it prioritiz-

DOI: 10.4018/979-8-3693-3542-0.ch009

es reflection and bridges the gap between theory and practice. Both types of action research are more focused on increasing the capacity and ability of future practitioners than contributing to a theoretical body of knowledge. (Ali, A. D.,2020)

Types of Action Research: There are two common types of action research: participatory action research and practical action research. O'Brien explains that by the mid-1970s four main types of action research had emerged: traditional; contextural (action learning); radical; and educational action research.

Traditional Action Research: These originated from Lewin's work and incorporates the concepts and practices of field theory, group dynamics, T-groups, and the clinical model. This traditional approach leans towards the conservative, and typically preserves the status quo regarding organisational power structures.

Contextual Action Research (Action Learning): These also known as action learning, involves reconstituting the structural relations among individuals in a social environment; is domain-based, in that it attempts to engage all affected parties and stakeholders; is holographic, as each participant comprehends the working of the whole; and emphasizes that par- ticipant's act as project designers and co-researchers.

Radical Action Research: These with its roots in Marxian dialectical materialism and the praxis orientations of Antonio Gramsci, focuses on emancipation and addressing power imbalances. Gramsci advocated the "renovation" and the "making critical" of the workers' common sense. Participatory Action Research, often found in liberationist movements and international development circles, and Feminist Action Research, both aspire to social transformation through an advocacy course of action to strengthen peripheral groups in society.

Educational Action Research: It has its beginnings in the writings of John Dewey, the American educational philosopher of the 1920s and 30s, who believed all professional educators should become involved in community problem-solving. Its practitioners operate largely out of educational institutions, and concentrate on developing curricula, professional development, and applying learning within a social context.

Participatory action research emphasizes that participants should be members of the community being studied, empowering those directly affected by outcomes of said research. In this method, participants are effectively co-researchers, with their lived experiences considered formative to the research process. This type of action research - an example would be a community of self-survey - seemed to be most effective for a limited range of problems. If was useful in disclosing particular and local facts (not general principles) which could provide examples for other communities.

Practical action research focuses more on how research is conducted and is designed to address and solve specific issues.

Diagnostic action research: It designed to produce a needed plan of action. The change agents would intervene in an already existing situation (for example, a race riot or anti-Semitic vandalism), diagnose the problem, and recommend remedial measures. Unless the proposed cures were feasible, effective, and acceptable to the people involved, however, this design of action was often wasted.

Empirical action research: It was primarily a matter of record keeping and accumulating experiences in day-to-day work, ideally with a succession of similar groups, such as boys' clubs. An inherent weakness of this procedure was that conclusions were drawn from experience with a single group, or with several groups differing in numerous ways, without test controls. Despite this handicap empirical action research could lead to the gradual development of generally valid principles as clinical medicine had already demonstrated. (Lewin K., 1920)

ADVANTAGES AND DISADVANTAGES OF ACTION RESEARCH

Advantages:

i. Action research is highly adaptable, allowing researchers to mold their analysis to their individual needs and implement practical individual-level changes.
ii. Action research provides an immediate and actionable path forward for solving entrenched issues, rather than suggesting complicated, longer-term solutions rooted in complex data.
iii. Done correctly, action research can be very empowering, informing social change and allowing participants to effect that change in ways meaningful to their communities.

Disadvantages:

- Due to their flexibility, action research studies are plagued by very limited generalizability and are very difficult to replicate. They are often not considered theoretically rigorous due to the power the researcher holds in drawing conclusions.
- Action research can be complicated to structure in an ethical manner. Participants may feel pressured to participate or to participate in a certain way.
- Action research is at high risk for research biases such as selection bias, social desirability bias, or other types of cognitive biases.

BACKGROUND

History and Illustration of Action Research: In schools, **action research** refers to a wide variety of evaluative, investigative, and analytical research methods designed to diagnose problems or weaknesses—whether organizational, academic, or instructional—and help educators develop practical solutions to address them quickly and efficiently. Action research may also be applied to programs or educational techniques that are not necessarily experiencing any problems, but that educators simply want to learn more about and improve. The general goal is to create a simple, practical, repeatable process of iterative learning, evaluation, and improvement that leads to increasingly better results for schools, teachers, or programs. Action research may also be called a *cycle of action* or *cycle of inquiry*, since it typically follows a predefined process that is repeated over time. A simple illustrative example: Identify a problem to be studied; Collect data on the problem; Organize, analyze, and interpret the data; Develop a plan to address the problem; Implement the plan; Evaluate the results of the actions taken; Identify a new problem; Repeat the process.

Figure 1. Steps involved in action research

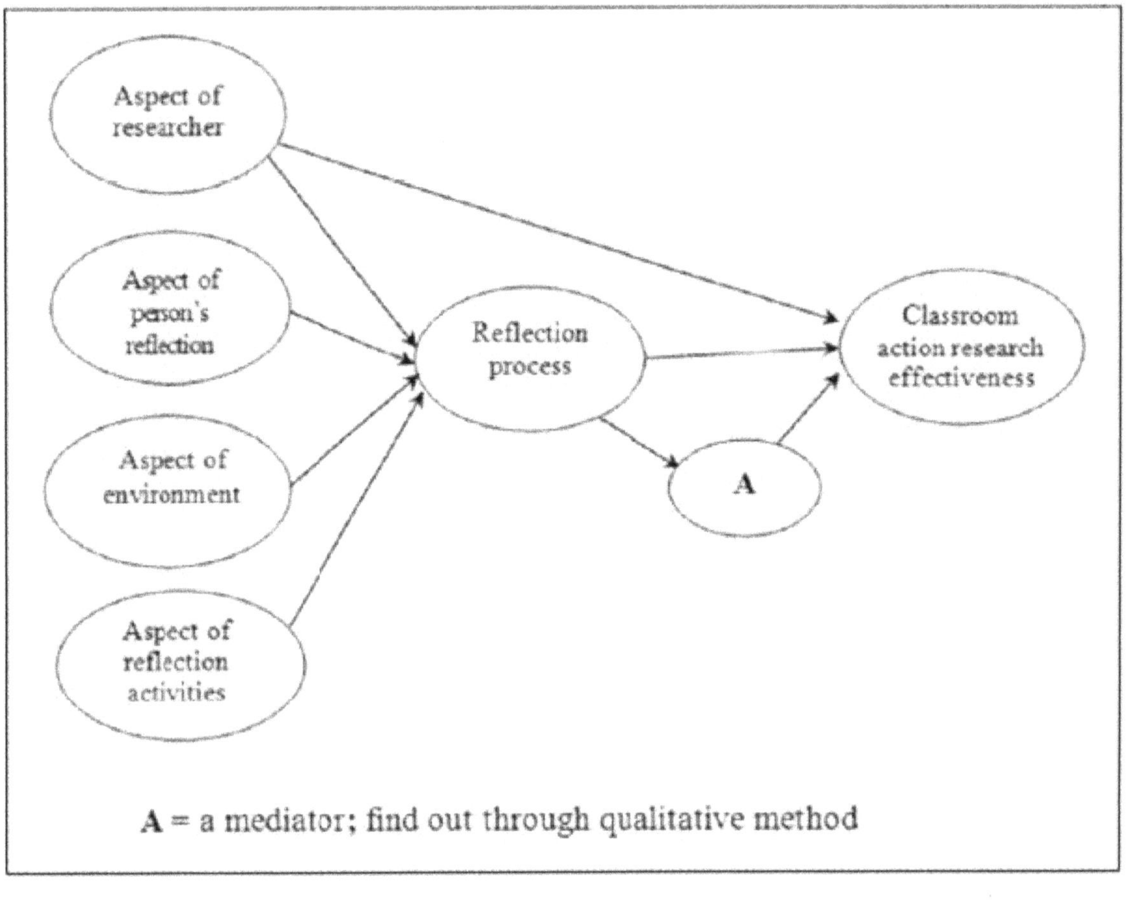

Action Research in Theory: Action research-based inquiry in educational contexts and classrooms involves distinct participants–students, teachers, and other educational stakeholders within the system. All of these participants are engaged in activities to benefit the students, and subsequently society as a whole. Action research contributes to these activities and potentially enhances the participants' roles in the education system.

Participants' Roles are Enhanced Based on Two Underlying Principles: Communities, schools, and classrooms are sites of socially mediated actions, and action research provides a greater understanding of self and new knowledge of how to negotiate these socially mediated environments; communities, schools, and classrooms are part of social systems in which humans interact with many cultural tools, and action research provides a basis to construct and analyze these interactions. In our quest for knowledge and understanding, we have consistently analyzed human experience over time and have distinguished between types of reality. Humans have constantly sought "facts" and "truth" about reality that can be empirically demonstrated or observed. Social systems are based on beliefs, and generally, beliefs about what will benefit the greatest amount of people in that society. Beliefs, and more specifically the rationale or support for beliefs, are not always easy to demonstrate or observe as part of our reality. Take the example of an English Language Arts teacher who prioritizes argumentative writing in her class. She believes that argumentative writing demonstrates the mechanics of writing best among types of writing, while also providing students a skill they will need as citizens and professionals. While we can observe the students writing, and we can assess their ability to develop a written argument, it is difficult to observe the students' understanding of argumentative writing and its purpose in their future. This relates to the teacher's beliefs about argumentative writing; we cannot observe the real value of the teaching of argumentative writing. The teacher's rationale and beliefs about teaching argumentative writing are bound to the social system and the skills their students will need to be active parts of that system. Therefore, our goal through action research is to demonstrate the best ways to teach argumentative writing to help all participants understand its value as part of a social system. The knowledge that is conveyed in a classroom is bound to, and justified by, a social system. A postmodernist approach to understanding our world seeks knowledge within a social system, which is directly opposed to the empirical or positivist approach which demands evidence based on logic or science as rationale for beliefs. Action research does not rely on a positivist viewpoint to develop evidence and conclusions as part of the research process. Action research offers a postmodernist stance to epistemology (theory of knowledge) and supports developing questions and new inquiries during the research process. (MacDonald B. 1978)

Focus of the Article

Action research as a paradigm, or model of practice, provides a framework for the ongoing improvement of practice. Within that framework of practice, many formal reports of action research have emphasized the improvement in student learning, or the changes needed in the curriculum, or the pedagogy. Less frequently, formal reports emphasized the changes that take place in the practitioner. Reflecting on changes in self is a significant part of action research because the practitioner is the person who identifies the problem, the intervention, the analysis, and whether or not to make changes in the future. Sometimes the changes that are needed are a new perspective by the practitioner on that practice. As one of our authors in this issue, Dresser recognized, "One of the most important findings in this study was that teacher candidates became cognizant of the impact of negative experiences on student academic achievement." It's not that the teachers needed a new curriculum, or needed to focus on a different part of practice, but rather, the teachers focused on their own perspective, adding to that focus an emphasis on the students' social-emotional learning needs as essential to improved academic learning.

Figure 2. Managerial approach of action research

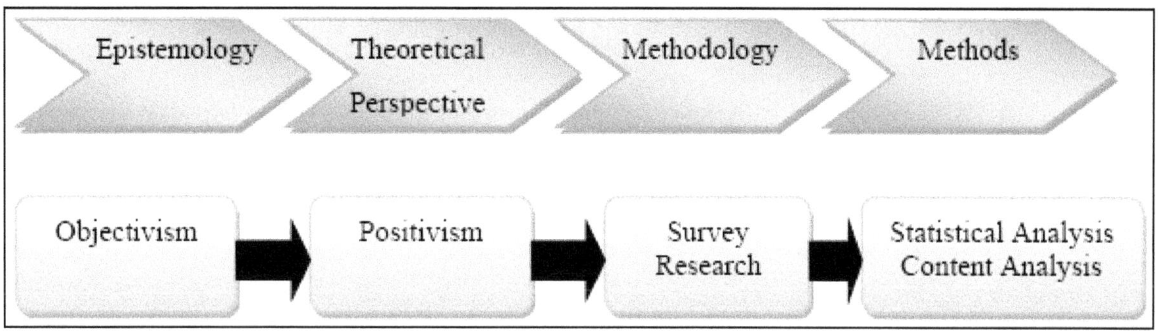

Action Research Modelling: Modeling the type of reflective practice expected from our students is one effective vehicle to teach both attitudes and skills. Observing a model allows preservice teachers to broaden their perspectives of teaching and provides opportunities to value reflectivity. Therefore, teacher educators should consciously examine the extent to which approaches to teaching action research demonstrate the type of reflection encouraged in their education students. (Madaus,G.F., 1989)

Review of Our Approach to Action Research: Action research was born when social scientists and practitioners, "concerned not only with the generation of scientific knowledge but also with its usefulness in solving practical problems, worked to bridge the gap between theory and practice". Kurt Lewin's work is generally taken as the starting point, when he wanted to use "research in a 'natural' setting to change the way that the researcher interacts with that setting". Lewin recognised the important role of participation in planned change processes, and constructed an action research theory, describing action research as ″proceeding in a spiral of steps, each of which is composed of planning, action and the evaluation of the result of action. This spiral or cyclical process involved a "non-linear pattern of planning, acting, ob- serving, and reflecting on the changes in the social situations". As action research is open-ended research in which techniques are examined systematically and scientifically, it starts with a concept, perception, or idea that has been developed, rather than starting with a fixed hypothesis.

SOLUTIONS AND RECOMMENDATIONS

Project Implementation and Analysis Modeling in Action Research: As a teacher educator for the past ten years it appeared that while the technical elements of action research were certainly mastered by most teacher candidates, many preserves teachers reacted negatively to completing a project while student teaching. For many, action research takes away from time spent preparing lessons, grading papers, or commiserating with peers. At best, projects rarely extend beyond fixing isolated problems. Yet, some students also gain a more holistic understanding of the interaction of the multiple variables in classrooms that are often messy and complex. A reflective model for such research attempts to connect received knowledge, such as facts, data, and theories, with experiential knowledge—what we know from our practical, professional experience—by a continuous process of reflection. (Adelman, C., 1993)

Figure 3. Effectiveness and casual models of reflection process of classroom teachers of action research

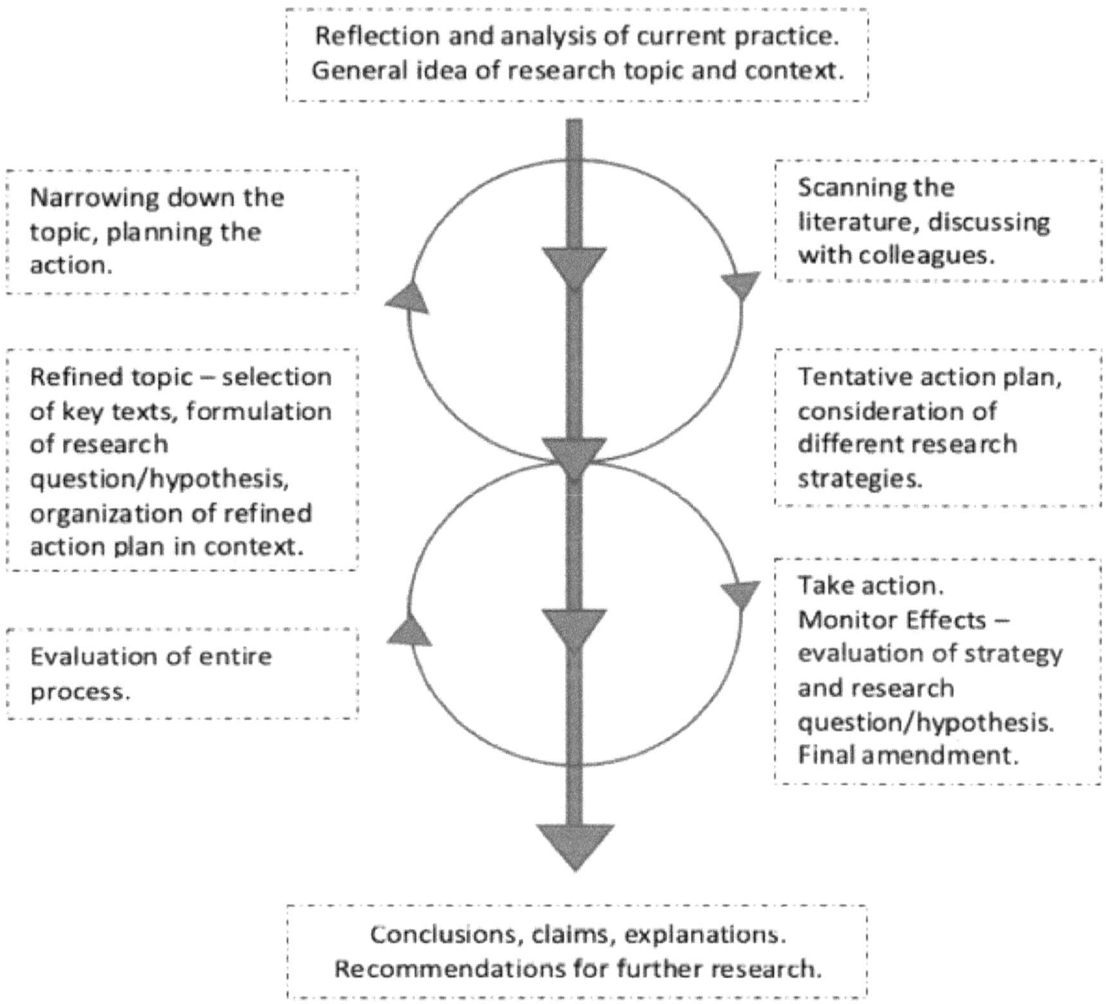

The Action Research Cycle: To illustrate the cycle, assume you are a teacher who has perceived a need or a problem with a class you are teaching (initial reflection)—that is, you have noticed that your students become nervous whenever they give an oral presentation. You carry out your plan (act) in light of your past experience of such situations, social landscape, and feedback from your students, and you collect data, using the techniques you planned. After the data is collected, you analyse your findings. For example: One of your students might mention that a particular strategy you taught them was very helpful. At the end of this cycle you reflect on what has happened so that experiential knowledge can feed back into received knowledge. In other words, you reconstruct your knowledge through reflection. For example, you might ponder how effective the changes you made in your teaching practices were, what you learned from the changes you made, or what barriers to change existed. Reflecting on barriers to change will likely lead you to continue the action research cycle. (Altrichter, H. & Posch, P. 1989)

Social Constructivism: The constructivist theory was framed in terms of individuals, disregarding that each person's development occurs in constant exchange with his or her social circumstances, such as immediate working relationships, the climate of the schools attended, and other environmental factors. Similarly, teachers' beliefs emerge from a complex of social and individual influences, including their own experiences as pupils, personality preferences, public educational theories, and the teachers' relationships with their students, colleagues, and superiors.

Multiple Forms and Role of Reflection: These models are used to reframe a problem or issue; then interventions are carried out which lead to outcomes which are analyzed further. This reflection-in-action model concedes there is very little or no separation of research from practice, and little or no separation of knowing and doing. Schön's model enhances the iterative and investigative nature of action research. This involves researchers asking themselves why they do the things they do, and why they are the way that they are. Although the term 'scientific' is not always clearly definable, Roberts states that research projects without "systematic reflection and learning" cannot be considered genuine action research projects. Reflection is therefore a critical component of action research. According to action research theory, "change does not come about as a result of spontaneous acts, but through reflection on and understanding of specific problems within their social, political, and historical contexts". Bailey out that "critical reflection is a form of analysis that not only explores how and why things happened but identifies the assumptions underpinning the analysis". Reflection in participatory action research is that moment where the research participants examine and construct, then evaluate and reconstruct their concerns. There are multiple forms of reflection that individuals engage in both formally and informally which are as follows.

Self-Reflection: Being able to conduct a self-reflection of practice provides teachers with a level of privacy and security to become vulnerable and admit failure that may not be as comfortable in a group setting. As discussed in the literature presented above, collaborative reflection requires a level of vulnerability that is not as direct when reflecting internally on your own practice.

Authentic Reflection: Although it was brought up, the discussion never continued to produce further data. The term authentic is used here as defined by the researcher. The term authentic reflection indicates, reflection that is unaffected by the prompting of peers or evaluators. An authentic reflection is one's own reflection of the process and outcomes without the influence of others. (Argyris, C. & Schön, D. A., 1976)

Informal Reflection: Another participant noted that conducting the reflection in real time was more meaningful. Participants used regularly scheduled data meetings, meetings outside of work, evaluation meetings, and team meetings to reflect on their goals with peers. Informal reflection in this study indicates reflection that takes place separately from the learning log. Being able to process through a problem or lesson with a partner provided an opportunity to explore different angles.

Collaborative Reflection: Teachers who are working to improve their reflective, have a lower probability of achieving success if working alone rather than collaboratively? The teachers felt comfortable getting feedback from their peers and in some cases thought they may be able to provide them more meaningful feedback. Embedding reflection into the regular routine of teachers would provide both a set time to reflect as well as opportunities for collaborative reflection that do not currently exist in any formal way. Using time that already exists, where peers are together provides opportunities to reflect collaboratively, such as Professional Learning Communities (PLCs), where the facilitation of peer reflection could naturally take place. It is more community-based, focused on building a network of similar individuals (e.g., college professors in a given geographic area) and compiling learnings from iterated feedback cycles. (Arlin, P., 1990)

Critical Reflection: It serves to contextualize systemic processes that are already ongoing (e.g., working retroactively to analyze existing school systems by questioning why certain practices were put into place and developed the way they did).[9-10]

Methodology: These responses from the school's primary evaluator indicate that there has been development of the learning log and reflection processes over the years since it has been included in the learning log process. Reflection is often an individual process that, that requires the autonomy to reflect in a singular way, while also being able to act on that reflection. A focus group participant explained that she is hesitant to reflect with peers, as it makes her vulnerable to show areas of weakness or failure. She stated, "I don't necessarily process with other people. It tends to be for me a lot of internal and just identifying things that were strengths or weaknesses, but it's not something I feel super comfortable making myself vulnerable about". This participant was somewhat of an outlier in the group, as all other participants identified talking with a peer as a preferred method of reflection. However, it should be considered that not all teachers will choose to reflect collaboratively, and there should be options to do so independently. Throughout the data collection both members of the focus group and the action committee discussed that feedback and collaboration from peers was a major factor in their daily reflection. Collaboration was a preferred method of reflection for four out of five participants on the action committee, with the fifth sharing a preference for collaboration, but was trying something new with media. *(Atkin, J. M. 1993)*

Data Collection: Based on the data gathered, teachers desire the feedback of others on their work. They want to improve and reflect on their practice, and are open to formal and informal channels of feedback. Access Another theme that continued to resurface from all members of the group was the limited or challenging access to the learning log document. This led to the assertion that procedures and access to reflection methods must be easy and consistent. If the system is too complicated or cumbersome, teachers are likely to avoid it, which limits visible reflection from the evaluator. The annoyance with a newly implemented online system for housing the learning log became clear throughout the process as early on as focus groups. This is one of the various reasons that were given for lack of participation in the formal reflection log, even though the log is where the evaluator provides observation feedback. During coding, the concern regarding access to the learning log was discussed by five out of six participants directly and was also closely tied into the codes for procedures and time, which were also mentioned by five out of six participants on the action committee. The primary evaluator in the school (the principal) writes in teachers' learning logs during observations, where she puts basic notes about the observation and what she saw, also asks probing questions about what was observed: where the teacher is going with the lesson, how he/she is collecting data, and other questions about the teacher's learning goal, etc. In theory, the teacher will check the log for observation notes and respond to those questions for a reflection. Being able to collect data and reflect on it or have regularly scheduled times to reflect with others keeps the goal and learning log visible and on track. Focus group participants noted that they only write in their log when they are told to do so, or when they know there is observation feedback. Between the cumbersome process and other timely priorities, the log is often put at the bottom of the list.

FUTURE RESEARCH DIRECTIONS

Data Analysis: Data from the present study provide evidence of the value of action research for both teacher educators and their preserves teachers. For teacher educators, action research self-studies provide insights into their curriculum and instructional methods and offer perspective into the complexities of examining teaching in systematic ways. Specifically, while my views on the merits of self-reflection and action research have been strengthened, my self-study also revealed a process that is demanding in both time and effort. Researchers argue that while elements of action research, such as data collection, are potentially daunting, teachers have the luxury of data surrounding them in multiple formats (e.g., tests, discussions, and questionnaires) and data access is relatively straightforward. He reports that, because teacher research typically follows less traditional statistical/scientific methodologies, researching data "story themes" can be accommodated in manageable ways. Data availability is certainly a convenience; however, proponents of teacher research grossly underestimate the complexity of teachers' lives and the enormity of their work. To assume that teachers will find or make the time to implement and analyze project components such as journal writing, video taping, or completing survey data analysis may be highly misguided. The choice of research design depends on the purpose(s) of research, and hence there are four types of research design which are: exploratory study, descriptive study, explanatory study, and emancipatory study. (Carr, W., 1993)

Exploratory Study: This type of studies focuses on investigating what is happening, asking questions, seeking new insights, assessing phenomena in a new light, as well generating ideas and hypothesis for future researches. An exploratory study is characterised as a flexible design. The exploratory study is conducted when there is no information is available or little information is known about how similar research has been conducted in the past. This study is useful for clarifying and understanding of an imprecise problem, it can be conducted based on three main ways are: a search of the literature, interviewing experts, and conducting focus group interviews.[12]

Descriptive Study: This study displays an accurate profile of persons, situations or event. This type requires collecting a lot of information about the situation that will be studied. The descriptive study may be flexible and /or fixed design. Descriptive study is conducted in order to determine and describe the characteristics of the variables in the situation. Therefore, the descriptive study aims to provide researcher a profile or describe aspects of the phenomena being researched at different levels such as individual, organisational, industry-oriented, and other perspective. This study is considered as a piece of, or a forerunner to exploratory research, and therefore it is necessary to have a clear picture of phenomena before conducting exploratory study.

Explanatory Study: This study seeks an explanation of a situation or problem being studied and not necessary to be in causal relationship, and explain of patterns relating to studied phenomenon. This study may be flexible and /or fixed design. Explanatory study investigates the relationship between variables of phenomenon in order to establish a causal relationships between variables. Other authors called this type of study as hypotheses testing. Hypotheses testing are usually conducted to explain the nature of the specific relationships, or indicate the difference among groups of independent variables, as well as explain the variance in the dependent variables or to predict outcomes. (Carr, W. & Kemmis S., 1986)

Result: The third assertion is that teachers value the freedom to make autonomous decisions about methods of reflection and evidence used for reflection. The code for autonomy was used when individuals discussed using an alternative method for reflection, trying something new for their reflection, and making individual decisions about evidence. Autonomy was mentioned by six out of six participants in various ways throughout the action committee meetings as well as during both focus groups. Efficacy was also discussed, primarily as it tied to the process of reflection, more so than connecting to the learning log itself. This discussion about taking authentic reflection whether done alone or collaboratively with peers, and translating it into the log was discussed by four out of six participants. However, having the autonomy to reflect in a way that is meaningful, loses some aspects of autonomy when the result is a prescribed format. Participant expressing frustration with the extra step of putting an autonomous reflection into a prescribed learning log. Participants stated that the exploration of different methods of reflection did make them feel more efficacious than typing directly into the learning log document. By exploring methods of reflection and collaborating with a peer, or in this case student teacher, the participant saw tangible evidence of improvement, which created an efficacious experience.

Purpose: The final assertion is connected to the purpose of the learning log within the evaluation process. Based on the data collected the learning log is used as a means of professional communication between the evaluator and the teacher; a place to share what is being done and what has been observed. The learning log does not hold significant value within the evaluation process, which brings about a question of its purpose. If the purpose and evaluator's use of the log is not clear to teachers, their motivation to set aside time sensitive tasks such as grading and planning to reflect, will be diminished. This uncertainty about the effects of the log on teacher self-efficacy is another indication that changes may need to occur within the learning log process to create clear messaging about the learning log and how it lives within the evaluation process.

Observation: After working extensively in diverse educational contexts and with a range of educators at all stages of their career, the author understood the importance of developing effective techniques and processes for critical reflection and the direct impact that the application of these skills and knowledge can have on practice and wider education outcomes. As a teacher-led researcher and researcher practitioner, the author has always valued evidence-based practice and strongly held the view that this can raise the status of the profession and awareness of the critical work of educators. However, it has been difficult to continually see government driven industry mandates and expectations applied with little regard to the resources and guidance teachers may require or the impact that this will have on teachers work and initial teacher education courses. (Clandinin D. J., 1989)

DISCUSSION

Some Ethical Considerations: Ethics is defined by Simons in Pring as "the search for rules of conduct that enable us to operate defensibly in the political contexts in which we have to conduct educational research" Chilisa highlights research ethics and legitimacy of knowledge saying "ethical issues in research include codes of conduct that are concerned with protection of the researched from physical, mental, and/or psychological harm. The codes of conduct to protect the researched include ensuring anonymity of the researched and confidentiality of the responses." Ethically speaking, there are two different positions that most researchers follow; the utilitarian and the deontological. The first study follows the deontological position as it sketches two case studies only and its results can't be generalized on each and every case study. On the other hand, the second study follows a utilitarian position. Pring sums up the meaning of utilitarianism saying "the utilitarians attach supreme importance to the creation of happiness for the greatest number." It will deal only here with those ethical issues related to the first study. First, the researchers of the present study didn't show that they have got the informed consent of the participants. Second, both researchers exploited their positions as university lecturers and didn't get permission of the university administration to conduct the research. Another unethical issue is publishing parts of the students' words without their permission. In addition, participants were not fully acquainted with the secondary use of data. What added insult to injury is that one of the researchers is conducting the research herself which is called the researcher's confirmation bias. Finally, the interpretations of the interview were not accurate as the transcribed audiotapes were not submitted back to the students to reflect, modify or change them. On the other hand, the second study has it ethical issues. First of all, the researchers of the present study manipulated the participants concerning their aims by using their authority as associate professors. (Clark C. & Yinger R., 1979)

They used the participants involuntarily to conduct their experiment. Second, the researchers did not get the consent of both the students and the university administration for conducting the study. Fourth, the interpretations of the interview conducted with the students were not accurate as the drafts of the transcribed interview were not submitted back to the students to modify, or change them. Moreover, participants of the present study were not knowledgeable about the purpose of the study and those who will have access to the data at present or at the future. Finally, one of the researchers of the present study is conducting the experiment himself and this is what is unethically called researcher's confirmation bias. From my own point of view, there are some unethical issues in the second study. First, the researchers should have enclosed the translated version of the Interview and the Listening Comprehension Test in English as appendices to reveal their honesty and accuracy in conducting research. Second, the researchers showed no proof the test validity. Finally, there is a reference to gender issue, 'she', 'he' and 'her' in the study report. This gender reference might be ethically problematic in terms of participants' confidentiality. (Corey S. M., 1952)

Qualitative Approach: This action research study was approached from a transformative worldview. I believe using a transformative lens for this research aided in understanding the need for change and the implications and consequences that come with change. Action research is a systematic inquiry conducted by teachers, administrators, or other invested party into a specific, actionable problem of practice; this approach worked best with the topic to examine the current learning log tool and develop an updated tool based on the results of the study. The issue, based on anecdotal evidence, is that teachers are not taking ownership of these logs and are completing them for compliance rather than to actualize professional growth. The district evaluation documents espouse the ideal of creating a teacher driven evaluation system that provides autonomy in goals and outcome; however, the autonomy, over time, has become the compliance of accomplishing a required task. Ideally the data collected through this process will enhance professional growth and improve teacher self-efficacy. Teacher evaluation meetings are conducted, therefore, it was necessary for each of the action research cycles to be completed as close to that deadline as possible. This study best lent itself to action research as the purpose was to design a more functional tool, based on the results of the study that uphold the intent of the learning log and maintain the fidelity of the evaluation system. Using a convenience sample was necessary as the researcher is a member of this organization and familiar with the site, policies, and procedures. Being a working employee in the district and school provided greater access to information and participants. Also, the unique nature of the evaluation system in the district made it an appropriate site for such a study. Participants. The initial fact finding stage of the research included two small focus groups of teachers from varying grade levels and disciplines within the school. Each focus group included four to five certified teachers currently working in the building. Participants were currently evaluated, certified teachers who have participated in a minimum of two evaluation cycles within the district. The transcripts from both groups were coded for themes that would then be used by the action committee. From the initial focus groups, an action committee of five teachers, including the researcher, were recruited to review the data and create a new reflection tool to supplement the learning log that is currently in place. Participants all volunteered, which provided a convenience sample to conduct the research. (Corey S. M., 1953)

The action committee was made up of three male and two female teachers (one being the researcher). There was one teacher who participated in the focus group and the first action committee meeting, but was unable to complete the study due to the birth of his child. During the initial phase, an interview was conducted with the building principal (primary evaluator) to gain a supervisory understanding of the use and purpose of the learning logs and how they are currently used and how they can be more effective in assisting with the evaluation process. To protect the integrity of the focus groups, administrators were not included. This information helped to understand the participant's baseline measure of self-efficacy. Data analysis took place in all stages of the research cycle. In the initial focus groups and principal interview, the discussions were transcribed by an external company and coded using in vivo coding, then further grouped into themes. This data was then used to inform the action committee as they determined interventions and revisions for the learning log. After each meeting of the Action Committee recordings were transcribed and coded using a set of in vivo codes from the conceptual framework then recoded using axial coding to determine themes and connections to the conceptual framework. This allowed for adjustments and discussion along the way regarding the themes that arose throughout the process as well as the identification of themes across the focus groups and action committee. Each of the codes were listed out in a spread sheet, then quantified by how many times they occurred throughout the first, second, and third committee meeting. For this reason, access and procedure were combined to identify the need for an improved procedure. Codes were not counted multiple times if they were a continuation of a participants comments or if they were stated multiple times within a participants marks. Potential next steps would be a larger implementation of the entire staff and over time throughout the district. The recommendations being made based on the data collected must be viable and accessible to all members of the staff, regardless of their content or position; for this reason, it was necessary to not only have a multi-disciplinary group serve on the committee, but also to test out the recommendations to ensure their compatibility in a wide range of classrooms. (Dewey J., 1938)

Trustworthiness: The use of multiple small focus groups to guide the work of the action committee provided an opportunity to consider issues that affect a larger population and be a basis for the newly designed tool. Using a collaborative process to revise and edit also allowed for multiple perspectives to be incorporated into the work. Similarly, recruiting group members from various disciplines and experiences established a diverse group with multiple perspectives to draw from. Member checking was attempted; however, only one participant responded with comments about the findings. This participant agreed with the findings of the committee's work as well as the recommendations found, Due to the time constrictions and end of the school year, member checking did not catch all participants before they left for the summer, resulting in the lack of responses. Experimental action research called for a controlled study of the relative effectiveness of various techniques in nearly identical social situations. Of all the varieties of action research, the experimental had the greatest potential for the advancement of scientific knowledge. Under favourable circumstances it could definitively test specific hypotheses. It was, however, the most difficult form of action research to carry out successfully.

Analysis: it is initially important to differentiate the terms 'reflection' and 'reflective practice'. Thus, reflection is associated with thinking and is judged to involve the cognitive processes of both 'problem finding' and 'problemsolving', concepts which continue to fascinate in cognitive psychology. This model of the 'epistemology of practice' was timely and well received within teacher education and research. According to Schön reflection-in-action acknowledges the tacit processes of thinking which accompany doing, and which constantly interact with and modify ongoing practice in such a way that learning takes place. Much of this may remain unconscious, tacit and unverbalised, though Loughran suggests that, in meeting unanticipated problem situations, reflection-in-action comprises reframing the problem and improvising on the spot so that the experience will be viewed differently. Reflection-on-action, on the other hand, is viewed as teachers' thoughtful consideration and retrospective analysis of their performance in order to gain knowledge from experience. Russell & Munby describe it succinctly as the 'systematic and deliberate thinking back over one's actions'. These two processes together, in Schön's terms, form the core professional artistry of the reflective practitioner. Subsequent research has been focused on determining that reflective practice exists, and identifying enabling and disenabling conditions which affect its use, and the means by which it might be fostered within all levels of the teaching. Schön's notion of reflective practice was a reaction against an instrumental notion of teaching where the teacher is a technician implementing others' knowledge in practice However, it and many other texts on reflection present an essentially rational, problem solving means of learning rooted as it is in cognitive psychology, with its emphasis on critical thinking. Johnston & Badley defined reflective practice as the 'acquisition of a critical stance or attitude towards one's own practice and that of one's peers'. The essential concern is the explication of meanings and understandings of teachers as they engage in actual practice across diverse teaching contexts. Critical theorists have extended Schön's categories adding 'reflection about action' as a means of ensuring that teachers reflect on the social, economic and political purposes and conditions of teaching and learning, as well as the school and classroom contexts. These notions continue to form the basis of much of the thinking about the nature of reflection and the development of effective reflective practitioners in the teaching profession. Reflective practice in these terms entails making conscious and explicit the dynamic interplay between thinking and action, what Shulman terms 'the wisdom of teaching', so that teachers may become thoughtful and learn from their work in the light of purposes and principles, which are 'moral'. Dewey considered reflection in practice as having a moral base, where professional actions would be treated as experimental, and the individual would reflect both on their actions and their consequences. The relationship of effectiveness to moral purposes of teaching remains a key area in the international debate on the nature of teachers' professionalism. Thus, while a reflective practitioner may be concerned to improve practice and to develop additional competence, what defines the effective reflective practitioner is more a set of attitudes towards practice based upon broader understandings of self, society and moral purposes than those which seek simply to increase efficiency in relation to 'delivery' and narrowly conceived achievement targets. (Freeman and J. Richards, 2020)

The declaration of the growing crisis in the education profession was sparked by an increased questioning of professional authority and infallibility. The shift to manage professional practice through measurable systems of accountability has been responded to with an acknowledgement that there is an increasing need for reflective practice and given the complex and multidimensional nature of education that this could be a more effective and authentic way to demonstrate evidence of practice. While it could be argued that processes involved in reflection could be regarded as antithetical to the more quantifiable and technocratic systems of managerialism. It is suggested that critical reflective practice can be seen as part of the same imperative, making professional practice more accountable through a systematic and ongoing scrutiny of the principles upon which it is based. Schön's 'epistemology of practice' model was discussed as the reflective practice theories were found to support why reflection is central to the professional work of teachers and educators. Through a systematic process of reflection teachers are able to develop an awareness of their impact in the classroom and document moments of illumination and action to demonstrate changed and improved practice. An evidence-base which supports notions of continuous professional development through systematic reflection can be combined with other forms of data gathering to broaden and deepen a teacher's portfolio of evidence. The role of critical reflection and its relationship to action research was established through Stenhouse's notion of the teacher-as-researcher. Hendrick's classroom and participatory action research approaches have been identified in particular, as a practical tool for critical reflection and a way for teachers to approach problem resolutions about their practice and educational issues. The complexities of action research have been deconstructed and consider the benefits, challenges and ethical issues. It has been established through a number of research projects and feedback from teachers that action research resonates well with the overall reflective requirements of impactful teacher practice and ensures credibility and reliability when questioning pedagogy. Evidence-based practice can have positive implications for teachers and educators in raising the status of the profession and in developing a voice of authority about complex educational issues. (Lange D. L., 1990)

Action research is highly recommended to teachers at all career stages and as part of an ongoing professional learning and development program to be included and embedded as part of teachers work. Research provides opportunities to challenges one's ideas and the theories that underpin these, to develop new knowledge, question understandings and contribute to a wider perspective on educational issues pertaining not only to individual classes, but whole school communities and the profession. In the contemporary educational climate, the action research model and extended collegial peer observation model proposed can facilitate critical reflection in a systematic, rigorous and authentic way, providing a practical and accessible approach for educators to assess, measure and provide evidence of practice. Exemplar key improvements in teaching and learning from the case studies mentioned using the action research model include: Increased teacher efficacy, knowledge and understanding of the high impact teaching strategies leading to improved change in pedagogy; collegial peer observation contributing to team building and a collaborative school culture; increased efficacy in the skills and knowledge to build evidence-based practice; enhanced student learning environment; and increased development of student metacognitive thinking processes. The opportunities for teachers and educators to consider the impact of their work in a systematic and critically reflective way and explore more deeply what they do, how and why is imperative. However, it is only with the appropriate support and resources that teachers will be able to engage appropriately with this work. If this is possible, it will have positive implications for future teaching practice and the development of the profession as teacher-led research becomes more accessible.

CONCLUSION

There are three reasons why reflective practice is increasingly being recognised as being essential to good teaching and having 'a central role in the learning life' of the effective teacher. The first concerns the nature of teaching. The assumption is that since teaching and learning are complex processes and since there is not necessarily one right approach, deliberating among competing versions of good teaching and recasting past understandings and current practices are likely to lead to improvement. Without a capacity to evaluate assumptions, teachers will be more inclined to remain 'prisoners of (their) programs' and, as a result, their professional effectiveness in circumstances which inevitably change over time will be decreased. The second is that engaging in reflective practice is a means of helping individuals towards greater self-knowledge and self-challenge – 'a useful way of achieving personal development' through, for example, an analysis of the personal values and theories that underlie teaching. Finally (and here the link with action research becomes sharper), reflective practice is considered to be central to the growth of teachers as inquirers who engage in collaborative research with others from inside and outside the school in generating knowledge of practice rather than finding themselves as objects whose role is to implement existing theory in practice. However, research continues to reveal that there is a continuum of reflective practice that exists within the profession. Teachers may reflect in differing ways at different times. It is important, therefore, to recognize the impact of teachers' positions in their career and life cycle, and the effects of the organizational and cultural contexts in which they work if opportunities for their professional growth to be maximized. Early evaluation of the impact of such approaches to teacher reflection has indicated that they not only provides the individual with opportunities for profound re-evaluation, but have also led frequently to powerful changes within the teachers' practices. While such claims clearly need further evaluation, documentation and validation, the current store of teacher journals and independent research theses, which are living testimony to this process, provide the basis on which the case for such a reflective journey may be built. Identifying reflection as the engine of action research processes, which are conceptualized more holistically, adds to the learning possibilities for all those involved, and the richness of what might otherwise be continuing adherence to restrictive action research paradigms and practices. Paradoxically, as policy makers in all countries exhort teachers to become lifelong learners in the ever more demanding and complex worlds of classrooms and schools, so the strictures which they place upon the use of time and the accompanying mechanisms for bureaucratic accountability increase. These often result in diminishing, rather than increasing opportunities for structured reflection through action research. They thus decrease teachers' capacities to raise standards of teaching, learning and achievement in circumstances where many children and young people are becoming disillusioned with schooling alongside raised expectations of teachers by parents and employers. (Lightbown P. and N. Spada. 2000)

The Mechanics of Action Research in Context Action Research: It provides a means to not only develop evidence-based practice, but also to formally or informally determine the impact of pedagogy on student learning, develop curriculum and program initiatives and respond to education policy and school reform. Challenging and questioning one's own practice is an expectation of the professional work of a teacher and an embedded part of the wider systemic culture of the school or education context. Many effective teachers will engage in a form of reflection as tool for self-assessment and evaluation. However, reflection in action research is a deliberate process and a key component of the methodology. It is characterised by a cyclical approach that is planned, systematic, iterative and critical, alternating between action and reflection, continuously refining methods and interpretations based on understandings developed in earlier cycles. While the focus of action research may be about one's own practice, it has the potential to significantly improve education outcomes holistically by incorporating change in a collaborative capacity with the common goal of improving practice. Educators have direct access to the research findings in order to implement immediate change as a result of engaging in systematic critical reflection. The research is situated in a localized context and focused on an identified local issue, acknowledging the importance of unique educational contexts, which may not always be resolved by findings that are generalisable.

Reflection: Reflective thinking, as defined by Dewey is "active, persistent, and careful consideration of any belief or supposed form of knowledge in light of the grounds that support it and the further consequences to which it leads". Here Dewey indicates that reflection is essential to questioning beliefs and accepting the consequences of that questioning. Brookfield identified those consequences, evolution and refinement of practice, as an essential component of professional development. Thus connecting to Dewey's belief that the process should result in growth of one's own practice. Mortari identifies reflection as playing a significant role in professional life as well as being necessary for aquiring professional expertise. The use of autobiographical writing is seen as a positive means of improving reflective learning. There is little in the literature regarding the use of written reflections as a means of building self-efficacy. Clarifying reflections brought greater understanding to the topic, enhancing understanding, by providing evidence teachers were able to connect their personal thinking to the group with evidence from experiences (personal or through fieldwork). Challenging occurs when a member of the group pushes or questions thinking in order to dig deeper into the reflection, and different thinking occurs when challenging results in a new perspective or understanding of the topic. In this study they found that the majority of responses were clarifying and providing evidence, relatively low level reflection skills, possibly as a result of new material or unfamiliarity with classmates. However, throughout the course the depth of reflection grew as teachers became more familiar with the content. This indicates that refection can be improved or enhanced with familiarity and knowledge.

Positivism Action Research Approach: It does not reject positivism altogether. It may make use of positivist research techniques (i.e., quasi-experimental design). On the other hand, positivism's major criticism for action research is its lack of objectivity. According to positivism, the researcher can and should separate himself from what is being researched. Thus, data collection and analyses processes would not be affected by a researcher's prejudices, acceptances or tendencies. However, in action research practitioners are also researchers. More to the point, positivism claims that insiders are also part of the phenomenon being investigated; therefore, they are too far from being objective. Therefore, it is possible for the action researcher to show tendencies that would have an impact on results. Third, a dual role will put too much burden on teachers that they may not overcome. "Doing research alone can be challenging due to lack of enough time, small numbers of subjects, too many variables to control for, difficulty randomizing subjects, inappropriate instruments or lack of support from administrators". Finally, a small number of participants is a threat to external validity. (Arnold, J., 2012)

Future Scope and Perspectives: Understanding of action research on the above-mentioned issues differ from the positivist paradigm's characteristics. According to action research, objectivity is not necessary in conducting a valid study since there are other ways to establish validity beyond objectivity. The action researcher should be an open-minded person who embraces democracy and applies it in his life. Therefore, relations between the action researcher and the teacher and students, parents, administrators, and other parties in education should be mutually honest, open, and equal. The action researcher attempts to prepare a clear and detailed description of how and why data were collected and share this information with participants. This kind of relationship will facilitate the data collection process for the action researcher and enable him-her to gain in-depth data. Also, action research views being an insider as an advantage rather than disadvantage because the insider is native to the setting and thus better able to establish connections needed to conduct a study (taking permission, arranging necessary equipment, reaching students, parents etc.); insiders have insights from the lived experiences. On the other hand, positivist education research is usually conducted by scientists who are outsiders and have never been an actual teacher. But the action researcher has to be aware that his own perspective is not intrinsically better than the perspective of others. The investigator has to be self-critical to avert any preventable impacts from his social, cultural, political, and educative background. The researcher should not act like an authority figure who assumes he knows better than the practitioners and tells them what to do and how to do it. S/he should be self-reflective in every step of the inquiry. This will enable him/her to develop meta-awareness about the process. Thus, the action researcher can continuously evaluate the process and himself. The researcher and teacher roles are constantly under review so that the researcher should be reflective on his own applications, ideas and acceptances. Action research has recruited the triangulation process to support validity. The teacher action researcher gathers data from different sources such as students, parents, administration and the community and in doing so embraces different perspectives, unlike the fundamentals of positivism. The data gathered from different sources compensate for each other's errors. Thus, action research seeks to attain a more complete picture for the researcher of the phenomena that eventually lead to better intervention. Although the action researcher has low control over the environment, this should not be a disadvantage since the teachers themselves are a variable in this environment. Their interaction with students and with class in general is dynamic, complex and changing; therefore, it cannot be stabilized and controled as with laboratory experiments. Therefore, action research is conducted in a natural teaching-learning environment rather than the artificial environments prepared for experiments.

Action Research, Post-Positivism and Interpretivism: Many characteristics of action research overlap with post-positivist and interpretivist paradigms, such as being against the superior status of research over participants and claiming to seek the improvement of social conditions by creating a more democratic, free and humanitarian social environment. Action research methods value devices such as the interview and observations to gather data, just as both the post-positivist and interpretivist approaches do. Also, triangulation techniques are widely used in action research as encouraged by the post-positivist perspective to provide a wider perspective and support validity. However, despite the aforementioned common grounds, only in action research does the researcher begin to conduct research as a practitioner and participant. Therefore, the interest of the practitioner is in the initiation of the research process. In post-positivist and interpretivist perspectives, the researcher can be an outsider (at least at the beginning) who seeks to learn and understand. As stated earlier, action research is educative—its main purpose is to improve educational practice. Therefore, results should have practical value (also known as workability). Action research is long term yet does not wait until its completion before applying an intervention. In doing otherwise, they would use their workability. On the other hand, post-positivist applications, such as anthropological studies, do not always seek practicality. Also, there can be very long intervals such as five-ten years between the beginning and end of studies. Everything may change before a study is finished; in this case an intervention with validity at the beginning may not be practical anymore. Since the classroom is a very dynamic environment, teachers need practical solutions almost every day. Therefore, the workability and flexibility of action research can be very beneficial to teachers who seek to improve their teaching. Furthermore, unlike interpretivist approaches and applications, action research makes use of standardized research techniques necessary to the purpose of a particular study. Another point between interprevitism and action research is the nature of facts. According to interprevists there are no extrinsic, objective facts—every single person has their own conceptualizations. These conceptualizations are molded by conditions and individual interactions with those conditions. In every situation the conceptualization of fact is subject to change. Therefore, the results and conclusions of interpretivist research are valid only for a certain time and people. An individual who has not been involved cannot benefit from results and conclusions of interpretivist research. However, action research claims that social facts are not completely knowable because social processes are dynamic and changing. However in this dynamic structure people reach consensus on some key issues for a relatively long time. Consensus and common values and applications enables an orderly social life. Action research is not as individualistic as the interpretivist paradigm. Also, people who are not involved in the research process can benefit from the results and conclusions of action research. (Atkin J. M., 1993).

REFERENCES

Adelman, C. (1993). Kurt Lewin and the Origins of Action Research. *Educational Action Research, 1*(1), 7–24. DOI: 10.1080/0965079930010102

Ali, A. D. (2020). Implementing Action Research in EFL/ESL Classrooms: A Systematic Review of Literature 2010-2019. *Systemic Practice and Action Research, 33*(3), 341–362. DOI: 10.1007/s11213-020-09523-y

Altrichter, H., & Posch, P. (1989). Does the 'Grounded Theory' Approach Offer a Guiding Paradigm for Teacher Research? *Cambridge Journal of Education, 19*(1), 21–32. DOI: 10.1080/0305764890190104

Argyris, C., & Schön, D. A. (1976). *Theory in Practice: increasing professional effectiveness*. Jossey-Bass.

Arlin, P. (1990). Wisdom: the art of problem-finding. In Sternberg, R. J. (Ed.), *Wisdom: its nature, origins and development*. Cambridge University Press. DOI: 10.1017/CBO9781139173704.012

Arnold, J., Edwards, T., Hooley, N., & Williams, J. (2012). Conceptualising teacher education and research as 'critical praxis'. *Critical Studies in Education, 53*(3), 281–295. DOI: 10.1080/17508487.2012.703140

Atkin, J. M. (1993). Developments in the Philosophy/Sociology of Science and Action Research. *Educational Action Research, 1*(1), 187–188. DOI: 10.1080/0965079930010111

Atkin, J. M. (1993). Developments in the Philosophy/Sociology of Science and Action Research. *Educational Action Research, 1*(1), 187–188. DOI: 10.1080/0965079930010111

Carr, W. (1993). Whatever Happened to Action Research? *Educational Action Research, 2*(3), 427–436. DOI: 10.1080/0965079940020310

Carr, W., & Kemmis, S. (1986). *Becoming Critical: education, knowledge and action research*. Falmer Press.

Clandinin, D. J. (1989). Developing Rhythm in Teaching: The narrative study of a beginning teacher's personal practical knowledge of classrooms. *Curriculum Inquiry, 19*(2), 121–141. DOI: 10.1080/03626784.1989.11075320

ADDITIONAL READINGS

Clark, C., & Yinger, R. (1979). Three Studies of Teachers Planning, Research Series 55. East Lansing: Michigan State University. Cole, A. (1997) Impediments to Reflective Practice. *Teachers and Teaching, 3*, 7–27.

Corey, S. M. (1952). Action Research by Teachers and the Population Sampling Problem. *Journal of Educational Psychology, XLIII*(October), 331–338. DOI: 10.1037/h0052459

Corey, S. M. (1953). *Action Research to Improve School Practices*. Teachers College Press.

Dewey, J. (1938). *Experience and education*. Collier Books. Gilpin.

Freeman and J. Richards. (2020). (pp. 79–96). Cambridge University Press.

Lange, D. L. (1990). A blueprint for a teacher education program. In Richards, J., & Nunan, D. (Eds.), *Second language teacher education* (pp. 245–268). Cambridge University Press.

Lewin, K.(1920) Die Sozialisierung des Taylorsystems, Praktischer Sozialismus, No.4.

Lightbown, P., & Spada, N. (2000). *How languages are learned*. Oxford University Press.

MacDonald, B. (1978). *The Experience of Innovation*. CARE,University of East Anglia.

Madaus, G. F. (1989). *Educational Evaluation: classic works of Ralph W. Tyler*. Kluwer.

Marrow, A. J. (1969). *The Practical Theorist the life and work of Kurt Lewin*. Basic Books.

KEY TERMS AND DEFINITIONS

Action Research: It is an inquiry which is carried out in order to understand, to evaluate and then to change, in order to improve educational practice

Transformative Research: According to Creswell, "contains an action plan that may change the lives of the participants, the institutions in which individuals work or live, and the researcher's life".

Reflection: It is considered as a process or activity that is central to developing practices.

Utilitarianism: Tarrant defines as "an ethical theory which has shown itself to be both flexible and comprehensive". It provided a number of ethics violation in research work.

Positivism: These dimensions of action research harm objectivity and therefore the validity and generalizability of the study.

Steps in Action Research: Koshy describes action research as: a constructive inquiry, during which the researcher constructs his or her knowledge of specific issues through planning, acting, evaluating, refining and learning from the experience.

Lewin's Model: It amalgamating research and action to enhance understanding and generate change, highlights that the introduction of action into the scientific model "by no means implies that the research needed is in any respect less scientific or 'lower' than would be required for pure science".

Chapter 10
Integrating Indigenous Knowledge Systems in Higher Education Institutions Using Progressive Environmental Action Research Activities

Tsebo Kgoto Matsekoleng
https://orcid.org/0000-0002-7577-3417
University of South Africa, South Africa

Tomé Awshar Mapotse
https://orcid.org/0000-0002-7234-2785
University of South Africa, South Africa

ABSTRACT

The chapter fills the gap in the literature void within Higher Learning Institutions by decolonising the curriculum through utilisation of progressive environmental action research (PEAR) activities to integrate Indigenous knowledge systems (IKS). This qualitative conceptual chapter employs perennialism theory to ground the chapter and the chapter is undergirded by a participatory paradigm. Because of the persistence of environmental issues, the purpose of this chapter is to share or highlight the applicability of PEAR activities in institutions of higher learning studies and close the gap on the death of literature. As such, this chapter will address How Indigenous Knowledge Systems can be integrated into Higher Learning Institutions' studies using progressive environmental action research activities.

1. INTRODUCTION

The inclusion of the Indigenous Knowledge System (IKS) in curriculum policy emerged and developed within an opposing of African Indigenous and Western knowledge systems (O'Donoghue, van Staden, Bhurekeni, Snow-Macleod & Ndlamlenze, 2024). Foreign culture, particularly Western culture, heavily influenced African development (Mnguni & Mapotse, 2023) in all spheres and walks of life,

DOI: 10.4018/979-8-3693-3542-0.ch010

and IKS was not spared. Indigenous knowledge systems are crucial in our daily lives, and schools and universities could serve as a foundation for the preservation of Indigenous practices in our society. As a result, curriculum decolonisation is critical to integrate Indigenous knowledge systems/practices through progressive environmental action research (PEAR) activities to make Higher Education relevant to African students conducting IKS research. There is a need to gain knowledge regarding the African continent, its histories, cultures, and civilisations, and textbooks must reflect IKS. South African universities must start asking deep questions that examine what decolonisation might look like in their teaching and how to achieve it (Du Plessis, 2021). Teaching and learning, as well as general operational and administrative practices that are Western-centric, alienate the African students from their immediate context (Hungwe & Mkhize, 2022). It is for this reason that Mbembe (2016) argues that:

> *Today the consensus is that part of what is wrong with our institutions of higher learning is that they are 'Westernized'. They are 'Westernized' in the sense that they are local instantiations of a dominant academic model based on a Eurocentric epistemic canon. A Eurocentric canon is a canon that attributes truth only to the Western way of knowledge production. It is a canon that disregards other epistemic traditions. It is a canon that tries to portray colonialism as a normal form of social relations between human beings rather than a system of exploitation and oppression (Mbembe, 2016: 32).*

It is against this background that Mapara (2022) argues that the indigenisation of the higher education curriculum is only possible if there is a willingness and a deliberate epistemic and cognitive shift on the part of the curriculum planners as well as implementers. It is, therefore important for universities to decolonise their curriculum structures. For instance, the University of KwaZulu-Natal introduced IsiZulu as a compulsory module for first-year students. Thus, the decolonisation of the curriculum should start by first acknowledging the richness of IKS and the knowledge that Indigenous student brings to the lecture hall. The implementation of PEAR activities, for instance, could serve as a foundation for the integration of IKS in the Higher Learning Institutions.

In accordance, to date, indigenous people across the globe have managed to live and survive using their Indigenous practices of treating various illnesses such as asthma and dysentery, where treatment mixtures are produced through indigenous technologies. Indigenous technologies such as Tshilo le Lwala (mill and stone) are used to extract substances from plants (Matsekoleng & Mapotse, 2023). Indigenous practices, in their nature, are hands-on and mind-on activities learned from generation to generation through observation and imitation, the Higher Learning Institutions could be supplemented by such freely available knowledge/practices. Kimbell (2008), for example, observed the construction of a dhow in Malawi. He was astounded by the manual selection, shaping, and fixing of the raw materials without the need for design or drawing, as well as the understanding of the timber's strength and how it was shaped and fixed.

Here children learned about indigenous architecture through observation and imitation. It is for this reason that IKS/practices should be infused into Higher Learning Institutions for students to relate to their everyday lives and further avail job opportunities that could address unemployment that persists to increase yearly. The integration of IKS into students' learning activities makes learning relevant to students' cultural knowledge and worldviews (Matsekoleng, Mapotse & Gumbo, 2022). On the foregoing argument, the integration of IKS and PEAR activities in Higher Learning Institutions could be realisable owing to the practicality of activities, including the utilisation of grass to make brooms, which thus minimises littering and/or pollution. In this case, PEAR refers to participatory, hands-on activities that can be used in Higher Learning Institutions to integrate IKS to address environmental issues, etc.

PEAR activities could fill gaps because higher education institutions have been unable to confront the severe social and environmental consequences of climate change (Lin, Stoltz, Aruch & Rappeport, 2021).

The discussion in this qualitative conceptual chapter covers the application of Perennial theory, the value of PEAR activities, the synergy of Perennial theory and PEAR activities, the integration of IKS, the relationship between IKS and Education for Sustainable development/Environmental Education (EE), and implications for integrating IKS in Higher Learning Institutions.

2. THE APPLICATION OF PERENNIALISM THEORY

Indigenous knowledge systems, PEAR, and Perennialism theory have a common understanding whereby students' contextual learning of knowledge is meant to be a lifestyle. The application of perennialism theory in Higher Learning Institutions by integrating IKS and PEAR is ideal for the relevancy of the curriculum and sustainability of Indigenous values in the global world. The perennial theory is founded on the belief that teachers should teach the things that are of everlasting relevance to all people everywhere (Nkhata et al., 2019). The biggest proponents of perennialism in education were Robert Hutchins (1954) and Mortimer Adler since this deals with the passing of culture and tradition from one generation to the next; hence, Higher Learning Institutions should be incorporated. For Indigenous people, the role of perennialism is to ensure the continuity of cultural heritage (Kasulwe, 2014). Maira and Kauka (2018:113) point out that there are some content areas in learning (knowledge, values, and skills) that are not bound to time or place. The researchers argue that the IKS content, which exhibits the conception of timelessness, must be part of the curriculum for Higher Learning Institutions. To preserve an ethnic group status quo, which includes Indigenous games, some African Indigenous communities prefer to apply perennial education to their next generation.

The diverse Indigenous games, such as the Morabaraba game, could serve as an important and necessary tool to decolonise the University curriculum, as the game requires students to use trash as objects to play the game and thus teach students about littering and recycling. Courses such as environmental management could utilise Indigenous games as hands-on and mind-on games in which Botho/Ubuntu is embedded. Further games allow students to learn and relate the constructed knowledge with their university curriculum. Indigenous games incorporate educational and environmental relevance to students and the incorporation of Indigenous games in teaching and learning promises to revitalise them again and make learning more meaningful, especially for Indigenous students (Matsekoleng et al., 2022). Students, while playing games, learn through the experience of engaging in the game and construct new knowledge, including rules. The application of perennialism theory adds value to PEAR activities.

3. METHODOLOGY

The qualitative conceptual chapter presents selected peer-reviewed articles. Researchers used various sites comprising Google Scholar and Eric to search for articles narrowed down to 10 years between 2014 and 2024. To gather relatable sources, keywords were used to search the information. These included "Environmental Education", OR "Practical IKS activities", OR "Challenges", OR "Barriers" OR "Obstacles", OR "Indigenous knowledge systems", OR "Indigenous knowledge", OR "Decolonisation",

OR "Africanising the curriculum", OR "Indigenising the university curriculum" OR "Decolonising curriculum in higher education institutions".

4. THE VALUE OF PROGRESSIVE ENVIRONMENTAL ACTION RESEARCH ACTIVITIES

University offers a variety of programmes to students, and the absence of Indigenous practices that could teach students about the value of firewood ashes in society leaves a gap in their qualifications for having environmental and IKS knowledge. For instance, the ashes from firewood could serve as manure for agricultural activities (Matsekoleng & Mapotse, 2023). The ash is freely available without any cost and is litter-free. Universities could improve their curriculum by collaborating with people in the community. For example, indigenous households fertilise their plants with chicken manure, cow dung, and crop wastes to improve yields. Furthermore, indigenous farmers frequently use indigenously created pesticide mixes for pest control, which explains why climatic risks and calamities are avoided (Kugara et al., 2022). The application of such PEAR activities addresses environmental issues, as in this instance of using natural resources to fertilise soil. One such example is the Maasai community in Kenya collaborating with local environmental educators. Together, they developed a curriculum that includes Maasai traditional knowledge of land management and conservation. PEAR efforts contribute significantly to environmental sustainability (Alam, 2021). It encompasses values and practices passed down through generations intending to preserve the natural environment (Wardhani, Arisanty, Nugroho, & Utami, 2024).

A broom can be made from freely available natural materials, such as grass, to sweep a house and yard (Matsekoleng & Mapotse, 2023; Rugedhla, Fidzani, & Moalosi, 2023, Matsekoleng 2022). Another major resource is grass, which is not only a primary source of food for animals but is also used as a building material, particularly in the thatching of huts (Sheya, 2014). Grass allows students to express their creativity while also increasing their understanding of the environment. Furthermore, students could sell the broom to subsist and minimise unemployment by hiring other people. These exercises can help to improve the university curriculum. Innovations such as the Mitticool refrigerator, which utilises ancient clay cooling techniques, demonstrate the necessity of combining IKS with modern technology to create long-term solutions for rural people (Kumar, 2024).

University students come with diverse potentials, experiences, and mindsets. They have many life experiences and prior knowledge on which to build new meaning. Critical thinking, problem-solving approach, and analytical skills are assumed to be present in higher education graduates (Basu, 2018). PEAR activities could accommodate students' diversity because the activities are practical with the purpose of empowering and dealing with the problem at hand. The PEAR activities in Higher Learning Institutions could decolonise the university programme by ensuring that students learn about Indigenous practices, including Lesolo/Nkelekele. In the practice of Nkelekele, it is believed that if there is insufficient rain or a total absence of rain during the rainy season, it is caused by the rubbish in the environment (Chauke et al., 2021). The practice offers students other options that are environmentally friendly for the rain to fall without using artificial methods. This may also add to the Indigenous weather forecast of predicting the weather based on clouds or sunshine. Indigenous people forecast the weather by observing the plants, birds, amphibians, reptiles, and insects in their environment. When making decisions, indigenous weather forecasting indications are critical. These decisions range from choosing which crops to plant in a specific season to deciding whether or not to fix a roof (Sebata, Dhlamini, & Ndlovu, 2024).

The PEAR activities are aligned with David Kolb's experiential learning theory because the theory proposes the assumption that learning is a process and not an outcome, driven by experience in a more holistic and integrative manner. Furthermore, learning requires the student to resolve conflicts through dialect and requires the individual to interact with their environment, thus creating knowledge (Matriano, 2020). The creation of knowledge could lead to the understanding of environmental issues and propose solutions to eradicate them. Further, PEAR activities incorporate five experiential learning elements: 1) direct, personal involvement, 2) student initiation, 3) pervasiveness, 4) student evaluation, and 5) the essence is meaning, as Rogers (1969) asserted. Learning is a student-centred approach that takes students learning away from traditional learning. Kolb's model describes that learning happens when knowledge is created through the transformation of experience (Matriano, 2020). One can argue that PEAR activities with the application of Perennial theory are important based on the following principles outlined by Revathi (2020:86-87):

- Learning is an active process in which the student uses sensory input and constructs meaning from it.
- Individuals learn to learn as they learn, and the key component of learning is motivation.
- Physical actions and hands-on experiences are necessary for learning, especially for children.
- Learning involves language, which influences learning and language and learning are inextricably intertwined.
- Learning is a social activity, and learning is associated with one's connection with teachers, peers, parents as well as casual acquaintances.
- It is not possible to absorb new knowledge without having some structure developed from previous knowledge to build on.
- Learning is not instantaneous. It takes time to learn.

Based on the above principles, PEAR activities encourage students to think independently, work collaboratively, and develop language communication skills, among other skills. For example, indigenous games are played with a variety of waste products. It is argued that constructing toys made from urban solid waste, such as the jaguar game, shuttlecock, and spinning top, was a rewarding experience that promoted creativity, teamwork, and environmental consciousness (Evangelista et al., 2024). As a result, excluding PEAR activities, such as indigenous games, from students' curricula may lead to cultural alienation. The introduction of novel pedagogical methods, such as indigenous games and activities, has the potential to improve the efficacy of this training process (Evangelista et al., 2024).

5. SYNERGY OF PERENNIALISM THEORY AND PROGRESSIVE ACTION RESEARCH ACTIVITIES

Both philosophies require students to take an active role in learning and building on factual knowledge to improve investigation and critical thinking skills (Klein & Merritt, 1994). Perennialism theory and PEAR activities place the student at the centre of the learning experience with the anticipation of increasing action competence. For instance, in the crocheting of plastic mats, students learn about littering, ways to address the high unemployment rate, and types of shapes, designs, and processes, to name a few. In turn, students share knowledge of addressing environmental issues as well. It thus individualises and contextualises students' learning experiences, helps students develop processes, skills, and attitudes, and focuses on knowledge construction, not reproduction (Basu, 2018). PEAR activities are action-oriented and contextualised, which leads to construction and experiential learning as advocated by perennialism theory. Experiential learning (as coined by David Kolb in 1984) is making meaning from direct experience. It plays a vital role in facilitating the process of creating knowledge, sense-making, and knowledge transfer in teaching, training, and development (Matriano, 2020).

It is against this background to argue that no meaningful learning can take place without the existence of perennialism theory and PEAR activities in the actual learning/practice. Learning, be it in the lecture room or outside, should integrate both and incorporate IKS learning/practice. The PEAR and perennialism activities are shaped around experimentation, debates, excursions, and research projects where students engage their hands, hearts, and minds (Matsekoleng, 2022). Further, the activities promote healthy living as students, while conducting activities, do exercise. The instructor and student both share the responsibility of learning. Perennialism learning can be facilitated using case studies or projects that closely align with real-life experiences (Basu, 2018). In the process of perennialism learning and PEAR activities, learning becomes interactive, collaborative, and authentic. For students to practice new skills, they must interact with their peers (Klein & Merritt, 1994). Perennialism learning and PEAR activities are cooperative, collaborative, and problem-based learning through the application of action research cycles comprising plan, observe, act, and reflect.

6. INTEGRATING INDIGENOUS KNOWLEDGE SYSTEMS

Indigenous weather forecasting, farming, food preservation (e.g., *Mageu*), and legal (e.g., *Lekgotla*), to name a few, are some of the IKS that the university can include in its syllabus for students to address their problems in their immediate environment. Incorporating Indigenous Knowledge Systems into curricula is a crucial step in advancing cultural diversity, educational significance, and student empowerment (Sumaiya, 2024). To accomplish Sustainable Development Goal 4, which calls for "inclusive and equitable quality education and promote lifelong learning opportunities for all," educational institutions and instructors must adjust to the diversity of their student cohort, especially for minority students who are likely to come from cultural backgrounds that differ from both the institution's and other major ethnic groups' (Fonua, 2021).

Over the last 35 years, there has been a groundswell in Australian school education to create a world-class Australian Curriculum (Michie, Hogue, & Rioux, 2023). The Queensland University of Technology Faculty of Indigenous Knowledge and Culture was established in 2024, with student enrolment beginning in 2025, to recognise and encourage Indigenous Australian brilliance (Queensland University of Technology, 2024). Efforts are underway around the world to integrate IKS into higher education institutions. New Zealand was the first to create a curriculum in their indigenous language (Te reo Māori) (Michie et al., 2023). These attempts are undertaken to make the curriculum relevant to Indigenous students while also preserving Indigenous practices.

Parent's rearing/philosophy is mutually coordinated through Ubuntu/Botho principles enshrined. Therefore, there could be a smooth integration of IKS in Higher Learning Institutions. Learning is experiential in many Indigenous societies, where repeated and continuous exposure and response to environmental situations generate the experience of how to manage them and permit responding with learning (Ford et al., 2020). The use of a donkey cart, for instance, explains the integration of IKS in the Higher Learning Institutions. The cart is made out of waste materials, and ropes are used to control domestic animals on the cart. This offers the opportunity for students to learn about the IKS of using such systems to transport goods (Matsekoleng, 2022). The use of donkey carts minimises air pollution as opposed to using a car. Further, the donkey cart exposes students to political and socio-economic issues, including the cost of petrol. Experiential and authentic learning is a result of student-centred, constructivist-based, and project-based methods or activities used in teaching (Matriano, 2020).

Students from Indigenous backgrounds in higher education require an educational environment that recognises and supports their cultural heritage (Sumaiya, 2024), and incorporating the Gadaa system into the curriculum broadens and diversifies the student's knowledge base (Jima, 2022). Bule Hora University introduced the Gadaa method into its curriculum in 2019, and Metu University followed suit (Jima, 2022). The Gadaa system is an Indigenous system that oversees Oromo social, economic, religious, and cultural matters (Debisa, 2022). The incorporation of IKS techniques into university curricula reveals gaps between Western and indigenous knowledge. Furthermore, it suggests that elements of knowledge can complement one another to achieve meaningful learning.

Supplementing the university curriculum with IKS is important because Indigenous people always interact with nature in their everyday activities. Daily, Indigenous people use plants for different purposes. For instance, trees, plant fluids, leaves/branches, stems, and roots offer Indigenous people shelter, food, medication, shoes, clothes, protection against strong winds, shade, roofing materials, weapons (knobkerrie), fences, instrumental music, fruits, and oxygen (Matsekoleng & Mapotse, 2023). This avails an opportunity for universities to sustain IKS and link it with the student's daily activities. The given scenario excludes the utilisation of Western materials, which proves the point that Indigenous students could benefit greatly from the integration of IKS into the curriculum. The downplaying of IKS does not help the Indigenous student but misses out on the innovative ideas of addressing problems that societies are facing, including diseases caused by environmental issues.

In the areas of Agriculture, Environmental Engineering, and Food Science Technology, there is a lot that the curriculum can benefit from when it comes to the incorporation of IKS. One widespread practice that the Indigenous community in Zimbabwe follows is to collect organic waste that is scattered around the homestead to create composites. This organic waste includes cow dung that is not dropped in cattle pens, as well as dead leaves of certain trees that quickly decompose. The manure that comes from such composites is used in the rainy season to feed crops (Mapara, 2022). The education system can help students connect with their cultural roots and develop the skills they need to navigate today's challenges by implementing Indigenous Knowledge Systems (Sumaiya, 2024).

6.1 Barriers to the Implementation of Indigenous Knowledge Systems in Higher Institutions

Integrating Indigenous Knowledge Systems into education is a difficult effort that meets multiple fundamental challenges. The challenges arise from historical, cultural, and epistemological gaps between indigenous knowledge and Western educational approaches (Sumaiya, 2024).

Banes and Cruz (2021) highlighted the difficulties that instructors encounter while integrating IKS into curricula, which hinder the comprehensive integration of IKS in the tertiary-level curriculum. The lack of a specific university policy/memorandum and monitoring system for incorporating IKS into the curriculum, teaching techniques, and pedagogies impedes the long-term integration of IKS in many topics. Other problems, such as a lack of reference materials/resources on IKS, as well as a lack of expertise, exposure, and respect for IKS, undermine the faculty's ability to incorporate IKS into instruction. Furthermore, Labonne (2024) mentions the following barriers: budgetary restrictions, staff proficiency, student diversity, university autonomy, the Indigenous/non-Indigenous divide, and relevance. Implementing PEAR activities can help to overcome some of the challenges, such as a lack of resources and a limited budget. The PEAR activities utilise natural resources and waste materials to promote effective learning.

7. RELATIONSHIP BETWEEN INDIGENOUS KNOWLEDGE SYSTEMS AND ENVIRONMENTAL EDUCATION

There is a strong link between IKS and Environmental Education in our day-to-day activities. For example, Indigenous peoples would collect plants and roots without causing environmental damage (Seroto, 2020). Some of the collected plants and roots (i.e., Umhlonyane/Lengana) are used to make medications to cure the flu, etc. Tree species are conserved for their medicinal value. Thus, this brings protection to the community and conserves floral species (Kugara et al., 2022). It is against this background that covertly, the relationship between IKS and Environmental Education exists, and it is evident that Indigenous people have made an immense contribution to sustaining the biophysical environment (Kanene, 2016). Accordingly, knowledge of Indigenous people offers information and perspective that complement traditional scientific and environmental observations (Dube & Munsaka, 2018).

The link is further evident in ornaments (e.g., wooden vases, toys, etc.) that Indigenous people make. Indigenous people are actively involved in the production of Indigenous products for their local communities, such as sculpted pieces, ceramics, textiles (weaving), dwellings, medicines, and food (Maluleke, 2019). These Indigenous products are transformed and passed down from generation to generation (Seroto, 2020). Learning about IKS is crucial because it allows students to study their heritage and, hence, keep their cultural identity (Sheya, 2014). For a longer period, the Indigenous practices have embedded environmental management in their activities, and it is evident in the practices of traditional healers that what comes from the soil eventually goes back to the soil in the form of compost. For instance, a broom made from grass after ageing can be used as charcoal or will be consumed by soil. Therefore, there is a relationship between IKS and Environmental Education. As such, integrating IKS in Higher Learning Institutions has implications.

8. IMPLICATIONS FOR INTEGRATING INDIGENOUS KNOWLEDGE SYSTEMS IN HIGHER LEARNING INSTITUTIONS

Indigenous knowledge systems are spiritual/cultural, and Western knowledge [Environmental Education] is academically based as such, in the learning process, students' hearts, brains, and body is engaged, so integration is possible (Matsekoleng, 2022). Seroto (2011: 37) claims that IKS was inextricably linked to social life, encompassing character development, physical aptitudes, and the learning of moral traits that are essential to adulthood. Indigenous knowledge systems are transgenerational, passed on verbally, visually, practically, and spiritually from generation to generation (Obiora & Emeka, 2015; Shava & Manyike, 2018) through the application of Perennial theory. Children learn by doing through exploration. Indigenous knowledge systems encompass traditional values, norms, attitudes, and skills that are learned at home (Sheya, 2014). Similarly, family members, including young and old generations, through imitation (models and methods), observation, and experiments practice IKS at home (Matsekoleng, 2022). Such practices help students to develop cognitive skills because experiential learning is continuous, cyclical, and lifelong (Matriano, 2020). The experiential learnings are grounded in the Perennial theory and IKS values. The PEAR Indigenous practices ensure that learning is student-centred, constructivist-based, project-based, experiential, multi-sensory, reflective, participatory, interactive, cooperative, collaborative, and active learning. Students must be engaged and involved in exploration, research, interaction, and creation so that they can adapt the skills and strategies for them to become responsible students and lifelong students (Matriano, 2020). Incorporating the principles of IKS into education enables students to grasp the interconnectedness between human actions and their impact on the environment and society (Kumar, 2024).

Figure 1 presents a stepped framework to implement the IKS/PEAR activities in higher institutions of learning. The framework can be implemented through seven steps by involving relevant stakeholders (i.e. community members, university staff, department of higher education, etc.), conducting curriculum review, training university staff, developing policies, piloting PEAR activities, developing IKS content, and lastly, integrating IKS content in the university curriculum.

Figure 1. PEAR activities framework

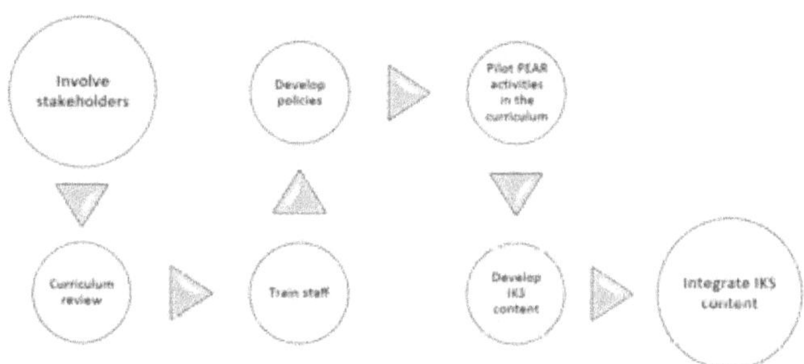

9. CONCLUSION

This chapter discussed the application of Perennial theory, the value of PEAR activities, the synergy of Perennial theory and PEAR activities, the integration of IKS, the relationship between IKS and Environmental Education, and the implications for integrating IKS in Higher Learning Institutions studies. The chapter fills the gap in the void of the literature within the Higher Learning Institutions of decolonising the curriculum by using PEAR activities to integrate IKS. A stepped framework is developed to implement the IKS practices in the university curriculum.

Indigenous practices, in their nature, embed a perennialism learning approach to help children acquire cultural heritage. The children, in their contextual learning experiences, also become the custodians of both the culture and traditional practices of their communities. Kanu (2017) observes that African Indigenous Education is anchored on five philosophical cannons namely: preparationism, hollisticism, communalism, functionalism, and perennialism. This study was based on the last listed philosophical canon. Indigenous practices in Higher Learning Institutions with the application of PEAR activities could close the irrelevancy of curriculum to Indigenous students. Because of irrelevancy, consequently, students are not inspired by the curriculum themes as students are primarily prepared for summative tests and examinations, which favour the memorisation of facts (Mentz & De Beer, 2019).

REFERENCES

Banes, G. G., & Cruz, K. B. D. (2021). The integration of Indigenous knowledge systems in the tertiary level curriculum of Benguet State University La Trinidad Campus. [formerly Benguet State University Research Journal]. *Mountain Journal of Science and Interdisciplinary Research*, *81*(1), 19–36. DOI: 10.70884/mjsir.v81i1.279

Basu, R. (2018). Constructivism Based Active Learning in Higher Education. *Shodh Sarita*, *5*(19), 24–28.

Chauke, O. R., Balotyi, T., Mapindani, A., Chauke, W. S., & Motlhaka, H. A. (2021). Rain petitioning as an indigenous agenda: Fusing ecological traditions with the modern IKS philosophy. *Turkish Journal of Computer and Mathematics Education*, *12*(6), 4946–4950. DOI: 10.17762/turcomat.v12i6.8741

Debisa, N. G. (2022). Building peace by peaceful approach: The role of Oromo Gadaa system in peace-building. *Cogent Social Sciences*, *8*(1), 1–17. DOI: 10.1080/23311886.2021.2023254

Du Plessis, P. (2021). Decolonisation of education in South Africa: Challenges to decolonise the University curriculum. *South African Journal of Higher Education*, *35*(1), 54–69. DOI: 10.20853/35-1-4426

Dube, E., & Munsaka, E. (2018). The Contribution of Indigenous knowledge to disaster risk reduction activities in Zimbabwe: A big call to practitioners. *Jàambá*, *10*(1), 1–8. DOI: 10.4102/jamba.v10i1.493 PMID: 29955265

Evangelista, R. D., Corrêa, D. H., Nunes, M. J., Figueiredo, J. S., Mascarenhas, S. W., Reis, M. da S., & Ferreira da Silva, D. (2024). Environmental Education in Promoting Sustainable Practices: Use of Indigenous Games and Play in Raising Awareness in Elementary School Students. *Journal of Social and Environmental Management*, *18*(12), e010376. DOI: 10.24857/rgsa.v18n12-099

Fonua, S. (2021). Enabling Sustainable Development by Embedding Tongan Knowledge into University Science Curricula. *Journal of Comparative & International Higher Education*, *13*(3), 157–179. DOI: 10.32674/jcihe.v13iSummer.3164

Ford, J. D., King, N., Galappaththi, E. R., Pearce, T., McDowell, G., & Harper, S. L. (2020). The resilience of indigenous peoples to environmental change. *One Earth*, *2*(6), 532–543. DOI: 10.1016/j.oneear.2020.05.014

Gumbo, M. T., & Williams, P. J. (Eds.). (2023). *Indigenous Technology Knowledge Systems. Contemporary Issues in Technology Education*. Springer., DOI: 10.1007/978-981-99-1396-1

Hungwe, J. P., & Mkhize, T. R. (2022). The Africanisation of universities in Africa: Reclamation of humanity and rationality. *Critical Studies in Teaching and Learning*, *10*(1), 136–152. DOI: 10.14426/cristal.v10i1.487

Hutchins, R. M. (1954). *Great Books: The foundation of liberal education*. Simon & Schuster.

Jima, A. O. (2022). Significance and restraint of indigenous knowledge inclusion in Ethiopian higher education curriculum: In focus Gadaa system. *Cogent Education*, *9*(1), 1–16. DOI: 10.1080/2331186X.2022.2046241

Kanene, K. M. (2016). 'Indigenous practices of environmental sustainability in the Tonga community of southern Zambia'. *Jàambá, 8*(1), 1–7. DOI: 10.4102/jamba.v8i1.331 PMID: 29955300

Kanu, I. A. (2017). The Philosophical Canons of Africa Indigenous Education. *Hummingbird: Journal of African Sustainable Development, 9*(8), 78-90. http://www.academicexcellencesociety.com/the_philosophical_canons_of_african_indigenous_education.pdf

Kasulwe, A. (2014). *African Indigenous Education*. https://patient.over-blog.fr/2014/07/african-indigenous-education-by-apollinaire-kaselwe.html

Kimbell, R. (2008). Indigenous knowledge, know-how, and design & technology. *Design and Technology Education: An International Journal, 10*(3), 8-10. https://openjournals.ljmu.ac.uk/DesignTechnologyEducation/article/view/1130

Klein, E. S., & Merritt, E. (1994). Environmental Education as a Model for Constructivist Teaching. *The Journal of Environmental Education, 25*(3), 14–21. DOI: 10.1080/00958964.1994.9941953

Kolb, D. A. (1984). *Experiential learning: Experiences as a source of learning and development*. Prentice-Hall.

Kugara, S. L., Kugedera, A. T. T., Sakadzo, N., Chivhenge, E., & Museva, T. (2022). The role of Indigenous Knowledge Systems in climate change. In Tshifhumulo, R., & Makhanikhe, T. J. (Eds.), *Handbook of Research on Protecting and Managing Global Indigenous Knowledge Systems* (pp. 1–23). IGI Global. DOI: 10.4018/978-1-7998-7492-8.ch001

Kumar, M. J. (2024). Forging Connections: Integrating Indian Knowledge Systems in Higher Education. *IETE Technical Review, 41*(3), 271–273. DOI: 10.1080/02564602.2024.2342625

Labonne, J. (2024). *Higher Education for Climate Change Mitigation: Sámi and Indigenous Knowledge Integration in Norwegian University Curricula* (Masters dissertation). Oslo: Oslo Metropolitan University.

Lin, J., Stoltz, A., Aruch, M., & Rappeport, A. (2021). Decolonization and transformation of higher education for sustainability: Integrating indigenous knowledge into policy, teaching, research, and practice. *Journal of Comparative & International Higher Education, 13*(3), 134–156. DOI: 10.32674/jcihe.v13iSummer.3255

Maira, J. N., & Kauka, E. O. (2018). Analysis of Ecopedagogy: Lessons from African Indigenous Education. *International Journal of Social Science and Human Research, 6*(2), 108–114.

Maluleke, R. (2019). *A pedagogy for technology education: An Indigenous perspective* (Doctoral thesis). Pretoria: University of South Africa.

Mapara, J. (2022). Indigenous Knowledge across the curriculum: An epistemic and cognitive shifting process. In Muyambo, T., Mapangisana-Hlatywayo, A., & Sithole, P. (Eds.), *Re-imagining Indigenous Knowledge and Practices in 21st Century Africa: Debunking Myths and Misconceptions for Conviviality and Sustainability* (pp. 41–68). Langaa Publishers.

Matriano, E. A. (2020). Ensuring student-centred constructivist and project-based experiential learning applying the Exploration, Research, Interaction and Creation Learning Model. *International Online Journal of Education & Teaching, 7*(1), 214–227. https://iojet.org/index.php/IOJET/article/view/727

Matsekoleng, T. K. (2022). *Challenges Confronting Technology Education Teachers in Raising learners' Awareness on littering in Gauteng Province* (DEd thesis). Pretoria: University of South Africa.

Matsekoleng, T. K., & Mapotse, T. A. (2023). Technology teachers' use of Indigenous Knowledge to integrate Environmental Education into Technology Education. In Gumbo, M. T., & Williams, P. J. (Eds.), *Indigenous Technology Knowledge Systems. Contemporary Issues in Technology Education* (pp. 283–295). Springer., DOI: 10.1007/978-981-99-1396-1_18

Matsekoleng, T. K., Mapotse, T. A., & Gumbo, M. T. (2022). The Role of indigenous games in education: A Technology and Environmental Education perspective. *Diaspora, Indigenous, and Minority Education*. Advance online publication. DOI: 10.1080/15595692.2022.2160436

Mbembe, A. J. (2016). Decolonizing the university: New directions. *Arts and Humanities in Higher Education, 15*(1), 29–45. DOI: 10.1177/1474022215618513

Mentz, E., & De Beer, J. (2019). The use of Cultural-Historical Activity Theory in researching the affordances of indigenous knowledge for self-directed learning. In J. De Beer (Ed.), *The decolonisation of the curriculum project: The affordances of indigenous knowledge for self-directed learning* (NWU Self-directed Learning Series Volume 2), pp. 49-86. Cape Town: AOSIS. https://doi.org/DOI: 10.4102/aosis.2019.BK133.03

Michie, M., Hogue, M., & Rioux, J. (2023). Two-Ways thinking and Two-Eyed Seeing as ways of implementing Indigenous perspectives in the science education curriculum. *Disciplinary and Interdisciplinary Science Education Research, 5*(23), 1–14. DOI: 10.1186/s43031-023-00084-3

Mnguni, J. N. C., & Mapotse, T. A. (2023). Examining the technological divide between Africa and the Western World: A case of South Africa. In Gumbo, M. T., & Mapotse, T. A. (Eds.), *Indigenous Technology Knowledge Systems: Decolonizing the Technology Education Curriculum* (pp. 221–233). Springer. DOI: 10.1007/978-981-99-1396-1_14

Nkhata, B., Mkandawire, S. B., Nachiyunde, K., Phiri-Nalube, P., Kaani, B., Mulenga, I. M., & Chipindi, F. M. (2019). Exploring selected Theories applicable to Educational Disciples and Social Sciences Research. *International Journal of Humanities Social Sciences and Education, 6*(12), 97–116. http://dx.org/10.20431/2349-0381.0612008

O'Donoghue, R., van Staden, W., Bhurekeni, J., Snow-Macleod, J., & Ndlamlenze, L. (2024). A Formative Study towards the inclusion of Indigenous Technologies and Knowledge practices in Science, Technology, Engineering, Arts, and Mathematics (STEAM) Curriculum settings. *Educational Research for Social Change, 13*(1), 34–47. DOI: 10.17159/2221-4070/2024/v13i1a3

Obiora, A. C., & Emeka, E. E. (2015). African Indigenous knowledge system and environmental sustainability. *International Journal of Environmental Protection and Policy, 3*(4), 88–96. DOI: 10.11648/j.ijepp.20150304.12

Revathi, T. (2020). Cooperative learning as a constructivist teaching-learning strategy. *International Journal of Pedagogical Studies, 8*(1), 84–91.

Rogers, C. R. (1969). *Freedom to learn*. Charles E. Merrill Publishing.

Rugedhla, M., Fidzani, L. C., & Moalosi, R. (2023). Indigenous technological knowledge for education in Zimbabwe. In Gumbo, M. T., & Williams, P. J. (Eds.), *Indigenous Technology Knowledge Systems: Decolonizing the Technology Education Curriculum* (pp. 235–251). Springer Nature Singapore. DOI: 10.1007/978-981-99-1396-1_15

Sebata, S., Dhlamini, N., & Ndlovu, M. (2024). Indigenous Weather Forecast Systems: A Case Study of Weather Forecast Indicators for Chapo Village, Ward 17, in Matopo Zimbabwe. *Journal of the Kenya National Commission for UNESCO*, *4*(2), 1–21. https://www.ajol.info/index.php/jkncu/article/view/281841. DOI: 10.62049/jkncu.v4i2.106

Seroto, J. (2011). Indigenous education in the pre-colonial era. In Booyse, J. J., le Roux, C. S., Seroto, J., & Wolhuter, C. C. (Eds.), *A history of schooling in South Africa: Method and context* (2nd ed., pp. 37–56). Van Schaik.

Seroto, J. (2020). Indigenous history and education before colonial times. In Seroto, J., Davids, M. N., & Wolhuter, C. C. (Eds.), *Decolonising education in the global south: Historical and comparative international perspectives* (pp. 40–58). Pearson.

Shava, S., & Manyike, T. V. (2018). The decolonial role of African Indigenous languages and Indigenous knowledges in formal education processes. *Indilinga*, *17*(1), 36–52. https://hdl.handle.net/10520/EJC-fe647aa43

Sheya, E. (2014). *Indigenous knowledge and Environmental Education: A case study of selected schools in Namibia* (Masters dissertation). Stellenbosch: University of Stellenbosch.

Sumaiya, B. (2024). Indigenizing Education: Integrating Indigenous Knowledge Systems for Cultural Relevance, Cognitive Justice, and Student Engagement. In Waris, A., & Tiwari, D. (Eds.), *Integration of Indian Knowledge System in Higher. Education, 16(1), 93-105*.

Wardhani, D. F., Arisanty, D., Nugroho, A., & Utami, U. B. L. (2024). Environmental Education Model Based on Local Wisdom of the Dayak Paramasan Tribe Indonesia. *Nature Environment and Pollution Technology*, 23(4), 2259–2272. DOI: 10.46488/NEPT.2024.v23i04.030

Chapter 11

Integrating Indigenous Technical Knowledge for Sustainable Agriculture and Horticulture:
Insights From India

Kushagra Sharma
https://orcid.org/0000-0003-1126-8697
Jawaharlal Nehru Krishi Vishwa Vidyalaya, India

Sapna Jarial
https://orcid.org/0000-0003-2660-9036
Lovely Professional University, India

Chethan Patil N. D.
https://orcid.org/0009-0001-2977-6422
Centre for Aromatic Plants, India

ABSTRACT

Indigenous Technical Knowledge (ITK) has immense potential for grassroots innovation in India, where numerous indigenous communities possess unique traditional knowledge on technologies. These technologies, comparable to modern systems, ensure self-sufficiency and sustainable practices in weather forecasting, seed germination, soil fertility, and pest control. Particularly regions inhabited by indigenous communities are, often overlooked in mainstream agricultural discourse, ITK plays a crucial role in their socio-economic development and environmental preservation. Despite the growing recognition of ITK, there remains a critical gap in integrating ITK with modern Holistic Knowledge Systems (HKS) to address contemporary agricultural challenges. This chapter explores ITK's impact on sustainable agricultural and horticultural practices, examining its integration with HKS and its role in sustainability, productivity, food security, biodiversity conservation, and resource management.

DOI: 10.4018/979-8-3693-3542-0.ch011

INTEGRATING ITK FOR SUSTAINABLE AGRICULTURE AND HORTICULTURE

Holistic knowledge systems (HKS) blend traditional, indigenous, formal, and informal systems. Interconnectedness, dependency, and a comprehensive view on element connections are stressed in these systems. HKS requires effective and sustainable solution to complex problems which requires integrating multiple knowledge systems (Baloyi, 2024). As a cultural perspective, traditional and indigenous knowledge systems emphasise reciprocal links and a comprehensive view of human-ecological interactions. Such systems illuminate sustainable behaviours and ecosystem-human interactions (Caudill, 2024). Systemic knowledge includes comprehending interpersonal norms, technological systems, and their linkages. Understanding context and relationships is crucial for managing knowledge and systems (Sheffield, 2005).

Context and relationships matter for managing knowledge and systems. Knowledge systems can be mechanical or holistic, thus it's best to examine their patterns rather than classify them (Ludwig & Poliseli, 2018). HKS may solve many issues, including sustainable development goals, due to their holistic nature which includes indigenous technical knowledge (ITK). These systems provide a framework for collaboration and action to achieve social goals (Bray & Dudgeon, 2020). ITK in education can support more holistic and contextual climate change measures (Mbah et al., 2021). Successful knowledge development and management in businesses demands a holistic approach (Zlatanović & Mulej, 2015). This includes latest technologies such as- Industry 4.0, which offers learning and help systems, however automated technologies' fragmented advice should not be overused. Keeping a broad perspective of how we learn and make decisions is crucial (Wotschack, 2023).

ITK includes local communities' generations-old wisdom, skills, and practices. ITK can improve agricultural outcomes and environmental sustainability when integrated with modern scientific research. Recent studies show how small-scale farmers have defied agricultural standards to meet subsistence demands (Mawere, 2014).

Necessity of ITK for Sustainable Agriculture and Horticulture

Modern agricultural practices, while boosting productivity, have also led to severe environmental and socio-economic challenges. Excessive dependence on synthetic fertilizers and pesticides has resulted in soil degradation, loss of beneficial soil microbes, and declining soil fertility (Dönmez, 2024). Monoculture cropping systems, which dominate global food production, have significantly reduced agrobiodiversity, making crops more vulnerable to pests, diseases, and climate change (Nair, 2023). The need for high-input farming has also escalated production costs, disproportionately affecting small-scale farmers who lack access to expensive agrochemicals and irrigation technologies (Mattu, 2017).

ITK offers practical, sustainable solutions to these challenges by promoting natural soil amendments, polyculture cropping, and traditional pest control strategies (Borah et al., 2023). Unlike industrial agricultural practices that prioritize short-term gains, ITK-based farming systems emphasize ecological balance, soil regeneration, and resource conservation (Sharma et al., 2022). The neglect of ITK not only erodes centuries-old farming wisdom but also threatens food security, particularly in regions vulnerable to climate change and resource scarcity (Pandey et al., 2020).

Plants are required for global nutrition, constituting 84% of the human diet. However, the dependence on a narrow spectrum of crop species is concerning, with only a fraction of the estimated 50,000 edible plant varieties used for sustenance (Nair, 2023). A mere 15 crop plants dominate 90% of the world's food energy intake, with rice, wheat, and maize collectively supplying over 60% of this essential dietary component, which leads to a number of ecological, social, and economic problems such as: eliminating diverse plant and animal species. Loss of biodiversity weakens ecosystems, thus, agricultural system becomes susceptible to pests, diseases, and climate change. Further, the same dominant crops deplete specific nutrients without regenerating the soil, thus making the system more chemical fertilizers dependent, which further degrades soil health. Focusing on a few staple crops farming can lead to the neglect of other essential crops that provide vital nutrients, resulting in imbalanced diets and malnutrition. Thus, it leads to the erosion of traditional agricultural practices and knowledge. As technology, urbanisation, and climate change affect traditional lives, knowledge erosion might lead to the loss of cultural heritage and basic survival skills. Young people move to cities in search of better jobs and are disinterested in crop-horticulture-livestock farming profession, which intensifies this knowledge gap, leaving the women and elderly with limited means to pass their expertise to the next generation. This has endangered the ITK.

Thus, to reverse this trend, technical documentation of local, sustainable agricultural practices that maintain ecological balance, enhance biodiversity, and support small-scale farmers is required to create a sustainable food system. While Indian farmers are encouraged to adopt modern agricultural technologies under the Digital India initiative, small and marginal farmers often face challenges in accessing these technologies due to limited land area, low per capita income, and insufficient government support. As a result, many farmers rely on ITK for crop production and farm management. This ITK, rooted in the experiences of indigenous peoples, evolves over time through experimentation and adaptation to changing conditions (Mattu, 2017). ITK holds profound importance, particularly in developing nations, where around 80% of the populace relies on age-old healing practices (Souravi & Rajasekharan, 2019). This recognition has spurred considerable interest in harnessing such wisdom for research and development, focusing on its implications for conservation and management strategies.

This chapter aims to explores in ITK's impact on sustainable agricultural and horticultural practices examining its integration with HKS and its role in sustainability, productivity, food security, biodiversity conservation, and resource management within various states of India.

Emerging Trends in ITK Integration

Over the past decade, there has been a growing interest in integrating ITK into mainstream agricultural research and policy frameworks. Several governments and institutions have initiated documentation and validation efforts, recognizing the potential of ITK in addressing food security and climate resilience (Pandey et al., 2017). Policies such as India's Organic Mission in the Northeast and the promotion of agroecological farming under Food and Agricultural Organisation (FAOs) guidelines have incorporated ITK-based practices to enhance soil health and reduce dependency on synthetic inputs (Borthakur & Singh, 2023).

Despite these positive developments, significant challenges remain. Many ITK practices are still undocumented, and scientific validation is often slow due to the complexity of traditional ecological knowledge systems (Caudill, 2024). Additionally, the lack of policy incentives, funding, and awareness has hindered large-scale adoption of ITK in modern farming systems (Sekhar, 2024). To bridge this gap, there is a need for participatory research involving local farming communities, policymakers, and agricultural scientists to ensure that ITK is integrated into contemporary agricultural extension programs.

ITK and Biodiversity in India

In India, approximately 800 food plant species are primarily utilized by tribal communities, with around 300 species thriving in the northeastern region alone. The northeastern states, often regarded as sleeping giants, hold a treasure trove of ITK waiting to be tapped. This rich repository of wisdom has the potential to fuel a second green revolution, driving "sustainable development" in rural areas while preserving the delicate ecological balance (Souravi & Rajasekharan, 2019). Despite their immense biodiversity and deep-rooted indigenous practices, much of this knowledge remains untapped and unknown to the outside world.

ITK originates from community, local, and rural sources and encompasses various agricultural practices, including weather forecasting, seed germination, soil fertility management, and pest and disease management for both plants and animals (De, 2021). A World Health Organization (WHO) survey reported 25% of prescribed medications are derived from plants, underscoring the enduring relevance of traditional medicine in today's world. ITK is deeply rooted in a spiritual connection to the natural environment and differs from scientific systems, yet there is a growing acknowledgment of the need to merge ancient knowledge with science (Jha et al., 2017). ITK have evolved through experiential learning and spontaneous experimentation, integrating into the cultural fabric of communities. Thus, ITKs are dynamic, adapted to local culture and environments, and prioritize risk reduction over profit maximization.

India is inhabited by various indigenous communities with diverse cultural, political, social, and economic values, is recognized as a biodiversity hotspot with prospect for agriculture based growth and development (Deo et al., 2021). Indigenous socio-economic farming practices such as *Apatani, Bun, Zabo, Zero Tillage*, and *Fruit-Based farming systems* emphasize minimal use of synthetic inputs and organic farming practices. Many states in India are gradually becoming organic states, with a majority of agricultural land declared as "Organic by default" due to minimal use of inorganic inputs.

Arunachal Pradesh state is known for its "biodiversity" and tribal populations heavily reliant on agriculture and animal husbandry (Borthakur & Singh, 2023). In Northeast India, farmers rely on ITK for sustainable organic crop cultivation (Pandey et al., 2020). Techniques such as embedding cucurbit seeds in fresh cow dung balls for optimal growth, using forest litter to enrich potato cultivation, and employing terraced methods for rice and maize to maintain soil health and prevent erosion are common practices (Sharma et al., 2020). ITK is important in nature conservation and sustainable resource management.

The preservation and proper management of bioresources are essential for human survival, yet overexploitation by a growing population has led to environmental degradation and challenges such as land degradation, biodiversity loss, soil erosion, landslides, floods, and drought. Indigenous knowledge-based "natural resource management systems" are considered impartial and promote equitable sharing of benefits derived from these resources. Local communities own and control large parcels of land and manage forests through community institutions (Chan, 2018).

Horticulture stands as a cornerstone of livelihoods in the hilly terrains of the northeastern states, where age-old management practices have sustained communities for generations (Ashwath et al., 2023). In Manipur state, local communities have long relied on certain plants for sustenance and as tools for forecasting weather patterns and foreseeing natural disasters. Embedded within their cultural beliefs, these plants serve as harbingers of environmental changes and guardians against negative occurrences (Refer to Table 1).

Table 1. Traditional uses and folklore of indigenous plants in Manipur

Plant Name	Common Names	Local Name (Manipur)	Significance/Beliefs	References
Agave americana Linn.	Agave, Century Plant	Kewa	Flowering pattern predicts direction of winds and storms; prevailing winds expected to blow from opposite direction of inflorescence with most flowers.	Hanglem et al., 2017
Alocasia indica Schott	Elephant Ear Plant	Yendem	Rarely flowers; blossoms considered ominous, foretelling difficulties and misfortune. Entire plant is removed from garden upon flowering.	Ashwath et al., 2023
Bambusa Spp.	Bamboo	Waa (Manipuri), Bans (Hindi)	Flowers associated with famine and rodent population increase. Apex crown shape believed to foretell family events; specific cutting customs to preserve prosperity.	Sharma et al., 2022
Brassica campestris Linn.	Mustard	-----	Symbolic in funeral rituals; mustard seeds sown on funeral sites symbolize deceased's societal contributions and potential rebirth.	Hanglem et al., 2017
Ficus rumphii Blume	Rumphf's Fig Tree	Khongnang	Falling branches believed to foreshadow death in direction of fallen branch. Crow nesting habits predict weather patterns.	De, 2021
Hibiscus cannabinus Linn.	Deccan Hemp	Sougri	Abundance of flowers predicts good rainfall. Violet-tinged leaves mark end of rainy season.	Ashwath et al., 2023
Mangifera indica Linn.	Mango	Heinou	Flowering patterns predict increased wind, storms, and heavy rainfall. Dark green leaves at season's start suggest favorable rainfall conditions.	De, 2021
Platycerium wallichi Hook.	Staghorn Fern	Saji-machi changkhrang	Dark green vegetative parts indicate impending rainfall. Dull coloration indicates lack of rainfall.	Ashwath et al., 2023
Quercus serrata Thunb.	Tasar Oak Tree	Uyung	Avoided in house construction due to belief it attracts lightning strikes.	Souravi & Rajasekharan, 2019
Terminalia tomentosa (Roxb) Weight and Arn	Indian Laurel	Mayokpha (Manipuri), Sadar/ Madar (Hindi)	Bark used in traditional heart ailment remedies. Cultural protocol for bark collection involves offering beetle nuts and leaves and requesting permission from tree. Only respectfully collected bark deemed suitable for medicinal use.	Hanglem et al., 2017
Acacia catechu Wild	-----	Kabokhajee	Produces durable reddish-black dye.	Ashwath et al., 2023
Achyranthus aspera Linn.	-----	Khujumpere	Valued for adhesive properties in dyeing, enhancing color vibrancy.	Ashwath et al., 2023
Amoora spectabilis Miq.	-----	Ungang	Yields pale scarlet dye.	Ashwath et al., 2023
Basella alba Linn.	-----	Urok sambal	Provides deep purple dye from leaves and stems, used in fabric dyeing and painting.	Ashwath et al., 2023

Despite their nutritional richness in vitamins and minerals, many indigenous and underutilized fruits (Table 1) remain untapped (Sharma et al., 2020). Wild fruits constitute a significant portion of the daily diet for tribal populations, are sold into local markets and serving as vital components for dietary diversity and nutrient security among rural communities year-round. Traditionally, minor fruits are consumed raw, transformed into beverages, pickled, or incorporated into various dishes. Some of these species hold medicinal properties, while others have commercial potential for tannin production, timber, dye, and phytochemical extraction (Singh et al., 2020).

The utilization of wild edible species holds promise for enhancing crop diversity through selective breeding and domestication. Additionally, wild species contribute to modern agricultural practices by providing a diverse gene pool for efforts aimed at crop improvement. The untamed species serve as staple food sources for indigenous populations and offer supplementary nutrition for non-local communities. The regular consumption of herbal remedies in the form of wild vegetables contribute to enhanced longevity and reduced prevalence of common ailments such as cancer, diabetes, and heart disease. It is plausible that the longer lifespans observed in rural and forest-dwelling populations, compared to their urban counterparts, can be attributed to their physically active lifestyles and daily consumption of medicinal vegetables (Deo et al., 2021).

Numerous studies have underscored the nutritional richness of natural edible plants, often surpassing that of conventional food crops.

Conservation Efforts and Environmental Insights

Internal and external factors present numerous obstacles to forestland sustainability. Internally, socio-economic pressures and environmental conditions are increasingly altering forest vegetation composition. Externally, interventions by actors outside the local economy and society further complicate these dynamics (Souravi & Rajasekharan, 2019). The inhabitants residing in 1000 to 4500 meters above mean sea level possess extensive knowledge of the diverse array of plants across altitudes. Their sustenance is intricately linked to the utilization of various wild plants, including notable wild vegetables such as *Paeonia emodi, Asparagus adscendens, Amaranthus viridis, Commelina maculata*, and *Diplazium esculentum.* Additionally, wild fruits like *Hippophae rhamnoides,* species of *Rubus, Fragaria nubicola, and Duchesnea indica* are staples in their diet. Beyond dietary needs, locals have identified numerous high-altitude plants to meet other requirements. Key cultivated crops include *Fagopyrum tataricum, Amaranthus paniculatus, Triticum aestivum, Oryza sativa, Phaseolus vulgaris, Glycine max, Macrotyloma uniflorum, Phaseolus mungo, and Pisum sativum,* along with vegetables such as *Brassica oleracea, Lycopersicon esculentum, and Solanum tuberosum* (Sharma et al., 2022). The abundance of medicinal plants in their natural surroundings, combined with traditional knowledge, facilitates the collection, processing, and trading of these invaluable resources. However, extensive utilization and significant biotic pressure threaten the existence of several essential plant species, including *Acorus calamus, Aconitum heterophyllum, Delphinium denudatum, Picrorhiza kurroa, Saussurea costus, and Zanthoxylum acanthopodium.* This underscores the urgent need for conservation efforts to protect these species from extinction.To grow crops various indigenous cultivation practices are followed.

Indigenous Agricultural Practices in Hills and Mountains of India

Jhum Cultivation and Shifting Cultivation: Northeastern hill farmers employ *Jhum* cultivation, a traditional slash-and-burn method practised from January to March. This approach maximizes the use of natural resources and minimizes cultivation risks, thereby supporting food security in hill regions. *Jhum* cultivation involves the sequential clearing, burning, and cultivation of land, allowing for the growth of a diverse range of crops, including cereals, legumes, oilseeds, root crops, cucurbits, vegetables, and fruits (Souravi & Rajasekharan, 2019). This diverse cropping system optimizes resource utilization and enhances crop diversity, ensuring a reliable food supply for local communities. In addition, farmers in hilly regions adopt alternative forms of shifting cultivation tailored to the varying topographies. Crops such as pulses, upland rice, vegetables, tubers, and root crops are strategically planted. Upper slopes are designated for crops like cassava and intercropped maize with soybean, while lower slopes are utilized for rice cultivation, maximizing the land's production potential (Gupta et al., 2012).

Moreover, double cropping of rice is practiced in the north easter region, where early rice varieties complement the main crop, significantly enhancing overall yield. Shifting cultivation also involves multiple cropping, with farmers planting between 8 to 35 different species on small plots of 2 to 2.5 hectares (Jha et al., 2017). This practice includes simultaneous sowing and sequential harvesting, which allows for efficient land utilization and continuous crop production. The crop mixtures provide essential nutrient cover, optimize resource use, promote biomass and nutrient recycling, and enhance soil quality. In Arunachal Pradesh's Jhum areas, farmers practice mixed cropping systems such as rice-maize, rice with cucurbits, rice with banana and citrus, and maize with French beans. These practices highlight the adaptability and resilience of traditional farming methods, showcasing their ability to sustain agricultural productivity in challenging environments (Pandey et al., 2020).

***Sustainable Rainfed Agriculture*:** In hilly terrains, rainfed farming predominates, except for a few valley pockets with irrigation. Crop selection and rotation are dictated by water availability, favoring crops with low water requirements. Legumes play a pivotal role in crop rotation, alongside erosion-resistant varieties like cowpea and *'kuth'.* Traditional crop choices prioritize nutrition, with protein-rich pulses and nutrient-dense grains like '*phaphra*' and '*chulai'* being common. Fruit-bearing plants like citrus, mango, and pomegranate are essential in households, providing seasonal vitamin-rich produce. To safeguard fields, thorny shrub biofences are employed (Sharma *et al*., 2022). Despite varied land characteristics, traditional rainfed farming persists, managing erosion through strategic crop rotation and terracing.

***Bun Cultivation Method in Meghalaya*:** The *Bun* cultivation system, prevalent in Meghalaya, involves growing crops on raised beds called *"Bun"* along the hill slopes. Typically, farmers cultivate crops consecutively for two years on these beds. The first year sees the cultivation of tuber crops like ginger, turmeric, potato, and sweet potato, followed by vegetables and upland paddy (Deo *et al*., 2021). This approach maximizes land productivity and optimizes resource utilization in hilly terrains.

***ZABO Farming System of Nagaland*:** The *ZABO* farming system, meaning "impounding of water," is an indigenous method practiced in Nagaland. It combines forestry, agriculture, and animal husbandry, emphasizing soil and water conservation (Sharma *et al*., 2022). The system features protected forest land, water harvesting tanks, cattle yards, and paddy fields. Innovative techniques, such as seepage control in paddy plots and the use of paddy husk on bunds, contribute to sustainable land management and crop production.

***Crop Rotation and Double Cropping*:** In the remote regions of Lahaul, situated along the Miar nallah in Himachal Pradesh, a unique crop rotation cycle unfolds (De, 2021). The agricultural calendar commences with the cultivation of barley, known for its ability to thrive in harsh conditions, during the initial year. Following this, buckwheat takes its place in the second year, benefiting from the enriched soil left behind by barley's heavy demand for organic manure. This sequential rotation not only optimizes soil fertility but also accommodates the limited availability of organic inputs, ensuring sustainable agricultural practices amidst challenging terrain.

Indigenous Maddim for Field Levelling*:Maddim*, a simple wooden apparatus, is utilized for leveling plowed fields in Uttarakhand. Adding weight, either through a heavy stone or a person, enhances levelling efficiency, a technique known as planking (Rawat *et al*., 2010). This method ensures optimal seed-soil contact, promoting robust crop germination, while also aiding moisture conservation in the fields.

***Animal-Powered Crop Threshing*:** Amidst the rugged terrain of the West Himalayan cold deserts, traditional methods of crop threshing employ the strength and endurance of animals, particularly *dzos*. In this age-old practice, a circular threshing floor is meticulously prepared, allowing *dzos* to tread upon the harvested crops without fatigue (Keerthana & Divakaran, 2021).Top of Form As the animals rhythmically circle the threshing ground, their movements are accompanied by traditional songs, infusing the laborious task with a sense of communal celebration. Meanwhile, a designated container ensures the collection of animal dung, preventing any contamination of the precious grains.

Innovative Horticultural Practices for Crop Cultivation and Preservation

From the age old tradition of manuring with organic materials sourced from livestock to the practice of burning grasses for fertilizer, local farmers exhibit a deep-rooted commitment to maintaining soil fertility. Their use techniques such as composite manure application, agricultural residue incorporation, and the strategic intercropping of legumes, all aimed at nurturing the land they depend on (Gupta *et al*., 2012). Moreover, traditional farming is exemplified in the meticulous construction of ridges and mounds for planting, a technique passed down through generations. This, coupled with the farmers' adept seed management skills, including the art of mass selection to cultivate resilient local crop varieties. Noteworthy, too, is the ingenious use of indigenous plant-derived pesticides, a testament to farmers' resourcefulness in pest management (Sharma *et al*., 2020). With leaves from specific trees and age-old preparation methods, they successfully fend off threats like the *'buluba'* and red ants, safeguarding the crops. In the face of modernization, the preservation and continued application of local horticultural wisdom stand as a tribute to the enduring legacy of traditional farming practices (Refer Table 2).

Traditional Agricultural Practices in High-Altitude Regions of India

***Root Expansion Techniques in Vegetable Farming*:** High up in Ladakh, Lahaul-Spiti, and Kinnaur, in northern India, farmers rely on a unique method called *"tokhre"* to prepare the soil, encouraging the lateral spread of roots in cabbage and various vegetable crops (Rawat & Kharwal, 2016). This method optimizes moisture and nutrient absorption; also effectively suppresses weed growth while safeguarding roots from potential damage.

***Grafting in Fruit Crops*:** In the Nubra Valley of Ladakh, north west India, communities practice the apricot grafting. Sweet scions are skillfully grafted onto wild bitter apricot stocks, employing specialized techniques like *Kalam* and *Dambu* (Sharma *et al*., 2022). The approach ensure the integrity of the graft union and facilitate successful grafting by providing essential moisture and protection against potential wind damage.

***Osmo-conditioning*:** Across Kinnaur, Spiti, and the upper zones of Shimla in Himachal Pradesh, garden pea emerges as a pivotal crop. Opting for sowing during October-November, preceding snowfall, results in superior and expedited germination compared to seeds planted in March or April (Souravi & Rajasekharan, 2019). Winter sowing allows sufficient time for seed physiological processes, leading to an early harvest and favorable economic outcomes.

***Modifying Microenvironment*:** Amidst the scenic Nubra Valley, grape growers thrive through innovative microclimate management. Basin temperatures are meticulously regulated using local materials like stone, grasses, and even textiles such as clothes or gunny bags (Lal & Verma, 2008). Grapes flourish in sun-kissed pockets, with pits filled with indigenous materials gently warming the sandy soil. Additionally, grapevines are shielded from adverse weather and animal interference through strategically designed protective coverings, resembling miniature greenhouses.

Bolstering Soil Health: In basin regions, a longstanding agricultural ritual involves fortifying the soil by burying animal bones. This natural approach enhances soil fertility by introducing phosphorus, fostering robust plant growth and sustainable agricultural yields (Souravi & Rajasekharan, 2019).

Table 2. Traditional agricultural techniques from different regions of India

Region	Indigenous Practice	Crop/Activity	Benefit/Outcome	Reference
Himachal Pradesh	Utilizing grafted mango pit; drying pit in sunlight for 3 months; burying burned leaves near mango tree	Mango cultivation	Decline in weeds, pests, and insects in the pit; promotion of flowering; mitigation of hopper populations	Sharma *et al.*, 2023
Tamil Nadu	Cultivating sunflowers between mango trees to attract honey bees; excising bark from mango trees on solar eclipse day	Mango and sunflower cultivation	Increased pollination and fruit production; prevention of premature flowering in mango trees	Sharma *et al.*, 2022
Andhra Pradesh	Covering mango fruit with paddy straw in closed room for uniform ripening	Mango cultivation	Uniform ripening of mango fruit	Mattu, 2017
Manipur	Using Pit Nursery Method for crop seedlings; maintaining water requirement in pits	Crop seedling management	Preventing evapo-transpiration loss; maintaining water supply for seedlings	Das *et al.*, 2012
Kerala	Burning dry leaves, twigs, and trashes for planting banana suckers; applying groundnut cake	Banana cultivation	Enhanced crop growth and yield	Jha *et al.*, 2017
	Intercropping banana with onion, tomato, cowpea, okra for 4 months; using Basella alba leaves in gunny bags	Banana cultivation	Pest management; diversification	
Meghalaya	Utilizing green leaf manures such as kolangi, *Agave* spp., and Ekka for grape cultivation	Grape cultivation	Soil enrichment; improved grape yield and quality	Sharma *et al.*, 2022
	Dissolving powdered groundnut cake in water for grape trees; thinning grape bunch tips	Grape cultivation	Enhanced fruit quality and yield; promotion of better fruit size	
	Nipping pineapple tips for increased fruit weight and size	Pineapple cultivation	-----	
Arunachal Pradesh	Cultivating various local cucurbit cultivars	Cucurbit cultivation	Diversification of crops	Sharma *et al.*, 2022
Bihar, Jammu, Maharashtra	Applying ash in onion nursery beds and fields	Onion cultivation	Improved bulb quality	Jha *et al.*, 2017

Traditional Postharvest Practices

Natural Apricot Drying at High Altitudes: In the arid highlands of places like Malling, Nako, and Pooh in Kinnaur district, apricots undergo a unique drying process while still on the tree. The exceptionally low relative humidity in these areas facilitates natural drying, resulting in apricots of exceptional quality (Rawat & Kharwal, 2016). Despite the availability of modern drying techniques such as checking, sulphuring, and mechanical dehydration, the apricots dried on the trees remain unparalleled in quality and flavor. The combination of low temperatures and dry weather not only enhances sugar accumulation in the fruits but also ensures a rich color and sweetness in the final product, making them prized and fetching prices two to three times higher than sun-dried apricots. This exemplifies the natural low-temperature drying process under ideal conditions.

***Sun-Dried Mango Slices*:** A common practice in lower regions abundant with in situ mango varieties involves slicing mature yet unripe mangoes and placing them in bamboo baskets. These baskets, known as *Changer* or *Chharolu,* are left in the open sun for drying. The resulting dried product, known as *Bukarian*, is stored in earthen pots and utilized as a souring agent.

***Ash-Assisted Drying of Cucumber and Pumpkin Seeds*:** In the Mandi district of Himachal Pradesh, a unique method is employed to dry cucumber seeds using wood ash. The ash and seeds are admixed; thereafter spread on plates, exposed to the sun (Sharma et al., 2022). This technique facilitates rapid drying by absorbing excess moisture from the seeds and preventing them from clumping together. The individual seed segregation increases the surface area, expediting the drying process.

***Sun-Dried Ginger*:** In the regions of Shillai and Renuka in Sirmour district, of Himachal Pradesh, ginger undergoes a traditional sun-drying process. Fresh ginger is first soaked in water and then peeled by rubbing against a gunny bag or placed in bamboo baskets (Sharma et al., 2020). The peeled ginger is then immersed in a lime solution and exposed to warmth to facilitate faster and more efficient fermentation, resulting in the prized dried ginger known as *Sonth.*

***Soil Pit Ripening for Bananas*:** Bananas encased in grass or cloth are ripened in soil pits, prolonging their storage period. Additionally, wrapping fruits and vegetables in moist gunny bags enhances their shelf life by harnessing the cooling effect generated through evaporation (Sharma et al., 2020).

Storage Techniques and Practices

***Preservation of Citrus Fruits*:** In the Himachal Pradesh, Mandi district, a unique method is employed to store ripe galgal fruit (*Citrus pseudolimon*). Pits are excavated in the fields, and the fruits are carefully arranged in alternating layers with ash (Rawat & Kharwal, 2016). The top is then sealed with a mixture of ash and soil. The ash's natural antifungal properties prevent spoilage, while the cool underground environment prolongs the fruit's shelf life significantly.

Another method for preserving citrus fruits involves utilizing wide-mouthed pitchers placed in underground pits. After loading the pitchers with fruits, they are covered with a thick layer of soil. This technique, popular in the *Nurpur area of Kangra*, Himachal Pradesh, to ensure citrus fruits remain fresh for up to a month without spoilage.

***Apple Preservation in Tribal Areas*:** In the tribal regions of Lahaul-Spiti and Kinnaur districts, Himachal Pradesh, apples are stored underground before winter sets in. Wooden crates filled with apples are carefully arranged in pits, with the mouths covered by thatch and soil (Lal & Verma, 2008). This method enables apples to be preserved until summer without any significant loss of quality. The incorporation of wooden boxes or pitchers within the pit serves as a natural cooling chamber, a concept which is a source of inspiration for modern zero-energy cool chambers.

Holistic Water Management Approaches

Water has been a vital resource shaping human civilization. In ancient times, semi-forest areas with water bodies beneath hillocks, known as *'mullai'* lands, were prime hunting grounds. This practice continues today, highlighting the enduring importance of water access (Sharma *et al.*, 2022). Water, known as *'Appu'* in Sanskrit, holds a revered status, believed to be as ancient as the Universe itself. The Rig Veda describes the world as "originally water without light," emphasizing its divine significance (Borthakur & Singh, 2023). Viewed as a source of peace, prosperity, and vitality, water was worshipped for its life-giving properties. From the Neolithic period to the Iron Age and early historic times, water bodies served multifaceted roles in society. They were utilized for irrigation, industry, trade, and religious ceremonies, underscoring their indispensable nature in human life. Indigenous water harvesting and storage methods have been integral since ancient times. Tailored to local conditions and annual rainfall patterns, these methods include *Jhalara, Talab/Bandhi, Bawari, Taanka, Ahar Pynes*, *Johads*, and more. Notably, archaeological sites like *Dholavira* reveal sophisticated water management systems, with interconnected reservoirs ensuring efficient water distribution (Sharma et al., 2020). Across civilizations, the stewardship of water resources has been paramount, reflecting its fundamental importance to human existence.

ITK in Management of Water in Northeast India

In Arunachal Pradesh, traditional water management systems have flourished among various tribes, showcasing innovative approaches to irrigation and conservation (Deo et al., 2021). The *Apatani's system*, practiced in *Ziro,* ingeniously harnesses both groundwater and surface water for agriculture (Sharma *et al.*, 2022). Utilizing terraced plots supported by bamboo frames approach, allows for efficient flooding and draining as needed, maximizing water utilization while minimizing erosion risks (Pandey et al., 2020). Similarly, the *Yetbung Lingang* and *Linkum structures*, employed by the Adi tribe in East Siang district, exemplify traditional water conservation techniques (Refer Table 3). Using locally available materials, these structures effectively regulate perennial streams to benefit agricultural lands. Moving to Assam, the *Bodo community's Dong-Bandh* system showcases a unique approach to water management along the Indo-Bhutan border. By constructing dongs over the *Pagladiya* River and creating intricate canal networks, water is efficiently channeled to agricultural fields, ensuring optimal irrigation. Additionally, Assam's *Pukhuri system* utilizes earthen canals to divert river water towards villages, providing a reliable "water source" for multipurpose (Souravi & Rajasekharan, 2019). Further east, Manipur, Meghalaya, and Tripura demonstrate innovative *bamboo drip irrigation methods*, particularly in the *Khasi* and *Jaintia* Hills (Sharma et al., 2020). By utilizing bamboo pipes to transport water from springs to fields, this system delivers precise irrigation directly to plant roots, minimizing water wastage. The adaptability of channel sections ensures efficient water distribution, catering to the diverse agricultural needs of the region.

These indigenous water management practices highlight the ingenuity and resourcefulness of local communities in optimizing water utilization for agriculture, paving the way for sustainable farming practices in the region (Borthakur & Singh, 2023). Rooftop rainwater harvesting stands as a widely embraced practice across the nation, particularly favored in hilly regions for its effectiveness in capturing and storing rainwater (Hanglem *et al.*, 2017). This method, commonly adopted by individuals, promotes self-sufficiency and cost-effectiveness by harnessing natural resources.

In Manipur, smaller lakes known as *"Kom"* supplement the major lakes and ox-bow lakes, enriching the water resources in districts such as Imphal, Bishnupur, Senapati, Tamenglong, and Thoubal (Hanglem et al., 2017). Additionally, public ponds, or *"Leikai Pukhri,"* serve as vital communal reservoirs, enhancing the region's water storage capacity. In Mizoram and Nagaland, indigenous practices play a significant role in water management. The *Chakhesang* tribe in Kikruma village, Phek district, employs the *"Zabo/Ruza"* method, which involves collecting and storing rainwater in catchments. This method integrates water conservation with forestry, agriculture, and animal care. Similarly, along the River *Mezii* in Kigwema, Nagaland, the *"Cheo-ozihi"* practice involves channeling water through bamboo pipes to irrigate terraces, showcasing the innovative use of natural materials for water management (Borthakur & Singh, 2023).

The utilization of Alder trees for soil fertility management exemplifies sustainable agricultural practices. By lopping the trees to maintain their height, their leaves and biomass are used to enrich the soil (Sharma *et al.*, 2022). In *Vangchhia* village, Mizoram, an ancient method of water storage in rocks, dating back centuries, was discovered. The local population ingeniously drilled holes in sandstone slopes to store water, sustaining themselves for extended periods (Hanglem et al., 2017). The rice-fish system on the *Apatani plate*au of Arunachal Pradesh is a holistic water management approach that integrates land, water, and farming systems (Sharma *et al.*, 2022). Developed by the local *Apatani* tribe, this system protects against soil erosion, conserves water for irrigation, and promotes paddy-cum-fish culture (Deo et al., 2021). Similar practices are observed in Manipur and Nagaland, where terraced fields accommodate small ponds for fish harvesting during paddy cultivation, ensuring year-round water conservation and multiple crop yields.

In Sikkim, various water sources such as springs *("Dhara"),* small rivers *("khola/kholsa"),* and shallow wells *("kuwa")* are crucial for providing water for agricultural and domestic purposes. These diverse indigenous water management practices underscore the rich and varied approaches across the region (Sharma *et al.*, 2022).

Table 3. Indigenous farming practices in soil and water management in India

Region	Practice	Technique/Method/Plant	Purpose/Effect	Reference
Sindhu Valley Civilization	Water harvesting structures	Reservoirs, dams, ponds, Chauka, Haveli system	Harvesting water	Bhan *et al.*, 2016
Various regions	Rainwater harvesting	Farm ponds, check dams, shallow wells, Bamboo drip method	Harvesting rainwater, controlling waterborne diseases	Sarkar *et al.*, 2015
Tamil Nadu	Micro-depression method	Neem, teak, mango trees	Controlling soil erosion, improving soil properties	Bhan *et al.*, 2016
Andhra Pradesh	Rainwater collection	Rolu method	Collecting rainwater	Sharma *et al.*, 2022
Himachal Pradesh	Water storage ponds	Chaal system	Drinking, irrigation	Sarkar *et al.*, 2015
Various regions	Soil amendment	Cow dung slurry, wood ash, rice husk, coco pit, paddy straw	Improving soil properties, moisture retention	Sarkar *et al.*, 2015

continued on following page

Table 3. Continued

Region	Practice	Technique/Method/Plant	Purpose/Effect	Reference
Uttarakhand	Water catchment reservoirs	Tals, Khals, Chals, Rou	Collecting water for domestic, agricultural use	Jha *et al.*, 2017
Assam	Water harvesting structures	Bari system	Harnessing water	Angadi *et al.*, 2021
Rajasthan	Open wells	Saza Kuva	Domestic, irrigation use	Angadi *et al.*, 2021
South India	Soil moisture evaluation	Aduthininapalai (Aristolochia bracteolacia)	Evaluating soil water	Sharma *et al.*, 2022
Various regions	Soil fertility improvement	Sheep/cattle penning, mixed cropping, intercropping	Improving soil fertility	Sarkar *et al.*, 2015
Tamil Nadu	Soil erosion control	Kolingi (Tephrosia purpurea), Nuna tree (Morinda tinctoria)	Controlling soil erosion, moisture retention	Jha *et al.*, 2017
Various regions	Organic liquid manure preparation	Neem, cow urine, asafetida waste, sand, cow dung	Improving soil productivity, pest control	Sharma *et al.*, 2022
Karnataka, Andhra Pradesh, Kerala	Soil conservation	Vetiver (Khus grass)	Managing land degradation, soil conservation	Angadi *et al.*, 2021
Maharashtra, Kerala, Assam	Soil improvement	Diancha (Sesbania sp.), Sun hemp (Crotolaria juncea)	Improving water holding capacity, soil properties	Deka *et al.*, 2017
Tamil Nadu, Kerala	Soil amendment	Poorvarasu (Thespesia populnea)	Mitigating water loss from soil	Bhan *et al.*, 2016
Tamil Nadu, Kerala, Karnataka	Soil improvement	Bagasse of sugarcane, Indian gooseberry leaves & branches	Improving saline soil	Deka *et al.*, 2017
Kerala, Hyderabad	Soil improvement	Tea quadrifolia, Cyanodan dactylon	Enhancing soil fertility	Sharma *et al.*, 2022
Tamil Nadu, Punjab, Haryana, Himachal Pradesh, Maharashtra	Soil improvement	Pirandai (Cissus quadrangularis)	Improving alkali soil properties	Bhan *et al.*, 2016
South India, Uttar Pradesh	Soil improvement	Diancha, Nut grass (*Cyperus rotundus*)	Ameliorating alkali soil	Angadi *et al.*, 2021
Various regions	Manure application	Cow dung, pig dung, sheep dung, goat dung	Improving soil fertility	Sharma *et al.*, 2022
Tamil Nadu, Karnataka, Gujarat, Bihar	Manure preparation	Cow dung, *Calotropis gigantea* leaves, neem cake powder	Improving soil properties	Angadi *et al.*, 2021
Bihar, North-East	Soil improvement	Water hyacinth compost or ash	Improving alkali, saline soil	Bhan *et al.*, 2016
Rajasthan	Green manure crops	*Tephrosia purpurea, Calotropis gigantea, Morinda tinctoria, Pongamia pinnata, Azadirachta indica, Thespesia populnea, Adathoda vasica*	Soil improvement	Sarkar *et al.*, 2015

ITK Practices for Sustainability

ITK provide a robust foundation for sustainable agriculture by minimizing external inputs, improving soil fertility, and enhancing biodiversity. Traditional intercropping techniques, such as the *"three Sisters method"* (a combination of maize, beans, and squash) or agroforestry systems in India, have demonstrated their effectiveness in maintaining ecological stability while ensuring food security (Deo et al., 2021). Similarly, natural pest management techniques(Table 4) —such as neem-based pesticides, cow urine applications, and wood ash for seed preservation—are eco-friendly alternatives to chemical-based solutions (Jha et al., 2017). Table 4 indicates the ITK regarding insect pest disease management in India.

Table 4. Indigenous pest and disease management practices in different regions of India

Region	Indigenous Practice	Crop/Activity	Benefit/Outcome	Reference
Punjab	Burning field debris after wheat harvest for pest control	Field crops	Pest population reduction	Adam & Apou, 2013
Uttarakhand	Burning dry pine leaves in fields; applying table salt for white grub control	Field crops	White grub control	Sharma *et al.*, 2022
	Deep plowing with indigenous plough for soil pest control	Field preparation	Pest and pathogen reduction in soil	
Kerala, Jharkhand	Mixture of wood ash/kitchen ash + farmyard manure for chewing and biting insect control	Field crops	Control of mouthpart insects	Gopi *et al.*, 2016
North-east, Uttarakhand, Madhya Pradesh	Burying severely infested crops such as mustard, cauliflower, cabbage for aphid control	Various crops	Aphid control	Gopi *et al.*, 2016
Bihar, Uttarakhand	Earthen mounds for potato tuber greening control	Potato cultivation	Tubercular greening control	Gopi *et al.*, 2016
Orissa, Kerala	Cow dung + cow urine compound for pesticide control, wilt symptoms, onion blight	Various crops	Pest and disease control	Adam & Apou, 2013
Uttarakhand, Orissa, Gujarat	Intercropping with Madira or Barnyard millet, Konri millet, mustard for insect-pest control	Various crops	Pest control through intercropping	Deka *et al.*, 2017
Chhattisgarh	Bicchu booti solution + cow urine for fungal disease control in vegetable crops	Vegetable crops	Fungal disease control	Roy *et al.*, 2015
Tamil Nadu	Smoking with edible/non-edible oil, mixing with ashes and neem powder for seed protection	Various seeds	Pest and disease control	Deka *et al.*, 2017
	Wood ashes for aphid pod borer control in vegetable crops; tobacco dust for aphid control	Vegetable crops	Aphid control	
	Hookah water for major and minor pest control in vegetable crops	Vegetable crops	Pest control	
Jammu & Kashmir, South India	*Chrysanthemum coronarium, Tagetes erecta* for nematode control in turmeric, tomato, chili, ginger	Various crops	Nematode control, soil enrichment	Gopi *et al.*, 2016
North-east	Liquid lime for Gummosis disease and caterpillar, trunk borer control in mandarin	Mandarin orchards	Disease and pest control	Jha *et al.*, 2017
Punjab, Haryana	Burning plant debris for insect harbor control	Vegetable, paddy crops	Pest control	Adam & Apou, 2013
Tamil Nadu, Kerala, North-east	Zero tillage practice for mycorrhizae root growth and nodulated frankia	Field crops	Soil health improvement	Sharma *et al.*, 2022

continued on following page

Table 4. Continued

Region	Indigenous Practice	Crop/Activity	Benefit/Outcome	Reference
Meghalaya, Jaipur, Punjab, Himachal Pradesh	Utilizing decomposed mulch for soil pathogen inhibition	Field crops	Pathogen inhibition	Roy *et al.*, 2015
Nagaland	Mixed cropping for pest and pathogen control	Various crops	Pest and pathogen control	Borah *et al.*, 2023
Indian farmer	Deep plowing after harvest for soil pest control	Various crops	Pest control	Borah *et al.*, 2023
Tripura, Meghalaya, Assam	Burning paddy husk, dry chili for rodent control in jhum fields	Field crops	Rodent control	Borah *et al.*, 2023
Kerala, Orissa	Fermented *Agave sissalana, Piper nigrum, Veronia amygdalina* for termite control	Soil	Termite control	Roy *et al.*, 2015
Punjab, Haryana, Telangana	Turmeric powder in plant irrigation for weevil and grain moth control	Various crops	Pest control	Roy *et al.*, 2015
Sikkim	Extract of *titey pati, banmara, Lantana camara* for aphid and white fly control in tomato and chili crops	Tomato, chili crops	Pest control	Jha *et al.*, 2017
Karnataka, North-east	Chilli seedlings in cow urine water for damping off prevention	Vegetable crops	Disease prevention	Adam & Apou, 2013
Kerala	Neem oil + detergent soap solution for hopper control in mango trees; neem powder + tobacco powder for nematode control in banana	Mango, banana trees	Pest and disease control	Deka *et al.*, 2017
Jorhat, South India	Cow dung + detergent soap solution for sooty mold disease control in mango trees	Mango trees	Disease control	Gopi *et al.*, 2016
Tamil Nadu, Kerala, Himachal Pradesh	Neem powder + tobacco powder solution for nematode control in banana suckers	Banana cultivation	Nematode control	Yamunarani *et al.*, 2023
Tamil Nadu, Kerala	*Calotrpis* spp. + Neem cake solution for white fly control in guava orchards	Guava orchards	Pest control	Yamunarani *et al.*, 2023
Tamil Nadu, North-eastern	Cotton seed with ash and cow dung slurry for seed protection and damping off prevention	Various seeds	Disease prevention	Yamunarani *et al.*, 2023
Assam, South India	Clay layer/cow dung ball in banana stalks to prevent spoilage and ripening	Banana cultivation	Pest control	Sharma *et al.*, 2022
Karnataka, Kerala, Tamil Nadu	Neem seed powder + cow urine + detergent solution for leaf miner pest control in citrus trees	Citrus orchards	Pest control	Borah *et al.*, 2023
South India	Neem fruit powder for nematode control in grape trees; maida flour + cow urine for sooty mold control in citrus trees	Grape, citrus trees	Pest and disease control	Yamunarani *et al.*, 2023
Karantaka, Mysore, Tamil Nadu	Leaves of Kasarka with cow dung for grub insect control in citrus trees	Citrus orchards	Pest control	Deka *et al.*, 2017
Assam, Maharashtra	Lime wash/lime soaked cotton for stem borer control in mandarin orange trees	Mandarin orange trees	Pest control	Jha *et al.*, 2017
South India	Aloe vera application for powdery mildew control in mandarin orange trees	Mandarin orange trees	Disease control	Sharma *et al.*, 2022
Tamil Nadu	*Cyanodon dactylon* leaf for tomato disease control	Tomato fields	Disease control	Adam & Apou, 2013
Orissa	Cow dung + Kochila seed powder + Kochila leaf compost for fruit and shoot borer control in brinjal	Brinjal fields	Pest control	Deka *et al.*, 2017

continued on following page

Table 4. Continued

Region	Indigenous Practice	Crop/Activity	Benefit/Outcome	Reference
Jharkhand	Cow urine + tobacco soaked powder for leaf and fruit disease control in cucurbits, cowpea, lady finger	Vegetable crops	Disease control	Deka *et al.*, 2017
	Rice starch + cow urine for insect control in lady finger, tomato	Vegetable crops	Pest control	Borah *et al.*, 2023
Bihar	Tobacco soaked water for shoot and fruit borer control in brinjal	Brinjal fields	Pest control	Borah *et al.*, 2023
Tamil Nadu	Cow dung slurry for rhinoceros beetle control in citrus trees	Citrus orchards	Pest control	Roy *et al.*, 2015
Madhya Pradesh	Dry mahua flower for *Scalopendra* spp. control in soybean	Soybean fields	Pest control	Gopi *et al.*, 2016
Tamil Nadu, Madhya Pradesh	Asafoetida solution for *Heliothis* sp. and other insect control in crops	Various crops	Pest control	Borah *et al.*, 2023
Orissa	Tobacco leaf solution for Heliothis armigera control	Various crops	Pest control	Jha *et al.*, 2017
	Neem leaf solution for various pests control	Various crops	Pest control	Adam & Apou, 2013
	Neem leaf + *Ipomea* bush solution for *Heliothis armigera*, spotted bollworm, armyworm control	Various crops	Pest control	Deka *et al.*, 2017
North-east, Orissa, Tamil Nadu, Bihar	Neem leaf solution for specific pests control	Various crops	Pest control	Roy *et al.*, 2015

ITK-based postharvest practices also play a crucial role in reducing food losses. Traditional storage methods, such as underground citrus storage in Himachal Pradesh and ash-assisted drying for cucumbers and pumpkins, minimize spoilage without requiring expensive infrastructure (Rawat & Kharwal, 2016). These techniques not only ensure year-round food availability but also contribute to household food security, particularly in remote rural areas (Sharma et al., 2020). By scientifically analyzing and optimizing these methods, policymakers and researchers can enhance their applicability in modern agricultural supply chains.

Geographical Indication Labels and Regional Horticulture

Geographical Indication (GI) labelling for northeastern India's rich horticulture assets opens up new possibilities. GI tags might revolutionise the region's economy. GI classification elevates these beautiful horticultural products to premium goods of quality and authenticity (Rajan et al., 2023). Farmers and artists benefit from this elevation, which opens export markets and boosts market value. However, while the potential is abundant, the current roster of GI-recognized goods remains scant. Urgent action is necessary to broaden this spectrum, necessitating grassroots awareness initiatives across sectors. Through concerted efforts, the abundance of northeastern India can be fully realized, enriching its cultural heritage and economic prosperity. Geographical indications serve as a crucial safeguard for traditional knowledge, granting collective rights that are vital for preserving indigenous wisdom (Kishore, 2018). India's vast biological and cultural diversity offers products derived from the biocultural knowledge of tribal communities and local groups. Beyond their commercial value, these goods hold the promise of bolstering local economies, fostering community unity, and safeguarding biodiversity, traditional knowledge, and cultural heritage. As market demand for biological resources and associated traditional knowledge grows, a new frontier of opportunities emerges for indigenous and local populations to reap benefits and enhance their livelihoods. Through initiatives like GI protection, iconic products like *Navara Rice, Kaipad Rice,* and *Pokkali Rice* are not only shielded but also empowered to leverage their market potential, uplifting the socio-economic fabric of rural communities (Rajan et al., 2023). While GI tagging ensures adherence to specific quality standards, maintaining consistency poses a formidable challenge. Variations in environmental factors and farming practices can impact product quality, necessitating robust quality control measures (Sharma *et al.*, 2022). Furthermore, the lack of modern infrastructure and technology in remote regions of northeastern India presents hurdles in production and postharvest processing. Inadequate facilities for sorting, packaging, and transportation undermine product quality and market competitiveness, highlighting the need for investment in infrastructure development and technological advancement. Additionally, addressing pest and disease control (Refer Table 4) in horticulture crops remains paramount, particularly in remote areas where access to effective control measures may be limited (De, 2021). By confronting these challenges head-on, stakeholders can unlock the full potential of GI-tagged products while ensuring sustainable agricultural practices and economic prosperity for local communities.

Protecting the Sacred Groves

Sacred groves, revered patches of forests dedicated to various deities and spirits, hold profound historical, cultural, and ecological significance. Protected by local communities for religious and ecological reasons, these sanctuaries offer insights into our ancestors' belief systems. In southern state Kerala, sacred groves are recognized as biodiversity hotspots, harboring a diverse array of flora and fauna. Environmentalists stress the importance of preserving these natural havens for future generations. References to sacred groves abound in India's religious texts, with mentions in epics like the *Mahabharata* highlighting their importance. Forest conservation holds a prominent place in Indian religion and culture, as evidenced by ancient myths and taboos surrounding these sacred spaces (De, 2021). D.D Kosambi's work sheds light on the cultural significance of sacred groves, illustrating how they were once revered as primordial sanctuaries, forbidden for human entry under penalty of transformation. Today, local communities continue to safeguard these forests, balancing conservation efforts with their socio-cultural and religious practices. Sacred groves stand as living testimonies to the enduring bond between humanity and nature.

Expanding Krishi Vigyan Kendras (KVKs)

KVKs are the knowledge lighthouse of Indian Council of Agricultural Research (ICAR) holds immense potential in the exploration, identification, characterization, and conservation of underutilized crop germplasm within their specific agroclimatic zones. Collaborating with research institutions is essential to develop standardized technologies for enhancing crop quality, processing, and the extraction of valuable secondary metabolites. Additionally, KVKs actively preserve ITK, foster partnerships with research bodies, promote cultivation, and raise awareness about the nutritional benefits of these crops among farmers, thereby stimulating a profitable market (Mattu, 2017).

Platforms for Knowledge Sharing and Collaboration

The landscape of platforms dedicated to preserving ITK is diverse and vibrant, offering valuable insights into traditional practices and techniques. The Honey Bee Network is celebrated for its innovative approach in safeguarding and sharing indigenous wisdom, while the Memory Banking Protocol serves as a repository for storing and disseminating traditional knowledge. Meanwhile, the United Nations Educational, Scientific and Cultural Organization's Management of Social Transformations (UNESCO-MOST) program spearheads international comparative research in social sciences, contributing to a deeper understanding of global societal dynamics. In addition, the inventory of ITK in Agriculture by ICAR, New Delhi, comprehensively documents traditional agricultural practices, preserving invaluable knowledge systems in the form of Compact Disc (Rwela, 2018).

Traditional Ecological Knowledge (TEK) and current Holistic Knowledge Systems(HKS) must be reconciled to solve modern agricultural problems (Pandey et al., 2017; Rivero-Romero et al., 2016). Examples show that TEK, based on long-term natural resource management observations and practices, may help ecosystems adapt and thrive, especially in the face of environmental disruptions like climate change (Tagliari et al., 2023; Kamakaula, 2024). Traditional knowledge, like rainfall prediction and weather patterns, has influenced farmers' agricultural decisions, emphasising the significance of incorporating it into current agricultural systems (Bheemappa et al., 2022).

Supporting traditional and modern agricultural methods with subsidies for sustainable inputs and incentives for innovative technology is essential for farming sustainability (Sekhar, 2024;Vijayan, 2022). Ecological intensification emphasises biological control to manage agroecosystems for sustainable agriculture (Doré et al., 2011). Green agricultural methods that address soil degradation, water shortages, and climate change are vital for resilient agricultural systems (Dönmez, 2024). ITK and experimental research can help sustain agricultural systems under complicated agro-environmental conditions (Allan et al., 2013).

CONCLUSION

The integration of Indigenous Technical Knowledge (ITK) with contemporary Holistic Knowledge Systems (HKS) is essential for addressing the complex challenges of modern agriculture. This chapter has highlighted the invaluable role of ITK in ensuring food security, conserving biodiversity, and promoting sustainable environmental management. Across India, traditional agricultural practices—ranging from localized greenhouse adaptations to indigenous water conservation methods—demonstrate a deep-rooted understanding of ecological balance and resource optimization.

By preserving and promoting ITK, we can enhance soil fertility, reduce dependency on synthetic inputs, and build climate-resilient farming systems. The conservation of underutilized crops, traditional postharvest techniques, and sacred groves further exemplify the interconnectedness between cultural heritage and sustainable agriculture. These time-tested practices not only support environmental sustainability but also empower smallholder farmers by reducing input costs and enhancing long-term productivity.

To fully realize the potential of ITK, it is imperative to bridge the gap between traditional knowledge and modern scientific research. This requires proactive policy support, scientific validation, and farmer-led participatory approaches to ensure that ITK is integrated into national and global agricultural frameworks. Strengthening agricultural extension services, incorporating ITK into educational curricula, and fostering cross-disciplinary collaborations will further accelerate this integration.

The journey toward sustainable agriculture is not a choice but a necessity. By harmonizing ITK with modern innovations, we can build resilient agricultural systems that honor traditional wisdom while addressing contemporary challenges. Moving forward, a collective effort from policymakers, researchers, and farming communities is needed to ensure that ITK remains a driving force for sustainable agriculture and food security. Through collaboration and knowledge-sharing, we can pave the way for a more inclusive, resilient, and ecologically sound agricultural future.

REFERENCES

Adam, K., & Apou, K. (2013). Indigenous traditional knowledge (ITK)-an useful tools in pests and diseases management. *Indian Agriculturist*, *57*(3), 185–190.

Allan, C., Nguyen, T., Seddaiu, G., Wilson, B., & Roggero, P. (2013). Integrating local knowledge with experimental research: Case studies on managing cropping systems in italy and australia. *Italian Journal of Agronomy*, *8*(2), 15. DOI: 10.4081/ija.2013.e15

Ashwath, M. N., Bodiga, D., Wagmare, B., Dechamma, N. L., Hosur, S. R., Toushif, P. K., & Dinesha, S. (2023). Threatened and near threatened underutilized edible fruit species of southern India for food security and diversifying agroecology. *Indian Journal of Ecology*, *50*(1), 19–31. DOI: 10.55362/IJE/2023/3847

Baloyi, T. M., Maphanga, T., Madonsela, B. S., Mongwe, Q. G., Malakane, K. C., Grangxabe, X. S., & Gqomfa, B. (2024). Indigenous Strategies for Managing Bush Encroachment in Rural Areas of South Africa. *Challenges*, *15*(3), 33. DOI: 10.3390/challe15030033

Bhan, S., Bharti, V. K., Bharti, S., & Bharti, S. (2016). Impactful Indigenous Technical Knowledge (ITK) for sustainable agriculture. *Journal of Soil and Water Conservation*, *15*(4), 349–355. DOI: 10.5958/2455-7145.2016.00023.0

Bheemappa, A., Shruthi, S. M., Maheshwari, K. D., & Biradar, N. (2022). Indigenous technical knowledge based rainfall prediction: A review. *Agricultural Reviews (Karnal)*, *43*(2), 205–210. DOI: 10.18805/ag.R-2347

Borah, N., Bey, B. S., & Deka, N. (2023). Indigenous technical knowledge (ITK) used in agriculture by selected ethnic communities of Assam. [IJTK]. *Indian Journal of Traditional Knowledge*, *22*(2), 264–272. DOI: 10.56042/ijtk.v22i2.46568

Borthakur, A., & Singh, P. (2020). Indigenous knowledge systems in sustainable water conservation and management. In *Water conservation and wastewater treatment in BRICS Nations* (pp. 321–328). Elsevier., DOI: 10.1016/B978-0-12-818339-7.00016-3

Caudill, C., Pulsifer, P. L., Thumbadoo, R. V., & Taylor, D. R. F. (2024). Meeting the challenges of the un sustainable development goals through holistic systems thinking and applied geospatial ethics. *ISPRS International Journal of Geo-Information*, *13*(4), 110. DOI: 10.3390/ijgi13040110

Chan, M. (2018). *Ten years in public health 2007-2017: report by dr margaret chan director-general world health organization*. World Health Organization., DOI: 10.1093/ww/9780199540884.013.249876

Das, A., Ramkrushna, G. I., Choudhury, B. U., Munda, G. C., Patel, D. P., Ngachan, S. V., ... & Kumar, M. (2012). Natural resource conservation through indigenous farming systems: Wisdom alive in North East India.

De, L. C. (2021). Traditional knowledge practices of North East India for sustainable agriculture. *Journal of Pharmacognosy and Phytochemistry*, *10*(1S), 549–556.

Deka, S., Nath, R. K., Sehgal, M., Ahuja, D. B., Kakoti, R. K., & Barbora, A. C. (2017). Indigenous technological knowledge (ITK) and practices in pest management of Assam. *Annals of Plant Protection Sciences*, *25*(1), 119–125.

Deo, M. R. C., Deo, C., Wangchu, L., Sarma, P., Singh, A. K., & Dobhal, S. (2021). Ethnobotany of indigenous (Traditional) vegetables in Adi tribe of East Siang region of Arunachal Pradesh, India. *Environment Conservation Journal*, *22*(3), 299–307. DOI: 10.36953/ECJ.2021.22335

Dönmez, D., Isak, M. A., İzgü, T., & Şimşek, Ö. (2024). Green horizons: Navigating the future of agriculture through sustainable practices. *Sustainability (Basel)*, *16*(8), 3505. DOI: 10.3390/su16083505

Doré, T., Makowski, D., Malézieux, É., Munier-Jolain, N., Tchamitchian, M., & Tittonell, P. (2011). Facing up to the paradigm of ecological intensification in agronomy: Revisiting methods, concepts and knowledge. *European Journal of Agronomy*, *34*(4), 197–210. DOI: 10.1016/j.eja.2011.02.006

Gopi, R., Avasthe, R. K., Kalita, H., Kapoor, C., Yadav, A., Babu, S., & Das, S. K. (2016). Traditional pest and disease management practices in Sikkim Himalayan region. https://doi.org/DOI: 10.23910/IJBSM/2016.7.3.1543

Gupta, A. D. (2012). Way to study indigenous knowledge and indigenous knowledge system. *Cultural Anthropology*, *373*, 393.

Hanglem, A., Thokchom, R., Singh, S. S., & Thokchom, J. (2017). Harnessing the indigenous knowledge base associated with horticultural enterprise for sustainable development of north eastern state, Manipur, India. *International Journal of Current Microbiology and Applied Sciences*, *6*(10), 3266–3274. DOI: 10.20546/ijcmas.2017.610.383

Jha, A. K., Verma, V. K., Deshmukh, N. A., Rymbai, H., Assumi, S. R., Devi, M. B., & Talang, H. D. (2017). Spices for Income Enhancement in NE Region: Needs and Focus. *Training Manual on Entrepreneurship and Skill Development on Market Driven Production and Processing of High Value Crops of NEH Region*, 32.

Kamakaula, Y. (2024). Ethnoecology and climate change adaptation in agriculture. *Global*, *2*(2), 473–485. DOI: 10.59613/global.v2i2.99

Keerthana, V., & Divakaran, A. (2021). A Study on Adoption of Indigenous Technical Knowledge Practices among Tribal Farmers in Kolli Hills. https://doi.org/DOI: 10.9734/ajaees/2021/v39i1130783

Kishore, K. (2018). Geographical indications in horticulture: An Indian perspective. *Journal of Intellectual Property Rights*, 159-166.

Lal, C., & Verma, L. R. (2008). Indigenous technological knowledge on soil and water management from Himachal Himalaya.

Ludwig, D., & Poliseli, L. (2018). Relating traditional and academic ecological knowledge: Mechanistic and holistic epistemologies across cultures. *Biology & Philosophy*, *33*(5-6), 43. Advance online publication. DOI: 10.1007/s10539-018-9655-x PMID: 30595613

Angadi, S. C., Mansur, C. P., & Kammar, M. R. (2021). Indigenous technical knowledge on soil and water conservation, soil health management and tillage techniques in Bagalkot district.

Mattu, V. K. (2017). Perspectives and challenges of apiculture for sustainable agriculture under changing climatic conditions. *Journal of Research: THE BEDE ATHENAEUM*, *8*(1), 130–142. DOI: 10.5958/0976-1748.2017.00018.2

Mawere, M. (2014). *Culture, indigenous knowledge and development in Africa: Reviving interconnections for sustainable development*. Langaa Rpcig.

Mbah, M., Ajaps, S., & Molthan-Hill, P. (2021). A systematic review of the deployment of indigenous knowledge systems towards climate change adaptation in developing world contexts: Implications for climate change education. *Sustainability (Basel)*, *13*(9), 4811. DOI: 10.3390/su13094811

Mbasalaki, P. K. (2023). *14 Building Coalitions Across Structural Borders As A Form Of Radical Intimacy And Kinship1*. The Sage Handbook of Global Sociology.

Nair, K. P. (2023). *Biodiversity in Agriculture: Sustainability of Soil, Soil Fauna and Soil Flora*. Springer Nature., DOI: 10.1007/978-3-031-44252-0

Nelson, M. K., & Shilling, D. (Eds.). (2018). *Traditional ecological knowledge: Learning from Indigenous practices for environmental sustainability*. Cambridge University Press., DOI: 10.1017/9781108552998

Pandey, N. K., Somvanshi, S. P. S., Prakash, O., & Kumar, S. (2020). Indigenous knowledge of agriculture and animal husbandry practiced by monpa tribes of tawang Arunachal Pradesh. *The Pharma Innovation Journal*, *9*(4), 1012–1016.

Pandey, V., Mittal, R., & Sharma, P. (2017). Documentation and application of indigenous traditional knowledge (itk) for sustainable agricultural development. *Asian Journal of Agricultural Extension Economics & Sociology*, *15*(3), 1–9. DOI: 10.9734/AJAEES/2017/31481

Rajan, S., Sharma, M., & Mishra, P. (•••). Geographical Indication for Horticulture Commodities in Northeastern India: Opportunities and Challenges. *Indian Horticulture*, *68*(6), 21–24.

Rawat, D. S., & Kharwal, A. D. (2016). Folk knowledge on indigenous practices in North-West Himalaya with special reference to Himachal Pradesh (HP), India. *Folk (Kobenhavn)*, *1*(8). Advance online publication. DOI: 10.1007/s10745-011-9396-9

Rawat, V. S., Rawat, Y. S., & Shah, S. (2010). Indigenous knowledge and sustainable development in the Tones Valley of Garhwal Himalaya. *Journal of Medicinal Plants Research*, *4*(19), 2043–2047. DOI: 10.5897/JMPR10.191

Rivero-Romero, A., Moreno-Calles, A., Casas, A., Melgoza-Castillo, A., & Camou-Guerrero, A. (2016). Traditional climate knowledge: A case study in a peasant community of tlaxcala, mexico. *Journal of Ethnobiology and Ethnomedicine*, *12*(1), 33. Advance online publication. DOI: 10.1186/s13002-016-0105-z PMID: 27539307

Roy, S., Rathod, A., Sarkar, S., & Roy, K. (2015). Use of ITK in plant protection. *Popular Kheti*, *3*(2), 75–78.

Rwela, A. G. (2018). Indigenous Knowledge Use in Horticulture for Sustainable Agriculture Development in Tanzania: Evidence from Nyandira Village in Morogoro.

Sarkar, S., Padaria, R. N., Vijayragavan, K., Pathak, H., Kumar, P., & Jha, G. K. (2015). Assessing the potential of Indigenous Technological Knowledge (ITK) for adaptation to climate change in the Himalayan and arid ecosystems.

Sekhar, M., Rastogi, M., Saikanth, D. R. K., Rout, S., Kumar, S., & Patel, A. K.Rajesh C M. (2024). Exploring traditional agricultural techniques integrated with modern farming for a sustainable future: A review. *Journal of Scientific Research and Reports*, *30*(3), 185–198. DOI: 10.9734/jsrr/2024/v30i31871

Sharma, A., Manpoong, C., Pandey, H., Gupta, C. K., Baja, Y., Singh, M. S., & Mounglang, C. C. (2023). A comprehensive update on traditional agricultural knowledge of farmers in India. In *Wild food plants for zero hunger and resilient agriculture* (pp. 331–386). Springer Nature Singapore., DOI: 10.1007/978-981-19-6502-9_14

Sharma, I. P., Kanta, C., Dwivedi, T., & Rani, R. (2020). Indigenous agricultural practices: A supreme key to maintaining biodiversity. *Microbiological advancements for higher altitude agro-ecosystems & sustainability*, 91-112. https://doi.org/DOI: 10.1007/978-981-15-1902-4_6

Sharma, P. K., Srivastava, S., & Chandauriya, M. (2023). Indigenous Knowledge and Traditional Practices for Water Resource Management in Rajasthan, India. In *Traditional Ecological Knowledge of Resource Management in Asia* (pp. 137–157). Springer International Publishing., DOI: 10.1007/978-3-031-16840-6_9

Sheffield, J. (2005). Systemic knowledge and the v-model. *International Journal of Business Information Systems*, *1*(1/2), 83. DOI: 10.1504/IJBIS.2005.007401

Singh, A. K., Janakiram, T., & Singh, J. (2020). Status of indigenous and minor vegetables Research "Way forward. *Indian Horticulture*, *65*(3).

Souravi, K., & Rajasekharan, P. E. (2019). Traditional Knowledge in Horticulture. *Conservation and Utilization of Horticultural Genetic Resources*, 621-636. https://doi.org/DOI: 10.1007/978-981-13-3669-0_21

Tagliari, M., Bogoni, J., Blanco, G., Cruz, A., & Peroni, N. (2023). Disrupting a socio-ecological system: Could traditional ecological knowledge be the key to preserving the araucaria forest in brazil under climate change? *Climatic Change*, *176*(2), 2. Advance online publication. DOI: 10.1007/s10584-022-03477-x

Vijayan, D., Ludwig, D., Rybak, C., Kaechele, H., Hoffmann, H., Schönfeldt, H. C., Mbwana, H. A., Rivero, C. V., & Löhr, K. (2022). Indigenous knowledge in food system transformations. *Communications Earth & Environment*, *3*(1), 213. Advance online publication. DOI: 10.1038/s43247-022-00543-1

Wotschack, P., Vladova, G., de Paiva Lareiro, P., & Thim, C. (2023). Learning via assistance systems in industrial manufacturing. an experimental study in an industry 4.0 environment. *Journal of Workplace Learning*, *35*(9), 235–258. DOI: 10.1108/JWL-09-2022-0119

Yamunarani, K., Sangeetha, S., Dhanushkodi, V., Chitra, K., & Dhanalakshmi, K. (2023). Application of Indigenous Technological Knowledge (ITK) for Plant Disease Management in Tamil Nadu, India. *Asian Journal of Agricultural Extension. Economia e Sociologia*, *41*(12), 310–315. DOI: 10.9734/ajaees/2023/v41i122334

Zlatanović, D., & Mulej, M. (2015). Soft-systems approaches to knowledge-cum-values management as innovation drivers. *Baltic Journal of Management*, *10*(4), 497–518. DOI: 10.1108/BJM-01-2015-0015

Chapter 12
Plan B Applications in Science Technology and Environmental Teaching

Canan Koçak Altundağ
https://orcid.org/0009-0006-2844-473X
Hacettepe University, Turkey

Sadık Taşkıran
https://orcid.org/0009-0002-9661-0518
Hacettepe University, Turkey

ABSTRACT

The aim of this study is to consider both face-to-face and distance education in effective activity planning during the educational process, and to create virtual alternatives for activities conducted in face-to-face education. Hybrid educational activities will be designed by planning the activities of 6 selected topics in science lessons to be implemented both in face-to-face and distance education. In recent years, situations have arisen where face to-face education may be disrupted due to reasons such as pandemics and earthquakes. Making alternative plans in education will ensure the progression of the process without interruption when transitioning to distance education. Although considered for mass distance education, hybrid activity plans will also be highly useful for students who cannot attend school for a period due to health problems.

INTRODUCTION

In our age, the face of the universe continues to be changed by science and technology, which develops at a rapid pace in parallel with it, while the new methods developed rapidly increase the power and richness of education and training. Because besides the great advances in science and technology, education is also developing as an applied science. In order for individuals to become constructive and innovative individuals, education introduces new methods and techniques to ensure the development of science teaching. In addition to equipping students with knowledge, it is necessary to train them as people who have acquired the skills and behaviours to solve the problems they face. Programs are being

DOI: 10.4018/979-8-3693-3542-0.ch012

developed that provide students with basic principles and concepts rather than information based on rote memorisation, that give them the habit of independent thinking by teaching subjects in an investigative way, that give importance to personal examination and research, and to laboratory studies that enable them to obtain information on their own. Within the framework of these programs, it is necessary to plan in accordance with the specific objectives of the course and to use laboratory tools and equipment in harmony with each other. Science is a branch of science that examines the nature and behaviour of all substances in the universe and thus uses the information obtained to meet the needs of humanity, peace and happiness. If science was not at the current level, the clothes, plastics, medicines, cosmetics used in daily life would not be as they are today. Science, which is an intellectual activity, is intertwined with human activities and other disciplines. Knowing science enables people to make sense of what is happening around them.

Education and teaching environments nowadays continue with face-to-face education. Face-to-face education is crucial for students' socialization, integration into daily life, and active engagement in education. In recent years, there have been negative developments globally or nationally that have disrupted education and teaching. Globally, the 'pandemic' affecting the world and nationally, earthquakes affecting Turkey have prevented face-to-face education. The negative events did not allow enough time to make additional plans and implement them, resulting in disruptions in education and teaching. In order to address these disruptions, a hybrid education model has been designed, and the necessity of having an alternative plan ready in the online environment when face-to-face education is disrupted has emerged. Our alternative plan will be readily available in case of adverse situations and education can continue without interruption. Our alternative plan includes online alternatives for activities included in the normal lesson plan. Our alternative plan can not only be used in case of disruptions to education and teaching but also as a hybrid education model that can be used to support face-to-face education being extended to the home environment. Our alternative plan, which we consider as Plan B in education, will replace the main plan in case of a general disruption in education and will create a hybrid education environment as an argument to support education in the normal process. In addition to a general disruption in education, if individual students who have health problems cannot attend school, they will be able to implement this plan and overcome the process without falling behind in their lessons. The inability to conduct face-to face education may lead to disruptions in education and teaching. The inability to carry out planned education and teaching on time can cause a decrease in student motivation. In subjects involving spiral education, the failure to learn a topic can decrease the learning motivation of a student who does not have prerequisite knowledge for the topic to be taught in other classes. In a scenario where education cannot be conducted on a general scale, immediately transitioning to online classes can instill a sense of creating an alternative when students encounter difficult situations. Incorporating individual students who cannot attend regular classes into alternative education can prevent the negative thoughts that falling behind in lessons may create for the students. The aim of this research is to ensure uninterrupted continuation of education when face-to-face teaching is not possible, by transitioning to remote education for middle school students. Our study, which includes alternative plans alongside face-to-face teaching plans, encompasses a functional planning that can be implemented collectively when transitioning to remote education or individually when remote education is required.

The research problem is formulated as follows: “Does making alternative remote teaching plans when face-to-face teaching is not possible contribute to the uninterrupted continuation of education?”

Which factors should be considered when creating an effective remote education plan?

What remote education approaches can be used to meet the needs of students with different learning styles and skills?

It is assumed that students' participation in face-to-face education will continue in the same manner in remote education.

Our study covers the attainment of learning objectives for six topics at the middle school level.

BACKGROUND

Science Education

Science education is a process aimed at teaching students the fundamental principles, concepts, and methods of the sciences. This process aims to develop scientific thinking skills, enhance the ability to understand and explain natural phenomena. Additionally, it aims to equip students with the skills to conduct scientific research and access knowledge. According to the nature of science theory in science education; the student understands how scientific knowledge is produced and the importance of the scientific method (Driver, 2008). Changes, which are an important part of the formation process of the universe, have been going on for billions of years. Science, which scientifically investigates and interprets the natural cycles taking place in the universe, is a discipline that continues to shed light on science, especially in the fields of matter and matter composition. They grasp that science is a part of life and that by understanding science, they can understand and even manage nature. Through experiments, observations, and evidence, they acquire scientific knowledge (Millar, 2005).

From ancient civilizations to today's societies, scientists living in each new year have added different dimensions to science and have supported developments that can be evaluated scientifically. Science education encompasses topics from various fields such as biology, chemistry, physics, and other branches of the natural sciences through an interdisciplinary approach. According to the Structural Theory, 'the student learns by connecting new knowledge to existing knowledge' (Ausubel, 1968). The student uses their existing knowledge to facilitate learning and enriches it with their experiences to ensure lasting learning. They understand concepts using analogies, models, and analogies (Novak, 1986). Experiences that not only provide theoretical knowledge but also stimulate the mind and enrich existing knowledge with new information play an active role in learning. The mind 'organizes its knowledge through a semantic network' (Bruner, 1966). Science education has seen the development of various approaches aimed at improving teaching methods. Some of these approaches include:

Activity-Based Learning: An instructional approach that encourages active participation of students and focuses on developing conceptual understanding. (Bybee, 2010) Students should actively engage in the process and make meaning of the information through the activities they perform, ensuring lasting learning.

Technology-Enhanced Learning: An instructional approach where technology is used to enhance learning and motivate students. (Prensky, 2001) Today, technology and education have become inseparable components. Whether in face-to-face or remote education, technology has become a stakeholder in education. Using technology will not only facilitate active learning but also concretize events that would otherwise be impossible to visualize with the capabilities of technology.

Experiential Learning: It is a fact accepted by all educators that experimental applications are indispensable for science teaching. This research has revealed that the best success in science education can be gained by learning based on experimental methods, but this can only be realized with well-trained teachers. In addition, it has been determined that students who receive laboratory supported science education are more successful in terms of 'cognitive and affective'. An instructional approach that assists students in connecting theoretical knowledge with practical experiences. (Kolb, 1984) Students should be able to achieve meaningful learning by applying existing theoretical knowledge into practice. In Turkey, science education is carried out in accordance with the curricula determined by the Ministry of National Education (MEB). These curricula aim to teach students fundamental concepts and principles in the field of science while also developing their scientific methods and research skills. Science education at the middle school level in Turkey involves conveying basic concepts and scientific methods in the field of science to students. This education covers topics selected from the branches of biology, chemistry, physics, and other fields within science. Science education typically aims to enhance students' scientific thinking and research skills using experimental and observational approaches. "The use of visual materials in science classes can enhance students' understanding level and make learning more effective (Yıldız, 2018)." Experimental applications include a multifaceted education in which basic concepts and scientific skills in science are acquired. Students obtain, interpret and increase their knowledge by observation, analysis of data collected by scientific methods and comparison with hypotheses. With experimental applications, it is aimed to provide a view of science within the framework of personal and world view.

Science teaching experiments contribute to the development of scientific process skills in students at any level. Experimental studies, observations, and analyses often play a significant role in science classes. When providing science education to students, the role of teachers is crucial. Teachers should create an effective teaching environment to increase students' curiosity and interest. Additionally, they should help students develop positive attitudes towards science and support their scientific thinking skills.

Science education also involves the effective use of technological tools. Students are provided with practical experiences in science classes through computers, tablets, experiment kits, and other technological tools. The effects of student-centered teaching approach in science education support students in gaining a deeper understanding of science topics and ensuring lasting learning (Yılmaz, 2020). This way, students not only learn science more effectively but also learn to use technology simultaneously. Research and practices conducted on science education in Turkey also provide important insights into the effectiveness and efficiency of science education. Findings from these studies contribute to the continuous improvement of science education methods and practices (Ministry of National Education, 2022). Today, educational technologies are impacting every aspect of life. Education and teaching have been particularly affected more than other areas, especially with the closure of schools worldwide and the initiation of distance learning practices during the Covid-19 pandemic. This process has accelerated the integration of education and technology like never before. New concepts such as distance learning, blended learning, and hybrid teaching have become popular in the process of integrating education and technology. When creating an alternative distance learning plan for science education in place of face-to-face teaching, several considerations should be taken into account. Preparing alternative distance learning plans for science education becomes crucial when face-to-face teaching is disrupted. An effective distance learning plan can minimize disruptions in teaching by ensuring that learning objectives are met, and student engagement and motivation are maintained.

Steps for Plan Preparation

Goal Setting: Set clear and measurable learning objectives for each lesson. Explain how the remote learning plan will achieve these objectives.

Content Selection: Choose engaging lesson content that aligns with the learning objectives. Utilize various resource materials (videos, texts, images, audio, interactive applications, etc.). Provide content that appeals to different learning styles.

Activities and Assessments: Plan various activities that encourage active student participation. Use different assessment methods to track learning levels. Be clear about providing feedback to students and sharing assessment results.

Communication and Interaction: Establish clear and regular communication with students and parents.

Provide opportunities for asking questions and giving feedback.

Utilize different communication channels (email, forums, chat platforms, etc.)

Infrastructure and Resources: Ensure that the necessary infrastructure and resources (internet access, computers, software, etc.) are available for implementing the distance learning plan. Provide access to additional resources if needed.

Support and Guidance: Provide necessary support and guidance to students and parents. Offer access to technical support and learning materials. Plan additional training and guidance programs if necessary.

Sample Distance Learning Plan Elements

Lesson Plan: Prepare a detailed lesson plan for each class. The plan should include learning objectives, content, activities, assessments, and a timeline.

Lesson Materials: Provide accessible access to the lesson materials you have prepared (videos, texts, images, audio, interactive applications, etc.).

Activity Guide: Prepare a detailed guide for the activities that students need to complete. The guide should include explanations of the activities, necessary materials, and a timeline.

Assessment Tools: Prepare assessment tools to track students' learning levels (tests, assignments, projects, etc.)

Communication Platform: Utilize a platform (email, forums, chat platforms, etc.) to communicate with students and parents.

Supporting Materials: Provide additional supplementary materials to students (lecture notes, study sheets, videos, etc.).

When preparing a distance learning plan, it is essential to consider the following:

- The plan should be focused on learning objectives.
- The plan should provide diverse and engaging content.
- The plan should cater to different learning styles.
- The plan should provide clear and open communication and interaction opportunities.
- The plan should rely on necessary infrastructure and resources.
- The plan should provide necessary support and guidance to students and parents.

Remote education is a method where learning materials and communication tools are delivered to students without the need for physical presence. In this model, the student and teacher may not be in the same physical environment, but interaction is facilitated through various digital platforms. This model provides educational materials to students through various digital tools and platforms. Moore (2013) states, "In distance education, the role of the moderator is to encourage student participation, guide discussions, and enhance interaction among students (Moore, 2007)." Historically, distance education, which began with materials sent by mail, has transitioned to being conducted through digital platforms with the widespread availability of the internet. Remote education offers advantages such as flexibility, overcoming geographical limitations, and accommodating different learning styles. However, it also brings along some challenges, such as the lack of face-to-face interaction and the requirement for technological infrastructure. According to the theory of independence, students need to possess independent study skills to be successful in distance education (Moore, 1983). Distance education will foster the development of students' self-regulation skills. While in face-to-face education, students are under the control of the teacher, in distance education, they must develop self-regulation and self-monitoring skills. They need to organize which learning tasks should be done when and how, independently. According to interaction theory, interaction between students and teachers plays a significant role in distance education (Garrison, 1993). Interaction should facilitate student engagement in a process where the teacher acts as a guide. Interaction should be fostered both between the teacher and students and among the students themselves, aiming to prevent the detachment from the social environment that face-to-face education provides in distance education.

Distance Education

Distance education offers students flexibility of time and location. Students can receive education anytime and anywhere they want. Distance education provides the opportunity for education without geographical limitations. This is a significant advantage for individuals facing difficulties accessing traditional educational settings, such as working individuals, homemakers, or people with disabilities (Holmberg, 2005). Catering to different learning styles: Distance education can utilize various instructional methods that appeal to different learning styles. Distance education platforms often allow students to progress at their own pace and access materials suitable for their learning styles. This can better cater to students' individual learning needs (Keegan, 1990).

Progress at their own pace: In distance education, students can progress at their own pace and manage their own learning.

More cost-effective: Distance education can be more cost-effective compared to traditional education.

Providing education opportunities to disabled students and students living in remote areas: Distance education can provide education opportunities to disabled students and students living in remote areas.

Disadvantages of Distance Education: Limited interaction: Interaction between students and teachers may be limited in distance education.

Lack of motivation: Students may struggle to stay motivated in distance education

Need for technological skills: Students in distance education need to have the necessary technological skills

Need for time management skills: Students in distance education need to have developed time management skills.

Lack of social interaction: In distance education, students may not have the opportunity to interact face-to-face with other students and teachers. Distance education reduces the opportunity for students to come together physically in a classroom, which can increase the risk of social isolation and limit interaction among students (Anderson, 2008).

Technical issues with distance education platforms or students' lack of access to appropriate technology can negatively impact the learning experience. This can be especially pronounced for students from low-income or rural areas (Hodges et al., 2020). Distance education should be seen not as an alternative to traditional education, but as a complement. When used alongside traditional education, distance education can provide students with more learning opportunities.

Blended Learning

Blended Learning is an educational model that combines traditional in-classroom instruction with online education. In blended teaching and learning, face-to-face instruction is supplemented with online elements. Teachers enrich the course content by providing additional resources such as documents, videos, and applications for students to use outside of class time. This model integrates traditional in-classroom teaching with online education to offer students a variety of learning experiences. In this model, students experience learning both in a face-to-face classroom environment and through online platforms. Bonk and Graham (2012) emphasize that blended learning provides students with learning opportunities tailored to various learning styles. With the increasing accessibility of technology for both students and teachers, it can be argued that many classrooms have to some extent adopted a blended structure. Blended learning offers students learning opportunities tailored to various learning styles and encourages effective use of technology.

Blended Learning Benefits

Increases student engagement: Blended learning provides students with more opportunities for active participation, making learning more engaging. While preserving face-to-face interaction in the traditional classroom environment, the blended learning model enhances communication among students through the opportunities offered by digital platforms. This can contribute to students' social and emotional development (Garrison & Kanuka, 2004).

Encourages personalized learning: Blended learning can offer a personalized learning experience by allowing students to progress at their own pace and manage their own learning. It enables students to personalize their learning processes (Vaughan, 2007).

Addresses different learning styles: Blended learning can employ various learning methods that cater to different learning styles. It provides students with personalized learning experiences where they can progress at their own pace and personalize their learning processes (Vaughan, 2007).

Provides time and location flexibility: Blended learning allows students to learn anytime and anywhere, offering them flexibility in time and location

Promotes the use of technology: Blended learning encourages students to conduct research, prepare presentations, and interact with each other using technology. By bringing together various learning tools, blended learning provides students with environments suitable for different learning styles, increasing access to materials tailored to visual, auditory, and kinesthetic learning styles (Bonk & Graham, 2006).

Blended Learning Application Examples

Online pre-class and post-class assignments: Students can watch online lessons in advance to acquire information about the topics to be covered in class and can complete online assignments after class to reinforce what they have learned.

Discussion forums: Students can share their ideas and interact with each other on online forums regarding the topics discussed in class.

Simulations and games: Students can use simulations and games to practice in an environment similar to real-life situations and enhance their skills.

Portfolios: Students can create online portfolios containing their work and projects demonstrating what they have learned.

Social media: Students can utilize social media platforms to acquire information about the topics covered in class and interact with each other.

Blended learning plays a significant role in education in the 21st century. This model provides students with opportunities not only offered by traditional education but also by online learning, thereby offering a more effective learning experience. However, blended learning also has some disadvantages. In blended education, students may lack access to technology or develop dependency on it. This can be a significant disadvantage, especially for students living in low-income or rural areas (Graham, 2013). For the effective implementation of the blended learning model, teachers need to be proficient in using digital tools effectively. This may increase the need for teacher training and ongoing support (Means et al., 2009). Maintaining students' motivation and discipline in blended education can be challenging. The lack of physical presence in the classroom and teacher supervision may lead to a lack of motivation and discipline issues among students (Kizilcec et al., 2013).

Hybrid Learning

Hybrid learning is an instructional approach that integrates the learning experiences of both online education and face-to-face instruction. In this teaching model, students benefit from both in-class interactions and continue learning through online platforms. Hybrid education provides flexibility for students while also maintaining face-to-face interactions, thereby supporting social and emotional development. "Hybrid education provides flexibility for students while also maintaining face-to-face interactions and offers potential to enhance student success (Garrison & Kanuka, 2004)." However, achieving success in hybrid learning requires a balanced use of technology and effective monitoring of student progress. Hybrid education is an innovative learning model that blends traditional face-to-face education with online learning. In this model, students have the opportunity to acquire knowledge and skills both in the classroom and online, independent of time and location constraints. As a result, the learning process becomes more flexible, personalized, and interactive, adapting to the needs and expectations of the modern world.

Basic Features of Hybrid Education

Flexibility: Hybrid education provides students with flexibility in terms of time and location, offering educational opportunities to individuals who may be deprived of traditional education due to busy work schedules or living in different geographical locations. As a result, students can receive education anytime and anywhere they want. The hybrid learning model allows students to determine their own learning pace and personalize their learning processes by providing flexibility (Graham, 2013).

Personalization: The hybrid education model, which caters to different learning styles and paces, enables each student to progress according to their individual needs and potential. Students can continue learning at their own pace and develop their skills in managing their own learning. Hybrid learning brings together various learning tools to provide students with environments suitable for different learning styles. This model enhances access to materials suitable for visual, auditory, and kinesthetic learning styles (Bonk & Graham, 2012).

Interaction: Hybrid education combines face-to-face interaction offered by traditional classroom settings with various tools provided by online platforms, resulting in a rich learning experience. "Hybrid learning preserves face-to-face interaction in traditional classroom settings while enhancing communication among students with the opportunities provided by digital platforms. This can contribute to students' social and emotional development (Garrison & Kanuka, 2004)." Students can interact with both their teachers and peers through online forums, discussion groups, and project work.

Accessibility: Hybrid education promotes equality in education by providing more access to education for disabled students and those living in remote areas compared to traditional education. As a result, every student is offered educational opportunities, ensuring equality of opportunity in education.

Benefits of Hybrid Education

Increased Student Engagement: Hybrid education provides students with more opportunities for active participation, thereby involving them more in the learning process. Consequently, students' motivation and interest in learning increase. "Hybrid learning combines traditional classroom interaction with digital technologies to offer students an enriched learning experience. This model enables students to shape their own learning journeys while benefiting from various resources (Vaughan, 2007)."

Personalized Learning Experience: Hybrid education offers each student a personalized learning experience by allowing them to progress at their own pace and manage their own learning. As a result, each student can maximize their potential.

Addressing Different Learning Styles: The hybrid education model, which incorporates various learning methods that cater to different learning styles, helps each student acquire knowledge and skills according to their learning style.

Utilization of Technology: Hybrid education contributes to the development of 21st-century skills by encouraging students to use technology for various purposes such as research, preparing presentations, and interacting with each other.

Cost-effectiveness: Hybrid education may be more cost-effective compared to traditional education. Particularly, the reduction in transportation and infrastructure expenses makes hybrid education a more economical option.

Some Applications Used in Hybrid Education

Online Pre- and Post-Class Assignments: Students can watch online lessons in advance to acquire information about topics to be discussed in class and complete online assignments after class to reinforce what they have learned.

Discussion Forums: Students can share their ideas on topics discussed in class and interact with each other in online forums.

Simulations and Games: Students can use simulations and games to practice in an environment similar to real-life situations and enhance their skills.

Portfolios: Students can create online portfolios showcasing their work and projects to demonstrate what they have learned.

Social Media: Students can use social media platforms to acquire information about topics discussed in class and interact with each other.

Blended Learning

Blended learning is a learning model that combines traditional face-to-face teaching methods with digital technologies. In this model, students experience both face-to-face interactions and online learning through digital platforms. Garrison and Vaughan (2008) emphasize that blended learning provides flexibility for students while preserving face-to-face interaction. "Blended learning combines face-to-face interaction with digital tools to provide students with various learning experiences (Whiteside & Lodge, 2019)." Blended learning not only offers flexibility and diverse learning experiences for students but also encourages the effective use of technological tools. For teachers, it provides opportunities to diversify course materials and better monitor student progress. In the 21st century, education must adapt to the changing world order and evolving technologies. In this context, blended learning emerges as a flexible, personalized, interactive, and accessible educational model that offers both the opportunities of traditional education and those of online education. In this study, we will examine the fundamental features, benefits, and applications of blended learning.

Blended Learning Key Features

Integrated Structure: Blended learning combines the social interaction and face-to-face communication provided by traditional education with the flexibility and accessibility offered by online education, creating a balanced learning environment (Graham, 2006; Garrison & Cleveland-Innes, 2005)

Personalized Learning: The blended learning model addresses different learning styles and paces, allowing each student to progress at their own pace and manage their own learning (Pichler et al., 2007; Ertmer & Ottenbreit-Hotz, 2010).

Increased Participation: Blended learning provides students with more opportunities for active participation, thereby involving them more in the learning process. This enhances students' motivation and interest in learning (Means et al., 2010; Hew & Brush, 2007).

Broad Access: Blended learning promotes educational equity by offering greater access to education for disabled students and those living in remote areas compared to traditional education (Simonson et al., 2013; Sharma & Mishra, 2015).

Benefits of Blended Learning

Enhanced Learning Experience: Blended learning enriches the learning process by combining the interaction provided by traditional classroom settings with the diverse tools offered by online platforms, making learning more engaging and enjoyable (Johnson et al., 2011; Oliver & Herrington, 2007).

Advanced Technology Use: Blended learning encourages students to use technology for various purposes such as research, presentation preparation, data analysis, and problem-solving, thereby contributing to the development of 21st-century skills (Ertmer, 2009; Hew & Brush, 2007). Easy access to the internet and online information sources in blended learning enables students to enhance their research skills. Skills such as analyzing different types of data, evaluating reliable sources, and synthesizing findings are critically important in various fields today.

Cost Advantage: Blended learning may be more cost-effective compared to traditional education. Particularly, the reduction in transportation and infrastructure costs makes blended learning a more economical option (Allen & Seaman, 2010; Pichler et al., 2007).

Some Applications of Blended Learning

Pre- and Post-Class Online Assignments: In blended learning, online lessons and assignments play a crucial role in enriching and reinforcing the learning process. This allows students to acquire information about topics to be covered in class beforehand and to reinforce what they have learned afterward by completing online assignments (Pichler et al., 2007; Ertmer & Ottenbreit-Hotz, 2010).

Virtual Discussion Spaces: Students can share their ideas and interact with each other about topics discussed in class through online forums, chat groups, and video conferences (Garrison & Cleveland-Innes, 2005; Johnson et al., 2011).

Simulations and Games: Online simulations and games enable students to practice in environments similar to real-life situations and enhance their skills (Oliver & Herrington, 2007; Sharma & Mishra, 2015).

Blended learning blends traditional and online learning methods, focusing not only on knowledge transfer but also on developing the skills students need in the 21st century. Active use of technology in blended learning plays a significant role in developing essential skills such as research, presentation, data analysis, and problem-solving. Thus, blended learning emerges as an educational model that prepares students for the future and equips them to face the challenges of the 21st century.

METHOD

The aim of science is to learn the structure of matter by examining it, and then to reach new substances by changing this structure and to make the daily life of human beings easier and healthier. It is not enough to see the perfect balance in nature and integrate with it. We should be able to ask ourselves questions about the working mechanism by analyzing nature. The answers we find will carry us to new horizons. First of all, science develops by asking questions. Instead of learning for an exam, we should prefer to learn because we are curious. Progress and development start with the question 'what if'. By researching and observing the subject of curiosity, a process that develops by revealing the problem and continues to produce solutions is reached. It is not correct to think of scientists as a person who only deals with solving problems and producing solutions. It is quite natural to find mistakes as well as the truths he/she finds while developing his/her work. Another scientist who comes after him/her will reach new researches, observations and different results by taking the truths of the previous study. The person who deals with science should know how to share his/her knowledge first of all. Science and technology today is both an individual and a social event. The work done or to be done will affect not only the scientist but the whole world. It is an indication that you can be a good scientist if you are a person who does not accept everything you are told as true, researches, asks, questions, closely follows technology, communicates well with people, and shares your findings with others.

This section presents the expert opinions on the current plan of our study and the alternative distance education plans prepared as part of this research. There are various teaching methods that can be used in distance education. The theoretical framework of this thesis will cover the following distance education methods:

Online Courses: These are courses that students can access and complete via the internet.

Live Webinars: These are real-time webinars that allow active participation from students.

Asynchronous Learning: A learning method where students can progress at their own pace, independent of time and location constraints.

Discussion Forums: Online platforms where students can interact with each other and share their ideas.

FINDINGS

This chapter includes a summary table of the teaching plans to be implemented in face-to-face instruction, the implementation of alternative distance education plans, and the links to the distance education applications. The activities proposed in the research were tried to be planned by considering the design cycle. With these activities, students are expected to make sense of knowledge, associate it with daily life, experience it and transform what they have learnt into practice under the guidance of the teacher. In addition, in the proposed activities, students are expected to identify a need or problem from daily life related to the topics addressed in learning-teaching practices. It is also aimed that these problems support students to develop original ideas. The teacher selects the appropriate one from the activities presented as plan A and plan B by comparing alternative ways and makes planning for the solution.

Activities:

- Absorption of Light
- Shadow Length Activity
- Mass and Weight Relationship
- Recycling Topic: "Separation of Waste at Source Activity"
- Electricity Unit: "Effect of Number of Bulbs on Brightness Activity"
- Periodic Table Topic: "Matching Elements and Symbols Activity"

Plans A and B were prepared to guide science teachers in the process of designing activities in line with the Science Education Program. The activities in the study were prepared by taking into consideration the field skills, conceptual skills, social-emotional learning skills and values in the Science Education Program and other components of the program. The activities were designed to be student-centred and based on experimental applications in accordance with the main features of the Science Education Program. Teachers should take into consideration the developmental characteristics, needs and interests of the students in their classes as well as the curriculum when preparing activities. From this point of view, the activities in Plans A and B were prepared as examples. Teachers are expected to make necessary adaptations on the activities according to the characteristics, needs and interests of the students. The activities were placed in Plans A and B in no particular order. Plans A and B included more than one example of activities for each domain skill, social-emotional learning skills and values. In addition, the researcher conducted experiments during the development of Plans A and B, and the qualitative and quantitative data obtained were analyzed.

Table 1. Plans A and B

7th Grade "Interaction of Matter with Light" Unit: "Absorption of Light Activity"	
Face-to-Face Education	Distance Education
Activities involve testing the absorption of light by objects of different colors by placing thermometers on fabrics of different colors, in water of different colors, and by popping balloons of different colors with a lens.	The temperature increase on fabrics of different colors is designed as a matching game on the web. https://phet.colorado.edu/sims/html/color-vision/latest/color-vision_all.html?locale=trhttps://wordwall.net/tr/resource/70877460
5th Grade "Propagation of Light" Unit: "Shadow Length Activity"	
Face-to-Face Education	**Distance Education**
For the Shadow Length Activity, a setup is designed where the length of the shadow is measured and compared as the object is moved closer to or further away from the light source. This allows for the conclusion that the shadow length increases as the object moves closer to the light.	A web tool is designed to demonstrate how the shadow length changes with the movement of the object and light. Students are provided with a link to access this tool. https://scratch.mit.edu/projects/986928555
7th Grade Force and Energy Unit: "Mass and Weight Relationship Activity"	
Face-to-Face Education	**Distance Education**
A section of the MTA (Mineral Research and Exploration) trip is planned to calculate the weight variations on different planets.	A web-based planet-weight matching activity is created. https://phet.colorado.edu/tr/simulations/masses-and-springs-basicshttps://phet.colorado.edu/sims/html/masses-and-springs/latest/masses-and-springs_all.html?locale=trhttps://sanaltur.mta.gov.tr/
Electricity Unit: "Effect of Number of Bulbs on Brightness Activity"	
Face-to-Face Education	**Distance Education**
An experiment is conducted where the number of bulbs is increased while keeping the battery count, wires, and switch constant, demonstrating that the brightness decreases.	An animation is created to show that increasing the number of bulbs, while keeping the battery count, wires, and switch constant, results in decreased brightness. https://phet.colorado.edu/tr/simulations/circuit-construction-kit-ac https://wordwall.net/play/69983/939/215
Recycling Topic: "Separation of Waste at Source Activity"	
Face-to-Face Education	**Distance Education**
A social responsibility project is created at school where plastic, glass, and paper waste are collected separately in large washing machine boxes.	Students are provided with a link to a web tool designed for zero waste, where they can perform a drag-and-drop activity to sort waste according to its type. https://www.cevkocочuk.org/ext/atiklari-yakala
Here is the translation for the periodic table activity	
Face-to-Face Education	**Distance Education**

continued on following page

Table 1. Continued

7th Grade "Interaction of Matter with Light" Unit: "Absorption of Light Activity"		
For the element and symbols matching activity, element names and symbols are prepared as magnets. The element names are attached to the board, and the symbols are placed on the table. Teams are formed, and when the timer starts, teams begin matching the elements with their symbols. The group that completes the first error-free match wins the game.	For the element and symbols matching activity, buttons containing element names and symbols will be prepared. The activity will be designed using web tools that provide feedback when correct matches are made. https://wordwall.net/tr/resource/69648584 https://bilimgenc.tubitak.gov.tr/periyodik-tablo	
Topic of the Activity Plan	**F.7.5.1. Absorption of Light**	**Activity 1 Plan A**
Lesson	**Science**	
Class	**7**	
Unit Title	Unit 5: Interaction of Light with Matter	
Topic	Absorption of Light	
Learning Outcome	F.7.5.1.1. Discover that light can be absorbed by matter as a result of its interaction with materials.	
Materials	**Different Colored Fabrics, Different Colored Waters, Thermometer (Plan A)**	

PRELIMINARY WORK

- Recall the information from previous lessons that white light actually contains all colors of visible light.
- Explain with illustrations why objects of different colors appear as they do. Provide information that each object absorbs all colors in white light except for its own color, which is reflected and therefore perceived as that color.
- Show with illustrations that white reflects all colors of light.
- Provide information that black absorbs all colors of light.

Part: How to Conduct the Experiment

- On a sunny day in the schoolyard, three fabric pieces made of the same material but in different colors are taken. The initial temperatures of the fabrics are measured and recorded.

Every 10 minutes, new temperatures are observed and recorded in the table.

Table 2. Temperature table

	White	Red	Black
Initial Temperature			
10th minute			
20th minute			
30th minute			
Difference			

Independent Variable:
Dependent Variable:
Constant Variable (Controlled):

30 ^{0}C

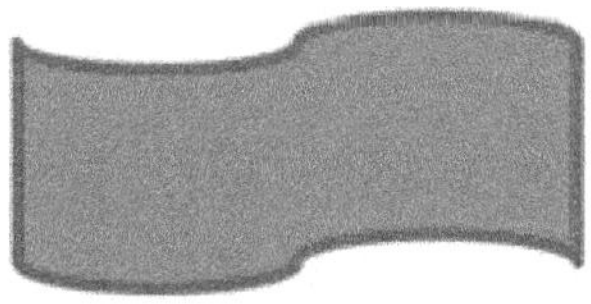

30 ^{0}C

30 ^{0}C

Figure 3. Representation of independent, dependent, and constant variables

C)

Every 10 minutes, new temperatures are observed and recorded in the table.

Table 3. Temperature table

	White	Red	Black
Initial Temperature			
10th minute			
20th minute			
30th minute			
Difference			

Independent Variable:
Dependent Variable:
Constant Variable (Controlled):
Part 2

RESULTS AND DISCUSSION

Finally, it is demonstrated that the balloon does not burst when the lens is held to the light-colored part but bursts when held to the dark-colored part. This is explained by the fact that the dark part absorbs more light, causing the balloon to burst.

Figure 4. Balloon and light absorption experiment

Table 4. Activity 1, Plan B

Topic of the Activity Plan	**F.7.5.1. Absorption of Light**	**Activity 1 Plan B**
Lesson	Science	
Class	7	
Unit Title	UNIT 5: Interaction of Light with Matter	
Topic	Absorption of Light	
Learning Outcome	F.7.5.1.1. Discover that light can be absorbed by matter as a result of its interaction with materials.	
Materials	Different Colored Fabrics, Different Colored Waters, Thermometer (Plan B)	

What Needs to be Learned: It should be understood that the amount of light absorbed by an object varies according to its color, and consequently, the temperature increases differently.

Materials: Different colored fabric pieces, different colored liquids, thermometer.

PRELIMINARY WORK

- Remind students from previous lessons that white light contains all colors of visible light.
- Explain with illustrations why objects of different colors appear as they do. Provide information that each object absorbs all colors in white light except for its own color, which is reflected and perceived as that color.
- Show with illustrations that white reflects all colors of light.
- Provide information that black absorbs all colors of light.

Activity: The link to the activity prepared in the web environment is provided. https://wordwall.net/tr/resource/70877460 Selecting Information that Matches the Visual.

Figure 5. Representation of how colors absorb light

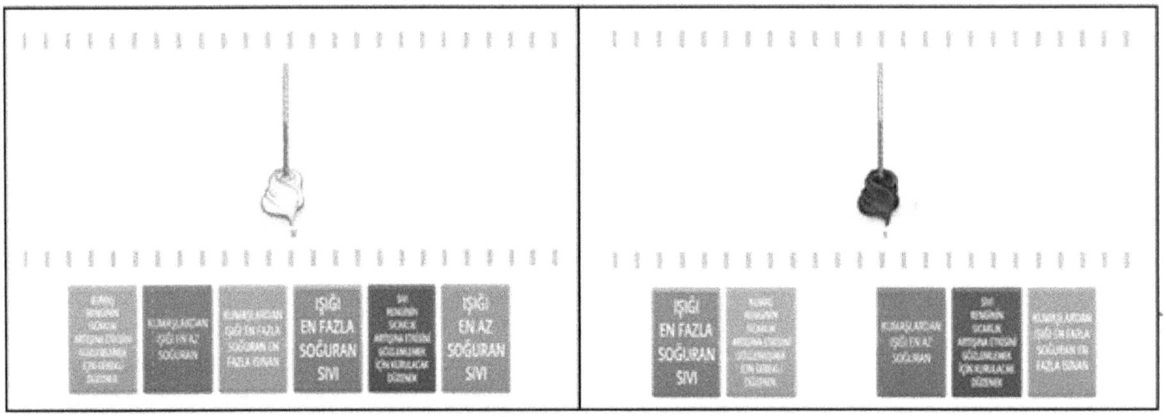

https://phet.colorado.edu/sims/html/color-vision/latest/color-vision_all.html?locale=tr The link is shared. It is demonstrated through the activity that white light contains different colors and that the combination of these colors forms white light again. It is specified that selecting the light color and filter is necessary to observe different color formations.

Figure 6. How light absorbs light to perceive color

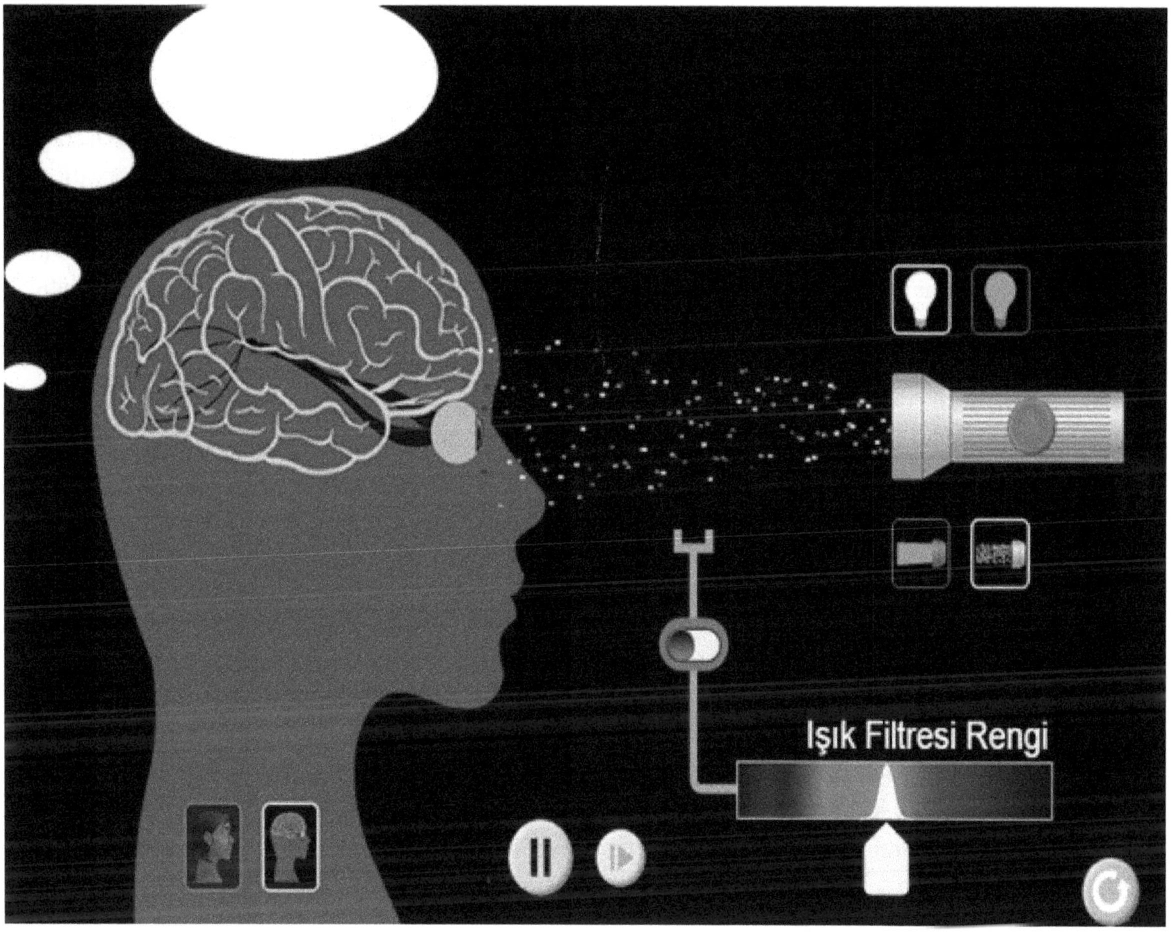

The activity is conducted simultaneously or asynchronously, depending on the available resources. It is expected that students will conclude that white light is formed from the combination of all colors.

The science area in Plans A and B aims to nurture students' natural curiosity and understanding of science phenomena in everyday life. Plans A and B aim to enable students to follow science-related events in daily life, to understand the functioning of natural laws and to enjoy science.

The activities in Plans A and B are designed to enable students to collect data through observations and to define the qualities of the observed object, event or phenomenon using these data. Observations made by students about the natural environment, especially starting from their immediate surroundings, are important for the acquisition of scientific process skills. Because observations are also a basic scientific process skill and are frequently included in other field skills. For this reason, students' acquisition of scientific observation skills is one of the important goals of Plans A and B.

It involves students grouping and labelling objects, events and phenomena according to their similarities or differences based on their observations using some qualifying features/variables. Students are expected to group the causes and consequences of science-related events or phenomena in their daily lives and name these groups. Classification skills are one of the skills that students acquire quickly by nature. This skill is also used in other skill areas.

The observation-based predictions in Plans A and B aim for students to make predictions about some situations using the observation data they have obtained and to think about the predictions made. This skill can be considered as the first step of using the language of science.

The predictions based on scientific data in Plans A and B are similar to the observation-based prediction skill, but they also refer to making predictions based on data obtained outside of observation or preliminary experiences. Science literacy requires understanding and using data-based expressions. In this way, it can contribute to the development of students' science capital. Data-based predictions also support the development of higher-order thinking skills. Data-based predictions about the use of natural resources, environmental problems and the development of living things can be used for this purpose.

In Plans A and B, it is aimed at designing and conducting simple experiments about the events or situations they are curious about. In this way, it is aimed that students enjoy the work they do in the field of science. This can be done through experiments that the student prefers or is curious about to increase his/her interest and motivation towards science.

Within the scope of inference-making skill in Plans A and B, the student's ability to express science-related situations based on data is included. The ability to make inferences about an event enables the formation of cognitive structures in the student's mind for the development of inductive and deductive reasoning skills. Therefore, this skill forms a bridge for the transition to higher level science domain skills.

Plans A and B involve students explaining science events or phenomena using models and making use of existing scientific models. This skill enables students to concretize abstract science events or phenomena and to make them suitable for students' age and developmental level.

The science activities in Plans A and B can be given in co-operation with different fields due to the scientific discipline in itself. Integration in this way can improve the content of science activities and contribute to the acquisition of skills. In this context, science activities can be made more effective and fun by associating them with other fields.

RESULTS, DISCUSSION, AND RECOMMENDATIONS

Today, the idea of becoming an information society, which is the primary goal of societies, is possible by raising new generations in a way that they can adapt to changes and developments. This can be achieved through student-centered environments where learning by doing and experiencing is realized. One of the teaching environments where learning by doing and experiencing is realized is science. In science teaching, laboratories are seen as an important factor in learning because they provide first-hand experience by observation and experimentation. Effective teaching and learning processes are significantly enhanced when students are actively involved in the instructional process. In face-to-face education, various activities can easily engage students in the learning process. When transitioning to remote education due to circumstances such as pandemics or natural disasters, or when individual students need to continue their education remotely due to personal health issues, providing alternative educational options allows students to participate in the learning process without losing valuable time. Having a pre-planned alternative in place ensures that education remains dynamic in situations where face-to-face instruction is disrupted. Additionally, these alternative methods can supplement in-person classes and positively impact academic achievement. Integrating the six topics and models developed in this study into all subjects and providing alternative links or QR codes in textbooks will facilitate easy navigation to the virtual alternative plans.

Active student participation and involvement are critical for an effective education process. Traditional face-to-face education allows for easy engagement through various activities and interactions. In cases where face-to-face education is disrupted due to pandemics, natural disasters, or individual health issues, alternative education models come into play. This approach ensures that students can continue their education without loss of time. Pre-planned and prepared alternative education plans provide flexibility and resilience, allowing education to proceed without interruption during unforeseen circumstances. This study developed and implemented six different alternative education models. The findings indicate that these models enhance student motivation and participation and positively affect academic performance. Comprehensive integration of the six alternative education models into all subjects and courses is recommended. Textbooks should include QR codes or virtual links to easily guide students to alternative education plans. Alternative education models should be developed for various subjects and their application areas expanded. These models should be designed to encourage active participation and increase interaction. The effectiveness and efficiency of alternative education models should be regularly assessed and improved as needed.

Alternative education models are essential tools that complement traditional education methods and ensure that education continues without interruption during challenging situations. Widespread implementation and application of the models developed in this study across different subjects will enhance the flexibility and resilience of the education system.

REFERENCES

Alkan, F. (2015). The role of technology in science teaching activities: Web based teaching applications. *Journal for the Education of Gifted Young Scientists*, *3*(2), 1–7. DOI: 10.17478/JEGYS.2015213531

Ausubel, D. P. (1968). *Educational psychology: A cognitive view*. Holt, Rinehart & Winston, Inc.

Bonk, C. J., & Graham, C. R. (2006). *Blended learning: In theory and practice*. Jossey-Bass.

Bonk, C. J., & Graham, C. R. (2012). *The hybrid learning imperative: Seven principles for implementing blended learning successfully*. Jossey-Bass.

Bruner, J. S. (1966). *Toward a theory of instruction*. Harvard University Press.

Bybee, R. W. (2010). Advancing STEM education: A 2020 vision. *Technology and Engineering Teacher*, *70*(1), 3035.

Ertmer, P. A. (2009). Blended learning in K-12 schools: A review of the literature. *Journal of Educational Technology & Society*, *12*(1), 25–35.

Ertmer, P. A., & Ottenbreit-Hotz, J. (2010). Blended learning in K-12 schools: A review of the literature. *Journal of Educational Technology & Society*, *13*(2), 25–35.

Garrison, D. R. (1993). A Cognitive Constructivist View of Distance Education: An Analysis of Teaching-Learning Assumptions. *Distance Education*, *14*(2), 199–211. DOI: 10.1080/0158791930140204

Garrison, D. R., & Cleveland-Innes, S. (2005). *Blended learning in online education: Theory, practice, and policy*. Jossey-Bass.

Garrison, D. R., & Kanuka, H. A. (2004). Blended learning in a K-12 school: One school's experience. *Journal of Asynchronous Learning Networks*, *8*(1), 1–17.

Graham, C. R. (2006). Blended learning: An effective model for K-12 education? *Educational Technology Review*, *52*(4), 233–255.

Graham, C. R. (2013). Blended learning: An effective model for K-12 education? *Educational Technology Review*, *59*(2), 233–255.

Jonassen, D., Spector, M. J., Driscoll, M., Merrill, M. D., van Merrienboer, J., & Driscoll, M. P. (2008). *Handbook of research on educational communications and technology: a project of the association for educational communications and technology*. Routledge.

Hodges, C., Moore, S., Lockee, B., Trust, T., & Bond, A. (2020). The difference between emergency remote teaching and online learning. *EDUCAUSE Review*, *27*(1), 1–9.

Holmberg, B. (2005). *Distance education: Culture and creativity*. Routledge.

Johnson, D. W., Johnson, R. T., & Smith, K. A. (2011). Cooperative learning: Returning to the classroom. *Educational Leadership*, *68*(7), 26–31.

Kizilcec, R., Sarı, H., & Uyar, D. A. (2013). The effectiveness of blended learning: A meta-analysis of literature. *Computers & Education*, *62*, 210–221.

Kolb, D. A. (1984). *Experiential Learning: Experience as the Source of Learning and Development*. Prentice Hall.

Means, B., Toyama, Z., & Vavrus, S. (2009). Implementing blended learning: Where the rubber meets the road. *The International Journal of Learning and Teaching in Educational Settings*, *1*(1), 3–24.

Moore, M. G. (1983). The individual adult learner. In Tight, M. (Ed.), *Adult Learning and Education* (pp. 153–168). Croom Helm.

Novak, J. D., & Gowin, B. D. (1984). *Learning How To Learn*. Cambridge University Press. DOI: 10.1017/CBO9781139173469

Prensky, M. (2001). Digital Natives, Digital Immigrants Part 1. *On the Horizon*, *9*(5), 1–6. DOI: 10.1108/10748120110424816

Sharma, R., & Mishra, S. (2015). Blended learning in higher education: A review of the literature. *International Journal of Learning and Teaching in Higher Education*, *36*(2), 181–201.

Vaughan, D. (2007). Blended learning: An effective model for K-12 education. *Journal of Asynchronous Learning Networks*, *11*(1), 1–10.

Vaughan, D. (2007). Blended learning: A framework for the effective integration of online and face-to-face learning. *International Journal of Learning and Teaching in Educational Settings*, *1*(1), 3–30.

Whiteside, K., & Lodge, J. M. (2019). Blended learning in higher education: A review of the literature. *International Journal on Teaching and Learning in Higher Education*, *30*(2), 228–242.

Yucel, A. S. (2006). E-learning approach in teacher training. *Turkish Online Journal of Distance Education*, *7*(4), 123–131.

KEY TERMS AND DEFINITIONS

Remote Teaching: A method of education where students and teachers are not physically present in the same environment, and classes are mostly conducted through the internet or other communication technologies.

Hybrid Teaching: In this method, traditional face-to-face teaching is combined with remote learning methods. While some students are physically present in the classroom, others can attend the same class remotely by connecting online.

Blended Teaching: Blended teaching combines different teaching methods and resources to provide students with various learning experiences. In this method, face-to-face interactive lessons are combined with online education, student-centered learning, and other pedagogical methods.

Mixed Teaching: Mixed teaching is an approach that combines traditional in-classroom education with online education. In this method, students benefit from both face-to-face classroom interactions and complete lesson materials and activities on online platforms.

Face-to-Face Lesson Plan: This plan is prepared for lessons conducted in a traditional classroom environment where students and teachers are physically present together. It usually follows a specific curriculum and includes elements such as lesson content, objectives, materials, activities, and assessment methods.

Remote Teaching Lesson Plan: This plan is prepared for lessons conducted through online platforms without students and teachers coming together. It includes elements such as the online platform for lesson delivery, communication tools, presentation of content, activities, assignments, and assessment methods. These lesson plans typically provide flexibility to support students' independent study as well.

ENDNOTES

1 Prepared from Hacettepe University Institute of Educational Sciences Graduate Project without Thesis.

Chapter 13
Pushing a Square Peg Into a Round Hole by Integrating IKS Into SA Technology Education Didactics

Tomé Awshar Mapotse
https://orcid.org/0000-0002-7234-2785
University of South Africa, South Africa

ABSTRACT

South Africa's Technology Education (TE) policy document encourages teachers to contextualize their teaching by infusing indigenous knowledge systems (IKS) into their lesson delivery. This chapter reports on a community engagement (CE) case with TE teachers from six selected schools in Limpopo Province of South Africa, where action research methods were applied. The community engagement project also known as the engaged scholarship and societal impact (ESSI) project was conducted with TE teachers in Mogoshi circuit. The southern theory underpinned the study and action research collaborative or cooperative paradigm undergirded the study. From these six schools, located in three villages, only eight TE teachers from five of the ESSI identified schools participated in a structured interview. The focus of the study was to inquire from the TE teachers how do they integrate IKS in their Technology classes during lesson delivery, after three years of their IKS emancipation through ESSI contact sessions.

INTRODUCTION

The Africanisation of Technology Education (TE) as a school subject has an agenda that far exceeds its mere importation into, and alignment with, the African context (Mapotse & Gumbo, 2019). Now lately a call to use indigenous knowledge systems (IKS) and Africanise knowledge is a demand which acknowledges the need to redress the damage done by ideologies such as apartheid in the South African context, and the general colonial domination and suppression of Africans (Msila & Gumbo, 2016). In South Africa (SA), the apartheid government embarked on a programme of destroying indigenous knowledge, while at the same time instituting laws that denied black school children modern Science and Technology Education: Africans were deemed fit for nothing than just certain forms of labour – manual

DOI: 10.4018/979-8-3693-3542-0.ch013

labour that would serve the most basic needs of apartheid's technocratic society (Serote, 1998). The colonizers of the education system portrayed African IKS as primitive, unscientific, sub-standard, and good-for-nothing (Kamwendo, 2017). The intention to silence IKS practice, on the part of the apartheid government in SA country, affected the training of teachers in both pre-and post-democratic dispensations. Despite numerous concerted efforts to destroy IKS, Shara (2016) asserts that modern Science is uncovering 'new' knowledge from old indigenous pearls of wisdom.

Teachers' didactical training in their field of specialization does not adequately include IKS, which should, in fact form part of every subject taught at the school level. The TE policy, for instance, has room for subject teachers to incorporate IKS in their lesson delivery. Teachers are free, in this regard, to invite a knowledgeable member of their community to join their class and lead the presentation of IKS during their lesson delivery. This can prove very helpful for TE teachers since indigenous people (the knowers) are creators of indigenous knowledge; they give it discourse and meaning based on and relating to, the experiences gained through interacting with their environment (the known), over time (Shava & Manyike, 2018). Most indigenous knowledge tends to wane away before it is recorded: its storage repositories are literally buried, as in most cases they are the elderly in our villages and rural areas. African homes are spaces where African bits of knowledge are created, with older people serving as the carriers of validated knowledge, while younger generations enjoy learning from them as apprentices (Mji, Kalenga, Ned, Alperstein & Banda, 2017). In many African countries, even though new indigenous knowledge is often uncovered, its application is not recognized within the formal schooling system, legal frameworks, or policy structures.

African scholars are not the only ones leading calls for the indigenization of the curriculum at the school level: Webb (2016) reports that several regions of the world, including New Zealand, Canada, the Philippines, and China, have embarked on indigenizing their curricula and education systems. All countries that are doing so, however, should prioritize their teachers' IKS didactics. The indigenization of the curriculum should be guided by the principles contained in countries' educational agenda, and this has prompted calls for an African Renaissance in education on this continent of Africa.

The call for an African Renaissance in education seeks to demonstrate how indigenous African epistemologies can be tapped as foundational resources for the cultural and socio-educational transformation of the continent, and how these epistemologies can be politically and economically liberating (Higgs, 2016). Besides a window of opportunity opening for indigenous African epistemologies to be liberating, they can spearhead the recognition of IKSs as true ethical pieces of knowledge which are to be valued. Ngulube (2017) believes that colonial literature does not have a good story to tell about indigenous communities, depicting them as inferior and unimaginative. Ngulube (2017) further stresses that the postcolonial literature continues to marginalize and neglect indigenous knowledge (IK). That marginalization of, and ignorance about, IK have paralyzed the ability of TE teachers to transfer related and relevant knowledge to their learners, as recommended by the subject policy. The TE teachers who participated in the study on which this chapter is based were members of a community engagement (CE) or engaged scholarship and societal impact (ESSI) initiative project that the author headed up in rural Limpopo Province of South Africa.

Maelene (2002) warns Africans never to forget their indigenous roots, even when subjected to the process of urbanization. TE teachers should follow their didactics, as enshrined in the technology policy pertaining to IK. Unfortunately, many Africans' greed for money (Maathai, 2009) has led them to forget who they are, and this has created a breeding ground for them to lose their indigenous orientation and roots (Mji, 2012), to the extent that they regrettably promote the ideologies of the colonizers. There was a perception, on the part of the early missionaries who first came to Africa wearing the hats of doctors/educators/ministers of religion, that they were entering a zone of pure ignorance regarding matters of civilization and knowledge (Mji et al., 2017). That is not the case. Many elders in local communities, to whom IK has been passed down over generations, are more knowledgeable about matters relating to IKS, while many TE teachers are ignorant in this regard. Gumbo and William (2014), who conducted focus group discussions as part of their study, exposed TE teachers' ignorance about the integration of indigenous technologies in their subject.

Pushing a square peg into a round hole is an English saying that refers to a person finding him/herself in a situation that is unsuited to his/her abilities or character (Kavanagh, 1978). TE teachers are square pegs since they cannot confidently deliver TE lessons that are infused with IKS. IKS is the round hole: unfortunately, TE teachers' abilities to deliver IKS lessons are limited due to a lack of exposure and the demonizing of IKS systems. TE teachers' ignorance of IKS and/or the devaluation of its role make it difficult for them to be part of the related discourse. This ignorance on the part of the 'square pegs' (TE teachers) unfortunately denies upcoming generations their cultural 'round hole' (IKS discourses). If TE teachers continue with this trend of ignorance, their learners will be prevented from cultivating the skills and knowledge that will enable them to solve the societal problems they are confronted with, be they indigenous or technological in nature. In Africa, IK is passed down from one generation to another through story telling or folklore not through a textbook. While the skills are transferred from one generation to the next through organised mentorship programmes.

BACKGROUND

There has lately been a gradually growing interest in IKS research trend due to its affluent and complex of both theory and practice based on the historical relationship between human advancement and the environment. Traditional research, has however, been more rigorous in the overview of IKS but Action research (AR) and Action Learning (AL) will unpack the interrelationship of IKS and Technology Education in details proportions. It should further be noted that both AR and AL concepts are constructive, innovative, and creative quality way to solve problems and to develop professionally while recognising human beings irrespective of their status in the society, forming relationships and regular reflections. Hence the TE teachers were chosen to be co-researchers in this study for quite some time just to make sure they integrate IKS in their lesson presentation as per national TE policy document. The TE teachers who are co-researchers or participants or interviewees in this study were engaged with IKS emancipation during ESSI for a period of three years which was split into contact sessions of one week per term (four months).

The IKS hypothesis maintain that, with such a conscious research, researchers can contribute towards the excavation, preservation, and knowledge generation of indigenous peoples. Taking lessons from the past for a individuals and communities to live a meaningful and beneficial life. The chapter impact on the teachers' professional development, to investigate and draw attention to IKS continuing practices from the past amongst indigenous peoples from which their learners come for the TE teachers to apply those practices in their lesson presentation. Scholars have been conducting research ever since time immemorial yet relatively few research practitioners are close to respond to the old Marxist question: "how can we change the world to become a better place for human beings?" This chapter aims at providing TE teachers with a practical way to conduct innovative and creative self-reflective integration of IKS in

TE classroom for progressive transformative and sustainable learning environments' towards answering the old global question. For the TE teachers to shape the future of the upcoming generation they must join their learners to the past indigenous ways of doing things. Learning can be made easier if learners can identify themselves with the curriculum.

The IKS has been developed by indigenous people since the beginning of civilization, including herbal medicine production, blacksmithing, woodcarving, textile-weaving and dyeing, leather works, fishing, beadworks, pottery making, architecture, agriculture breeding, metal work, salt production, gold-smithing, copper-smithing, leather-crafting, soap-making, bronze-casting, hunting and war weapons, canoe-building, brewing, glassmaking, agriculture, mining, and interpreting the milky-galaxy, climate change and weather signs. Skill transfer was done through mentorship and story-telling folklore from one generation to the next. Most of these civilisation components has been adopted by many countries in their education systems and are taught within the classroom to produce engineers, technologists, bakers, artists (in all forms), plumbers, electricians, paediatricians', architects, builders, astronomers, educators, bankers, biologists, carpenters, florists, horticulturalists, geologists, gardeners, wedding planners, judges, lawyers, jewelists, philosophers, psychotherapists', lexicographers, farmers, navy, armies, merchants, scientists, meteorologists, cartographers, postmen, zookeepers, veterinarians, undertakers, jockeys, politicians, etc. All these listed professionals have been shaped through many centuries by incorporating some form research in each field. This chapter was a study with TE teachers to make sure they don't downplay IKS in the classroom with their learners.

The next sections address the research problem, conceptualization of the study, IKS didactics within TE, the theoretical framework, research methodology, findings, recommendations for future integration of IKS into the TE classroom, and the study is finished with a conclusion.

STATEMENT OF THE PROBLEM

The main research question is, 'How can the action research practitioner make a square peg (teachers) round, to fit into a round hole (IKS content and TE curriculum)?' In this instance, the action research practitioner was an ESSI project leader from a higher education institution (HEI).

Both the main research question and the sub-questions (see Table 1) were formulated to underscore the advantages of having a square peg fit into a round hole: achieving the perfect fit is, after all, envisaged by the African Renaissance (Mbeki, 1998). Table 1 lists the sub-questions posed during the structured interviews with TE teachers. These questions served as data-collection instruments during the study. The responded were probed to support their answers.

Table 1. Research sub-questions posed during the structured interviews.

1	**Do we have only one way to learn IKS during TE practices?**
2	Can diversely people from all cultures contribute to IKS?
3	Does this believe hold water: technologists are creative, and elders are IKS resourceful?
4	In IKS does Science and Technology complement each other?
5	IKS and Technology were they there before science?

continued on following page

Table 1. Continued

1	**Do we have only one way to learn IKS during TE practices?**
6	Is IKS part of the social and cultural traditions of every community?
7	Your take on the statement is: IKS transferable, Science experimental, and Technology practical?
8	Why IKS rely mostly on the excavation, preservation, and knowledge generation of local people?
9	How did a strong Renaissance agenda strengthen IKS research approach?
10	Is there a way in which the indigenous knowledge contributes to creating the historical experiences of ordinary local people?

These sub-questions sought to determine how TE teachers were exposed to IKS during their pre-and in-service training, by their district facilitators. The sub-questions stated in Table 1 guided the study, as the author used structured interviews as the data-collection instrument. It was assumed that most TE teachers had never been exposed to IKS during their pre-service (and worse, during in-service) workshops. The teachers' facilitators did not cover IKS integration with TE didactics either, which made it seem that those teachers had been placed in a situation unsuited to their abilities since they did not know how to integrate IKS during their lesson delivery. Engaging with these communities of TE teachers was the least that could be done to encourage them not to look down on IKS, but instead to promote it. It was through such engagement, using action research methodologies, that their love of indigenous knowledge was rekindled, this statement responds to the main research question. The ensuing section unravels the terms mostly used in this study.

CONCEPTUALIZATION

During the TE didactic situation teachers should make sure they enable learners to solve their community or society problems technologically. If learners can acquire the technological skills, they will be able to confront their societal challenges and extend human abilities as they further improve life of their society, they find themselves in. On the other hand, indigenous knowledge (IK) is being transferred verbally from one generation to another, not recorded per se. This IK prepare learners to solve problems they encounter in life as they grow towards adulthood. It is incumbent on the teachers during their didactics to transfer both TE skills and IK to capacitate learners to confidently face the world and its challenges while navigating through life by solving life problems through societal and cultural knowledge.

Didactics

Skatkin and Lerner (1970–1979) view didactics as methods of instruction meant to achieve interconnected activities, between the teacher and learners, to facilitate the acquisition of knowledge and skills, the formation of a worldview, and the development of aptitudes. Didactics is a theory of education that specifies what elements of the culture people have accumulated, to transmit as educational content, to deliver to educated individuals. Didactics is a theory of teaching and learning, and the process of designing educational programs. It involves selecting content, structuring learning experiences, and assessing student understanding.

Merriam-Webster Dictionary (n.d), https://www.merriam-webster.com/dictionary/didactic first gave us the etymology of didactics before it provides its meaning. *Didaktikós* is a Greek word that means "apt at teaching." It comes from *didáskein*, meaning "to teach." Something didactic does just that: it teaches or instructs. *Didactic* conveyed that neutral meaning when it was first borrowed in the 17th century, and still does; a didactic piece of writing is one that is meant to be instructive as well as artistic. Parables are generally didactic because they aim to teach a moral lesson. *Didactic* now sometimes has negative connotations, too, however. Something described as "didactic" is often overburdened with instruction to the point of being dull.

A Technology class can be used as a platform to teach learners about diverse cultures since they themselves come from multi-dimensional, multi-linguistic, multi-cultural, and pluri-religious backgrounds. TE didactics offers an opportunity to tap into those elements of culture which learners have accumulated over time, thanks to their diverse traditions. TE is thus not only a mental undertaking but also a hands-on endeavour, where the aim is to help learners acquire new skills and knowledge, including of IKSs.

Technology or Technology Education

Notably, the definition and development of TE have been greatly affected by the Western cultural conception of the subject, which confines it to the conventional model of industry (Mapotse & Gumbo, 2019). As these authors further emphasize, 'this linear approach to Technology has unfortunately denied the global South, which is predominantly indigenous, to value IKS' (Mapotse & Gumbo, 2019). Technology as a subject became part of the national curriculum in South Africa after democracy in 1994. It involves the use of knowledge, skills, values, and resources, to meet people's needs and wants, by enabling them to develop practical solutions to problems, taking social and environmental factors into consideration (Department of Basic Education, 2011). In line with this study, the definition of Technology from the author's lenses can be read as follows:

> Technology is the application of both African indigenous and Western bits of knowledge, including the technologies within an indigenous community, to resolve societal problems technologically whilst enhancing creativity and innovation as well as reviving traditional cultures, values, and aptitudes, by preserving both the environmental and communal influences.

Technology is about humankind's purposeful mastering and creative use of knowledge and skills in products, processes, and approaches to better control their environment (Gumbo, 2019). Technology teachers engage learners in hands-on projects to solve design problems using a plethora of skills including investigation, design, making, evaluation, and communication (Blose & Gumbo, 2024). Technology includes accumulating experiences generated by indigenous people and integrating with culture to fulfil life's needs, solving problems while maintaining sustainability and a harmonious relationship with nature (Hill et al., 2020; Umoh & Jacob, 2020).

Indigenous Knowledge Systems or Indigenous Knowledge

Indigenous Knowledge Systems (IKS) encompass the wisdom, knowledge, and practices of indigenous people, developed over time through direct interaction with the environment and nature (*My* Courses, 2024). South Africa has infused indigenous knowledge into its school curriculum, and this is evident in a subject such as Technology. IKSs have been included as part of the human rights principles undergirding the school curriculum, as enshrined in the South African constitution. Key aspects of indigenous knowledge include people (indigenous people), context (Spatio-temporal), culture, language, knowledge, practices, and dynamism (Shava, 2013). This study has inspired the researcher, as an action research practitioner, to redefine IK in the following manner:

Indigenous knowledge refers to the bits of knowledge from indigenous communities the world over which were acquired over a lengthy period and transferred from one generation to another, in a variety of ways, to preserve cultural practices whilst observing the community's evolution into Technology practices, due to the industrial revolutions.

Different scholars from diverse part of the globe define IK differently, but highlighted here are those that speak to the context of the present study:

- Zidny et al (2021): IK is broadly described as indigenous people's knowledge of a particular culture in the local community [from Indonesia].
- Mji et al. (2017): IK is an essential transformative instrument that could enhance the beingness and becoming of Africans [from both SA and Zambia].
- Koehler (2017): IK is a global concept referring to knowledge that arises out of the respective cultural, historical, and social backgrounds of individuals, in other words, prior knowledge is primarily shaped by the family and community [from Germany].
- Tapfuma (2017): IK is a local knowledge that is unique to a community or society that has been developed over time [from Zimbabwe].
- Lutomia and Bello-Bravo (2017): IK is a context-specific knowledge that communities have developed themselves for centuries, and which allowed them to live in their environment for often long periods of time [from the USA].

In Canada, the land is associated with indigenous knowledge, whereas in South Africa land is equated to wealth; its redistribution and ownership constitute a hot potato in parliament and present a headache for the recent government of national unity. Conversations concerning land in a settler society in Canada relate to the relations of people to place, the fundamentals of sustenance, the massive changes wrought by capitalism and colonization, and pressing questions around environment and indigeneity (Corbett & Gereluk, 2020). Across the centuries, the indigenous peoples of sub-Saharan Africa maintained complex practices for preserving and transmitting knowledge from one generation to the next. This knowledge was practical, social, moral, artistic, historical, and religious. Practical skills spanned a wide range of activities, from the production of food and shelter to techniques for treating disease. In West Africa, griots were historians, storytellers, praise singers and poets, acting as repositories of oral tradition (Unisa, n.d).

Synonymous terms used for 'indigenous knowledge' include local, native, or traditional wisdom. IK is dynamic and has been improving over time, in keeping with people's latest desires, needs, and wants. While the advancement of technological devices and machinery may render some IKS practices obsolete, the fact remains that, in many parts of the world, those IKS practices are still applicable, practiced, and valued. IKS can still be applied and practiced in the education milieu, more so by TE teachers, for instance working with beads, or recycling plastic for making mats/hats, freeing a bird artefact from a cow's horn, creating ornaments from wires, etc.

IKS DIDACTICS WITHIN TECHNOLOGY EDUCATION LESSONS

In the introduction, it was highlighted that most teachers cannot deliver TE lessons that integrate IKS, as required by the technology policy document, due to their ignorance of IKS. The technology policy document (Department of Basic Education, 2010) gives teachers leeway to infuse IKS in their teaching, by reiterating: 'Wherever possible, learners should be made aware of different coexisting knowledge systems. They should learn how indigenous cultures have used specific materials and processes to satisfy needs and become aware of indigenous intellectual property rights.' Even though the policy urges TE teachers to try to transfer indigenous culture to their learners, sadly many are unable to do so. The study on which this chapter is based was undertaken to encourage teachers to transfer IKS skills to their learners, and to determine whether, by engaging TE teachers through ESSI, the former can achieve the means to effect such skills transfer. Blose (2019) suggests that teachers lack IKS understanding and are not aware that IKS is included in the Curriculum and Assessment Policy Statements document (CAPS).

There are myriad issues that hinder Technology teachers to infuse the integration of IKS in their lesson delivery as highlighted by a few scholars. The lack of availability of teaching support resources prevents the introduction of IKS into the school curriculum (Blose, 2019). The obstacles, such as the stigma attached to culturally specific environmental ethics, the requirement of proof and experimentation, the use of a foreign language in schools, formal education, the loss of fear for gods and spirits in nature, centralized curriculum, democracy, and political biases, exacerbate the impediment of IKS in the school curriculum (Opoku & James, 2020). Maluleke (2019) asserts that factors such as teachers' inability to use indigenous objects, the use of a single evaluation tool for all learners, a negative attitude toward IKS, and TE teachers' inability to incorporate culturally appropriate pedagogy in the teaching of design skills, obstructs IKS integration. In addition, due to attitude, lack of understanding of IKS by learners, and insufficient IKS awareness, TE teachers encounter challenges in incorporating indigenous Technology into the curriculum (Gumbo, 2015; Kpan, 2020).

The is a concerted effort by some African scholars to unpack IKS even though a lack of clarity on what defines IKS has hampered its applicability (Dube & Munsaka, 2018) because teaching IKS is difficult for many teachers since it involves new learning (Jacobs, 2015). Likewise, Samuels (2003) highlights that older people are reluctant to reveal details, that IKS teaching, and evaluation procedures are shallowly established, and that there is an insufficient insight into the process of transition, personal bias, cultural bias, and insufficient insight into IKS and Western knowledge. The indigenous knowledge's mindset, vocabulary, holistic nature, contextual nature of IKS, undocumented format, diverse, unstructured, responsive, invalidated nature of IKS, insufficient resources, lack of a coherent stakeholder empowerment strategy, intellectual property rights, heterogeneous classes, and non-acceptance of IKS all create obstacles to indigenous knowledge's use in a classroom for teaching and learning.

It is reasonable to pronounce that education systems of African states still mirror colonial education inherited from former colonial education systems. Whilst Africa has a dual mode of education delivery (i.e., colonial education and postcolonial operation), Higgs (2016) argues that education systems in Africa in the 21st century must operate in both postcolonial context and global context. South African teachers often lack appropriate pedagogical content knowledge to infuse indigenous knowledge in TE classrooms (Mapotse, 2020). Learners often experience the world they live in and a school subject that is being taught to them as two different cultures – African and Western cultures, (Jegede & Aikenhead, 1999), especially the indigenous knowledge in TE.

The TE teachers are needing to play an influential role for learners to embrace their IKS traditions and cultures.

THEORETICAL FRAMEWORK

The emphasis is on valuing IKS by considering the theory of multiple intelligences, which should guide learning (Gumbo, 2019). This means that TE teachers should not raise Western knowledge above any other knowledge forms but should try to diversify their teaching so that it better suits both indigenous and non-indigenous learners. Africa offers the world certain views (ontologies) and knowledge systems (epistemologies) with the potential to have a significant and transformative impact on both theory and practice (Phillips, 2017), particularly in the TE field. Transformative and postcolonial paradigms and theories informed the progress of this study and underpinned its context with TE teachers. In this action research study of ESSI, the cooperative/collaborative inquiry paradigm undergirded the emancipation activities with TE teachers, whilst southern theory was used to underpin the progress of the study.

Cooperative inquiry, also known as collaborative inquiry, was first proposed by John Heron in 1971 and later expanded by Peter Reason; the main idea of a cooperative inquiry is to research 'with' rather than 'on' people (Mapotse, 2012). Reason (1999: 211) points out that in cooperative inquiry we say that knowing will be more valid—richer, deeper, more true to life and more useful—if these four ways of knowing [*Experiential knowing, Presentational knowing, Propositional knowing, and Practical knowing*] are congruent with each other; if our knowing is grounded in our experience, expressed through our stories and images, understood through ideas which make sense to us, and expressed in worthwhile action in our lives. It emphasizes that all active participants (TE teachers, in this case) are fully involved in the action research spiral and circular activities, as well as in decision-making as co-researchers. This study reflects on the responses obtained during structured interviews about IKS within TE didactical activities, undertaken through the cyclical and participatory processes of planning, acting, observing, and reflecting on TE teachers' empowerment. As such, the collaborative inquiry paradigm created a path for participants to multitask as facilitators and as dedicated co-researchers in the research process. Cooperative inquiry is a modality within the broad family of action research approaches (Coghlan, 2010). As action research, its focus is on generating practical outcomes and knowledge which distinguishes it from other reflective approaches (Coghlan, Brady, O'Leary & Hynes, 2024). Heron and Reason (2008) define cooperative inquiry 'as a form of second- person action research in which all participants work together in an inquiry group as co-researchers and co-subjects. In this cooperative inquiry, AR facilitator research a topic with TE teachers based on their own experience of TE policy, to make sense of some aspect of integrating IKS in their TE classroom and further develop new and creative ways of looking at that integration and to learn how to act to change things by contextualising their teaching through the infusion of IKS within the lesson delivery. The cooperative inquiry paradigm and southern theory created room for TE teachers to look at IKS from the unique perspective of ownership.

The southern theory promotes multicentred social science perspectives and critiques, as social science produces many forms of knowledge, and is relevant to democracy (Wahyudi, 2014). Since southern theory produces many forms of knowledge which are relevant to democracy, it was deemed appropriate that indigenous knowledge should be among the knowledge produced in conjunction with the participating TE teachers. The southern theory challenges the fact that some pieces of knowledge are displaced to the periphery by the northern theory (Gumbo, 2017). The term 'Southern Theory' can be loosely defined as a conversation that questions universalism in the field of social theory. It critiques the dominance of Western models by identifying alternative thinkers and particular issues that have been overlooked in the academies. It seeks to empower thinkers beyond the trans-Atlantic metropolitan centres to focus on experiences in their place. Africans, women, indigenous people, and all individuals whose identity is reduced to 'other' have been routinely excluded from literacy, schooling, jobs, and publication venues that would otherwise legitimate their knowledge (Collins, 2013). This theory is opposed to a linear approach to knowledge construction because its principles embrace other forms of knowledge, which make it relevant for the teaching of Technology in the South African context (Gumbo, 2019). The didactics of TE teachers' views are crucial in this democratic era, as these were not catered for during the segregationist dispensation.

RESEARCH METHODOLOGY

Most HEIs have identified ESSI as a vital component of their operational needs. The ESSI enabled the researcher to gather TE teachers in Limpopo province, for a week once every four months, to address all their challenges within the TE themes per term, and to stress the role of IKS throughout the contact sessions. This exercise offers the university (through its lecturers) an opportunity to add African voices in promoting IKS within the education curriculum. There is an existential and humane need today to decolonize the curriculum in Africa, by means of postcolonial education systems that can reclaim indigenous African voices through curriculum reform and the transformation of educational discourse (Higgs, 2012; Ramose, 2003). Odora-Hoppers (2001) supports this view that African voices now seek to contribute to the quest to return humanism to the centre of the educational agenda, and she dares educators to view the African child-learner not as a bundle of Pavlovian reflexes, but as a human being who is culturally and cosmologically located within authentic value systems. TE teachers/educators are no exception in contributing to the quest to return humanism to the educational agenda, hence they formed part of this study's research methodology.

The research methodology was guided by Connell's (2007) vision of southern theory. Gluck (2015) draws attention to 'southern' as referring to relationships of authority, exclusion and inclusion, and hegemony between the actors in the centre and those on the periphery. During apartheid, the colonizers made sure the curriculum for teachers' didactics would not cater to any IKS components. Teachers' training colleges were meant to produce teachers who looked down on their own cultural practices, observations, and knowledge. Based on southern theory, teachers were under the authority of the colonizers, and their training excluded African IKS and anything they could identify with, culturally speaking. The hope was that taking TE teachers through ESSI activities would open their eyes to the authoritarianism of the colonizers, who sought to enforce their hegemony by demonizing IK and deciding what knowledge to include or exclude. This research methodology served as an umbrella term, referring to both the research methods and study design. Structured interviews were used to collect data in this qualitatively designed action research undertaking.

Population and Sampling

The marginalization of grades 7, 8, and 9 TE teachers, and the lack of IKS integration during their lesson delivery, took centre stage in this study. TE teachers were purposively selected because they teach Technology subject and volunteered to be part of the ESSI team. Senior phase TE teachers were selected from three primary schools and three secondary schools located in three villages in Limpopo province (each of those villages has one primary and one secondary school). Teachers from one primary school in one village did not participate in the structured interviews.

Data Collection and Organization

The data collected through an action research cycle from eight Technology/TE teachers is reflected in Table 2. Structured interviews were used to collect data from TE didactics in the field of IKSs by using an interview schedule, which is a set of questions posed in a predetermined order (Bertram & Christiansen, 2020). Data were collected during the third term of the year, during weekly ESSI contact sessions with the teachers. Table 2 displays the question statements put to the participants during the structured interviews, as well as their collective responses. After securing a time slot with the TE teachers, the researcher visited their schools to conduct the interviews, when all eight interviewees/participants/co-researchers were asked to respond to question statements 1 to 10 (see Table 2). The researcher read a question to the interviewee, who had to choose an answer from three options provided, where A = Agree, UN = Uncertain and DA = Disagree. Those three answers determined the degree of consensus with the statement.

Table 2. Didactics data-collection instrument: TE didactics versus IKS integration

Didactics Questions		Participants' responses		
NO	STATEMENTS	A	UN	DA
1	There is only one way to learn IKS during TE practices	1	3	4
2	Diverse people from all cultures contribute to IKS	7		1
3	Technologists are creative, and elders are IKS resourceful	7	1	
4	Science and Technology do not complement each other in IKS		1	7
5	IKS and Technology were there before science	7	1	
6	IKS is part of the social and cultural traditions of every community	8		
7	IKS is transferable, Science is experimental, and Technology is practical	8		
8	IKS does not rely on the excavation, preservation, and knowledge generation of local people	1		7
9	The IKS research approach has been strengthened by a strong renaissance agenda	8		
10	Indigenous knowledge contributes to creating the historical experiences of ordinary local people	8		

The participating TE teachers' responses are consolidated in Table 2. Data about the infusion of IKS didactics were collected from the participants during 2019, with their ESSI having begun in 2017 (i.e., a three-year emancipation journey). In 2020, this cohort was to be certified with attendance certificates and mandated to help TE colleagues within their cluster to teach Technology with confidence, and every chance of success. As of now, SARS-2/Covid-19 pandemic and the national lockdown have delayed the conferring of such certificates until further notice.

Data Analysis and Interpretation

In this study, qualitative content analysis/textual analysis was used to analyze the data collected during the interviews. Both analyses refer to content as text, and the process involved in the analysis is the same (Bezuidenhout & Cronjé, 2014). The theorist McKee (2003) refers to text as anything we produce as an 'interpretation of something's meaning'. Figure 1 was used for data analysis, flowing from Table 1, to reflect details of the data collected during the action research process.

Bertram and Christiansen (2020) point out that, in structured interviews, questions may require closed responses (often Yes/No answers, Agree/Disagree, one-word answers, or answers where limited information is given, such as a list of schools taught at or attended), open-ended responses (often where the participants express their opinions, give their impressions or are asked to supply information that will be essentially different from other participants' answers). The interview schedule used for the purposes of this study did not contain any open-ended questions.

Figure 1. Analysis of data collected.

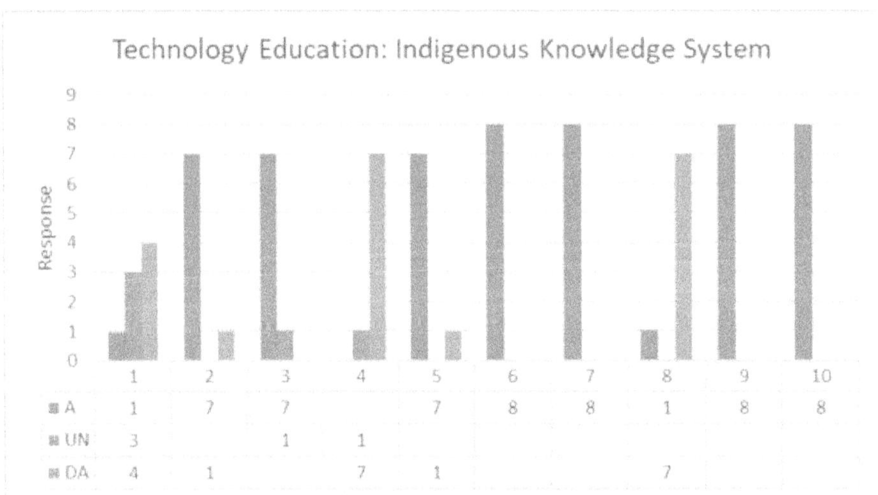

Of the eight participants, one argued that there is only one way in which to learn IKS during TE practice, three were uncertain and the remainder disagreed with statement 1. In statement 2, seven participants agreed that diverse people from all cultures contribute towards IKS, and only one disagreed. In statement 3, seven of the participants agreed that technologists are creative and that elders are IKS resourceful, and only one disagreed. Seven participants disagreed with the statement that Science and Technology do not complement each other in IKS, and one was uncertain when it came to statement 4. In statement 5, seven participants agreed that IKS and Technology preceded Science, while only one participant disagreed. All participants, in terms of statement 6, agreed that IKS forms part of the social and cultural traditions of every community. As for statement 7, all participants also agreed that IKS is transferable, Science is experimental, and Technology is practical. Only one participant agreed with statement 8, that IKS does not rely on the excavation, preservation, and knowledge generation of local people, while the remaining participants disagreed. All the participants agreed with statement 9, that the IKS research approach has been strengthened by a strong Renaissance agenda. Statement 10, which questions whether IK contributes towards creating the historical experiences of ordinary local people, resonated with all the participants, as they were all in agreement.

RESEARCH FINDINGS

During the ESSI contact sessions, the participants were divided in such a way that IKS learning in a Technology class was more ubiquitous irrespective of their context. During a weekly contact session of ESSI with the TE teachers, who can be described as participants or co-researchers, a desire for, and love of, IKS was rekindled. The TE teachers were made aware that diverse people contribute equally to their own knowledge systems. They strongly agreed that Science and Technology complement each other in IKS. Interestingly, all the participants agreed that IKS is part of the tradition of every community, thus they could differentiate the role of science subject from that of Technology subject were aware that IKS is transferable. It is heart-warming to note that an overwhelming majority of participants agreed that IKS is reliant on the excavation, preservation, and knowledge generation of local people. All participants acknowledged that a strong Renaissance agenda could contribute to imploring an IKS research approach, and all supported the view that IK contributes towards creating the historical experiences of ordinary local people.

DISCUSSION

It seems that partnerships between universities and local communities, through ESSI, have borne positive results in respect of the latter's emancipation. The ESSI project managed to encourage TE teachers to infuse IKS during their Technology lessons. Teachers are ready to accept their learners' cultures and traditions, to give the latter an opportunity to share their cultural differences in the TE class, and to help learners learn from one another. Technology teachers will command the respect of both learners and their parents if they make IKS part of their lesson presentation. TE teachers should promote the African Renaissance agenda, as this will make their learners grateful for who they are and for their shared destiny.

Many studies around the Didactics of Technology, which sought to determine whether IKS can be infused into TE curricula, have had somewhat disappointing results. As Gumbo and William (2014) found, many teachers who failed to incorporate IKS in their teaching did so out of ignorance. This study has proved otherwise: ignorance was the result of a lack of knowledge about how to infuse IKSs into teaching practice. Once TE teachers are empowered, they will be ready and able to integrate IKS into their lesson delivery. The CE or ESSI participants had been through various action research spiral cycles during a three-year journey to unpack numerous Technology-related themes and resolve (as a team) the challenges individuals had encountered within those themes.

The systematic integration of technology in the teaching and learning process foster a population that leverages twenty-first century resources, and IKS is part of this development (Gumbo &William, 2023). It is through incorporating TE in school curricula where IKS is shared within the school and local communities that leaners and teachers can benefit their society with functional modern technology that respond to the societal needs.

RECOMMENDATIONS

Every country should strive to develop on the firm foundation of multilateralism and further set to be in the front as well as the centre of the global quest to build "pluriversal" world. Escobar (2018), coin the term "pluriverse" as referring to a world that is more multipolar, more inclusive, just and equitable, a world that respects the potential and contributions of all countries to human progress. A "pluriversal" world embraces knowledge from all sources and cultures and harness them to build a more humane community with a shared future (Ngcaweni, 2024). Such a shared world is the antithesis of the idea of unipolarity, espoused and practiced by the major powers of the global North. Skills development that can help TE teachers to improve lives and livelihoods of their learners is when they can identify with the curriculum and integrate the IKS in their lesson delivery.

- The following are proposed to direct future research, policy, and practice of integrating IKS in during TE lesson presentation from '*My* Courses Editor' (2024, January):

  1. ***Incorporating traditional materials and techniques*:** Using traditional materials and techniques in design projects can be a way to honour Indigenous knowledge and culture. This could involve using materials such as natural fibres, clay, or wood, and employing traditional methods of weaving, carving, or pottery-making.
  2. ***Enhance sustainability***: IKS often encompass sustainable practices that have been used for centuries to maintain balance between humans and their environment. By applying these principles to design, we can create solutions that are more environmentally friendly and sustainable. For instance, in architecture, studying and applying indigenous building techniques could lead to designs that are better adapted to the local environment, use materials more efficiently, and have a smaller ecological footprint.
  3. ***Create Culturally Sensitive Designs:*** IKS can be used to create designs that are more culturally sensitive and relevant. For instance, in graphic design, elements drawn from indigenous art and symbolism can make designs more resonant with local cultures. Similarly, product designs can be adapted to fit better with traditional lifestyles and practices.
  4. ***Foster Community Involvement*:** Incorporating IKS in design projects often involves working closely with indigenous communities. This can help foster community involvement in the project, leading to designs that are more accepted by the community and meet their needs more effectively.
  5. ***Encourage Local Economy:*** Applying IKS may involve sourcing local materials or working with local artisans, which can have positive impacts on the local economy. For instance, using traditional materials or techniques can provide employment and income for local communities, and promote the preservation of traditional crafts.
  6. ***Encourage Innovation:*** IKS can lead to innovation by providing new perspectives and ideas. For instance, indigenous knowledge can inspire novel solutions to design problems, or suggest new uses for materials and techniques.

7. ***Preserve and Promote Cultural Heritage:*** Including IKS in design can help preserve and promote indigenous cultures and traditions. For instance, using indigenous patterns in textile design can help keep these traditions alive, while also raising awareness and appreciation of the culture

- The following are proposed to direct future research, policy, and practice of integrating IKS in during TE lesson presentation from 'Blose and Gumbo' (2024, June):

1. By integrating IKS, design scenarios could consider indigenous contexts.
2. Learners can engage in learning activities that stimulate their critical and creative thinking on sustainability.
3. Learners can learn a great deal from each other because they possess knowledge from their historical and cultural contexts.

The two authors also develop a framework to integrate IKS in TE for sustainable development. This framework consists of three layers that is: Curriculum, Social environment, and Methodologies. These layers are only distinguishable but inseparable.

Figure 2. A framework for integrating Indigenous knowledge systems in technology education for education on sustainable development (Blose & Gumbo, 2024)

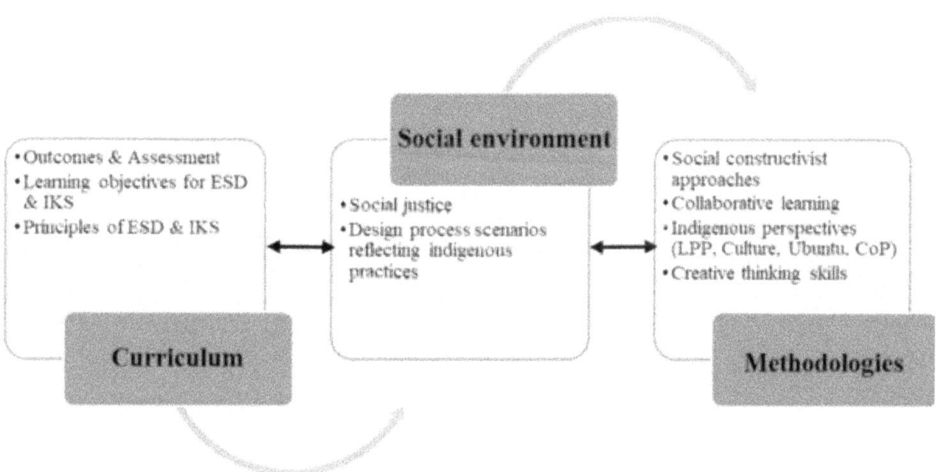

The framework for teaching TE intends to capacitate teachers to integrate IKS in their classroom practices.

- Quest for preserving and promoting IKS cultural heritage and IKS integration in TE classroom by involving Higher Education Institutions from the author perspective.

The IKS integration in SA is supported by the Technology policy unfortunately from the previous studies TE teachers are not implementing that to their fullest ability. The TE teachers still under the spell of their colonisers since they too still demonise and look down upon community IKS practice and observances. Figure 2 display how best can IKS integrated in TE classroom and furthermore levelling the plain field of treating IKS knowledge with respect. Any interaction in the form community engagement

(CE) or engaged scholarship and societal impact (ESSI) project run by higher education institutions (HEIs) will bridge the recognition of IKS as valid knowledge pass down from one generation to the next, through folklore, storytelling, mentoring, observation, or Action Research methodologies.

Figure 2 outline the quest for preserving and promoting IKS cultural heritage and IKS integration in TE classroom by involving Higher Education Institutions.

Figure 3. Indigenous knowledge systems integration framework within technology education

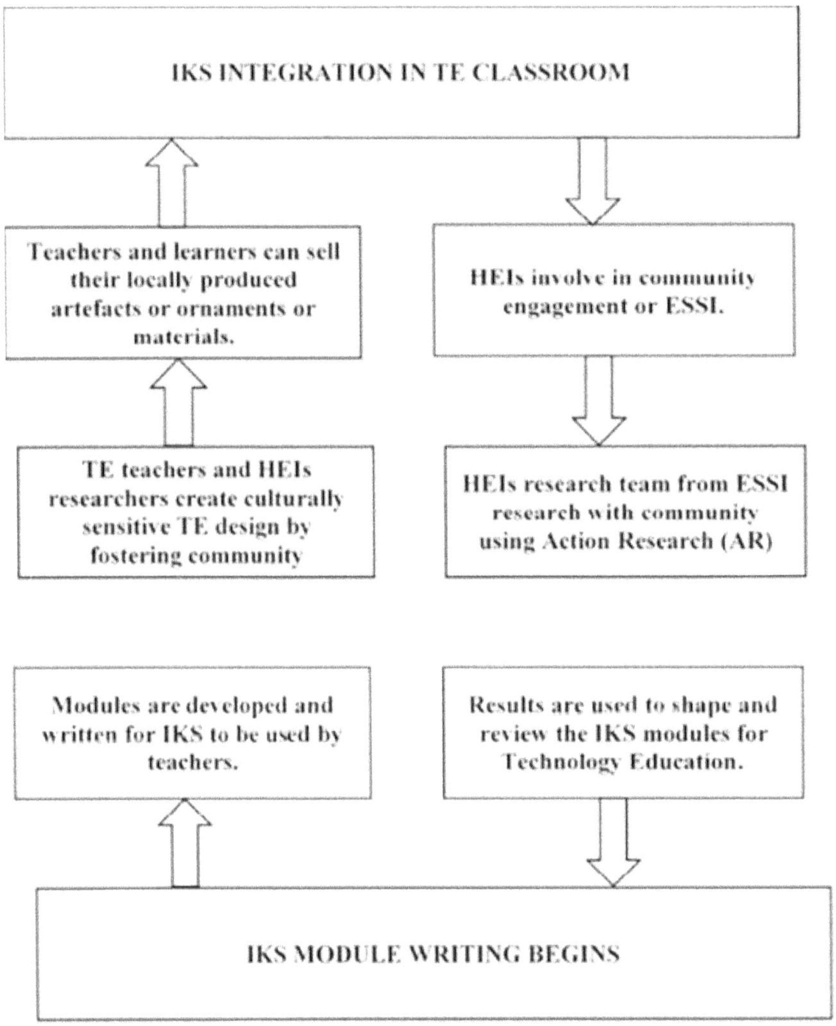

This research journey by the ESSI team from HEIs will allow the possibility of developing module laced with IKS substantial content that community based. The CE/ESSI project has the power to revive love for culture, traditional practices, and indigenous event celebrations. In this case learners will enjoy their learning as they can identify with the curriculum since it hit home, and they can identify with it. What is more mutual and beneficial is that the learners through teachers' guidance can be innovative and creative in designing products that con be sold. This will boast the economy of the local community.

As the TE scholar my recommendation to teachers to be able to integrate IKS in their field of specialization during their lesson presentation will be possible if, the Higher Education Institutions interested in promoting their community indigenous knowledges, cultures and traditions can have a special module to orientate teachers on how to integrate IKS across diverse field.

In South Africa (SA), colonialism and apartheid suppressed indigenous cultures. Apartheid introduced by white-minority government, was a policy based on separate development for different racially categorised (Seda, 2024). The system of education in SA is still suffering the legacy on both colonialism and apartheid since IKS is still looked down at and more so the chains of demonising IKS has not been completely slacken off. The developed framework from this study has the potential for preserving and promoting IKS cultural heritage and IKS integration in TE classroom as it encourages partnership between Higher Education Institutions, local communities, and Technology Education teachers. This partnership will make it easy to push the square peg into a round hole.

CONCLUSION

From the pre-democratic era in SA to the present, IKS has survived and avoided its extinction from ordinary local communities, despite the uncalled-for ideologies imposed by the colonizers. The Technology teachers who participated in this study realized that they could serve as good conduits for transferring IKS to their learners, by acting as the middleman and inviting IKS knowers from the community to share their knowledge with the learners in class. The process of inviting older people to TE class will, in a way, preserve the available indigenous knowledge and serve to pass it on to the next generation. The imperative to indigenize the curriculum has become a practice the world over. In Africa, learners at secondary schools and students' tertiary institutions are not only calling for the indigenization of the curriculum but also its Africanisation. Doing ESSI with teachers will help to recognize the value of IKS for both TE teachers and learners. TE teachers should advocate for an African Renaissance in education. African scholars need to prioritize knowledge generation, to focus on developing more literature on IKS. Before our indigenous knowledge and technologies are lost forever, they need to be recorded for future generations. The education ministries of different countries should support both universities and their nations' teachers in the promotion of IKS policies and resources.

The current study managed to fit a round peg in a round hole, by equipping TE teachers with the know-how of designing a TE lesson infused with IKS. TE teachers can be better positioned to make their learners aware of the different coexisting knowledge systems at their disposal. As southern theory underpinned the study, and collaborative inquiry informed it, the researcher, as an action research practitioner, was pleased to leave that region of Limpopo with Technology teachers having been professionally empowered to integrate IKS in their lesson delivery. This study was anchored on the ideas of Tanzania's founding president, the late Mwalimu Julius Nyerere, who argued for the indigenization of his country's education system. Nyerere (1968) argued that the purpose of education was to cultivate skills and knowledge that would enable a society to function into the future. His ideas of a type of education that is relevant to African contexts still resonate with biting poignancy nearly over five decades since he first stated them (Shara, 2016). This study contributed an IKS framework to integrate TE in a classroom. Once this framework in called, "Indigenous Knowledge Systems integration framework within Technology Education" can successfully be implanted and yield the desired results then it will therefore be labelled as the IKS integration model.

REFERENCES

Bertram, C., & Christiansen, I. (2020). *Understanding research: an introduction to reading research* (2nd ed.). Van Schaik.

Bezuidenhout, R.-M., & Cronjé, F. (2014). Qualitative data analysis. In du Plooy, F., Davis, C., & Bezuidenhout, R.-M. (Eds.), *Research matters* (pp. 228–251). Juta.

Blose, P. (2019). *The integration of indigenous graphics knowledge and skills to enhance Grade 9 learners' understanding of graphic designs in Technology Education*. MEd Dissertation: University of South Africa, Pretoria.

Blose, P., & Gumbo, M. T. (2024). Developing a Framework for Integrating IKS in Technology Education with Sustainable Development Principles. *African Journal of Research in Mathematics, Science and Technology Education*, 1–13. https://doi.org/DOI: 10.1080/18117295.2024.2356416

Coghlan, D. (2010). Seeking Common Ground in the Diversity and Diffusion of Action and Collaborative Management Research Methodologies: The Value of a General Empirical Method. In W. Pasmore, A.B.(Rami) Shani, & R. Woodman (Eds.), *Research in Organizational Change and Development*, 18:149–181. Bingley: Emerald

Coghlan, D., Brady, V., O'Leary, D., & Hynes, G. (2024). Exploring vulnerability and risk in an action research writing group: A cooperative inquiry. *Educational Action Research*, *32*(1), 25–42. DOI: 10.1080/09650792.2022.2084133

Collins, P. H. (2013). Critical intervention in western social theory: Reflections on power and southern theory. *Decentering Social Theory. Political Power and Social Theory*, *25*, 173–182.

Connell, R. (2007). *Southern Theory: The Global Dynamics of Knowledge in Social Science*. Polity Press.

Corbett, M., & Gereluk, D. (Eds.). (2020). Introduction: Why do rural matters in Canadian education? In M. Corbett & D. Gereluk (Eds.), *Rural teacher education: connecting land and people*. Singapore: Springer Nature. [Online]. Available at: https://doi.org/ (Accessed on 10 April 2020).DOI: 10.1007/978-981-15-2560-1

Department of Basic Education (DBE). (2010). Curriculum and Assessment Policy Statement (CAPS) Technology, grades 7–9. Final draft [Online]. Available at: https://www.education.gov.za/CAPS.asp (Accessed on 15 November 2011).

Dube, E., & Munsaka, E. (2018). The contribution of indigenous knowledge to disaster risk reduction activities in Zimbabwe: A big call to practitioners. *Jàambá*, *10*(1), 1–8. DOI: 10.4102/jamba.v10i1.493 PMID: 29955265

Escobar, A. (2018). *Design for the Pluriverse: Radical Interdependence, Autonomy, and the Making of Worlds. United State*. Duke University Press. DOI: 10.1215/9780822371816

Gluck, A. (2015). De-Westernisation: Key concept paper. United Kingdom" the University of Leeds.

Gumbo, M. T. (2015). Indigenous technology in Technology Education curricula and teaching. In Williams, P. J., Jones, A., & Buntting, C. (Eds.), *Contemporary Issues in Technology Education: The future of Technology Education* (pp. 57–75). Springer.

Gumbo, M. T. (2017). Indigenous knowledge in the teaching of technology: a quest for transforming Technology education. In M. T. Gumbo & V. Msila (Eds.), *African voices on indigenization of the curriculum: insights from practice*, pp. 119–154. Wandsbeck: Reach.

Gumbo, M. T. (2019). Integrating forms of knowledge and technology. In Gumbo, M. T. (Ed.), *Teaching Technology: Intermediate to Senior Phase* (pp. 20–39). Oxford University Press.

Gumbo, M. T. (2019). What is technology? In Gumbo, M. T. (Ed.), *Teaching technology: Intermediate to senior phase* (pp. 2–19). Oxford University Press.

Gumbo, M. T., & Williams, J. P. (2014). Investigating Grade 8 Technology teachers' pedagogical content knowledge in the Tshwane district of Gauteng province of South Africa. *International Journal of Educational Sciences*, *6*(3), 479–488. DOI: 10.1080/09751122.2014.11890159

Gumbo, M. T., & Williams, J. P. (Eds.). (2023). An Introduction to Indigenous Technology Knowledge Systems. In M. T. Gumbo & J. P. Williams (Eds.), *Indigenous Technology Knowledge Systems: Decolonizing the Technology Education Curriculum*, Part I, pp (v)-(xvii). Singapore: Springer.

Heron, J., & Reason, P. (2008). Extending Epistemology within a Cooperative Inquiry. In Reason, P., & Bradbury, H. (Eds.), *The SAGE Handbook of Action Research* (2nd ed., pp. 367–380). Sage. DOI: 10.4135/9781848607934.n32

Higgs, P. (2012). African philosophy and the decolonisation of education in Africa: Some critical reflections. *Educational Philosophy and Theory*, *44*(2), 22–34. DOI: 10.1111/j.1469-5812.2011.00794.x

Higgs, P. (2016). The African Renaissance and decolonization of the curriculum. In Msila, V., & Gumbo, M. T. (Eds.), *Africanising the curriculum: indigenous perspective and theories* (pp. 1–15). Sun Media Metro.

Hill, R., Adem, ., Alangui, W. V., Molnár, Z., Aumeeruddy-Thomas, Y., Bridgewater, P., Teng , M., Thaman, R., Adou Yao, C. Y., Berkes, F., Carino, J., & Xue, D. (2020). Working with indigenous, local, and scientific knowledge in assessments nature and nature's linkage with people. In *Current opinion in environmental sustainability* (Vol. 43, pp. 8–20). Elsevier B. V., DOI: 10.1016/j.cosust.2019.12.006

Jacobs, K. R. (2015). *The classroom implementation of indigenous knowledge in the science curriculum by science teachers in the Western Cape province*. DEd Thesis: University of Cape Town, Cape Town.

Jegede, O. J., & Aikenhead, G. S. (1999). Transcending cultural borders: Implication for Science teaching. *Research in Science & Technological Education*, *17*(1), 45–66. DOI: 10.1080/0263514990170104

Kamwendo, G. H. (2017). Foreword. In Ngulube, P. (Ed.), *Social, cultural, and educational considerations of indigenous knowledge in developing countries* (p. xix). IGI Global.

Kavanagh, K. (Ed.). (1978). *The South African concise Oxford dictionary*. Rhodes University.

Koehler, C. (2017). Crucial role of indigenous knowledge in formal education systems. In Ngulube, P. (Ed.), *Social, cultural, and educational considerations of indigenous knowledge in developing countries* (pp. 60–79). IGI Global. DOI: 10.4018/978-1-5225-0838-0.ch004

Kpan, M. A. (2020). *Indigenous knowledge systems and the curriculum of basic schools in the Nadowli district of the upper west region*. MA Dissertation: University for Development Studies, Tamale.

Lutomia, A. N., & Bello-Bravo, J. (2017). Community of practice and indigenous knowledges: a case study of empowering women in processing Shea butter using scientific animations. In Ngulube, P. (Ed.), *Social, cultural, and educational considerations of indigenous knowledge in developing countries* (pp. 226–243). IGI Global. DOI: 10.4018/978-1-5225-0838-0.ch012

Maathai, W. (2009). *The challenge for Africa: a new vision*. Random House.

Maelene, M. (2002). Implications of indigenous African beliefs and customs for health education and prevention of HIV/AIDS. *Nutrition Update*, *26*(8), 14–17.

Maluleke, R. (2019). *A pedagogy for technology education: an indigenous perspective*. DEd Thesis: University of South Africa, Pretoria.

Mapotse, T. A. (2012). The teaching practice of senior phase Technology Education teachers in selected schools of Limpopo Province: An Action Research Study. DEd Thesis: University of South Africa, Pretoria.

Mapotse, T. A. (2020). Developing Technology teachers through Action Research: An African perspective. In Gumbo, M. T. (Ed.), *Decolonization of Technology Education: African indigenous perspectives* (pp. 331–350). Peter Lang.

Mapotse, T. A., & Gumbo, M. T. (2019). Technology Education as a school subject. In Gumbo, M. T. (Ed.), *Teaching Technology: Intermediate to Senior Phase* (pp. 62–77). Oxford University Press.

Mbeki, T. (1998). The African Renaissance, South Africa, and the world. South African Deputy President Thabo Mbeki speaks at the United Nations University 9 April 1998.

McKee, A. (2003). Textual analysis: A beginner's guide. *Metro Magazine*, *127/128*, 138–149.

Merriam-Webster. (n.d.). Didactic Definition and Meaning. In *Merriam-Webster.com dictionary*. Retrieved September 10, 2024, from https://www.merriam-webster.com/dictionary/didactic [Didactic Definition & Meaning - Merriam-Webster].

Mji, G. (2012). The health knowledge utilised by rural older Xhosa women in the management of health problems in their home situation, with a special focus on indigenous knowledge, DEd Thesis: University of Stellenbosch, Cape Town.

Mji, G., Kalenga, R., Ned, L., Alperstein, M., & Banda, D. (2017). Indigenous knowledge exclusion in education systems of Africans: impact of beingness and becoming an African. In Ngulube, P. (Ed.), *Social, cultural, and educational considerations of indigenous knowledge in developing countries* (pp. 36–59). IGI Global. DOI: 10.4018/978-1-5225-0838-0.ch003

My Courses Editor. (2024, January). Applying Indigenous Knowledge Systems (IKS) in Design Project. https://mycourses.co.za/applying-indigenous-knowledge-systems-iks-in-design-projects/

Ngcaweni, B. (2024, March). BRICS building a "pluriversal" world. *UBUNTU: Stronger Together, 31*, 34-35, 110. South Africa's Public Diplomacy in action: Department of International Relationship and Cooperation (DIRCO).

Nyerere, J. (1968). *Ujamaa: essay on socialism.* Oxford University Press.

Odora-Hoppers, C. A. (2001). Poverty, power, and partnerships in educational development: Post-victimology perspective. *Compare: A Journal of Comparative Education*, *31*(1), 21–38. DOI: 10.1080/03057920020030144

Opoku, M., & James, A. (2020). Challenges of teaching Akan (Ghana) culturally specific environmental ethics in senior high schools: Voices of Akan Indigenous Knowledge -holders and biology teachers. *Southern African Journal of Environmental Education, 36*, 33–50. DOI: 10.4314/sajee.v36i1.4

Phillips, A. (2017). *Afrophobia, a result of colonialism and global capitalism.* University of South Africa.

Msila, V., & Gumbo, M. T. (Eds.). (2016). *Africanising the curriculum: Indigenous perspectives and theories.* African Sun Media.

Ngulube, P. (Ed.). (2016). *Handbook of research on social, cultural, and educational considerations of indigenous knowledge in developing countries.* IGI global.

Ramose, M. B. (2003). I doubt, therefore African philosophy exists. *South African Journal of Philosophy*, *22*(2), 110–121. DOI: 10.4314/sajpem.v22i2.31364

Reason, P. (1999). Integrating Action and Reflection through Cooperative Inquiry. *Management Learning*, *30*(2), 207–226. DOI: 10.1177/1350507699302007

Samuels, K. G. (2003). *Perceptions and knowledge regarding indigenous knowledge and Environmental Education in the Natural Science curriculum.* DEd Thesis: University of South Africa, Pretoria.

Seda, O. (2024). *Sho Madjozi: The pop star using traditional culture to shape a fresh identity for young South Africans. UBUNTU: Stronger Together, 31, 106-107, 110. South Africa's Public Diplomacy in action: Department of International Relationship and Cooperation.* DIRCO.

Serote, M. W. (1998). Initiatives for the protection of rights of holders of traditional knowledge, indigenous peoples, and local communities. *Proceedings of the World Intellectual Property Organisation conference*, Geneva, July 23 & 24.

Shara, S. (2016). The ownership of global knowledge. In Msila, V., & Gumbo, M. T. (Eds.), *Africanising the curriculum: indigenous perspective and theories* (pp. ix–xvii). Sun Media Metro.

Shava, S. (2013). The representation of indigenous knowledges. In Stevenson, R. B., Brody, M., Dillon, J., & Wals, A. (Eds.), *International handbook of research on Environmental Education.* Routledge.

Shava, S., & Manyike, T. V. (2018). The decolonial role of African indigenous languages and indigenous knowledges in formal education process. *Indilinga*, *17*(1), 36–52.

Skatkin, M. N., & Lerner, I. A. (Eds.). (1970-1979). *The great Soviet encyclopaedia* (3rd ed) [Online]. Retrieved from: https://encyclopedia2.thefreedictionary.com/Didactics (Accessed on 9 March 2020).

Tapfuma, M. (2017). Preservation of indigenous knowledge systems in Zimbabwe: some ethical considerations. In Ngulube, P. (Ed.), *Social, cultural, and educational considerations of indigenous knowledge in developing countries* (pp. 244–258). IGI Global. DOI: 10.4018/978-1-5225-0838-0.ch013

Unisa [University of South Africa]. (n.d). Indigenous knowledge systems & education. Retrieved from https://www.unisa.ac.za/sites/corporate/default/Unisa-History-and-Memory-Project/Themes/All-themes/Indigenous-knowledge-systems-&-education [Indigenous knowledge systems & education] (Accessed on 8 August 2024).

Wahyudi, R. (2014). 'Democratic' online courses as 'globalized communication' in English language teaching and applied linguistics: A proposal. *Journal of Global Literacies, Technologies, and Emerging Pedagogies*, *2*(3), 248–260.

Webb, P. (2016). Indigenous knowledge and science education: What knowledge, whose knowledge and how do we bridge the gap? In Msila, V., & Gumbo, M. T. (Eds.), *Africanising the curriculum: indigenous perspective and theories* (pp. 177–194). Sun Media Metro.

Zidny, R., Sj str m, J., & Eilks, I. (2021). A multi-perspective reflection on how indigenous knowledge and related ideas can improve Science Education for sustainability. *Science & Education*, *29*(1), 145–185. DOI: 10.1007/s11191-019-00100-x

Chapter 14
South African Preservice Teachers' Experiences About Traditional Environmental Practices:
A Participatory Action Research Method

Tsebo Kgoto Matsekoleng
https://orcid.org/0000-0002-7577-3417
North-West University, South Africa

Anette Hay
https://orcid.org/0000-0002-2210-7218
North-West University, South Africa

ABSTRACT

The richness of Indigenous Knowledge Systems (IKS) received the attention of scholars such as Govender and Mutendera (2020) in the research arena. Traditional weather forecasting through observation of clouds, birds, and the sun provides the richness of IKS as Indigenous people use it to predict weather. As such, this chapter explores South African preservice teachers' experiences about traditional environmental practices by utilising the Participatory Action Research (PAR) method. Bandura's social learning theory grounded the study. The PAR process, which had two cycles, was used to generate data from preservice teachers' reflective assignments. The thematic analysis method was used to analyse generated data. Preservice teachers doing a compulsory sustainability module were purposively selected for the study. The chapter revealed that South African preservice teachers know traditional environmental practices that are crucial to protecting and conserving the environment and can integrate with subject themes.

DOI: 10.4018/979-8-3693-3542-0.ch014

1. INTRODUCTION

This chapter aims to explore South African preservice teachers' experiences about traditional environmental practices through the application of the Participatory Action Research (PAR) method. Participatory action research has three main principles: (1) it aims to transform injustices, (2) researchers become activists through direct involvement in socio-political processes, and (3) social actors become co-researchers by identifying needs and potential problems, gathering information, analysing, and making decisions. Participatory action research methodologies are cyclical, recursive, transdisciplinary, and transformative (Hensler, Alatorre Frenk & Merçon, 2023). The method therefore emphasises that traditional practices are processes that give people an opportunity to share knowledge. In the process, hands-on and mind-on activities could be utilised to foster effective traditional practices. Participatory action research could serve as a suitable methodology to implement traditional practices which teacher could use in their classrooms to raise awareness of environmental issues including global warming. Participatory action research is a philosophy, a methodology, a theory of learning, and a facilitation process (Zuber-Skerritt, 2018) in favour of the collaborative nature of learning in traditional practice and social learning theory through experimentation.

Albert Bandura's social learning which was established in 1977 ground the chapter owing to its experimentation. The theory grounds the chapter as it shares elements with PAR and traditional practices of being experimental, it is, therefore, important to explore teachers' experiences with traditional practices. Albert Bandura's social learning theory is a theory of the learning process and social behaviour which proposes that new behaviour can be acquired by observing and imitating others (Cilliers, 2021). In PAR and traditional practices, engagement is key to come with implementable solutions for addressing identified issues. It is for this reason, that Albert Bandura's theory of social learning is based on the assumption that humans are flexible and can change through direct observation (Feri & Maemonah, 2022). The behavioural change can be influenced by various internal and external factors such as imitation which learners can learn traditional practices from their teachers.

2. NATURE OF LEARNING

The nature of learning can happen in a formal or informal structure. For instance, schools/colleges/universities are classified as formal learning institutions. In these institutions, children are taught/learn about environmental practices and to some extent about traditional practices. This gives children an opportunity to learn traditional practices from adults through experimentation as advocated by social learning theory. In State institutions, learning is based on the prescribed standard curriculum (Govender & Mutendera, 2020). Learning at state institutions follows a specified standard curriculum (Govender & Mutendera, 2020). Home/community/village could be referred to as an informal structure of learning, where knowledge is transferred from one generation to another using folktales etc. However, learning is not prescribed. Formal and informal structures of learning endorse continuous learning and influence how children learn. Teaching and learning have evolved into an ongoing process (Amoyedo-Peter, 2023). Accordingly, social learning theory processes such as observational methods are reviewed by the relevant committee within the structured institutions and follow certain steps to execute the plan through experimentation and reflection. However, social learning theory and Indigenous Knowledge Systems (IKS) advocate cooperative learning to take action and produce practical knowledge of mitigating environmental issues including pollution. Elders use proverbs to guide children about traditional practices, and schools use systematic rules to mould children and transfer knowledge. Both practices intend to teach children about life and nature. Knowledge is generated and understood in the context of relationships (Hare, 2021). Therefore, in social learning theory, PAR and IKS relationships are established through the action of observing or experimenting within immediate environments. Such that learning/knowledge cannot happen/be generated without experimentation or observation to effect behavioural change in the PAR setup of applying IKS and social learning theory. The interconnectivity of PAR, IKS and social learning theory in a classroom setup could help in achieving Sustainable Development Goal 4.

3. THE APPLICATION OF SOCIAL LEARNING THEORY IN INDIGENOUS KNOWLEDGE SYSTEMS

Social learning theory contextualises IKS learning/practice where children learn from people within their space. For example, children in their homes learn to cook traditional dishes including *Kgodu* (Pumpkin porridge). Contextualised learning uses an immediate environment that could enable children to experience learning, foster cooperation, and make children creative and critical thinkers. Children observe adults and their surroundings to formulate new behaviours about traditional practices. The inclusion of traditional practice in the school curriculum will ensure quality education where Sustainable Development Goal 6 is achieved.

The social learning theory and IKS have a common understanding whereby people explore learning-based imitations, observing, and modelling to know the phenomena based on their experiences. The application of the social learning theory in the IKS and PAR are ideal for the sustainability of traditional values in the global world. It is founded on the belief that one constructs knowledge from one's experiences and mental structures (Misra & Misra, 2020). Children learn from being active children in collaborative settings; and from constructing new ideas and concepts based on their prior knowledge and experiences. Participatory action research in this instance is connected to IKS and social learning theory through the processes of observing to construct new ideas.

4. TRADITIONAL PRACTICES RELATING TO PARTICIPATORY ACTION RESEARCH AND SOCIAL LEARNING THEORY

Traditional practices and PAR relate to social learning theory through engagement processes of observing people while carrying out a task. Albert Bandura's social learning theory identifies learning as a dynamic interaction between people, environment, and behaviour. Engagement in a social context involves a dual process of making meaning (Firmansyah & Saepuloh, 2022). New patterns of behaviour in the social learning system can be learned by direct experience or observation of the behaviour of others (Bandura, 1971). In accord, PAR and IKS use the immediate environment to learn through collaborative efforts. The social learning theory and IKS describe how humans gain knowledge and learn. As a result, it has a direct educational application. As social learning theory suggests humans generate knowledge and meaning through their experiences (Bada & Olusegun, 2015).

Indigenous Knowledge Systems practices are practical and involve hands, mind, body, and heart to transfer learning to children. For instance, agroforestry has the potential to restore degraded ecosystems (Viswanath & Lubina, 2017) and agroforestry addresses Sustainable Development Goal 15 (life on land) of sustaining the land for future generations. Indigenous people use existing or dug springs/fountains to harvest water, which teaches children to interact with and respect nature. Indigenous Knowledge Systems practices have enormous potential to contribute to children's moral upbringing and to teach Education for Sustainable Development/Environmental Education concepts to create an awareness of climate change, and global warming (Govender & Mutendera, 2020). As well, Zuber-Skerritt (2015) alludes that PAR involves Body/Brain/Mind interactions for learning. Indigenous Knowledge Systems and PAR relate in so many ways. This does not relegate social learning theory from relatability, as social learning theory also uses models to foster traditional practices (Amoyedo-Peter, 2023). The relatability of traditional practices, PAR, and social learning theory offer an opportunity to close theoretical learning with an advance of practical learning.

Traditional practices infuse the Ubuntu/Botho principle of working together in society to address problems communities encounter. Participatory action research is social and we learn together (Zuber-Skerritt, 2015). In social learning theory, learning occurs by observing and imitating others to alter the behaviour. Collaborative learning is promoted in traditional practices, social learning theory and PAR, which is ideal for providing children with good quality education (Sustainable Development Goal 4).

A variety of teaching methods such as role-playing, games, observation, demonstration, imitation, inquiring, self-teaching, and learning and peer teaching can be used to transmit societal norms and values to the younger generation (Edinyang, 2016). Prior knowledge and observation play a crucial role in learning. Social learning theory is said to use imitation, observation, and modelling to model new behaviour and new behaviour is moulded through the learning processes. Participatory action research uses observational evidence and draws on prior knowledge to create knowledge (Zuber-Skerritt, 2015) working with participants. As well, Indigenous people use observation to learn from their surroundings. Figure 1 presents the authors' contribution to traditional practices that relate to PAR and social learning theory. In Figure 1, the authors explore the relatability of traditional practices, PAR and social learning theory in relation to their philosophy, relationship, approach and learning/rearing. The table elements of fostering participatory learning processes in PAR, traditional practices and social learning theory.

Figure 1. PAR and social learning theory

	Traditional practices	PAR	SLT
Philosophy	Ubuntu/Botho	Co-operation	Imitation
Relationship	Mutual respect	Collaboration	Observe
Approach	Holistic	Holistic	Holistic
Learning/rearing	Active	Active	Active

5. VALUE OF PARTICIPATORY ACTION RESEARCH AND INDIGENOUS KNOWLEDGE SYSTEMS IN SUSTAINABLE DEVELOPMENT GOALS

The PAR process always involves an event where participants can share their learning with others, both to recognise their learning and to educate and encourage the wider community (Zuber-Skerritt, 2015). Figure 2 presents the PAR processes where preservice teachers shared their learning experiences. Preservice teachers through engagement in their activities come across new things and reflect on the experiences gained. This is because preservice teachers are the centre of exploring learning.

In the classroom, learning can point toward several different teaching practices as IKS enhances classroom engagement (Suarta et al., 2022). In the most general sense, it usually means encouraging children to use active techniques (experiments, real-world problem solving) to create more knowledge and then to reflect on and talk about what they are doing and how their understanding is changing (Bada & Olusegun, 2015). Children build new knowledge based on their prior experience. Participatory action research and IKS ensure children are active participants in their learning, which promotes Sustainable Development Goals such as Sustainable Development Goal 6 "Ensure availability and sustainable management of water and sanitation for all". In the process, children develop skills that are needed to sustain the environment for future generations.

Relationships are at the heart of PAR since all learning and development take place in small reflective groups underpinned by values such as trust, care, and respect (Zuber-Skerritt, 2015). Indigenous Knowledge Systems relationships are built through Botho where community members care for each other. While building a relationship the social learning theory uses imitations and observation to relate with those around them to learn new things.

6. RESEARCH METHODOLOGY

This section discusses the research design and methods of the study.

6.1 Research Design

This qualitative PAR case study reports on findings derived from 747 out of ±2197 preservice teachers who were enrolled in the Environmental Management for Sustainability 312 compulsory module. The aim was to explore South African preservice teachers' experiences with traditional environmental practices by utilising the PAR method.

6.2 Recruitment of Participants

Purposive sampling was used to select the third-year preservice teachers from North-West University according to the phase they teach e.g., foundation; intermediate-; senior- and further education and training (FET) phases, because the study needs different contributions from all phase-pre-service teachers. The participants are aged between 18 – 22 years. Preservice teachers completed the assignment in groups in a PAR cyclic process of 48 groups. Researchers extracted data from 46 groups because two groups did not include their reflections in the assignment. The groups derived from 747 out of ±2197 preservice teachers as the focus was on the Potchefstroom campus owing to accessibility to one of the researchers.

6.3 Participatory Action Research Cycles

The PAR process unfolded with two cycles in this chapter. Both cycles had three phases each one focusing on a certain aspect. In cycle one of phase one, preservice teachers were recruited by an independent person from the Geography Education subject group. The researchers established a relationship with the pre-service teachers by employing the strengths, weaknesses, opportunities, and threats (SWOT) strategy as suggested by Zuber-Skerritt (2015: 19). Using a SWOT analysis helped the action learning groups to consider possible implications for the project. For phase two, a mapping exercise activity was carried out where researchers explained O'Donoghue's (1993) model for the environment and how Education for Sustainable Development aims to develop competencies that empower individuals to act in complex situations sustainably (UNESCO, 2017). Subsequently, the participants mapped out the environmental issues in their local area by drawing a mind map using systems thinking as an example of sustainability competency. Lastly, in phase three preservice teachers reflected on what they had learnt during their activity. For cycle two, in phase one preservice teachers collectively designed the research processes and tools to analyse the mapping exercise activity. In phase two, preservice teachers held informal group discussions and lastly, in phase three preservice teachers reflected and discussed the possible competencies obtained during the exercise. Figure 2 depicts the PAR cyclic process.

Figure 2. PAR cyclic process

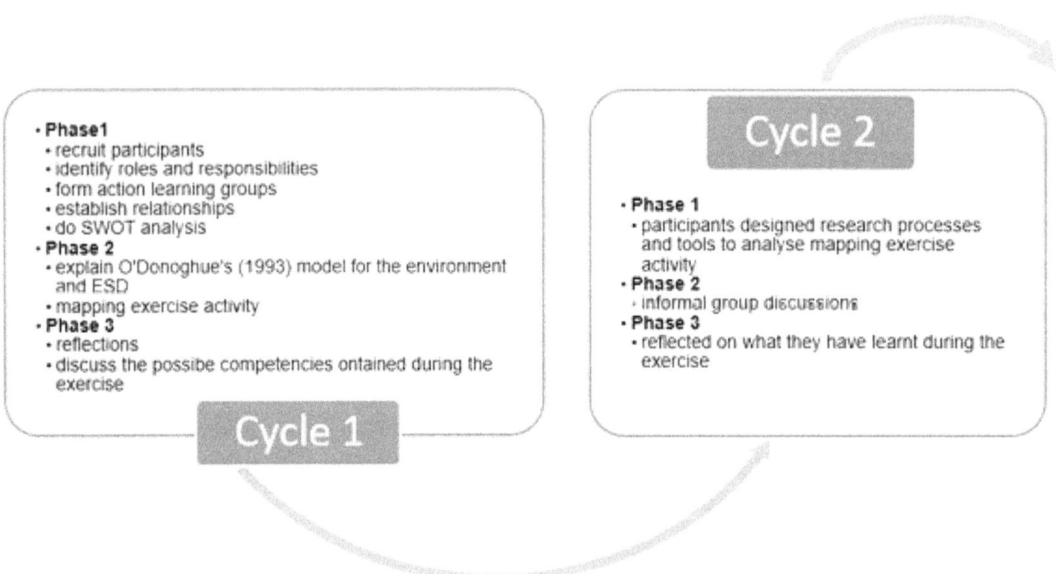

6.4 Data Collection Instruments

Written assignments were used to generate data in both cycles of the PAR process. The assignment was divided into three questions. The first question was based on a lesson plan, the second on learning activity design, and the third on incorporating IKS into Environmental Education. The third question had reflective questions on the incorporation of IKS into Environmental Education. The question allowed preservice teachers to voice out their unique voices in the learning processes. Thus, this chapter reports on the reflections of the assignment based on question three owing to the richness of the data.

6.5 Data Analysis

Figure 2 presents PAR processes that were followed to analyse the generated data. The study used thematic data analysis to analyse the generated data from the reflections of the participants who had completed the mapping exercise. The participants reflected on their experiences on how to integrate IKS into their subjects and specific subject themes. The thematic analysis method was used to identify, organise, and interpret data by establishing codes, categories, and themes (Saldana, 2016). This was done in steps in an inductive process. The inductive analysis involves coding data without attempting to fit themes into the researchers' beliefs about the research (Braun & Clark, 2006). ATLAS.ti version 23 was used to analyse data. Excerpts from the reflections are used to capture preservice teachers' reflections. Emerged themes from the data are displayed in Figure 3. Themes emerged through the coding processes and categorisation of the coded data in search for themes. Similar ideas that emerged were categorised into a specific theme displayed in Figure 3. Other themes were turned into subthemes as shown in Figure 3 in column one.

Figure 3. Themes in coded data

Traditional farming practices • Agroforestry • Rotational Grazing	Rainwater harvesting	Traditional Environmental conservation	Traditional agricultural and land management

7. RESEARCH RESULTS AND DISCUSSIONS

Results captured here are derived from preservice teachers' reflections on their assignments to answer this research question *"What are the South African preservice teachers' experiences about traditional environmental practices"*.

7.1 Traditional Farming Practices

Traditional environmental practices have been part of Indigenous people for years from one generation to the next one. This includes traditional farming practices to manage soil, land, and water through agroforestry, rotational grazing, etc. The practices are learnt through observation and experimentation. Within the social learning theory, one might learn new behavioural patterns by watching other people or by direct observation (Bandura, 1971). The ensuing subthemes emanated from the preservice teacher's reflections as discussed.

7.1.1 Agroforestry

Indigenous people have embraced agroforestry because they understand its ability to enhance their lives by generating income, assuring food security, providing medicine, firewood, and handicraft materials, and enhancing the microclimate (Gonçalves, Schlindwein & Martinelli, 2021). In the practice of agroforestry for food security, Indigenous people use the practice of drying up edible leafy vegetables for rainy days.

> *"..the agroforestry practice of the indigenous people as a way to support food security, cultural heritage, and environmental conservation."* (Group 26).

> *"..indigenous tribes frequently rely on agriculture for a living. Agroforestry allows them to diversify their crops and boost their food security and profit.*" (Group 29).

Further, Indigenous people use the agroforestry system at home to grow medicinal plants for preservation purposes including African wormwood. Medical plants are grown at home to have access to traditional medicine for when a need arises. "*...the importance of conserving medicinal plants like Aloe vera and Artemisia afra Jacq. ex Willd (African wormwood), emphasizing their value not just as biological resources but as vital components of Indigenous knowledge and cultural heritage*" (Group 33). Conservation of the medical plants was noted when society faced the COVID-19 pandemic such as African wormwood which is believed to boost antibodies. Overharvesting of the African wormwood became a concern among traditionalists which could result in scarcity of the plant. In addition, the plantation of the Marula tree within home yards for instance serves the purpose of providing fruits, medicine, firewood, and shadow. Thus, the Marula tree is deemed to be crucial as it offers multiple benefits and live longer.

> *"The trees provide shade, protection against erosion, and increase soil fertility by providing organic matter. It also promotes biodiversity through a habitat to create for various species of flora and fauna."* Group 42.

Group 2 of the preservice teachers allude that "*Indigenous agroforestry methods often emphasise sustainability and resilience, intending to preserve ecosystem health…*" Other Indigenous practices including crop rotation and Indigenous seed-saving practices help preserve the environment

> "*Using crop rotation is also good for the soil because it helps it to be fertile, it reduces soil rotation, and it positively contributes to sustainable living, and it improves the environment in a lot of ways*" (Group 18).

> "*Indigenous seed-saving practices are rooted in deep knowledge of ecosystems and genetic diversity, which is crucial for sustainable agriculture. These practices also hold cultural importance in preserving traditions and connection to the land*" (Group 20).

This emphasises that Indigenous people sustain traditional practices that benefit their wellbeing and the environment knowing the importance of co-existing with nature. Teachers' reflections stress the importance of traditional practices within a community. The inclusion of the practices in a classroom could benefit the learners in learning other environmentally friendly practices.

7.1.2 Rotational Grazing

The rotational grazing practice is one of the practices that Indigenous people continue to use. For instance, as seasons change Indigenous people also move domestic animals such as cows around to find grazing bushes.

"We need to protect the environment and that we can do so by using rotational grazing. This is also teaching the learners how to use the resources they have more sustainably, by rotating the grazing for the livestock." Group 11.

Grazing animals, such as cattle, sheep, goats, or bison, should be utilised as a sustainable way to enhance the land, absorb carbon, and reduce concerns regarding animal welfare (Grandin, 2022). The practice reduces soil erosion as animals continually move from one bush to another. Rotating grazing cattle or other livestock on a cover crop rotated with cash crops, such as corn or soy, will improve both soil health and "reduce dependence on expensive, fragile fertiliser supply chains" (Grandin, 2022: 1). It is for this reason that the preservice teachers believe that,

"*....rotational grazing is a sustainable practice. It lets you connect with Indigenous wisdom as well as help you actively contribute to environmental conservation*" (Group 7).

This further indicates traditional practices are sustainable for our fragile environment, which pre-service teachers consider ideal for conserving the environment. "*...environmental sustainability has been going on for ages and ages and there are various ways to sustain the environment such as moving cattle around grazing areas and even planting trees. This ensures that the grass is not used up by the cattle because as the cattle move to other grazing areas, the grass grows back and ensures that we never run out of trees and food supplies because trees are being planted. By doing the above mentioned, the environment is conserved and protected.*" Group 2. Rotational grazing also provides an opportunity for people to learn and connect with Indigenous wisdom where intergenerational occurs through observation and imitation to preserve the practice.

"This ancestral wisdom helps us understand sustainable land management" (Group 35).

7.2 Rainwater Harvesting

For decades, Indigenous people have used rainwater harvesting practices for their survival. Water could be harvested using rooftops, aloe leaves, and bow-shaped stones among other techniques, and thus "*....conserves water and lessens dependency on groundwater or municipal sources*" (Group 23). SDGs build on existing knowledge, as Indigenous people know about the conservation and sustainability of water management. The practice indicates that IKS could save us from environmental degradation in today's lives. One of the techniques Indigenous people have used over the years to have clean water is rainwater harvesting. The practice contributes largely to sustainable development in terms of providing 'free' water to many people (Owusu & Asante, 2020). Not only does the practice provide free water but also the practice addresses the scarcity of water among community members.

"*A good example is collecting rainwater for gardens...*" Group 46

According to the above reflections, the United Nations (UN) SDGs have 17 goals, two of which expressly address water. They include Goal 6, "Ensure availability and sustainable management of water and sanitation for all", as well as Goal 14 "Conserve and sustainably use the oceans, seas and marine resources for sustainable development". Both goals emphasise avoiding water pollution, enhancing and restoring water quality, and safeguarding and restoring water-related ecosystems worldwide (Huang et al., 2021). Further, preservice teachers highlight that "*Rainwater collection can eventually result in cost savings, depending on where you live and water usage*" (Group 23). Rainwater harvesting is a good source of water for bathing, washing, and other household uses, particularly in places where it is challenging to get to public water supply (Owusu & Asante, 2020). Harvesting of rainwater also reduces households' expenses on irrigation, cooking, and laundry. Rainwater collected from the rooftop within households reduces costs because there will be no transportation of water in channels and no treatment needed for the water as it is naturally harvested. Indigenous Knowledge Systems such as rainwater harvesting point out that Indigenous practices have a role to play in our modern societies to address environmental issues and alleviate costs for water purifications.

> *"It encourages learners to value and apply this knowledge in their own lives through practical actions such as collecting rainwater and creating water purification systems"* Group 47

7.3 Traditional Environmental Conservation

Traditional environmental conservation is one of the crucial IKS practices for sustaining the environment. The IKS offers practical skills to protect the environment, for example, Indigenous peoples would collect plants and roots without causing environmental damage (Seroto, 2020). Woods have been used to make fire without destroying the plant/tree but by harvesting the branches, twigs, and stem. Some Traditional practices are about conserving natural resources from depletion, including taboos that are a valuable way of conservation. For example, taboos protect some edible animals and plants because it is taboo for one to eat his or her totem animal (Govender & Mutendera, 2020). Preservice teachers from Group 33 reflect "*...sustainable harvesting guarantees that these resources will be available for generations to come by promoting conservation activities*". In addition, "*We carefully rotate animals across meadows, ensuring their welfare. This stimulates pasture growth, reduces erosion, and helps soil retain moisture.*" Group 35. Therefore, traditional environmental conservation is crucial to conserve the extinction of rare types. In this instance, PAR and social learning theory can be integrated into IKS practices to foster the preservation of natural resources. It is understood that "*... traditional medicine is a sustainable method of environmental conservation as well as an important part of Indigenous culture*" (Group 34). The practice of traditional medicine African wormwood further stresses the importance of intergenerational learning and its co-existence with nature.

7.4 Traditional Agricultural and Land Management

Traditional agricultural and land management are Traditional practices of interacting with nature without causing any harm from composting to ploughing. For instance, the use of cows to plough the land and the application of cow dung to fertilise the soil. These exclude the use of machines and chemicals that could cause pollution. Hence, Group 28 of preservice teachers aver that traditional agricultural and land management are "*...useful strategies to maintain a healthy environment and sustainable agriculture, such as crop rotation, organic composting, and intercropped fields....*" Group two reiterates, "*Environmental sustainability has been going on for ages and ages and that there are various ways to sustain the environment such as moving cattle around grazing areas and even planting trees*". Traditional agriculture practices are conservative and localised in the Indigenous community (Patel, Sharma & Singh, 2020)—the application of the Traditional practices aimed to manage the land through rotational grazing, etc. Management of the land not only improves the Traditional agricultural practices but also "*...improves food sovereignty for Indigenous populations and attain environmental sustainability by fusing these traditional methods with contemporary farming methods*" (Group 28). Further Indigenous agricultural practices play a crucial role by offering sustainable farming practices including

"Indigenous seed-saving practices are rooted in deep knowledge of ecosystems and genetic diversity, which is crucial for sustainable agriculture. These practices also hold cultural importance in preserving traditions and connection to the land" (Group 20).

8. IMPLICATIONS

Traditional practices have the potential to save the environment from degradation and can be used to enhance the school curricula. Traditional practices including rotational grazing require young kids to learn skills to sustain the practices. For example, children are shaping their behaviour by observing the most relevant adults especially teachers in school and parents at home (Amoyedo-Peter, 2023) and ignorance of the practices could mean there will be minimal intergenerational learning. In society, parents within the family, characters on mass/social media, friends within their peer groups, religion, other members of society, and the school can influence children either negatively or positively. Children pay attention to some of these agents of socialisation and imbibe the behaviours exhibited (Edinyang, 2016). Social learning theory helps us understand how our environment and the people around us shape our behaviour and learning experiences. It helps explain how individuals develop new skills and behaviours by paying attention to the behaviour of others and then trying to reproduce that behaviour themselves (McLeod, 2024).

Participatory action research, social learning theory, and IKS can contextualise traditional practices, and promote collectiveness in the school curriculum. Indigenous people over the years used traditional practices to teach young kids about environmental conservation including rainwater harvesting. It is for this reason that learning is centred and knowledge is co-constructed through engagement as presented in Figure 1. Learning, on the other hand, is a relatively permanent change in behaviour resulting from activities, training, or experiences (Amoyedo-Peter, 2023). The application of PAR, social learning theory, and IKS in school curricula could add value to enhancing the curricula.

9. CONCLUSION

This chapter explored South African preservice teachers' experiences with traditional environmental practices from the perspective of the PAR approach. It also discussed the nature of learning, the application of social learning theory in IKS, traditional practices relating to PAR and social learning theory, and the value of PAR and IKS in the Sustainable Development Goals.

The chapter responded to the research question *"What are the South African preservice teachers' experiences about traditional environmental practices"* It is established in this chapter that preservice teachers know traditional practices that could conserve the environment from degradation. The practices include traditional framing practices (agroforestry and rotational grazing) rainwater harvesting, traditional environmental conversation and traditional agricultural and land management. These practices have been used in indigenous communities for centuries to sustain the environment. The awareness of the practices is important for intergenerational learning because traditional practices are learnt through observation and experimentation for preservation. The chapter fills the gap in the literature on traditional environmental practices from the perspective of PAR.

REFERENCES

Amoyedo-Peter, C. (2023). Albert Bandura's Social Learning Theory: The significance. *The Prodigious Musicologist: Celebrating Professor Agatha Ijeoma Onwuekwe*, 194-204.

Bada, S. O., & Olusegun, S. (2015). Constructivism learning theory: A paradigm for teaching and learning. *Journal of Research and Method in Education*, *5*(6), 66–70. DOI: 10.9790/7388-05616670

Braun, V., & Clarke, V. (2006). Using thematic analysis in psychology. *Qualitative Research in Psychology*, *3*(2), 77–101. DOI: 10.1191/1478088706qp063oa

Cilliers, E. J. (2021). Reflecting on Social Learning Tools to Enhance the Teaching-Learning Experience of Generation Z Learners. *Frontiers in Education*, *5*(606533), 606533. Advance online publication. DOI: 10.3389/feduc.2020.606533

da Silva, C., Pereira, F., & Amorim, J. P. (2023). The integration of Indigenous knowledge in school: A systematic review. *Compare: A Journal of Comparative Education*, *54*(7), 1210–1228. DOI: 10.1080/03057925.2023.2184200

Edinyang, S. D. (2016). The significance of social learning theories in the teaching of social studies education. *International Journal of Sociology and Anthropology Research*, *2*(1), 40–45.

Feri, M., & Maemonah, N. H. (2022). Analysis of social attitudes of elementary school learners with the application of Albert Bandura's learning theory. *Jurnal Pendidikan Dasar*, *14*(2), 149–157. DOI: 10.17509/eh.v14i2.41081

Firmansyah, D., & Saepuloh, D. (2022). Social learning theory: Cognitive and behavioral approaches. *Jurnal Ilmiah Pendidikan Holistik*, *1*(3), 297–324. DOI: 10.55927/jiph.v1i3.2317

Gonçalves, C. B. Q., Schlindwein, M. M., & Martinelli, G. C. (2021). Agroforestry Systems: A Systematic Review Focusing on Traditional Indigenous Practices, Food and Nutrition Security, Economic Viability, and the Role of Women. *Sustainability (Basel)*, *13*(20), 11397. DOI: 10.3390/su132011397

Govender, N., & Mutendera, G. (2020). Teachers' and custodians' views and dilemmas arising thereof regarding the integration of indigenous knowledge in the primary school. *Alternative*, *16*(4), 356–368. DOI: 10.1177/1177180120970935

Grandin, T. (2022). Grazing cattle, sheep, and goats are important parts of a sustainable agricultural future. *Animals (Basel)*, *12*(16), 2092. DOI: 10.3390/ani12162092 PMID: 36009682

Hare, J. (2021). Indigenous Synergies for Decolonizing Action Research. *The Canadian Journal of Action Research*, *21*(3), 4–16. DOI: 10.33524/cjar.v21i3.561

Hensler, L., Alatorre Frenk, G., & Merçon, J. (2023). Participatory action research. *Handbook Transdisciplinary Learning*, *6*, 257–266. Advance online publication. DOI: 10.1515/9783839463475-027

Huang, Z., Nya, E. L., Rahman, M. A., Mwamila, T. B., Cao, V., Gwenzi, W., & Noubactep, C. (2021). Integrated water resource management: Rethinking the contribution of rainwater harvesting. *Sustainability (Basel)*, *13*(15), 8338. DOI: 10.3390/su13158338

McLeod, S. (2024). Albert Bandura's social learning theory. *Simply Psychology. London*. https://www.simplypsychology.org/bandura.html

Meyiwa, T., Letsekha, T., & Wiebesiek, L. (2013). Masihambisane, Lessons Learnt Using Participatory Indigenous Knowledge Research Approaches in a School-Based Collaborative Project of the Eastern Cape. *South African Journal of Education*, *33*(4), 1–15. DOI: 10.15700/201412171329

Misra, V., & Misra, P. K. (2020). Implications of constructivist approaches in the classrooms: The role of the teachers. *Asian Journal of Education and Social Studies*, *11*, 17–25. DOI: 10.9734/ajess/2020/v7i430205

O'Donoghue, R. K. (1993). Environment, development, and environmental education: An environmental policy and curriculum initiative in formal education. Howick: Share-net.

Owusu, S., & Asante, R. (2020). Rainwater harvesting and primary uses among rural communities in Ghana. *Journal of Water, Sanitation, and Hygiene for Development : a Journal of the International Water Association*, *10*(3), 502–511. DOI: 10.2166/washdev.2020.059

Patel, S. K., Sharma, A., & Singh, G. S. (2020). Traditional agricultural practices in India: An approach for environmental sustainability and food security. *Energy, Ecology & Environment*, *5*(4), 253–271. DOI: 10.1007/s40974-020-00158-2

Pooley, S. (2021). Fire in African landscapes. In *Oxford Research Encyclopedia of African History*. http://dx.doi.org/DOI: 10.1093/acrefore/9780190277734.013.984

Saldana, J. (2013). *The Coding Manual for Qualitative Researchers*. Sage.

Seroto, J. (2020). Indigenous history and education before colonial times. In Seroto, J., Davids, M. N., & Wolhuter, C. C. (Eds.), *Decolonising education in the global south: Historical and comparative international perspectives* (pp. 40–58). Pearson.

Suarta, I. M., Noortyani, R., Yarsama, K., & Adhiti, I. A. I. (2022). The role of teachers' Indigenous Knowledge and Cultural Competencies in enhancing students' engagement and learning outcomes. *Journal of Ethnic and Cultural Studies*, *9*(1), 244–264. DOI: 10.29333/ejecs/1025

Viswanath, S., & Lubina, P. A. (2017). Traditional Agroforestry Systems. *Agroforestry: Anecdotal to Modern Science*, 91-119. http://dx.doi.org/DOI: 10.1007/978-981-10-7650-3_3

Zuber-Skerritt, O. (2015). Participatory action learning and action research for community engagement: A theoretical framework. *Educational Research for Social Change*, *4*(1), 5–25. DOI: 10.1108/978-1-78769-537-520191017

Zuber-Skerritt, O. (2018). An educational framework for participatory action learning and action research. *Educational Action Research*, *26*(4), 513–532. DOI: 10.1080/09650792.2018.1464939

Chapter 15
The Role of Indigenous Participants in Participatory Action Research:
An Application of a Decolonial Approach

Monicca Thulisile Bhuda
https://orcid.org/0000-0002-8506-9562
University of Mpumalanga, South Africa

Mamokete Venolia Setshego
https://orcid.org/0000-0002-1925-5949
North-West University, South Africa

Motheo Koitsiwe
North-West University, South Africa

ABSTRACT

Indigenous peoples were formerly treated more as passive study subjects than active participants in research, demonstrating that the purpose of the research was to marginalise rather than empower them. Therefore, Participatory Action Research (PAR) with indigenous people is a decolonial approach which aims to achieve a better balance between research participants' input, decision-making, and power with research teams. This study examined the role of Indigenous participants in PAR using a Qualitative Secondary Analysis method based on desktop research which was sourced from academic works with related studies. Using thematic analysis, various themes were developed from existing literature to complete the scope of this study. This study's discussion offers examples of useful implementation of PAR with indigenous participants in two case studies. It further makes a case that since it is founded on the ideas of participation and action, PAR is suitable to be applied in studies involving indigenous participants who have historically been colonised and had no voices in research about them.

DOI: 10.4018/979-8-3693-3542-0.ch015

INTRODUCTION

Participatory Action Research (PAR) as explained by Blumenthal (2011) is an approach of inquiry built in collaboration with people directly concerned with the issue at stakes. Priorities and perspectives of these people with lived experience are essential and central in initiating a project. Chen et al (2010) argued that their engagement within the process is at the core of the endeavour. While researchers have too often positioned Indigenous peoples as passive research subjects, the PAR paradigm situates them as full subjects invested with agency. Community-based participatory research at its theoretical roots implies a process of shared decision-making, ownership and co-construction between researchers and the communities where the research is to take place (Cornish et al., 2023).

Historically, Cornish et al (2023) indicated that studies involving Indigenous peoples have not guaranteed benefits to their communities or acknowledged indigenous worldviews. Researcher oppression and uneven power relations, wherein Indigenous communities are treated as passive study subjects rather than independent, self-governing actors, have been and still are encountered by Indigenous communities. Consequently, PAR is a methodological approach that presents itself as a project with roots in a decolonisation framework, particularly in Indigenous situations (Guillemin et al., 2016). PAR is appealing in these kinds of situations because of its dedication to assisting people that have historically been underrepresented in the research community and to promoting initiatives that would directly impact their circumstances. Understanding knowledge generation through PAR is also compatible with publications that discuss Indigenous paradigms and methodology and highlight the significance of epistemological heterogeneity (Bhuda & Koitsiwe, 2022).

It was indicated by Datta (2018) that some researchers have resorted in conducting PAR without the involvement of communities as participants who should benefit from research. This has been challenging, especially because Indigenous communities are custodians of Indigenous Knowledge which researchers need to source. Datta (2018) emphasizes the necessities of community engagement, ensuring respect for traditions, perspectives, and ethical considerations inherent in IKS. Therefore, it is uncommon to find PAR undertaken in isolation from the communities it aims to understand and serve (Guillemin et al, 2016; Datta, 2018). This study therefore examines the role of Indigenous participants in PAR. The study's examination is based on the argument that the research was primarily conducted on Indigenous communities for the benefit of the researchers, with little or no consultation or benefit for the communities participating (Guillemin et al, 2016). Thus, PAR serves to deconstruct the Western research paradigm and is inclusive toward an Indigenous understanding of knowledge. Based on the above discussions, the study's objectives are as follows:

1. To investigate the role of Indigenous participants, have in PAR.
2. To explore how PAR benefits Indigenous participants.
3. To discuss how PAR contributes to decolonising research.

Defining PAR

According to Kemmis et al. (2014), PAR is an umbrella for research approaches, methodologies, and frameworks that employ systematic inquiry in close cooperation with people impacted by the topic under study with the goal of action or change. PAR places a high priority on co-constructing research through collaborations between scientists and stakeholders, locals, or other people who have firsthand experience and knowledge of insider (Jagosh et al., 2012). In other words, PAR involves people who are not trained in research but who either belong to or represent the interests of the people the research is focused on.

PAR aims to place power in the research process in relation to those who will be most impacted by a program. The idea is for the participant and the researcher to work together as equal partners (Stern, 2019). The process of collective, self-reflective inquiry that stakeholders engage in is a crucial component of this involvement. It aims to better understand and enhance the situations and practices in which they are involved. Since action is a part of this process, the individuals or communities it affects should ideally have more influence over their lives (Halliday et al., 2019).

The goal of PAR, according to Hallidayet al. (2019), is to promote participation, empowerment, and community development. According to Cornish et al. (2023), PAR acknowledges and values the fact that individuals live in social, political, and economic contexts. Strongly value-oriented, PAR aims to address important concerns pertaining to the well-being of individuals, their communities, and the larger ecosystem in which we all live. As such, research participants are not study subjects; rather, they are active researchers who take part in every stage of the research process (Kemmis et al., 2014).

Kemmis et al. (2014) further highlighted that the PAR process assists in restoring a person's capacity to be a creative player in the world by including them actively in important decision-making. As a result of communal inquiry's promotion of knowledge ownership in PAR, the research process becomes less mysterious and creates more space for the development of trust. Brydon-Miller et al. (2020) argued thatin order to create emancipatory social change, PAR entails the involvement and leadership of individuals who are struggling with challenges. These individuals do methodical study to produce new knowledge. The important factors for designing a PAR project are outlined as follows: building relationships, establishing working procedures, creating a shared understanding of the problem, observing, gathering, and producing materials, collaborative analysis, planning, and action are the six fundamental building blocks for PAR project design that are introduced in the primer's core (Nadeau et al., 2022).

Indigenous Communities as Participants in Research

The Indigenous participants or communities refers to a unique group of people that practices unique traditions and have social, economic, cultural and political traits that are different from the main society (Simpson & Mendenhall, 2022). They are connected to their ancestral land because it has fundamental importance for their collective physical and cultural survival as peoples. Leung and Minkler (2004) stated Indigenous participants or communities are in most cases among the society that is neglected when it comes to service delivery, therefore, they use their Indigenous knowledge and the natural resources available to them for their livelihood; and that is how some of their Indigenous Knowledge is formed. Thus, Indigenous participants in this study refers to Indigenous Knowledge holders and laymen in Indigenous communities who are aware of the traditions and cultures in that particular area.

For many years the Indigenous communities were considered as passive subjects of the research rather than as researchers. However, in PAR, people who would otherwise be called informants became core-searchers in a process of collaborative knowledge generation (Keane et al., 2017). PAR is a process wherein participants and researchers jointly create objectives and methods, take part in data collection and analysis, and apply the findings in a way that will increase critical awareness and encourage positive changes in the lives of those involved changes that are under the direction and control of the participating group or community (Balakrishnan & Claiborne, 2017).

Moreover, PAR involves collaboration of two different parties which is a community with lived experience of a social issue and professional researchers who contribute relevant knowledge, skills, resources and networks (Cornish et al, 2023). In contrast to the traditional western approach which in most cases suck knowledge from the Indigenous participants without solving any social issue in their communities; PAR is not a research process driven by the authoritative to generate knowledge for scientific progress, or knowledge for knowledge's sake; it is a process for generating knowledge for action and knowledge through action, in service of goals of specific communities (Cornish et al., 2023).

For a long time, Indigenous participants were considered to be illiterate by the standard of the Western and their knowledge is emphasized as primitive. Owusu-Ansah and Mji (2013), support by asserting that the intelligence of the Indigenous people/participants accentuates the practical, interpersonal and social domains of functioning and is quite distinguished from the cognitive intelligence that dominates and embraced by Western. However, as indicated by Israel et al. (2005), the researchers all over the world have found an interest in IKS and have comprehended its significance.

In most of the research conducted in Indigenous communities about IKS, the participants are the providers of the knowledge, which makes them the researchers, not passive subjects. Indigenous participants in research, should be involved in the research project from the beginning and acknowledged as researchers and owners of the study. According to Nadeau et al. (2022), PAR seeks a better balance of input, decision-making and power between research participants and research teams and values participants' knowledge.

Therefore, the Indigenous participants are active in PAR by articulating their views, wishes, concerns and research questions that deem important to their cultural context and experiences (Owusu-Ansah & Mji, 2013). To elaborate, they contribute their knowledge, perspective, and lived experiences to shape the research questions, methodology and outcomes. Furthermore, Caxaj (2015) argued that the Indigenous participants perform as a voice of their Indigenous community concerning the issue that is researched.

Maditsi and Bhuda (2023) indicated that Indigenous participants provide insight concerning a matter using their Indigenous worldviews and that assist the professional researchers to come up with the solutions that will be accepted within the cultural context of that certain community. To clarify, Indigenous participants are every so often the experts of their culture, therefore, they bring vital cultural knowledge and insights that are important for understanding the community's context, history, values, and ways of knowing (Bhuda & Koitsiwe, 2022). Moreover, the Indigenous participants safeguard the Indigenous knowledge of the community. They ensure that the Indigenous knowledge is properly used on the outcome of the research and the outcome respect and protect the intellectual property and cultural heritage.

Despite this, the importance of Indigenous participants is overlooked. For instance, there are much research done with and on Indigenous people (participants), however there are few studies that has outlined the role of Indigenous participants in research (Blumenthal,2011). The research conducted is in most cases about the Indigenous knowledge they possess concerning an interest of research (Bhuda & Marumo, 2022). Therefore, it is significant to conduct this study because it acknowledges the agency and contribution of the Indigenous participants after being marginalized for so long; the study can contribute to efforts to decolonize research by challenging dominant research paradigms and fostering more equitable partnerships and in general, the significance of this study is ethically imperative, contributing to inclusive, equitable, and respectful research practices that prioritize Indigenous perspectives and promote social justice (Kemmis et al., 2014).

METHODOLOGY

Qualitative Secondary Analysis refers to the process of utilizing previously obtained qualitative data to address an alternative research topic. When it comes to patient populations who are hard to reach, secondary analysis of qualitative data offers a chance to optimize data utility (Irvine, 2022; Ruggiano & Perry, 2019). Qualitative Secondary Analysis (QSA) of secondary data that was sourced in this study focuses mainly on studies such as community-based participatory research, decolonising research, PAR, and Indigenous research. The primary sources of information for the data gathering method were literary sources, which included books, book chapters, and published peer-reviewed journal articles. A range of platforms, including Google Scholar, JSTOR, Research Gate, and institution repositories, were searched in order to guarantee a thorough examination of the resources that were available.

Thematic analysis was used to analyse data collected from above mentioned sources. Clarke and Braun (2017) explained thematic analysis as a method for analyzing qualitative data, such as transcripts or interviews, to identify patterns, themes, and ideas. Therefore, the analytical procedure entails both electronic and physical documents evaluated by reading and comprehending the information provided to get codes that generated themes, then the themes were reviewed and named. The study aims to explore the ways in which Indigenous people engage in PAR. This is significant because researchers who ignored IKS have marginalized and maligned indigenous people and their knowledge (Bhuda, 2024). Thus, the themes that were generated include 1) Indigenous people in PAR 2) PAR as a benefit to Indigenous participants 3) PAR approach and decoloniality.

The study aims to explore the ways in which Indigenous people engage in PAR. This is significant because researchers who ignored IKS have marginalized and maligned indigenous people and their knowledge (Bhuda, 2024). Therefore, to build positive relationships with indigenous people and accomplish the objectives of the research at hand, researchers must employ the proper methodology in their work, such as PAR . Themes were generated in this study using thematic analysis which Clarke and Braun (2017) explained as a method for analyzing qualitative data, such as transcripts or interviews, to identify patterns, themes, and ideas. Researchers employ thematic analysis as a qualitative research method. In order to read, comprehend, and expand upon the information provided by electronic and physical documents, the procedure entails evaluating both types of materials. The themes that were generated are as follows: 1) Indigenous people in PAR 2) PAR as a benefit to Indigenous participants 3) PAR approach and decoloniality.

Indigenous People in PAR

There is a difference between doing research on Indigenous people and with Indigenous people. In a case of PAR, the research is normally conducted with and on Indigenous communities. Therefore, both non-indigenous researchers and Indigenous participants enter into a PAR as equal partners, whereby both parties develop a proposal and come into an agreement; however, the research should only proceed if its focus and processes align with the priorities and needs of Indigenous peoples (Dadich et al., 2019). Thus, Indigenous participants are involved in PAR to share a light on Indigenous values and principles that exist within their communities and also to state their needs and priorities that necessitates research.

They stand in a gap to ensure that the non-indigenous researchers do not conduct research that will violate their culture and ensures that their ethics are followed.

As stated by Caxaj (2015), Indigenous values and principles are able to illuminate on ethical and methodological considerations outside of the purview of standard academic institutes. As results, the outcome of the research will still hold the essence of the culture and meaning of their knowledge (Bennett, 2019). The involvement of Indigenous participants denounces the monopolistic control of academia over the research process, specifically, when the analysis and interpretation of research findings must take account of Indigenous science, which is based on experiential and humanistic interpretation, rather than academic needs (Bennett, 2019). Consequently, the primary goal of participant action research will be achieved which is to create the positive social change with involvement of the Indigenous participants who are able to state the community's need and are equipped to bring that change (McHugh & Kowalski, 2009).

PAR as a Benefit to Indigenous Participants

PAR has been advocated as the relevant approach that is suitable for working with Indigenous communities because it can center the local knowledge and involve Indigenous participants fully on research (Caxaj, 2015). PAR focuses on achieving certain goals and by achieving that goal will benefit the people of area of study, including the participants. Simpson and Mendenhall (2022) assert that PAR has been used as an instrument to understand and develop solutions to the problems faced by the Indigenous communities.

In addition, empowering Indigenous participants is one of the main goals in PAR, for it is process that builds a reciprocal relationship between the Indigenous participants and researchers and also emphasizes the researcher responsibility towards the participants (Datta et al., 2014). Indigenous participants involvement in research gives them an understanding of the process of research, therefore it equips them to advocate a change (Simpson & Mendenhall, 2022). Moreover, Dudgeon et al, (2020) affirms that the study of Peltier (2018) indicates that the PAR provided a catalyst for action and positive changes and sustained changes in the health service and greater empowerment for health workers.

Furthermore, the involvement of Indigenous participants creates a position for them to have a voice on a matter researched and also on other matters that are not part of the study. For instance, Khupe and Keane (2017), stated that the PAR that they were conducting in Mqatsheni in KwaSani Municipality, Sisonke District in the KwaZulu-Natal Province, was supposed to be project that involved partnering with communities in identifying ways of bringing together IK and science to benefit communities.

However, Khupe and Keane (2017) ended up also addressing the issue of food insecurity that was raised by the community. Therefore, the investigation of practical concerns of Indigenous people, enable the account of the voiceless to be heard (De Oliveiria, 2023). Moreover, the involvement of Indigenous participants in the research process builds their analytical and critical skills, in doing so returns controls, and power to the community (Hoar et al, 1993). In addition, data analysis segment ensures Indigenous participants that the identified themes at the beginning of the study, which might be significant issues has been addressed; therefore, the research would have an impact at the policy level and speak on behalf of Indigenous participants (Datta et al., 2014). De Oliveiria (2023), supports that Participants Action Research has been a tool used to reach policymakers more easily and represents a more effective means of informing them on current social issues.

Formerly, researchers would enter Indigenous communities armed with goodwill in their front pockets and patents in their back pockets (Dadich et al., 2019). To be exact, the researchers would conduct research with the Indigenous communities, pretending to help them, only to steal their knowledge and claim the ownership of it. Conversely, in PAR, the Indigenous participants are acknowledged and are co-owners of the knowledge (Datta et al., 2014).

PAR Approach and Decoloniality

Decolonisation is a process that has its roots in the idea of self-determination. It aims to reduce the possibility of oppression and micro aggressions by redefining power dynamics and alliances between Indigenous and non-Indigenous peoples and institutions. Heleta (2018) state that research acknowledges the agency of Indigenous communities and their ownership of research data. It necessitates a paradigm change that emphasizes Indigenous epistemologies while valuing and complementing other epistemologies. It entails incorporating the promise of Western research methodologies while enhancing Indigenous values, knowledge, and traditions (Tavernaro-Haidarian, 2019).

Decoloniality, then, is a key idea in postcolonial thinking that centers the struggles of marginalized individuals and groups, frequently in the global South, in order to undermine long-standing power systems (Chaka, 2020). It challenges the prevailing ideas that continue to oppress individuals who are impacted by them, such as racism, sexism, colonialism, and imperialism (Sefa-Dei & Cacciavillani, 2024). By redistributing economic resources and making room for other points of view, decoloniality envisions a society free from these oppressive structures and strives to promote equity for everyone (Ndlovu-Gatsheni, 2015). In addition, it aims to reverse the effects of colonialism by redefining identity, culture, and community in ways that transcend or bridge geographical boundaries.

Understanding decoloniality is crucial to comprehending the state of affairs in many regions of the world today. It challenges prevailing narratives and power structures that have disadvantaged some communities by taking a critical stance toward knowledge systems that are controlled by the West (Ndlovu-Gatsheni, 2015; Mignolo, 2020). Decolonial theory acknowledges a variety of difficulties resulting from the histories of multiple colonizations, but it also concentrates on establishing more fair practices and dynamics in order to pave new paths towards social justice (Mignolo, 2020). Decoloniality emphasizes the necessity for us to rethink various socialization strategies that result in transformation rather than just offering criticism (Omodan & Dastile, 2023).

Based on the above explanation, this study contends that PAR is a means of promoting inclusivity and teamwork that subverts the prevailing power structures that decoloniality proposes, based on the logical reasoning that emerges from both concepts (Omodan & Dastile, 2023). By enabling people to collaborate to develop a solution, it helps to balance the power dynamics between individuals facing an issue. Gill et al. (2012) explained that the act of collaborating facilitates the creation of alternative channels for the production of information, giving previously marginalized or silenced voices a chance to be heard and, consequently, influence the final solutions selected. This method provides an alternate means of creating tactics that can aid in the gradual dismantling of oppressive structures, even though it cannot eliminate them completely. PAR proposes an alternative form of power structure, based on the ideas of decolonisation. It suggests an inclusive and collaborative pattern as an alternative to a hierarchical one, allowing people who are having problems to join together and find solutions (Fernández, 2022).

Additionally, this approach promotes involvement from all parties, establishing forums where all opinions can be heard equally and assisting in the group's collective identification of workable solutions (Kidd & Kral, 2005). In the end, this transforms the colonial notion of power into something new and more equal by giving the oppressed people the opportunity to take the lead in developing solutions that are good for everyone. Balakrishnan and Claiborne (2017) argued that since power supremacy removes the possibility of further subjugating the studied, this further demonstrates that the inclusive and collaborative process of PAR is a methodical means to decolonize the research process. Thus, one may consider PAR to be a decolonial research methodology.

Vaughn and Jacquez (2020) stated that PAR serves as a strategy that provides voice to IKS and the perspectives of Indigenous peoples is demanded by the decolonisation project in response to this. This is a response to the ways in which colonization has affected Indigenous rituals and belief systems. From a research perspective, it also emphasizes the necessity of decolonial methodologies that capture reality as experienced by these Indigenous populations. Additionally, Bartlett et al. (2007) developed a framework (Figure 1) that is an Aboriginal-focused approach/methodology for decolonising research that looks at the experiences of Indigenous peoples in Canada. The six steps of reasoning, enabling, facilitating, experiencing, accepting, and implementing decolonising research are included in it, in sequence of application (Bartlett et al., 2007). The process of implementing an Indigenous research methodology to question conventional Western research methods in order to reformulate underlying assumptions and methodologies is of special importance here.

Figure 1. Depiction of framework for Aboriginal-guided decolonising research (Bartlett et al., 2007).

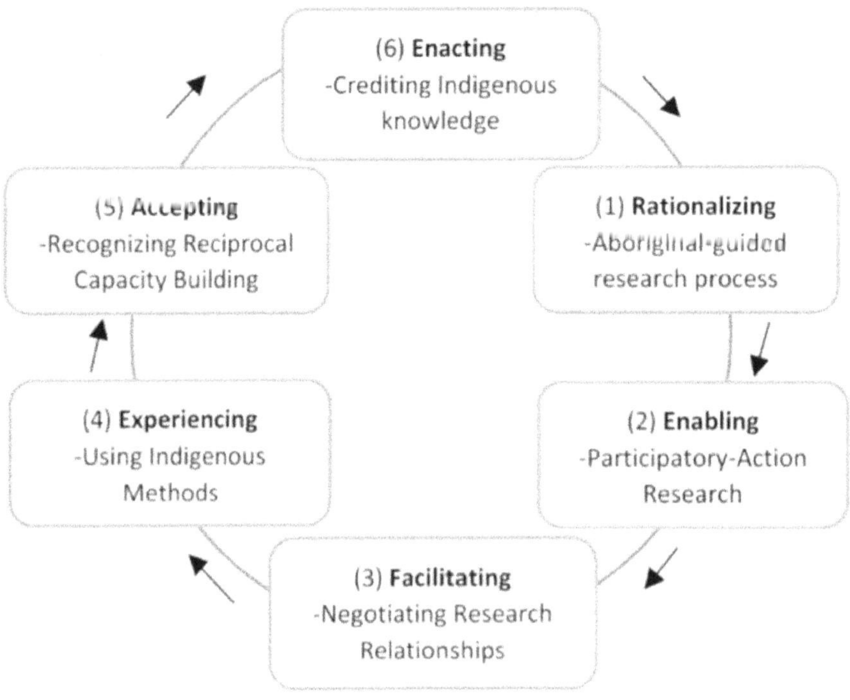

Case Studies with a Participatory Research Approach

This study makes example of case studies by two researchers who employed PAR. The study first makes an example of a case study by Khupe (2014) who conducted a study using Participatory Research Approach in Mqatsheni rural KwaZulu-Natal, South Africa by (Khupe, 2014). The second study was conducted by Mwangi et al. (2023) with the two communities (Maasai and Samburu people) in Kenya about the nature conservation.

Participatory Research Approach in Mqatsheni Rural KwaZulu-Natal

Khupe (2014) conducted participatory science curriculum research which focused on possible integration of Indigenous Knowledge and Science education. This study provides examples pertaining to community access, research goals, data production procedures, ethics, and results. Through her efforts, Khupe (2014) promotes the use of PAR as a vital instrument for gathering Indigenous knowledge and fostering social change among Indigenous communities.

- *Study area*

Mqatsheniis located in the KwaZulu-Natal province's KwaSani Municipality, Sisonke District. The area as explained by Khupe (2014) is located roughly 30 kilometers north of the tiny urban community of Underberg in the southern UKhahlamba mountains. In Mqatsheni, the majority language is IsiZulu. Although English is the language of teaching in the local schools, very few people speak it outside of the classroom. The usage of isiZulu is so great that it frequently replaces English in classrooms. In Mqatsheni, cultural values are very prevalent. Furthermore, the region is governed by a Traditional Council of Elders, or induna, which is led by a chief. The elders are in charge of preserving culture and knowledge and ensuring that it is passed down to future generations.

- *Research approach, design and paradigm*

In order to find and document IK in Mqatsheni for use in science teaching and learning, as well as to help transform IK research by employing techniques that acknowledge the significance of language and culture in the lives of community members, Khupe (2014) conducted an interpretive research study. The research was a component of a larger initiative on indigenous knowledge, science curricula, and development, which involves working with communities to find ways to combine science and IK for the good of the community.

Secondly, keeping with the goal of the study, the Mqatsheni community and university academics collaborated through a transformative participatory design. Recognizing the connection between knowledge and power, transformative participatory research aims to establish a shared-power posture between community members and university researchers so that both parties can actively contribute to knowledge development. The method aims to bring people and their everyday experiences from the periphery of epistemology to the center of the knowledge generating process by placing communities at its center (Hall, 1992).

Furthermore, relationships are recognized as essential to the research process in an indigenous paradigm, which means that people and information are not viewed as "objects" (Hart, 2010). Four key indigenous methodology tenets are summarized by Louis (2007). First, the network of connections that indigenous peoples have with their physical and social environments is characterized by relational accountability. In the field, the researcher is accountable to all of these relationships in addition to being in charge of building them. Respectful representation is the second principle. The researcher considers how she portrays herself, community members, the events, and phenomena in the study in light of this principle.

The third principle, which deals with reciprocal appropriation, stems from the understanding that social science research fundamentally entails appropriation and, as such, necessitates that the procedure be advantageous to all parties. This suggests that the researcher must provide stakeholders with ongoing input and seek their advice on how to make the process advantageous for all parties. Fourth, it is necessary to recognize that the study focuses on local cultural customs in order to comprehend rights and regulations (Louis, 2007). Participants in the community own the knowledge they give; thus, the research process should be cooperative and follow a path that reflects a balance of power between community and university researchers.

The study's methodology was chosen in part because of the participants' primarily oral culture, the significance of history, the role of the Traditional Council, and the role of the Zulu language as a medium of the participants' culture. Khupe (2014) conducted collaborative research using the Ubuntu principles. The individuals' preferred language was used throughout the investigation. The methodological framework for this investigation was based on ubuntu and indigenous research, which included collaborative and participatory research. Mqatsheni people's knowledge and the worldview that supports it were identified with the aid of these frameworks.

- *Data collection stages*

Understanding the circumstances in Mqatsheni helped Khupe (2014) dispel many of her preconceptions, misconceptions, and stereotypes during the data collection stage, which was carried out through observations and interviews. It also promoted courteous interactions and improved comprehension of the data.

- *Ethics*

Khupe's (2014) study received approval from the university's ethical committee. Consent forms were signed by participants at the start of the investigations. None of the community members requested anonymity or to be given false identities, which is contrary to what is commonly believed in traditional research. They even consented to the research papers using their photos. But real ethical problems go beyond formal agreements. Decisions about ethics were made both during and after fieldwork. Khupe's (2014) initiative received approval from the university's ethical committee. Consent forms were signed by participants at the start of the investigations. None of the community members requested anonymity or to be given false identities, which is contrary to what is commonly believed in traditional research. They even consented to the research papers using their photos showing that they trusted the researcher and wanted lasting relationships with her (Keane, 2008).

PAR (PAR) for Community-Based Conservation in Kenya

The study of Mwangi et al. (2023) is another example of PAR research conducted with the two communities (Maasai and Samburu people) in Kenya about the nature conservation. The qualitative research was conducted in Mbirikani Group Ranch (MGR) and Lekurruki Community-based conservation (CBC). The researcher intended to engage all the community members at all stages of planning conservation projects through active participatory approaches. They argue that the top-bottom approach has been used when it comes to participation of the communities in policies whereby communities participate in the last stage of decision-making which is implementation (Li, 2023). As a result, the projects fail after the donors stop funding the project; therefore, there is need for a bottom-up approach to participation for a successful community approach to wildlife conservation programs.

- *Methodology*

The study used qualitative research methods which include observations, case-study and the thematic analysis and cross-case synthesis was used to analyze data (Clarke & Braun, 2017). Synthesis of case studies is necessary to build a body of knowledge from individual cases (Cruzes et al., 2015). The primary and secondary data was used for data collection tools to explore community needs to achieve full potential in conservation of wildlife, starting with conceptualizing and particularizing the problem and moving through several interventions and evaluations (Mwangi et al., 2023).

- *Results*

1. Mbirikani Group Ranch

The data collection with the Maasai people was done in 3 months, whereby the researcher has the translator to assist. According to the data collected, the challenges faced by community towards conservation of wildlife were human wildlife conflict leading to loss of crops and livestock, low rates of compensation that caused dissatisfaction, low tourism revenues experienced due to the Covid-19 pandemic, lack of involvement by the youth. The researchers observed that children under 18 years and women are either partially participating or not participating entirely in conservation due to their culture. Therefore, the researcher urges that the cultural hindrance should be addressed because everyone is significant in assuring the conservation of nature (Mwangi et al, 2023).

2. Lekurruki CBC

The study by Mwangi et al. (2023) explained that this community is made up of the Maasai and Samburu people and they live in the heart of Laikipia County between mountain and neighboring other CBCs and Samburu community. Community was engaged in conservation programs through employment in schools and ranger security while others worked in the Lodge. The children under 18 years were not allowed to be involved in conservation program and young people were not interested claiming that it is an agricultural activity. Women were not supposed to speak in the community development meetings held every quota of year.

- Discussions

The results of the study by Mwangi et al. (2023) indicated that culture plays a significant role in building society and for the progress of the community. Most of the African cultures don't allow women to be part of the decision-making and leadership hence the Maasai and Samburu women on both case-studies are not actively participating in conservation programs. Errico (2021) affirmed that women have a very limited and unsecured access to the land and natural resources and are excluded in decision-making concerning them. Furthermore, Errico (2021) explained that some of the factors that continue to pose the challenges to access and control the land and natural resource are, the discrimination of women by putting up the unfavorable laws within and outside community, lack of participation in relevant decision-making institutions and processes, lack of their recognition in productive role (Ndonye et al., 2021). That calls for re-evaluation of some of the African cultures concerning women which is one of the actions that need to be considered on this PAR research.

The study also indicates that young people and children were not actively involved in conservation of nature in those areas of study. Young people are disinterested, and children are forbidden since they were taught about conservation in school. In order to make the constitution appealing to young people and enable them to assume responsibility when the time comes, researchers intended to use the PAR study (Mwangi et al., 2023). Through this PAR study, the community is empowered to run its conservancy with occasional managerial assistance when need arises. Community needs are increasing, and new approaches are necessary for development and sustainability to be achieved. The most important thing is a change of policy which will allow both women and young people to be fully active and the gender guidance are being developed to include the role of the other voices, so that everyone in a community might participate in conservation of nature (Greiner, 2012). This study demonstrated the pillars that build a successful PAR study namely, collaboration, empowerment, contextual understanding, inclusion, flexibility, action and ethics (Mwangi et al., 2023).

The Challenges of Decolonizing PAR in Indigenous Contexts

Nadeau et al. (2022) explained that one of the main principles of the decolonisation effort is that the legacy still exists in the nations that started the colonisation process and that it is now a component of a worldview that systematically favors Western people. Bourgeois et al. (2024) argued that as a result, decolonisation endeavor finds common ground with numerous other social groups that oppose the current quo and have radical beliefs. Using the PAR paradigm does not ensure a decolonising strategy, as the research is beginning to recognize. PAR initiatives run the danger of being impeded by the structures in which they are executed (Fernández, 2022).

Given the persistence of colonialism within institutions, many have outlined significant obstacles to research that is truly decolonising: standard terminology, grant agency and academic traditions and requirements, technology accessibility, and resource availability tend to limit the integration of Indigenous knowledge and the empowerment of Indigenous communities. According to Gobena et al. (2023), PAR is essentially ingrained in Western paradigms, even if it is a component of an alternative research movement. However, PAR that takes a critical stance holds great promise as an investigative approach in Indigenous contexts because of its goal of advancing emancipatory justice and community and research subject participation (Aliverti et al., 2021).

According to a study by Gobena et al. (2024), scholars working in Indigenous communities and the Global South have been charged with historically exploiting Indigenous or local knowledge without giving due credit or acknowledgment. They are charged with insensitivity to Indigenous and local cultural norms, conducting research studies that do not benefit local communities, and unauthorizedly disclosing sensitive information about Indigenous and local populations (Prior, 2007).

Moreover, McGowan et al. (2020) was concerned that the Indigenous communities are forced to begin with a posture of caution and skepticism when it comes to participating in research because of a history of exploitative research techniques, which breeds mistrust (McGowan et al., 2020).Decoloniality in research has not been adequately used and limited in the Global South, according to a study by (Gobena et al., 2023) as opposed to nations like Canada, New Zealand, and Australia. Furthermore, research approaches informed by IKS are rarely discussed in studies conducted in the Global South which that has to change as higher education needs to be decolonized (Emnet, 2021). This is because of colonial powers that are still favored and promoted by Higher education institutions and researchers

Decolonising knowledge and upending hegemonic power in all of its manifestations must be a key component of a decolonial PAR practice since the coloniality of power produces colonial thinking. In addition to producing unique patterns of expression and thought that can contest or reject Western epistemologies linked to the colonial gaze and disembodied research objectivity (Tuck & Yang, 2014), decolonial thought can foster and support the radical imagination of oppressed people.

CONCLUSION

Using a decolonial approach, this study investigated the role of indigenous participants in PAR by focusing on three objectives and research questions which were attempted to be answered. The study's findings indicate that PAR is an approach that enables Indigenous communities to be part and owners of the research process as they are treated as participants and not objects. PAR as an approach corrects the injustice that has been taking place in Indigenous research where participants have been exploited and violated. Moreover, PAR further contributes to the decolonisation of research by serving as a strategy that provides voice to Indigenous people and their knowledge during the research process. Case studies have been presented in this study as examples of how PAR was carried out by researchers. Understanding the roles of the researcher and participants who are indigenous people in PAR required the use of this example. Thus, this study concludes that indigenous people are essential to PAR because they are participants and indigenous knowledge holders who build strong bonds with researchers who carry out research that benefits indigenous people as well. By receiving PAR recognition, academics help decolonise research and eliminate western approaches that failed to acknowledge indigenous people and their expertise.

REFERENCES

Aliverti, A., Carvalho, H., Chamberlen, A., & Sozzo, M. (2021). Decolonizing the criminal question. *Punishment & Society*, *23*(3), 297–316. DOI: 10.1177/14624745211020585

Balakrishnan, V., & Claiborne, L. (2017). Participatory action research in culturally complex societies: Opportunities and challenges. *Educational Action Research*, *25*(2), 185–202. DOI: 10.1080/09650792.2016.1206480

Bartlett, J. G., Iwasaki, Y., Gottlieb, B., Hall, D., & Mannell, R. (2007). Framework for Aboriginal-guided decolonizing research involving Métis and First Nations persons with diabetes. *Social Science & Medicine*, *65*(11), 2371–2382. DOI: 10.1016/j.socscimed.2007.06.011 PMID: 17689163

Bennett, M. (2019). A Review of the Literature on the Benefits and Drawbacks of participatory action research. *First Peoples Child & Family Review*, *14*(1), 109–122. DOI: 10.7202/1071290ar

Bhuda, M., & Koitsiwe, M. (2022). The Importance of Underpinning Indigenous Research Using African Indigenous Philosophies: Perspectives From Indigenous Scholars. In *Handbook of Research on Protecting and Managing Global IKS* (pp. 223–248). IGI Global. DOI: 10.4018/978-1-7998-7492-8.ch013

Bhuda, M. T. (2024). Decolonising and indigenising African Indigenous research: Why Indigenous ways of knowing, being and living matter in research. In *Higher Education Transformation in Africa* (pp. 50–69). Routledge. DOI: 10.4324/9781003528319-4

Bhuda, M. T., & Marumo, P. (2022). Ubuntu Philosophy and African IKS: Insights from decolonization and indigenization of research. *Gender & Behaviour*, *20*(1), 19133–19151.

Blumenthal, D. S. (2011). Is community-based participatory research possible? *American Journal of Preventive Medicine*, *40*(3), 386–389. DOI: 10.1016/j.amepre.2010.11.011 PMID: 21335275

Bourgeois, R., Karuri-Sebina, G., & Feukeu, K. E. (2024). The future as a public good: decolonising the future through anticipatory participatory action research. *foresight, 26*(4), 533-549.

Brydon-Miller, M., Kral, M., & Ortiz Aragón, A. (2020). Participatory action research: International perspectives and practices. *International Journal of Qualitative Research, 13*(2), 103–111. DOI: 10.1177/1940844720933225

Caxaj, C. S. (2015). Indigenous storytelling and participatory action research: Allies toward decolonization? Reflections from the peoples' international health tribunal. *Global Qualitative Nursing Research*, *2*, 2333393615580764. DOI: 10.1177/2333393615580764 PMID: 28462305

Chaka, C. (2020). Translanguaging, decoloniality, and the Global South: An integrative review study. *Scrutiny2, 25*(1), 6-42.

Chen, P. G., Diaz, N., Lucas, G., & Rosenthal, M. S. (2010). Dissemination of results in community-based participatory research. *American Journal of Preventive Medicine*, *39*(4), 372–378. DOI: 10.1016/j.amepre.2010.05.021 PMID: 20837290

Clarke, V., & Braun, V. (2017). Thematic analysis. *The Journal of Positive Psychology*, *12*(3), 297–298. DOI: 10.1080/17439760.2016.1262613

Cornish, F., Breton, N., Moreno-Tabarez, U., Delgado, J., Rua, M., de-Graft Aikins, A., & Hodgetts, D. (2023). Participatory action research. *Nature Reviews. Methods Primers*, *3*(1), 34. DOI: 10.1038/s43586 023-00214-1

Cruzes, D. S., Dybå, T., Runeson, P., & Höst, M. (2015). Case studies synthesis: A thematic, cross-case, and narrative synthesis worked example. *Empirical Software Engineering*, *20*(6), 1634–1665. DOI: 10.1007/s10664-014-9326-8

Dadich, A., Moore, L., & Eapen, V. (2019). What does it mean to conduct participatory research with Indigenous peoples? A lexical review. *BMC Public Health*, *19*(1), 1388. DOI: 10.1186/s12889-019-7494-6 PMID: 31660911

Datta, R. (2018). Decolonizing both researcher and research and its effectiveness in Indigenous research. *Research Ethics*, *14*(2), 1–24. DOI: 10.1177/1747016117733296

Datta, R., Khyang, N.U., Khyang, H.K.P., Kheyang, H.A.P., Khyang, M.C. & Chapola, J. (2014). Participatory action research and researcher's responsibilities: an experience with an Indigenous community. *International Journal of Social Research Methodology*, *18*(6), 581–599. Advance online publication. DOI: 10.1080/13645579.2014.927492

De Oliveira, B. (2023). Participatory action research as a research approach: Advantages, limitations and criticisms. *Qualitative Research Journal*, *23*(3), 287–297. DOI: 10.1108/QRJ-08-2022-0101

Dudgeon, P., Bray, A., Darlaston-Jones, D. and Walker, R., (2020). Aboriginal participatory action research: an indigenous research methodology strengthening decolonisation and social and emotional wellbeing.

Errico, S. (2021). Women's Right to Land Between Collective and Individual Dimensions. Some Insights From Sub-Saharan Africa. *Frontiers in Sustainable Food Systems*, *5*, 690321. DOI: 10.3389/fsufs.2021.690321

Fernández, J. S. (2022). Decolonising participatory action research in community psychology. *Decolonial enactments in community psychology*, 29-51.

Gill, H., Purru, K., & Lin, G. (2012). In the midst of participatory action research practices: Moving towards decolonizing and decolonial praxis. *Reconceptualizing Educational Research Methodology, 3*(1).

Ginn, C. S., & Kulig, J. C. (2015). Participatory action research with a Group of Urban First Nations Grandmothers. *International Indigenous Policy Journal*, *6*(1), 1–16. DOI: 10.18584/iipj.2015.6.1.4

Gobena, E. B., Hean, S., Heaslip, V., & Studsrød, I. (2024). The challenge of western-influenced notions of knowledge and research training: Lessons for decolonizing the research process and researcher education. *Journal of Ethnic & Cultural Diversity in Social Work*, *33*(5), 276–287.

Greiner, C. (2012). Unexpected consequences: Wildlife conservation and territorial conflict in northern Kenya. *Human Ecology: an Interdisciplinary Journal*, *40*(3), 415–425. DOI: 10.1007/s10745-012-9491-6

Guillemin, M., Gillam, L., Barnard, E., Stewart, P., Walker, H., & Rosenthal, D. (2016). "We're checking them out": Indigenous and non-Indigenous research participants' accounts of deciding to be involved in research. *International Journal for Equity in Health*, *15*(1), 1–10. DOI: 10.1186/s12939-016-0301-4 PMID: 26772174

Halliday, A. J., Kern, M. L., Garrett, D. K., & Turnbull, D. A. (2019). The student voice in well-being: A case study of participatory action research in positive education. *Educational Action Research*, *27*(2), 173–196. DOI: 10.1080/09650792.2018.1436079

Hart, M. A. (2010). Indigenous worldviews, knowledge, and research: The development of an Indigenous research paradigm. *Journal of Indigenous Social Development*, *1*(1A).

Heleta, S. (2018). Decolonizing knowledge in South Africa: Dismantling the 'pedagogy of big lies'. *Ufahamu. Journal of African Studies*, *40*(2).

Irvine, A. (2023). First steps in qualitative secondary analysis: Experiences of engaging with the primary research team. *International Journal of Social Research Methodology*, •••, 1–13. DOI: 10.1080/13645579.2023.2267836

Israel, B. A., Eng, E., Schulz, A. J., & Parker, E. A. (2005). Introduction to methods in community-based participatory research for health. *Methods in community-based participatory research for health, 3*, 26.

Jagosh, J., Macaulay, A. C., Pluye, P., Salsberg, J. O. N., Bush, P. L., Henderson, J. I. M., Sirett, E., Wong, G., Cargo, M., Herbert, C. P., Seifer, S. D., Green, L. W., & Greenhalgh, T. (2012). Uncovering the benefits of participatory research: Implications of a realist review for health research and practice. *The Milbank Quarterly*, *90*(2), 311–346. DOI: 10.1111/j.1468-0009.2012.00665.x PMID: 22709390

Keane, M. (2008). Science education and worldview. *Cultural Studies of Science Education*, *3*(3), 587–621. DOI: 10.1007/s11422-007-9086-5

Keane, M., Khupe, C., & Seehawer, M. (2017). Decolonising methodology: Who benefits from indigenous knowledge research?. *Educational Research for social change, 6*(1), 12-24.

Kemmis, S., McTaggart, R., Nixon, R., Kemmis, S., McTaggart, R., & Nixon, R. (2014). Introducing critical participatory action research. *The action research planner: Doing critical Participatory Action Research*, 1-31.

Khupe, C. (2014). *Indigenous knowledge and school science: Possibilities for integration* (Doctoral dissertation, University of the Witwatersrand, Faculty of Science, School of Science Education).

Khupe, C., & Keane, M. (2017). Towards an African education research methodology: Decolonising new knowledge. *Educational Research for Social Change*, *6*(1), 25–37. DOI: 10.17159/2221-4070/2017/v6i1a3

Kidd, S. A., & Kral, M. J. (2005). Practicing participatory action research. *Journal of Counseling Psychology*, *52*(2), 187–195. DOI: 10.1037/0022-0167.52.2.187

Leung, M. W., Yen, I. H., & Minkler, M. (2004). Community based participatory research: A promising approach for increasing epidemiology's relevance in the 21st century. *International Journal of Epidemiology*, *33*(3), 499–506. DOI: 10.1093/ije/dyh010 PMID: 15155709

Li, Y. (2023). The Influence of Top-Down Mode and Bottom-up Mode to National Innovation and Entrepreneurship. *Journal of Innovation and Development*, *2*(3), 158–161. DOI: 10.54097/jid.v2i3.7640

Louis, R. P. (2007). Can you hear us now? Voices from the margin: Using indigenous methodologies in geographic research. *Geographical Research*, *45*(2), 130–139. DOI: 10.1111/j.1745-5871.2007.00443.x

Maditsi, M. E., & Bhuda, M. T. (2023). Decolonising and Humanising Pedagogies in South African Postgraduate Education: Lessons From IKS. In *Digital Preservation and Documentation of Global IKS* (pp. 282-302). IGI Global.

McGowan, K., Kennedy, A., El-Hussein, M., & Bear Chief, R. (2020). Decolonization, social innovation and rigidity in higher education. *Social Enterprise Journal*, *16*(3), 299–316. DOI: 10.1108/SEJ-10-2019-0074

McHugh, T. L. F., & Kowalski, K. C. (2009). Lessons learned: Participatory action research with young Aboriginal women. *Pimatisiwin*, *7*(1), 117–131.

Mignolo, W. D. (2020). On decoloniality: Second thoughts. *Postcolonial Studies*, *23*(4), 612–618. DOI: 10.1080/13688790.2020.1751436

Mwangi, R., Maina, S. M., & Munene, M. (2023). Participatory action research (Par) for Community Based Conservation in Kenya. *Africa Design Review Journal*, *1*(4), 480–496.

Nadeau, L., Gaulin, D., Johnson-Lafleur, J., Levesque, C., & Fraser, S. (2022, December). The challenges of decolonising participatory research in indigenous contexts: The Atautsikut community of practice experience in Nunavik. *International Journal of Circumpolar Health*, *81*(1), 2087846. DOI: 10.1080/22423982.2022.2087846 PMID: 35979584

Ndlovu-Gatsheni, S. J. (2015). Decoloniality in Africa: A continuing search for a new world order. *Australasian Review of African Studies*. *The*, *36*(2), 22–50.

Ndonye, H. N., Mulwa, A., & Kyalo, D. N. (2021). Stakeholder engagement and performance of community based conservation projects: A case of Laikipia Conservation Region Conservancies. *Journal of Economics and Sustainable Development*, *12*(4), 41–50.

Omodan, B. I., & Dastile, N. P. (2023). Analysis of participatory action research as a Decolonial research methodology. *Social Sciences (Basel, Switzerland)*, *12*(9), 507. DOI: 10.3390/socsci12090507

Owusu-Ansah, F. E., & Mji, G. (2013). African indigenous knowledge and research. *African Journal of Disability*, *2*(1), 1–5. DOI: 10.4102/ajod.v2i1.30 PMID: 28729984

Peltier, C. (2018). An application of two-eyed seeing: Indigenous research methods with participatory action research. *International Journal of Qualitative Methods*, *17*(1), 1609406918812346. DOI: 10.1177/1609406918812346

Prior, D. (2007). Decolonising research: A shift toward reconciliation. *Nursing Inquiry*, *14*(2), 162–168. DOI: 10.1111/j.1440-1800.2007.00361.x PMID: 17518828

Ruggiano, N., & Perry, T. E. (2019). Conducting secondary analysis of qualitative data: Should we, can we, and how? *Qualitative Social Work: Research and Practice*, *18*(1), 81–97. DOI: 10.1177/1473325017700701 PMID: 30906228

Sefa-Dei, G. J., & Cacciavillani, A. (2024). ActualisingDecolonisation: A Case for Anti-Colonising and Indigenising the Curriculum. *Journal of Philosophy of Education*, qhae036.

Simpson, J. E., & Mendenhall, T. J. (2022). Community-based participatory research with Indigenous youth: A critical review of literature. *Alternative*, *18*(1), 192–202. DOI: 10.1177/11771801221089033

Stern, T. (2019). Participatory action research and the challenges of knowledge democracy. *Educational Action Research*, *27*(3), 435–451. DOI: 10.1080/09650792.2019.1618722

Tavernaro-Haidarian, L. (2019). Decolonization and development: Reimagining key concepts in education. *Research in Education*, *103*(1), 19–33. DOI: 10.1177/0034523719839750

Tuck, E., & Yang, K. W. (2014). Unbecoming claims: Pedagogies of refusal in qualitative research. *Qualitative Inquiry*, *20*(6), 811–818. DOI: 10.1177/1077800414530265

Vaughn, L. M., & Jacquez, F. (2020). Participatory research methods–choice points in the research process. *Journal of Participatory Research Methods*, *1*(1). Advance online publication. DOI: 10.35844/001c.13244

Woldegiorgis, E. T. (2021). Decolonising a higher education system which has never been colonised. *Educational Philosophy and Theory*, *53*(9), 894–906. DOI: 10.1080/00131857.2020.1835643

Chapter 16
Tracking the March from Traditional Knowledge to Geographical Indication:
A Case Study of Stone Craft of Jabalpur

Rama Sirpurkar
RDVV, Jabalpur, India

Shashikala Gurpur
Symbiosis Law School, Pune, India

Richa Dwivedi
Symbiosis Law School, Pune, India

ABSTRACT

Traditional knowledge is the wisdom passed on from generation to generation within a community which forms part of their cultural identity. This community wisdom held by indigenous people is the source of many modern inventions and innovations in various fields of human development. Lack of representation, recognition and protection to this body of knowledge makes it susceptible to misappropriation and misuse. In an effort to create strong indigenous knowledge system it is essential that they are protected through robust legal framework. Intellectual property is one of the tools to guard this traditional knowledge against unauthorized exploitation. This paper aims to study the interface between traditional knowledge and intellectual property. Jabalpur Stone Craft is identified as an expression of traditional knowledge of craftsmen carrying this art for generations. The chapter studies the progression of this craft from traditional knowledge to a commercialized and well-protected piece of art. Even though the craft has been protected as a geographical indication, this protection is limited and focuses on the art itself. This paper studies the traditional knowledge aspect of this stone craft and proposes approaches to protect the underlying wisdom of cultural expressions. To encapsulate, this paper aims at fostering the relationship between traditional knowledge and intellectual property laws through the case study of stone craft of Jabalpur.

DOI: 10.4018/979-8-3693-3542-0.ch016

1. INTRODUCTION

Traditional knowledge is the wisdom passed on from generation to generation within a community, forming part of their cultural identity. There is no accepted definition of traditional knowledge at the international level but it can be identified as knowledge resulting from intellectual activity in the traditional context and includes know-how, practices, skills, and innovations.(WIPO, n.d.) Indigenous knowledge can be defined as the process through which the natives of an area have built a relationship with their natural environment. (Semali& Kincheloe, 1999) This knowledge is developed over time and tied to a certain community. It includes traditional technologies in various areas like agriculture, hunting, medicine, ecology, climate, celestial navigation, and craft skills. The Report of the International Council for Science Study Group on Science and Traditional Knowledge characterizes Traditional Knowledge as 'A cumulative body of knowledge, know-how, practices and representations maintained and developed by peoples with extended histories of interaction with the natural environment. These sophisticated sets of understandings, interpretations and meanings are part and parcel of a cultural complex that encompasses language, naming and classification systems, resource use practices, ritual, spirituality, and worldview.' (International Council for Science, 2002)

Moving further, the United Nations Declaration on the Rights of Indigenous People (UNDRIP) and World Intellectual Property Organization (WIPO) define indigenous knowledge as including both, traditional knowledge (TK) as well as traditional cultural expressions (TCE). The term "traditional cultural expressions" is used by WIPO to refer to "any form of artistic and literary expression" in which traditional culture and knowledge are embodied. They are transmitted from one generation to the next, and include handmade textiles, paintings, stories, legends, ceremonies, music, songs, rhythms and dance." (Zuckermann et al., 2015) It is clear that traditional knowledge and indigenous knowledge are interwoven and can be used alternatively. WIPO thus gives international recognition to indigenous knowledge through traditional cultural expressions (TCE) and accords legal protection in various forms through TRIPS.

2. IDENTIFICATION OF THE STONE CRAFT OF JABALPUR

The stone craft of Jabalpur is carried out by artisans living near the banks of the Narmada river in Bhadaghat. Bhedaghat is a town and a Nagar Panchayat in Jabalpur district, Madhya Pradesh, India. It is situated on the banks of river Narmada, 20 km away from Jabalpur. The exact map location of the place is 23.13° North, 79.8° West and 408 mts from the sea level.[1] This place is famous for the Dhuandhar falls (Waterfall) and marble rocks. The major tourist attraction here is these waterfalls. The tourists have encouraged the Marble Craft here. These marble rocks are world famous and have attracted entrepreneurs from Rajasthan. The extension of these rocks in between Jabalpur and Katni is being quarried. The marble from these areas is exploited for its off-white, fine-grained, banded attributes. (Indian Minerals Yearbook, 2011) Recently, Bhedaghat has been nominated as the site to be included in the coveted list of UNESCO's world heritage sites. (The Economic Times, 2012) In this backdrop, it is essential to mention that the marble and varieties of soft stone quarried in the region and stone carving skills are prolific in this area.[2] Stone carving is a very famous craft of Jabalpur.

2.1 History of the Stone Craft in Madhya Pradesh

The stone crafts of Madhya Pradesh demonstrate the highly acclaimed tradition of stone carving that commenced in the medieval era with royal patronage. Singi, Mandsaur, Ratlam, and Narsinghagarh are known for their stone craft in Madhya Pradesh. The artisans of Ghosipura in Gwalior are praised for their dexterous sculptors, considered the descendants of Vishwakarma. Artisans in Madhya Pradesh create two types of idols – two-dimensional relief work and three-dimensional relief work. Gwaliorspecializes in jalli work, while Tikamgarh and Jabalpur are famous for decorative items.Tribal communities have a significant contribution in the field of stone crafts. The rich history of India showcases great stone carvings in various historical monuments. The city of Gwalior is known for its exquisite jaali work, while Tikamgarh and Jabalpur are famous for animal and bird figures. Bhedaghat is known for its soft marble rock handicrafts, and Balaghat is renowned for its green stone handicrafts. The artisans draw inspiration from nature to create lifelike stone images for forts, palaces, and temples.

Stone crafts of Madhya Pradesh have their distinct style, combined with the style of tribes. The craft has been practised for generations, with 300-400 craft people practising it.

2.2 History of Jabalpur Stone Craft

To understand the nature and background of the stone craft of Jabalpur a peek into the historical facts and dialogue with the craftsmen was felt essential. Bhedaghat, located to the west of Jabalpur, is a site that offers a unique view of river Narmada, falling from a height. Bhedaghat has marble rocks along the banks of the Narmada River, which are limestone cliffs rising 30 meters above the river waters? This site is known internationally for its archeological wealth that dates since pre historic times. (Indian Minerals Yearbook, 2011) It has a resource of limestone, marble, iron ore and refractory clay. Marble and varieties of softstone are quarried in the region and stone carving skills are prolific in Jabalpur and Bhedaghat. The dolomite marble is quarried at Jabalpur, Balaghat, Chindwara, Bijawar and Sabalghat areas. Most of the local population is engaged in a craft that is almost 60 to 80 years old. The sculpting and carvings are done on the marble that is quarried from the nearby region. It has 300-400 craft persons who practice this craft. The craft is practiced here since generations. They have inherited the traits which kept on modifying with time. The craft here reveals the culture and tradition of the region. The making of idols seems to be inspired by the idols in the temple of Chousath Yogini. It is a temple of 10th century situated in Bhedaghat. These hereditary craftsmen belong to the Visvakarma caste, some are Jharia and Adivasi and some are Muslims who have learnt the craft. The objects made from stone carving are of religious significance i.e. the idols of Gods and Godesses. Certain type of utensils and artistic decorative items also form the part of the craft practiced here. It is evident that the craft has been carried out at this particular place since time immemorial.

2.3 Stone Craft Sustainability and Challenges

The interplay of sustainability and handicrafts presents both challenges and possibilities for artisans. Globalization has brought about mass-produced alternatives that hinder the commercial viability and cultural significance of many traditional crafts practices (Shafi, 2022). Addressing the impact of these developments on rural artisan communities is crucial. The call for sustainable practices includes environmental, economic, and social factors, promoting innovative approaches to enhance their products while preserving authenticity (Kesaboyina et al., 2024). Over the time the artisans have mastered stone craft which has impacted socioeconomic fabric of many rural communities. According to the UNESCO Report on Cultural Practices (Halder & Sarda, 2021), families uphold significant cultural traditions and ancestral knowledge. Several studies emphasize that these crafts preserve communal rituals, social hierarchies, and identity and transcend economic significance (Beel et al., 2017). However, influx of international alternatives has led to a decline in demand for these items, posing a threat to the livelihoods of these artists and exacerbating their socioeconomic vulnerabilities (Guha, 2022; Singh & Alam, 2024). Hence, it is crucial to acknowledge the cultural significance of stone crafts to develop sustainable practices that safeguard artisans' livelihoods and preserve cultural heritage. The Indian handicraft sector is a living testament to the country's diverse cultural heritage, reflecting centuries of tradition, skill, and creativity (Datta & Bhattacharyya, 2016; Grobar, 2019). It showcases the most intricate facets of the nation's cultural landscape, with each region presenting unique techniques and styles. These artisanal practices, handed down through generations, are a testament to India's deep-rooted connections to its past and traditions (Majeed, 2018; Patel et al., 2023; Upadhyay & Jain, 2019).

The Indian handicraft sector is a powerhouse in the Indian economy after agriculture (Yadav et al., 2022). Indian handicrafts are classified as informal sector (Agarwal et al., 2023). The informal sector is the second largest employer in India after agriculture. There were 6.572 million people involved in 2005–06, which increased to 7.617 million in 2010–11. (Ahmad Bhat & Yadav, 2016). The Planning Commission (2012) reported that out of the 7.617 million individuals, approximately 24.7% are from Scheduled Castes (SC), 2.3% from Scheduled Tribes (ST), and 47.4% are female (Ahmad et al., 2016). This data indicates that the handicraft sector significantly empowers women and minority communities in India who are the integral part of indigenous communities. Social capital, loss of traditional skills (Alam & Singh, 2022, 2024), and changing community structures are obstacles that may weaken artisans' identity and the cultural relevance of traditional crafts (Banks, M. 2014). Traditional handicrafts are expensive and time-consuming to make compared to machine-made products (Shafi et al., 2021). A closer analysis shows that lack of government intervention, poor legal frameworks, market changes, globalisation, and industrial competition have gradually declined local knowledge and skills. (Devid Deshmukh et al, 2024)

2.4 Identification of Ways to Protect Stone Craft of Jabalpur as Traditional Knowledge

To build an Indigenous knowledge system, two factors play an important role: first, the creation of robust provisions for the protection of traditional knowledge, and second, the creation of subject-specific action research to make such protection feasible.

This research identifies the Stone Craft of Jabalpur[3] as the culmination of the traditional knowledge of the artisans.[4] A previous study (Sirpurkar&Gurpur, 2014) carried on by the researchers was action research to find out the feasibility of marble craft of Bhedaghat as a geographical indication. This research through a detailed review of literature concluded that many arm chair studies were carried out on the topic of intricacies of Geographical Indications of Goods(Registration and Protection) Act, 1999.[5] However, these studies only provided reasons and measures to be taken for a product to be registered as GI and did not present the steps to be taken or already taken for their actual implementation. No published research provided for checklists, processes, or possible frameworks for the registration of specific products as GI. The literature was thus inadequate. (Sirpurkar&Gurpur, 2014). The research created a case for the registration of the marble craft of Bhedaghat as a geographical indication. It is important to refer this research as it successfully proved that the marble craft of artisans from Bhedaghat was a definitive product for registration as geographical indication.[6]

Moving further, this paper endeavours to examine the traditional knowledge aspect of this stone craft. The paper, in the next segment, tries to review if this stone craft of Jabalpur constitutes traditional knowledge of the craftsmen and if geographical indication is sufficient way to protect and preserve this community wisdom. This is done through an action research.

3. OBJECTIVE OF THE RESEARCH

The objective of this research is to identify Stone craft of Jabalpur as traditional knowledge of the artisans and find best possible way of its protection to enable creation of indigenous knowledge system. To achieve the above objective, action research is used. The research focuses on understanding the issue, collecting the data, making a SWOT analysis and drawing comprehensive conclusions to support preservation of communal wisdom. The research also highlights the significance of action research in preserving the traditional knowledge.

4. LITERATURE REVIEW

Before diving into the core of this research, it is pertinent to understand the theoretical framework which forms the base for lucid analysis of the empirical data. Understanding the concepts also facilitate explicit discussion and fuller conclusions. The following sections discuss development of concept of traditional knowledge, interface of traditional knowledge and intellectual property over a period of time and investigate the importance protecting traditional knowledge through intellectual property. It also enumerates various other studies aimed at understanding the interplay of handicraft and traditional knowledge and their protection through Geographical indication.

4.1 History and Development of Concept of Traditional Knowledge

Traditional knowledge is deeply rooted in the practices and cultures of indigenous people. Indigenous people and their traditional knowledge were recognized at an international level some seventy years ago. (Chernela, 2004). In 1945, when the United Nations was established, its Charter stated, "the purposes of United Nations are …to achieve international co-operation…… in promoting and encouraging respect for human rights and for fundamental freedoms for all without distinction as to race, sex, language and religion."(United Nations, 1945) This was followed by Declaration of Human Rights, adopted in 1948. Article 1 states that,"All human beings are born free and equal in dignity and rights."(United Nations, 1948) Article 2 states that, "Everyone is entitled to all the rights and freedoms set forth in this Declaration, without distinction of any kind, such as race, colour, sex, language, religion, political or other opinion, national or social origin, property, birth or other status. Furthermore, no distinction shall be made on the basis of the political, jurisdictional or international status of the country or territory to which a person belongs, whether it be independent, trust, non-self-governing or under any other limitation of sovereignty."(United Nations, 1948) These two documents formed the basis for creating an environment of equality and non-discrimination among the people of the world community. In 1953, International Labour Organization published the first report on indigenous peoples, 'Indigenous Peoples: Living and Working Conditions of Aboriginal Populations in Independent Countries.' ILO further adopted Indigenous and Tribal Populations Convention no 107, which recognized their right of ownership over their traditional territories. This formed the first international juridical instrument to address indigenous peoples and their rights. This convention was later replaced in 1989 by the 'Convention Concerning Indigenous and Tribal Peoples Independent Countries (ILO Convention no.169). (Zappalaglio, 2013) This Convention in its Preamble recognizes, 'the aspirations of these peoples to exercise control over their own institutions, ways of life and economic development and to maintain and develop their identities, languages, and religions, within the framework of the States in which they live.' (International Labour Organization [ILO], 1989)

This was a turning point in the movement of recognition of Indigenous people and their rights over their way of life, their own developed means of living, and, of course, their wisdom developed over a period of time. This can be drawn from the fact that even if this treaty does not explicitly mention traditional knowledge and traditional cultural expressions as is understood in today's relevance, Article 23(1) provides that:"handicrafts, rural and community-based industries, and subsistence economy and traditional activities of the peoples concerned, such as hunting, fishing, trapping and gathering, shall be recognized as important factors in the maintenance of their cultures and in their economic self-reliance and development."(Zappalaglio, 2013)

All this development in the '70s and early 80s focused on recognizing Indigenous people and creating a conducive environment for the protection of their rights that were very closely essential. It was still far from conceptualizing Traditional Knowledge. However, these developments at international and local level led to recognition of a corpus of knowledge that formed an integral part of indigenous people's existence. This knowledge, without sufficient recognition and protection, came to be easily misappropriated. The unawareness on the part of indigenous people, made such kind of theft effortless without any action being taken. In this regard, the first and most famous case is represented by the song "El Condor Pasa", belonging to the Andean folk tradition, turned into a bestselling hit by Simon and Garfunkel in 1970. This provoked the protestations of the Bolivian government, while the indigenous communities remained silent.(Halewood, 1999)

In no time the concern of misappropriation grew to other areas as well like craftsmanship, agriculture, environment etc.(Antons, 2009) the indigenous rights came to be recognized at the UN level around 1982, when these issues became structurally part of the activity of the UN with the creation of the UN Working Group on Indigenous Populations within the frame of the UN Economic and Social Council. This became the largest forum dedicated to the protection of indigenous population's human rights. (International Labour Organization [ILO], 1989) However the concrete conceptualization of traditional knowledge came with the Convention on Biological Diversity(CBD). This international treaty was signed at Rio de Janeiro on 5th June 1992 which aimed at developing strategies for conservation of biodiversity and sustainable use as well as fair sharing of biological resources.(Convention on Biological Diversity, 1992) Article 8(j) of the convention provides that each contracting party shall, "Subject to its national legislation, respect, preserve and maintain knowledge, innovations and practices of indigenous and local communities embodying traditional lifestyles relevant for the conservation and sustainable use of biological diversity and promote their wider application with the approval and involvement of the holders of such knowledge, innovations and practices and encourage the equitable sharing of the benefits arising from the utilization of such knowledge, innovations and practices."(Convention on Biological Diversity, 1992) This article is particularly important both from a substantive and terminological point of view.

For the first time, it is explicitly recognized that local communities are the depositaries of a body of knowledge manifested through their traditional lifestyle that can be important for the protection and conservation of biodiversity and environment.(Frater, 2009)

The development of the concept of traditional knowledge began from initially acknowledging its existence as a body of knowledge held by indigenous people and then by initiating attempts to accord protection to it. The importance of protection was realized as it was being misappropriated explicitly and attempts were being made for its commercial exploitation. Even though traditional knowledge came to be considered and recognized through various international as well as local instruments, robust framework for its protection in all its varied forms still remained uncharted.

The Convention for the Safeguarding of the Intangible Cultural Heritage, adopted by the UNESCO General Conference in 2003 that came into force from April 2006(UNESCO, 2006) recognized traditional knowledge in theform of Traditional Cultural Expressions (TCE's). The parallelism between the concepts of traditional knowledge and Intangible Cultural Heritage (ICH)(Convention Concerning Indigenous and Tribal Peoples in Independent Countries, 1989) is made clear through the definition of ICH provided by the convention,

> "Intangible Cultural Heritage means the practices, representations, expressions, knowledge, and skills (…) that communities, groups and, in some cases, individuals recognize as part of their cultural heritage. This intangible cultural heritage, transmitted from generation to generation, is constantly recreated by communities and groups in response to their environment, their interaction with nature and their history, and provides them with a sense of identity and continuity, thus promoting respect for cultural diversity and human creativity."

This definition is important from the point of view of interface between traditional knowledge and intellectual property,[7] as cultural heritage is defined to be intangible which takes us to the point that it is eligible for IP like protection.

United General Assembly in September 2007, adopted Declaration on the Rights of Indigenous Peoples (UNDRIP). Many articles of this Declaration mention traditional knowledge as well as traditional cultural expressions in various forms. Article 31(1) being important reads,

> *'Indigenous Peoples have the right to maintain, control, protect and develop their cultural heritage, traditional knowledge and traditional cultural expressions, as well as the manifestations of their sciences, technologies and cultures, including human and genetic resources, seeds, medicines, knowledge of the properties of fauna and flora, oral traditions, literatures, designs, sports and traditional games and visual and Performing arts. They also have the right to maintain, control, protect and develop their intellectual property over such cultural heritage, traditional knowledge, and traditional Cultural expressions.'*

The paragraph is very interesting for at least four reasons. First, it shows how the terminology "traditional knowledge" has become standard in international law; second, it mentions TK and TCEs together, emphasizing the continuity and the proximity, rather than the clear distinction, between the two concepts; third, it gives numerous examples of different manifestations of the two concepts, thus emphasizing their complexity and width; fourth and final, it explicitly recognises that TK can be object of IP protection. (Zappalaglio, 2013)

4.2 Traditional Knowledge and World Intellectual Property Organization

The World Intellectual Property Organization ("WIPO") was established in 1967 by the Convention Establishing the World Intellectual Property Organization, signed in Stockholm, in order to encourage creative activity, to promote the protection of intellectual property throughout the world and to ensure administrative cooperation among the Unions. (Convention Establishing the World Intellectual Property Organization, 1967) Between the end of the 70' and the beginning of the 80', WIPO worked on the issue of folklore together with the United Nations Educational, Scientific and Cultural Organization. Model Provisions for National Laws on the Protection of Expressions of Folklore against Illicit Exploitation and Other Prejudicial Actions were adopted in 1982 under the auspices of WIPO and UNESCO. This was looked upon by member countries as a model for creating a sui generis system to protect folklore in their domestic structure. (Stoll & von Hahn, 2008) WIPO carries out its work on traditional knowledge through intergovernmental committee on intellectual property and genetic resource, traditional knowledge and folklore (IGC). It began operating in 2001on the basis of the results of fact-finding missions conducted in 1998 and 1999.(Antons, 2009) IGC is mandated to work on draft of instruments to protect traditional knowledge, traditional cultural expressions and genetic resources. The 25th session of the works of the IGC, held in Geneva (Switzerland) from 15 to 24 July 2013, has discussed, among the others, two draft articles, one on the protection of TK and another on the protection of TCEs. In addition, it has presented a "Consolidated Document Relating to Intellectual Property and Genetic Resources"(WIPO, 2001) WIPO was thus instrumental in building an environment for protecting traditional knowledge. This contribution helped in crafting the concept of traditional knowledge in its present form. WIPO's strong connections with other institutions such as UNESCO and WTO, as well as its action aimed at spreading the culture of IP law and researching all aspects of TK, make it a key factor in the development of the concept of TK at the international level. (Convention Concerning Indigenous and Tribal Peoples in Independent Countries, 1989)

4.3 Traditional Knowledge and World Trade Organization

World Trade Organization does not directly legislate on traditional knowledge, but it is mentioned in the Trade-Related Aspects of Intellectual Property Rights (TRIPS) Agreement. (World Trade Organization, 1994) The Doha Declaration(WTO, 2001) which aimed at bringing in reforms to WTO law in 2001, provides in para 19 that work in the TRIPS Council on implementation issues should cover: the relationship between the TRIPS Agreement and the UN Convention on Biological Diversity (CBD); the protection of traditional knowledge and folklore; and other relevant new developments that member governments raise in the review of the TRIPS Agreement.(World Trade Organization [WTO], 1994) TRIPS agreement does not provide for protection of traditional knowledge. It is believed that the patentability criteria laid down in Article 27 (3) (Agreement on Trade-Related Aspects of Intellectual Property Rights, 1994) of the agreement should be extended to traditional knowledge which will give it a robust form of protection. It has been majorly advocated by countries like India, China, and the EU that the protection of geographical indication provided under article 23 of the agreement be extended to traditional knowledge in order to bring it within its ambit certain products embodying expressions of traditional knowledge.(WTO, 2001) There has been however no progress in this direction until 2013. (WTO Ministerial Conference, 2013) Making traditional knowledge a part of WTO will have a major impact as traditional knowledge is most vulnerable of misappropriation. This protection is particularly important when traditional knowledge is presented and exploited as a product in the world market.

This brings us to the conclusion that the concept of traditional knowledge developed over a long period of time, and gradually, the international community began realizing the importance of its protection. This protection is still fragmented, and a robust form of protection to traditional knowledge is anticipated to create a strong indigenous knowledge system.

4.4 Importance of Protecting Traditional Knowledge and the Role of Intellectual Property Rights

Traditional knowledge is generally unscripted and passed on orally or through practice. The know-how and the skills are, therefore, susceptible to misappropriation as this use of traditional knowledge is not acknowledged or even remunerated under the present scheme in place. This might deprive the indigenous people of their intellectual wealth that has been treasured by them for centuries. The role of intellectual property is to protect creativity and intellectual effort. The knowledge or wisdom deep-rooted in the rich culture and traditions of Indigenous people, devoid of any legal protection, is exploited by private corporations to reap tremendous profits. This biopiracy is detrimental to the rights of indigenous people.

This is where the interface between traditional knowledge and intellectual property comes to the fore. The expression of traditional knowledge comes to be protected through intellectual property laws.

Protecting traditional knowledge and creating an indigenous knowledge system is possible only through intellectual property laws. The nature of traditional knowledge is such that it is held by a community as opposed to an individual. It becomes difficult, therefore, to identify the owner of such rights. Community rights are best protected through geographical indications as being the intellectual property of the community involved in creating or manufacturing a particular product. This product can be the expression of traditional knowledge held by these communities for generations together. Geographical indications are granted to a community of manufacturers/producers. The GI tag is given to the products whose qualities are attributable to a particular geographical region. The qualities intrinsic to the product are due to the geographical region in which they are produced /manufactured. The protection through GI helps to prevent third parties from using the GI tag without the authorization of producers/ manufacturers to whom the protection is granted. The purpose of protecting traditional knowledge is thus, to foster economic growth and enhance the art practiced and developed through generations. This purpose can be efficiently achieved only when the expression of traditional knowledge in various forms is protected. The intangible nature of traditional knowledge makes it the subject of intellectual property rights.

4.5 Other Studies on Various Handicrafts from India

A study carried out on Shivarapatna stone craft (Tanima Chanda & Aneesha Sharma, 2024) revealed how this stone craft is a traditional craft carried out through generations. It talked about the sustainable practices that would help in preserving the traditional wisdom. It is thus important from the point of present study. Another paper (Ruchi Pant, 2015) studied how geographical indication is best tool for protecting traditional knowledge. The research is important as it takes up five case studies to prove that GIs can help traditional knowledge holders get increased recognition for their traditional knowledge and higher benefits from its commercialisation through full benefit capture. A study on challenges to revival of handicrafts in India, (Ramratan Guru, 2024) observed that protecting the traditional art and organizing the handicraft sector contributes to promote and revive the rich heritage of the country. In impact of GI on Craftsmen of Varanasi, Uttar Pradesh, (Arobinda Kumar Sarkar, 2019), the researcher concluded that, in an era where brand value draws sales, GI is the best possible brand-creator and holds potential for increasing value for artisan communities. A study on Protection of Traditional Handicrafts under Indian intellectual property laws, (J Sai Deepak, 2008) suggests that blend of

Geographical indications and anciallary rights is way of encouraging ingenuity in traditional crafts.

Most of the studies are carried out for recognition, propagation and revival of traditional handicrafts in India. The study of this material helps to understand the best possible ways and approaches to protect stone craft of Jabalpur.

5. RESEARCH QUESTIONS

This study tries to track the journey of stone craft of Jabalpur from traditional knowledge to geographical indication through following research questions:

i. Whether Stone Craft of Jabalpur be considered as Traditional Knowledge?
ii. Whether protection of Stone Craft of Jabalpur as Geographical Indication is sufficient to protect traditional knowledge of Artisans?
iii. How does action research help in preserving the traditional knowledge of artisans of Jabalpur Stone Craft?

6. METHODOLOGY

The study aims to answer these questions through a scientific and comprehensive SWOT analysis. This analysis will provide a more holistic approach to the understanding of the traditional knowledge aspect of the stone craft of Jabalpur. To conduct this research, the SWOT analysis is carried out to understand and evaluate several aspects of internal, external factors that impact stone craft of Jabalpur as the traditional knowledge of artisans. It includes, Internal factors (strengths and weaknesses) and External Factors (opportunities and threats) (Knierim & Nowicki, 2010). This methodology provides a clear and detailed plan for doing the study, including the design, techniques for collecting data, and the analytical methodologies used.

This research is both doctrinal as well as non-doctrinal. The doctrinal part of the research helps to evaluate the concept of traditional knowledge as defined under various texts as elaborated earlier. It is important to analyze the definition and scope of traditional knowledge so as to answer the first research question.

The non doctrinal or empirical research helps us to collect and document firsthand information from the craftsmen themselves that ensures a profound understanding, enabling the collection of numerical data and insightful perspectives.

For the study, action research has been specifically chosen as it enables investigation as well as intervention in preserving traditional knowledge. Action research helps in achieving practical outcomes. In the case of Jabalpur stone craft, Action research has helped in documenting the traditional knowledge by directly engaging with the artisans. The documentation and interaction have helped in identifying the challenges that the artisans have faced in preserving the craft. Finally, it has helped in developing and implementing solutions.

The sample for the study comprises of all the craftsmen of Bhedaghat engaged in stone craft, residing in Bhedaghat NagarParishad, district of Jabalpur, Madhya Pradesh. India. Random sampling is used to ensure diversity of responses. Data collected from the interaction with craftsmen revealed that out of the total craftsmen working there, around 15% were female artisans and 40% were male artisans. The data was collected mainly through unstructured interviews and observation method. The interaction with the craftsmen was mainly in the vernacular.

Table 1. SWOT questions for stone craft artisans of Bhedaghat

Empty Cell	Positive	Negative
Internal Factors	Strengths	Weaknesses
	S1: Availability of natural resources	W1: Lack of modern techniques
	S2: Low labour costs and skilled artisans	W2: Lack of training programmes for developing skills
	S3: Income and employment generation	W3: Declining interest among youth
External Factors	Opportunities	Threats
	O1: Growing demand for products at trade fairs and exhibitions	T1: Increasing cost of transport and raw materials
	O2: Preference for naturalness over artificiality	T2: Interference of middlemen and profit reduction
	O3: Government interventions	T3: New generation moving to cities for better job opportunities post covid

Figure 1. SWOT methodology

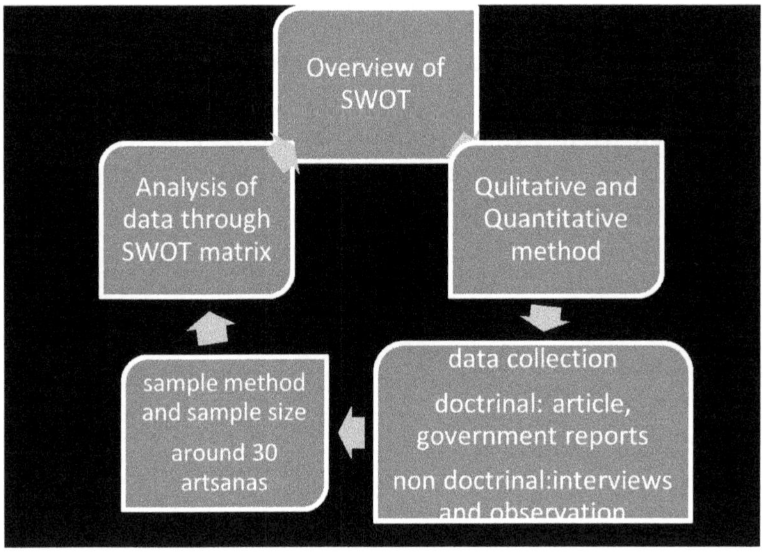

7. SWOT MATRIX ANALYSIS FOR STONE CRAFT ARTISANS OF BHEDAGHAT

SWOT analysis, a methodical approach is employed to assess strengths, weakness, opportunities and threats that are associated with a corporate endeavour. It offers a systematic way of understanding all the factors that lead to success or downfall of a corporate. Employment of this method provides best judgment to the artisans about their strengths, weaknesses, opportunities and threats to make their craft more sustainable. It will help them in multi criteria decision making to improve market position, competitiveness and preserve their generational art. The SWOT questions were meticulously selected after in depth literature review of similar handicrafts.

7.1 Strengths

7.1.1 S1 Availability of Natural Resources

The stone craft carried out at Bhedghat is mainly done using the soapstone available at the banks of river Narmada. Ramesh (38/Male/Craftsman at Bhedaghat) says that the stone is easily available to them near their place of work. The raw material is procured by the craftsmen according to their need and the market demand. The basic stone that has been used in most of the stone crafts of Madhya Pradesh is soapstone. This particular stone is used mainly for shaping and carving idols. The work of craftsmen near Bhedaghat is mainly of carving beautiful designs and making decorative items out of it.

7.1.2 S2 Low Labour Costs and Skilled Artisans

During the interaction with the artisans, Sadanand (Male/43/Craftsman at Bhedaghat) told that he and his younger brotherare into the crafting business since they were 15 years old. As children they had seen their grandfather and father creating the art pieces. They never felt the need to go out of Bhedaghat for better opportunities. They were trained by their father and mastered the art with practice. They also told that they never felt the need to hire outside labour. So there no labour cost involved in the process of creation of art pieces. They are into this art since generations and the skills have been learnt by most of them from their elder generation.

7.1.3 S3 Income and Employment Generation

Stone art plays a vital role in the socio-economic development of tribal communities, providing livelihoods to thousands of artisans, especially in rural areas. The art form preserves cultural heritage and contributes to sustainable development and empowerment of marginalised communities. Vinod Dahiya (Male/40/Craftsman at Bhedaghat) and his wife Sunita(Female/34/Craftswoman at Bhedghat) practice this stone craft and it is their family occupation. The family earns income only through the soap stone work.As the variety of products ranges from small goods to big ones, the cost also ranges from Rs. 10/- to Rs.10,000/-. The retailers sell the products in local shops and showrooms in Jabalpur. It was told that the goods also are sent to various cities across India. The products like idols of average size when purchased in wholesale are available at Rs.500/- to Rs.800/-. But the same is sold for Rs.1200/- in big showrooms. The craftsmen thus suffer a loss. The average income range of the craftsmen working here is between Rs .Eighty thousand to five lacs per annum.

7.2 Weakness

7.2.1 W1 Lack of Modern Techniques

The craftsmen narrated the exclusivity of the craft as it is completely handmade craft. The designs are developed by the local craftsmen only and that the soft marble specialty is unique to the place. Marble carving is done in a workshop where the craftsman draws an outline of the design on the boulder. The artisans remove unwanted material using a hammer, chisel, and drill to create the final figure. Small chisels and drills are used for fine detailing. Sunita (Female/34/Craftswoman at Bhedghat) narrated that they do not have any machines or any modern techniques for creating their craft. They use simple tools and this takes a lot of time in creation of the final piece. She stated that she creates the small decorative pieces like photo frames, name plates and can make only 6 to 8 pieces in a day.

7.2.2 W2 Lack of Training ProgrammesFor Developing Skills

Sudhir (Male/28/Craftsman at Bhedaghat) says that the craft that he engages in is learnt through the earlier generation. He learnt it from his parents mainly by observation and some specific skill were thought to him by elders more experienced in crafting. He learnt carving through other elder craftsmen in the community.

7.2.3 W3 Declining Interest Among Youth

It was observed through the interaction that many craftsmen were concerned about the fact that their sons and daughters lacked interest in carrying out the traditional craft. They felt that the craft does not fetch enough money and it takes a lot of time in creation of bigger pieces which would cost higher. The new generation expressed that the sale of the craft was seasonal i.e during December to March the sales increased while during monsoon due to closure of the place, they suffered no income period.Naresh (Male/17/Craftsman at Bhedaghat) says that he migrates to Jabalpur during the off season to fetch extra income for the family and was looking for better jobs in the city.

7.3 Opportunities

7.3.1 O1: Growing Demand for Products at Trade Fairs and Exhibitions

The product range available in this particular craft are small animal figures like elephants etc. with carvings on them, kitchen utility items, idols small and big, flower vases, tables, crockery etc. The workers also make certain specific utensils like pata for making chapattis and marble grinders etc. Sheikh Sehjad,(Male/ 43/Craftsman at Bhedaghat) one of the artist and owner of Karim Arts in Bhedaghat, revealed that government invites these craftsmen to the trade fairs and exhibitions with their products. During these exhibitions, products are exchanged between the craftsmen from Agra and Jaipur. This opportunity helps for the betterment of their craft.

7.3.2 O2: Preference for Naturalness Over Artificiality

Sheikh Sehjad,(Male/ 43/Craftsman at Bhedaghat) also mentions that the demand for their products like idols and carved pieces of marble is more in big cities than actually at Bhedaghat. The appreciation to their created craft is because city crowd prefers naturalness and value the effort and beauty of their work. The prices that they get are five times more than in Bhedaght. It was also told that they even receive orders for customized pieces at such exhibitions and fairs. It definitely improves their income and gives better prospects.

7.3.3 O3 Government Interventions

While talking about the government initiatives or support, it was told that training camps used to be organized for training of these artisans. But the same has not been organized since last many years post covid. It was told that training was carried out by training centers. Some of the craftsmen are in the process of registering an association comprising of 20-25 artisans. They planned to register a society of some of the craftsmen under the M.P Society RegistrikaranAdhiniyam, 1973. The advantages that they pointed out if such association is formed were mainly to revive the dying art, to train more and more persons in the art as well to create recognition for themselves and the art. The association would be able to represent the interests of these artisans before the government authorities so that the Government takes interest in their activities. The craftsmen unanimously stated that registration of stone craft of Jabalpur as geographical indication has immensely added to its brand value. This in turn has helped them get better prices at the fairs and exhibitions.

7.4 Threats

7.4.1 T1: Increasing Cost of Transport and Raw Materials

The marble required for this stone craft of Jabalpur is a hard marble which is a metamorphic rock composed of recrystallized carbonate minerals mostly calcite or dolomite. The marble available here is dolomite, which costs Rs. 2000-4000/- per tone. Earlier the Bhedaghat craftsmen used to get the soapstone from the riverside which required no added costs. However,sincea few years High Court has banned this practice has put a stay order to break stones from marble mountains. Therefore, the craftsmen now have to buy soapstone from local dealers. They buy 5 to 10 kgs of raw material which is utilized in 10 days. Earlier the price was Rs.10 to 12/- per Kg which then increased to Rs.18 to 20/- per Kg. This has affected the pockets of poor craftsmen.

7.4.2 T2 Interference of Middlemen and Profit Reduction

Vinod Dahiya (Male/40/Craftsman at Bhedaghat) narrated that how the interference of middlemen has affected the profits and increased the overall making cost. This has surged in the post covid era. An investigation into the economic conditions of the craftsmen revealed that most of them earned hand to mouth, except for some who have shops as well as factories for the production of Marble Craft goods. Some of the craftsmen work for merchandisers and have to sell the items at very low cost.

7.4.3 T3 New Generation Moving to Cities for Better Job Opportunities Post Covid

The fear of next generation moving out of Bhedaghat in search of better job opportunities is expressed by almost all craftsmen. This trend is seen increasing post the covid times.

7.4.4 T4 Lack of Knowledge about Protecting the Stone Craft

Sheikh Sehjad,(Male/ 43/Craftsman at Bhedaghat) expressed concern over the fact that most of the craftsmen are poor and earn a living only through the stone craft. However, they lack awareness about the importance of protecting and preserving the craft. Even though the stone craft has been granted GI protection, it value is not known to most of the artisans.

8. INFERENCES AND DISCUSSION

It is pertinent to note that the historical brief about the stone craft and interaction with the craftsmen presented through SWOT Matrix, revealed that:

1. The stone craft carried in Bhedaghat, Jabalpur is an age old practice carried on since pre historic times
2. The skill and knowledge of crafting has been passed on to the craftsmen from their forefathers
3. The artisans are unable to garner the much sought attention to their craft and are economically deprived [8]

4. The craftsmen fear the craft being misappropriated and sold in bigger marketplaces at exorbitant prices
5. Most of the craftsmen lack awareness about preserving and protecting their craft. They are unaware of the fact that protection of geographical indication to the stone craft will help in adding brand value to their art

These conclusions lead to the answer to the first research question that Stone craft of Jabalpur is a fit case of traditional knowledge of artisans. The previous research (Sirpurkar&Gurpur, 2014) concluded that the craft attributable to the geographical area and owing its characteristics to the skill of artisans working there can be well protected as Geographical Indication. This suggestion was successfully implemented with Jabalpur Stone Craft being registered as a geographical Indication on 31/03/2023.[9] This research (Sirpurkar&Gurpur, 2014), was the part of supporting documents that were filed in the Geographical Indications Registry(CGPDTM, 2022)to make a out a case for registration of Jabalpur Stone Craft as geographical indication. It is listed as a reference in the Geographical Indications Journal[10]where Jabalpur Stone Craft is entered as a registered Geographical Indication. This is one of the ways of protecting traditional knowledge of artisans. It is inferred that awareness about the importance of GI protection has to be created among the artisans.

The stone craft is the expression of the traditions and culture[11] of the indigenous[12] people residing in Bhedaghat, Jabalpur. This craft is practiced by a community at large. Looking at the nature of this craft, the best-suited protection that could be sought was through intellectual property law in general and protection through GI in particular. The journey of a traditional stone craft from being the source of livelihood to the craftsmen that was on the verge of extinction [13]to a product of handicraft possessing the GI tag is incredible. Post GI era for the stone craft has added brand value to the craft. The economic benefit of such recognition and added value has proved to be encouraging to the local craftsmen as gathered from interaction with the stone craft artisans at Bhedaghat. It is also evident from the above discussion that GI protection is not sufficient and a better protection in the form of TK or TCEs is sought for the benefit of the craftsmen.[14]

8.1 Exploring the Traditional Knowledge Aspect of Jabalpur Stone Craft

As discussed in the preceding section, stone craft of Jabalpur was found eligible to be protected as geographical indication. This protection enhanced the value of the product and accorded certain rights to the craftsmen. The paper though acknowledges the importance of GI protection intends to dive deeper and analyse the traditional knowledge aspect of the stone craft. The examination of the documents supporting the registration of stone craft of Jabalpur (Geographical Indications Registry, n.d.)explicitly describe this craft as being carried over for generations by the artisans. This craft is the expression of culture and traditions of the local community as is evident from the study of its historical background from the SWOT analysis carried out through interaction with local craftsmen. This inspires the researchers to look at the craft form traditional knowledge perspective.

8.2 Protecting Traditional Knowledge through Intellectual Property: Scope and Limitations

Traditional knowledge being abstract in nature cannot be protected as such. It has to have a form and needs to be expressed to become eligible for protection. As discussed earlier the term traditional knowledge has been recognized and defined at various levels but there aren't any substantial provisions made for its protection. It has always been protected in the form in which it is expressed. For example, the Jabalpur stone craft is protected as geographical indication which is the articulation of traditional knowledge and know how held by the artisans. It is coherent that traditional knowledge or the wisdom of indigenous people is protected in the form it is expressed with the appropriate intellectual property protection. Traditional knowledge as such is vast and indefinite, protecting which is consequently difficult. Intellectual property in its modern forms protects creations through patents, utility models, industrial designs, copyrights, trademarks and the like. The expression of traditional knowledge also comes to be protected through one of these kinds of intellectual property. IP tools developed as a means to provide a temporary monopoly to an inventor or creator as an incentive by rewarding their innovative and creative outputs. The notion that IP is a system to 'protect' intellectual or creative endeavor is thus a metaphor for the ability to prevent others from acts of misappropriation and to enable the owner of the subject matter to exclusively benefit from that invention or creation for a fixed term.[UNCTAD], 2014) For this pattern to work the expressions of traditional knowledge must satisfy the criteria of protection laid down in each case. The major problem lies in the contrasting features of IP on one hand and TK as a customary right on the other. This can be asserted from the fact that traditional knowledge is passed on from generation to generation while IP protection in cases of patents and copyrights is time-bound. TK and TCEs[15] are not novel in the sense that they embody a technology that was created possibly ages ago and has been passed on and would not constitute a novel technology for purposes of patents or utility models or a new seed in the case of plant variety protection. Some TK may be spread more or less widely in the public and might even be documented in publications, hence would not fulfill the basic criteria to receive patent protection. Geographical indications and collective trademarks protect a mark or a name, rather than the underlying TK or TCE. (The Convention on Biodiversity and the Nagoya Protocol: Intellectual Property Implications, 2014) Although protection as geographical indication can save the creation from being misappropriated. Another issue is that of unknown or collective ownership. As traditional knowledge is a collective wisdom of indigenous people, identifying ownership becomes impossible. These limitations, however, can be overcome by a possibility to claim collective authorship or to let an institution function as a substitute for unknown authors under copyright laws. Similarly, it could be possible that patents are given to an institution that represents a collective of inventors. The literature on existing national legislation and experience shows that solutions to the above limitations cannot be developed through amending solely the existing IPR but through a combination of it with *sui generis* options. (The Convention on Biodiversity and the Nagoya Protocol: Intellectual Property Implications, 2014) At the same time, it is very difficult to amend the national IP laws as the countries are members of the WTO TRIPS agreement and various WIPO IPR treaties with IP-related obligations. The members are boundto meet certain international standards and limitations in formulating their IP laws.

It is thus essential in this context to understand another aspect of traditional knowledge, i.e., Traditional Cultural Expressions (TCEs). Traditional cultural expressions (TCEs), also called "expressions of folklore", may include music, dance, art, designs, names, signs and symbols, performances, ceremonies, architectural forms, handicrafts and narratives, or many other artistic or cultural expressions.(WIPO, n.d.)

Traditional cultural expressions:

- May be considered as the forms in which traditional culture is expressed
- Form part of the identity and heritage of a traditional or indigenous community
- Are passed down from generation to generation

TCEs are integral to indigenous and local communities' cultural and social identities, embody know-how and skills, and transmit core values and beliefs. Their protection is related to the promotion of creativity, enhanced cultural diversity and the preservation of cultural heritage.(WIPO, n.d.) Even though protecting TCEs is equally difficult as protecting traditional knowledge, they enlarge the scope of TK and give TK a definite form.[16]TCEs will also protect TK directly rather than protecting just the name, mark, or any typical form of expression.

Returning to Jabalpur Stone Craft, it is observed that the Jabalpur stone craft definitively falls with the definition of TCEs. The stone craft of Jabalpur is an integrated heritage of the craftsmen that brings within its ambit the traditional knowledge and traditional cultural expressions held by them for generations. Delving into protection of TCEs, there is no uniformity among the various jurisdictions regarding the legal protection of TCEs. TCEs can sometimes be protected by existing systems, such as copyright and related rights, geographical indications, appellations of origin and trademarks. For example, contemporary adaptations of folklore are copyrightable, while performances of traditional songs and music may come under the WIPO Performances and Phonograms Treaty and Beijing Treaty (WIPO Performances and Phonograms Treaty (WPPT), 1996) on Audiovisual Performances. Trademarks can be used to identify authentic indigenous arts, as the Maori Arts Board in New Zealand, Te Waka Toi, has done. Some countries also have special legislation for the protection of folklore.[17]However, this will also lead to a similar situation as in the case of protecting traditional knowledge.

8.3 Deliberations on Protecting Traditional Knowledge and Traditional Cultural Expressions and its Effect on Stone Craft of Jabalpur

The Intergovernmental Committee of the WIPO on Intellectual Property and Genetic Resources, Traditional Knowledge and Folklore (IGC) undertook text-based negotiations to finalize an agreement on an international legal instrument for the protection of traditional knowledge (TK), traditional cultural expressions (TCEs), and genetic resources (GRs). (WIPO Intergovernmental Committee, 2024) To further this objective, WIPO adopted a treaty on Intellectual Property, Genetic Resources, and Associated Traditional Knowledge on 24th May 2024 at the WIPO headquarters in Geneva, which addresses the interface between intellectual property, genetic resources, and traditional knowledge. This is the first WIPO Treaty to include provisions specifically for Indigenous peoples as well as local communities. Once it enters into force with 15 contracting parties, the Treaty will establish in international law a new disclosure requirement for patent applicants whose inventions are based on genetic resources and/or associated traditional knowledge.The IGC in its Mandate 2024-2025(WIPO Intergovernmental Committee, 2024) resolves and agrees without prejudice to work on the protection of genetic resources (GRs), traditional knowledge (TK) and traditional cultural expressions (TCEs), with the objective of finalizing an agreement on an international legal instrument(s), without prejudging the nature of outcome(s), relating to intellectual property, which will ensure the balanced and effective protection of TK and TCEs. (WIPO Intergovernmental Committee, 2024) Adoption of the treaty on Intellectual Property, Genetic

Resources and Associated Traditional Knowledge in May 2024 was in fulfillment of this very mandate. It has further resolved to work on text-based negotiations, with a primary focus on narrowing existing gaps and reaching a common understanding on core issues that include, as applicable, inter alia, definitions, beneficiaries, subject matter, objectives, scope of protection, and what TK/TCEs are entitled to protection at an international level, including consideration of exceptions and limitations and the relationship with the public domain. This seems to be a very promising agenda of the IGC which will shed clarity on the dark areas of protection of traditional knowledge and TCEs. A very interesting step taken by the IGC also includes establishment of ad hoc expert groups to address specific legal, policy, or technical issues[18]

The Committee will use all WIPO working documents, including The Protection of Traditional Knowledge: Draft Articles (WIPO/GRTKF/IC/47/14) and The Protection of Traditional Cultural Expressions: Draft Articles, (WIPO/GRTKF/IC/47/15) as well as any other contributions of Member States, such as conducting/updating studies covering, inter alia, examples of national experiences, including domestic legislation, impact assessments, databases, and examples of protectable subject matter and subject matter that is not intended to be protected; and outputs of any expert groups established by the Committee and related activities conducted under the capacity-building and technical assistance program of the Secretariat the Traditional Knowledge Division. The Secretariat is requested to continue to update studies and other materials relating to tools and activities on databases and on existing disclosure regimes relating to GRs and associated TK, with a view to identifying any gaps and continuing to collect, compile and make available online information on national and regional sui generis regimes for the intellectual property protection of TK and TCEs. (WIPO, n.d.)

This Mandate of IGC seems truly promising and if worked successfully will go a long way to protect the interests of indigenous people.

As far as the position in India is concerned, India has been the pioneer of protecting traditional knowledge at various levels. It is the first country to have created Traditional Knowledge Digital Library (TKDL),[19] a compendium of 200,000 medicinal formulations that deter patent claims on such knowledge. It serves as the prior art in patent search all over the world. .India's TKDL is proving a powerful weapon in the country's fight against erroneous patents, sometimes referred to as biopiracy(Gupta, 2011) this library is, however, restricted to traditional knowledge of medicine and mother medicinal formulations. The Patents Act in India does not permit patent protection to traditional knowledge, thus providing a safeguard against monopolizing this wisdom of indigenous/local people. The art forms or craft embodying traditional knowledge is usually protected in India through copyright or geographical indication as the case may be. However, it is pertinent to note that India has always valued ancestral wisdom and cultural heritage and has taken steps to protect and preserve it. This approach towards traditional knowledge in India helps in creating a strong knowledge system. While India has a rich heritage of TCEs it protects them through different IP laws and lacks a sui generis law on the same. However, India has always championed for sui generis laws for protection of TCEs at IGC.

8.3.1 Effect of Protecting Stone Craft of Jabalpur as Traditional Knowledge (TK) and /or Traditional Cultural Expressions (TCEs)

1. Recognizing and protecting stone craft of Jabalpur as traditional knowledge and /or traditional cultural expression would give it a robust and permanent form of protection. This in a way would give recognition to this stone craft and similar other crafts as they embody the community wisdom of the indigenous people.

2. The problem of misappropriation of the art and cultural heritage would be addressed if stone craft is protected as TK and/or TCEs. Such kind of protection would ensure that benefits of using the skills of artisans are well rewarded.
3. TK protection would also help propagating the craft in international market, value addition and brand recognition.

The SWOT analysis of stone craft of Bhedaghat and the doctrinal study of this craft being protected as traditional knowledge reveal that the sector is rich in tradition, but faces challenges in contemporary times. To overcome the challenges the research suggests:

1. Ensuring access to better education and opportunities for new generation of artisans
2. Effective resource management and better opportunities to bigger marketplaces
3. Creating awareness among the artisans about using the tool of GI protection to promote, protect and preserve their craft.

This can be better ensured through according protection to the craft as traditional knowledge

8.3.2 Preserving Traditional Knowledge through Action Research

In this study, action research was employed which has directly contributed to preserving the traditional knowledge of Jabalpur Stone Craft. As discussed in the methodology, Action research has helped in the systematic documentation of the practices of the craftsman, which was possible only through proper and systematic interaction with the artisans. The documentation created now serves as a permanent record for future generations and provides evidence of the intellectual property of the craftsman.

Apart from the documentation of the intellectual property, this action research has also helped them support the community by seeking legal protection for their craft through geographical indication.[20]Linking the successful registration of Jabalpur stone craft as a geographical indication has provided evidence for the policymakers as a model to protect similar traditional knowledge.

The action research helped in aligning the traditional craft with modern market requirements by showcasing the same in exhibitions, markets, etc by maintaining their cultural integrity. Finally, this has helped the craftsman to overcome their economic challenges by promoting their traditional knowledge and its cultural significance.

CONCLUSION

The vastness of traditional knowledge is evident from the fact that it has its impact on varied fields of human existence. It has been the breadbasket that has fed the world; it has provided medicines to heal the world, it has also provided ways of sustainable management of resources including biodiversity. It has in its realm wisdom relating to languages, art forms, music and dance. It provides insights on celestial navigation, ecology, climate change, agriculture and technology. Yet today, this repository of conventional understanding is in danger of being marginalized. If we aim to create a robust knowledge system today, it is vital that it is based on the traditional wisdom held by our ancestors. Creating indigenous knowledge system that not only benefits the indigenous people but also guides the development of modern technology and know-how is attainable through promising action research. As aptly quoted earlier Jabalpur Stone craft is a paradigm of how traditional knowledge or traditional cultural expression can be modernized, protected, enhanced and promoted through intellectual property rights. The action research thus aides in creation of indigenous knowledge systems. It is further concluded that for an action research to fructify it has to be backed by well established legal framework. The enduring efforts at the international level aim to protect traditional knowledge underlying the manifestation, unlike the present state where traditional knowledge is protected in fragmented forms. The expression is protected while the wisdom gone into its creation remains open to exploitation.

This action research demonstrates that Jabalpur Stone craft is a fit case of traditional knowledge being protected through its embodiment in the art. It highlights the role of community engagement in preserving Traditional knowledge. Artisans played an important role in preserving and practicing the knowledge and they have had an active participation in protecting their craft as GI.It is the indigenous people who are the custodians of traditional knowledge. With the changing times it is essential that their participation in the wider world is ensured through awareness and capacity building. This has to be ensured that they understand the value of their wisdom and are aptly benefited by their exploitation. This will not only disseminate their community wisdom for a higher purpose but also will help in preserving it. Through action research on the relationship between intellectual property and traditional knowledge, we will serve the purpose of creating a conducive environment for the Indigenous communities to grow and preserve their wisdom.

The transformation of oral traditions into documented knowledge, combined with legal protection through geographical indication, ensures that this cultural heritage will be preserved for future generations while protecting the economic interests of the artisan community.

REFERENCES

Antons, C. (2009). The international debate about traditional knowledge and approaches in the Asia-Pacific region. In Antons, C. (Ed.), *Traditional knowledge, traditional cultural expressions and intellectual property law in the Asia-Pacific region* (pp. 40–51). Wolters Kluwer.

Chanda, T., & Sharma, A. (2024). *Shivarapatna stone craft in Karnataka - Exploring the transformative role of innovation in traditional handicraft and cultural identity*. ResearchGate. DOI: 10.29121/shodhkosh.v4.i2CDSDAD.2023.595

Chernela, J. (2004, December 3). The rights of indigenous people. Retrieved from http://janetchernela.outfoxing.com/rights-of-indigenous-peoples.html

Food and Agriculture Organization. (2001). *International treaty on plant and genetic resources for food and agriculture*. Retrieved from ftp://ftp.fao.org/docrep/fao/011/i0510e/i0510e.pdf

Frater, L. (2009). Protecting indigenous knowledge: The role of the convention on biodiversity and intellectual property rights (Working Paper Series No. 52). The Centre for Business Relationships, Accountability, Sustainability and Society.

Geographical Indications Registry. (n.d.). GI Search Version 2.0. Government of India. https://search.ipindia.gov.in/GIRPublic/Application/Details/710

Government of India. (2012). *Indian minerals yearbook 2011 (Part-II): Marble* (50th ed.). Ministry of Mines, Indian Bureau of Mines.

Government of India. (n.d.). Application details of Jabalpur stone craft. Retrieved from https://search.ipindia.gov.in/GIRPublic/Application/Details/710

Gupta, V. K. (2011, June). Protecting India's traditional knowledge. *WIPO Magazine*. Retrieved from https://www.wipo.int/wipo_magazine/en/2011/03/article_0002.html

Guru, R. (2024). A study on challenges to the revival of handicrafts in India. *ResearchGate*. Retrieved from https://www.researchgate.net/publication/383661483_A_Study_on_Challenges_to_Revival_of_Handicrafts_in_India

Halewood, M. (1999). Indigenous and local knowledge in international law: A preface to sui generis intellectual property protection. *McGill Law Journal. Revue de Droit de McGill*, *44*, 953–996.

Indian Bureau of Mines. (2012, October). *Indian minerals yearbook 2011: Marble (Part-II, 50th ed.)*. Ministry of Mines, Government of India. Government of India. Geographical Indications Registry. Application details of Jabalpur stone craft (Application No. 710).

International Council for Science. (2002, March). *Science and traditional knowledge: Report from the ICSU Study Group on Science and Traditional Knowledge*.

International Labour Organization. (1989). *Convention No. 169 - Indigenous and Tribal Peoples Convention*. Retrieved from https://normlex.ilo.org/dyn/normlex/en/f?p=NORMLEXPUB:12100:0:NO:P12100_ILO_CODE:C169

Office of Controller General of Patents. (2022). Designs and Trademarks. *Registered geographical indications*. Department for Promotion of Industry and Internal Trade, Ministry of Commerce and Industry, Government of India. https://ipindia.gov.in/registered-gls.htm

Office of Controller General of Patents. (2022). Designs and Trademarks. *Geographical indications journal No. 166*. Department for Promotion of Industry and Internal Trade, Ministry of Commerce and Industry, Government of India.

Pant, R. (2015). Protecting and promoting traditional knowledge in India: What role for geographical indications? *IIED*. Retrieved from https://www.iied.org/16576iied#:~:text=Sustainable%20markets -,Protecting%20and%20promoting%20traditional%20knowledge%20in,what%20role%20for%20 geographical%20indications%3F&text=Geographical%20indications%20(GIs)%20can%20help,and%20 promotion%20of%20traditional%20knowledge

Sanders, D. (1980). *Background information on the formation of the World Council of Indigenous Peoples*. Centre for World Indigenous Studies.

Sarkar, A. K. (2019). Impact of geographical indication (GI) on craftsmen of Varanasi in Uttar Pradesh. *NBS*. Retrieved from https://www.nbsc.in/MediaGallery/Impact%20of%20Geographical%20Indication %20(GI)%20on%20Craftsmen%20of%20Varanasi%20in%20Uttar%20Pradesh.pdf

Semali, L., & Kincheloe, J. L. (1999). *What is indigenous knowledge?: Voices from the academy*. Routledge.

Sirpurkar, R. J., & Gurpur, S. (2014). Protection of traditional knowledge through geographical indications: A case study of marble craft of Bhedaghat, Jabalpur. *American International Journal of Research in Humanities, Arts and Social Sciences*, 6(1, 2, 3). Retrieved from https://www.academia.edu/7371551/ ISSN_Print_2328_3734_ISSN_Online_2328_3696_ISSN_CD_ROM_2328_3688

Stoll, P. T., & von Hahn, A. (2008). Indigenous peoples, indigenous knowledge and indigenous resources in international law. In von Lewinski, S. (Ed.), *Indigenous heritage and intellectual property: Genetic resources, traditional knowledge and folklore* (2nd ed.). Kluwer Law International.

The Economic Times. (2012, December 18). Bhedaghat may make it to UNESCO list soon.

UNESCO. (2003). *Convention for the safeguarding of the intangible cultural heritage*. Retrieved from http://unesdoc.unesco.org

United Nations. (1945). *Charter of the United Nations*.

United Nations. (1948). *Universal declaration of human rights*. Retrieved from https://www.un.org/en/ about-us/universal-declaration-of-human-rights

United Nations. (1992). *Convention on biological diversity*. Retrieved from www.cbd.int

United Nations. (1994). *Convention to combat desertification*.

United Nations Conference on Trade and Development. (2014). The Convention on Biodiversity and the Nagoya Protocol: Intellectual Property Implications, Chapter 5 Protection of Traditional Knowledge. https://unctad.org/system/files/official-document/diaepcb2014d3_ch5_en.pdf

United Nations Conference on Trade and Development. (n.d.). The Convention on Biodiversity and the Nagoya Protocol: Intellectual Property Implications, Chapter 5 Protection of Traditional Knowledge. Retrieved from https://unctad.org/system/files/official-document/diaepcb2014d3_ch5_en.pdf

World Intellectual Property Organization. (1967). *Convention establishing the World Intellectual Property Organization*. Retrieved from https://www.wipo.int/treaties/en/text.jsp?file_id=283854

World Intellectual Property Organization. (1996). *WIPO performances and phonograms treaty*. Retrieved from https://www.wipo.int/treaties/en/ip/wppt/

World Intellectual Property Organization. (2001). *Intellectual property needs and expectations of traditional knowledge holders: WIPO report on fact-finding missions on intellectual property and traditional knowledge (1998-1999)*. WIPO.

World Intellectual Property Organization. (2012). *The protection of traditional knowledge: Draft articles* (WIPO/GRTKF/IC/21/4). Retrieved from https://www.wipo.int/edocs/mdocs/tk/en/wipo_grtkf_ic_21/wipo_grtkf_ic_21_ref_facilitators_text.pdf

World Intellectual Property Organization. (2012). *The protection of traditional cultural expressions: Draft articles* (WIPO/GRTKF/IC/22/4). Retrieved from https://www.wipo.int/edocs/mdocs/tk/en/wipo_grtkf_ic_22/wipo_grtkf_ic_22_ref_facilitators_text.pdf

World Intellectual Property Organization. (2019, August 9). WIPO Director General quote on Indigenous Day [Tweet]. X (formerly Twitter).

World Intellectual Property Organization. (2023). *Assemblies of member states of WIPO, 64th series of meetings, agenda item 15(v) report on intergovernmental committee on intellectual property, genetic resources, traditional knowledge and folklore (IGC), decision*. Retrieved from https://www.wipo.int/export/sites/www/tk/en/igc/docs/igc-mandate-2024-2025.pdf

World Intellectual Property Organization. (n.d.). *Traditional cultural expressions*. Retrieved from https://www.wipo.int/tk/en/folklore/

World Trade Organization. (1994). *Agreement on trade-related aspects of intellectual property rights*. Retrieved from https://www.wto.org/english/docs_e/legal_e/27-trips_01_e.htm

World Trade Organization. (2001). *Doha ministerial declaration*. Retrieved from https://www.wto.org/english/tratop_e/dda_e/dohaexplained_e.htm

Zappalaglio, A. (2013). Traditional knowledge: Emergence and history of the concept at international level. *University of Oxford*. Retrieved from https://www.researchgate.net/publication/271207052_Traditional_Knowledge_Emergence_and_History_of_the_Concept_at_International_Level

Zuckermann, G., et al. (2015). ENGAGING - A guide to interacting respectfully and reciprocally with Aboriginal and Torres Strait Islander people, and their arts practices and intellectual property [PDF]. Australian Government: Indigenous Culture Support.

KEY TERMS AND DEFINITIONS

Traditional Knowledge: Traditional knowledge is the knowledge including skills, practices and know-how that is developed, sustained and passed on from generation to generation in a community, which forms the part of oits cultural identity. For the purpose of this paper traditional knowledge is the wisdom of craftsmen of Bhedaghat practicing stone craft which they are engaged into from generations. They are practicing, preserving and protecting the craft for future generations and making a living out of it.

Action Research: For the purposes of this paper action research is the kind of research that is based on understanding the concepts of traditional knowledge and geographical indication and applying those to the empirical data collected from the craftsmen of Bhedaghat. The combined study is used to draw inferences and propose the best possible way to preserve traditional knowledge of craftsmen and create an indigenous knowledge system.

Stone Craft: Stone craft for the purposes of this paper is the art created from stone by the artisans living at the banks of river Narmada in Bhedaghat, Jabalpur

Jabalpur: A district in the state of Madhya Pradesh situated at the banks of river Narmada. It is the nearest place to Bhdaghat where the stone craft artisans reside.

Geographical Indication: A geographical indication (GI) is a sign used on products that have a specific geographical origin and possess qualities or a reputation that are due to that geographical origin. In order to function as a GI, a sign must identify a product as originating in a given place.

ENDNOTES

1 For GeoHack coordinates: Bhedaghat coordinates: 23.13°N 79.80°E

2 Application Details of Jabalpur Stone Craft; https://search.ipindia.gov.in/GIRPublic/Application/Details/710

3 Jabalpur, a city in state of Madhya Pradesh, India is situated at the banks of river Narmada. A small town situated in the district known by the name of Bhedaghat, is characterized by presence of magnificent Marble Rocks by the side of river Narmada.Bhedaghat in Jabalpur is known for its fine stone craft made out of soft marble stones found on Narmada banks.

4 Thiscase study is the best example of how traditional knowledge came to be protected through geographical indication

5 The Act for protection and registration of geographical indications in India

6 After almost nine years from publication of this research, Sirpurkar R.J and Gurpur S. 2014, AIJRHASS, Issue 6, Volume 1,2,3, March –May 2014https://www.academia.edu/7371551/ISSN_Print_2328_3734_ISSN_Online_2328_3696_ISSN_CD_ROM_2328_3688on 31st March 2023, the marble craft of Bhedaghat was given the GI tag by the Indian Geographical Indication Registry. It came to be registered as Jabalpur Stone Craft.Theconclusionof the above referred research manifested and the objective of the research to acknowledge the creativity of artisans by adding value to their art was successfully achieved.

7 Intellectual property is the mode of protection of traditional knowledge

8 This situation has improved a bit after Jabalpur stone craft was registered as geographical indication (around the beginning of 2023)

9 Application Details of Jabalpur Stone Craft; https://search.ipindia.gov.in/GIRPublic/Application/Details/710

10 Geographical Indications Journal NO. 166, Novenmber 30, 2022https://search.ipindia.gov.in/GIRPublic/Application/Details/710

11 This is evident from the figures that are created through this craft

12 The artisans residing here are practicing this art for generations using the naturally available resources which in this case is the soft marble stone

13 Many craftsmen moved to urban areas owing to meager wages

14 Answering the second research question

15 Traditional Cultural Expressions, elaborated in later part of the chapter

16 Form in which traditional knowledge is expressed is ultimately the subject matter of IP protection

17 Available on the website of GI Registry https://search.ipindia.gov.in/GIRPublic/Application/Details/710

18 The expert group(s) will have a balanced regional and gender representation and use an efficient working methodology. Accredited representatives of Indigenous Peoples as well as local communities will be invited to participate, in line with past practice of the Committee.

19 It is a database containing 34 million pages of formatted information on some 2,260,000 medicinal formulations in multiple languages.

20 https://search.ipindia.gov.in/GIRPublic/Application/Details/710

Chapter 17
Using Indigenous Knowledge System and Participatory Action Research to Reframe Western Epistemology in Research

William Chakabwata
https://orcid.org/0000-0002-4224-5239
University of South Africa, South Africa

ABSTRACT

The chapter averred that the use of positivism in research in higher education has proved to be of disservice to indigenous communities because of the assumptions inherent in that paradigm. The paradigm assumed that knowledge has to be verified scientifically by using methods derived from modernity and then it must be applied in universal setting. This approach is being challenged by the use of participatory action research which transforms participants from being objects of research to empowered communities who can demonstrate agency in changing their communities. The chapter explores a number of theoretical lenses to discuss the importance of PAR which include communicative theory, ubuntu, decolonial, and education for emancipation by Paulo Freire. The chapter calls for the use of PAR in higher education because of its potential for democracy and empowerment of communities.

INTRODUCTION

This chapter will advance ways in which indigenous knowledge system and participatory action research may be used to reframe western epistemic systems in research in Africa. Kolawole (2022) explained that local knowledge (LK) is as old as humanity. LK emerged as mankind grappled to survive in their environment and it assumed the guises of medicine, veterinary medicine, music, metallurgy, knowledge of plants among many other forms. Studious investigations in local knowledge began around 1900 and motivation on the subject grew exponentially around 1970s. The burgeoning interests in LK was accounted for by the growing climate crises in the Sahel and droughts in the 1990s which were linked to unsustainable environmental practices such as emission of fossil fuels gasses which are destructive to the ozone layer into the atmosphere and failure of the green revolution to address issues of poverty and

DOI: 10.4018/979-8-3693-3542-0.ch017

hunger. There was also academic influences on the rise of LK such as the emergence of post-modernism, in the 1980s and 1990s and its critical approach towards modernist claims of absolute nature of knowledge, rationality and normativity (Zhang, 2019; Medina and Boharquez, 2020). In recent times there is a burgeoning scholarship on the urgent need to delink higher education curriculum from western epistemic ideology (Barnes, 2018; Mpoe and Swartz, 2019; Lakso and Adu, 2024; Tegama and Fox, 2023; Molebatsi, 2006). It is for that reason that a motivation for Participatory Action Research (PAR) is advanced as an alternative, due to its democratic nature and the way it turns participants from being objects of research to subjects, who are empowered and capable of transforming their communities. The chapter is underpinned by a number of theoretical lenses which included decolonial theory, communicative action, Freirean thinking on critical consciousness and Ubuntu lens. The theoretical assumptions of PAR are also explored. Discussion on the value of PAR in higher education is submitted as well as the suggested recommendations Decolonising of research methodology has been associated with the work of Smith (1999) who presented a feminist and critical censor of Western positivistic world view to research. The same author submitted a counter to Western approach to research using a perspective from the colonised people who seek to preserve cultures, indigenous people lives, and ecosystems. This innovative work used Kaupapa Maori as a case to argue for the use of culturally relevant research ethics and methodology.

This clarion call has assumed numerous and diverse guises such as calls for decolonizing the curriculum, African renaissance discourse, a call to Africanise higher education and a call to delink research from western ideological premises. This demand for decolonizing higher education has also resonated with some scholars in the global north and in the United States who have averred that the university is replete with "colonised space" and that it is critical that the university become a site of ongoing struggle to decolonize the curriculum (Parker et al., 2017, p. 233). Apparently, the challenge to create higher education that promotes social justice and which is inclusive as noted in the preceding events in the United States has assumed universal proportion with the Rhodes Must Fall protests in South Africa. Science was employed throughout the colonial period to extend the colonial empires. There was a universal disregard for non-western scientific methods, which allowed a Eurocentric bias to grow in the field of science. Science was also utilized as a weapon for imperial expansion and it was used to detect minerals, allocating land on racial basis and to control outbreak of diseases such as small- pox. Multiple scientific advances were realised under the ambit of colonialism in the fields such as engineering, chemistry, geology and mineralogy. Science gave an impetus to imperialism, giving western nations an edge to cultures that they wanted to subdue (Harrison, 2005; Roy, 2014; Joseph, 1987). The claims of positivism to present knowledge and research in a value -free way that is universal began to come under serious interrogation by the publication of the social construction of reality and the structure of scientific revolutions (Berger and Luckmann, 1966; Kuhn, 1970). The field that participatory action research has focused on in the past include race ethnicity, disability studies, poverty and marginalization.

Background

It is important to deal with the nomenclature that is associated with this chapter by addressing the meaning of indigenous people. The oxford dictionary defines indigenous as one born in a specific place (Allen, 1990). Cunningham and Stanley (2003) expounds that the term has applications to a group of people who have lived in a place for a thousands of years as opposed to those who resided there for hundred years or so. There are numerous terms that have been employed to characterise indigenous people in diverse places, which included Aborigines in Australia, Metis and Inuit populations in Canada,

Maori in New Zealand or the term 'Tangata Whenua' or 'People of the Land'. This term has also been applied to refer to people whose cosmology is perceived to be unity between people and nature. This is a direct contrast with Judeo – Christian cosmology where God is perceived to be above, inhabiting heaven (Cunningham and Stanley, 2003). This also differs from the eastern perspective which seeks to access extraordinary powers internally through meditation. Indigenous people view people as constituting an unbroken connection between people and nature, the land, seas, rivers and forests.

Sarivaara et al., (2013, p369) indicated that the term indigenous is a contested term, however the authors postulated a definition as a " a construction that is applied to certain people in certain communities". In addition, it is also asserted that indigenous people are those who lived in a place for many years before colonisation, athough the definition may have its own limitations. Definitions of indigeneity have political, cultural and social connotations indicating an identity of a people. Burgeoning work on indigenous people is motivated by a desire to realise self- determination for the people. There are international proclamations on the rights of indigenous people, which include International Labour Organisation which defined indigenous people as people who dwelt in any area prior to colonialism and the formation of state borders. In addition, ILO No 169, acknowledges the rights of indigenous people to protection of their language, culture and environment (Joona, 2012). The global estimation of indigenous people around the world is 478 million people around the world (Planent Anomaly, 2023).

Smith (1999) explicate the term indigenous people has a recent etymology that is associated with American Indian Movement (AIM) and Canadian Indians and it came to the fore in the 1970s. It is a concept that highlight the struggles for self -determination of some of the world's most colonised people. The term has taken root due to the way it foregrounds the struggle for self- determination among the colonised people in Africa, New Zealand, Australia, Latin America and United States. Although, it is acknowledged that the term is not representing a homogeneous category, it still found to subsume the collective aspirations of colonised people in the international arena. Indigenous people share experiences of land alienation, colonisation of their cultures and denial of the right to self -determination. In spite of the official of colonialism, the colonising powers continues to shape the destiny of many people in the subaltern states.

The United Nations General Assembly in 1993, promulgated International Decade for World's Indigenous People 1995-2004. In addition, the General Assembly pronounced a theme for this decade as "Indigenous People: Partnership in Action". The pronounced intention of this declaration was to realise the following objectives:

- Implementation of projects by intergovernmental agencies and national agencies that seek to profit indigenous people
- The securing and enforcement of rights of indigenous people
- Protection of the rights of indigenous people

Modernism was marked by englightenment and scientific invention in which knowledge was supposed to be scientific and open to verification and has universal application. Postmodernism holds that knowledge is created and structured in certain historical environments and among certain cultures and is generated by human beings and has different value. Power determines which knowledge is produced and shared or suppressed (Kolawole, 2022).

Magni (2016) defined indigenous knowledge (IK) as traditional or localised knowledge that is imbedded in communities which subsumes " cultural traditions, values, beliefs, and worldviews of local people". Indigenous knowledge is associated with a group cultural identity and is therefore fundamental to the survival of the community. Siambombe et al., (2018) noted that indigenous knowledge among the BaTonga people in Zimbabwe included use of taboos, totems, and sacred places to protect the environment and to ensure sustainability. Sarkhel (2016) indicated that indigenous knowledge is constitutive of fisheries, food storage, selection of seeds in times of draughts, predicting weather patterns and draught, veterinary treatment for livestock, religion, arts, astrology and handcrafts.

Indigenous Knowledge system is perceived to be comprehensive and reflects a dynamism between the people and their environment. Although the implementation of LK in research is not without its own challenges, it is widely acknowledged that it has the potential to inform policy formulation at local level and illuminate other forms of research (Kolawole, 2022). It is broadly accepted among different workers that there are insurmountable challenges to implement development projects at local level without taking cognisance of LK. Kalawole (2014) adumbrated some of the reasons that will make LK continue to be relevant in contemporary discourse as:

- An predisposition by indigenous people for local technologies that are user friendly and environmentaly friendly and easy to adapt
- Limited financial resources among indigenous communities which makes them inclined to use local innovations which they can employ to address local challenges
- The need to preserve identity which is amplified by practices that uphold LK in the face of challenges from modernity and globalisation
- It helps communities to experiment with novel approaches to solving problems, as they try non-linear approaches
- Dependency of indigenous communities on local ecosystems makes their dependency on LK important

The voices of indigenous communities is emerging and the best way forward is to create a platform that facilitate engagements between communities and researchers. This platform is best provided through PAR. LK is created from the indigenous communities through story -telling, public testimony, theatre, and or dance. One of the chief struggle of the indigenous people is to reclaim the history of indigenous people. The process of reclaiming the history of colonised people involves probing the premises of Western epistemic system which assumed that there is a single universal history for humanity that can be compiled objectively and around the motif of binary categories of the coloniser and the colonised. In addition Western way of life with the ability to write were upheld as a heuristic for civilization. Western epistemic research was used to objectify the people who were different as the other. Smith (1999) averred that human history was also assumed to be Western patriarchal history. Furthermore, decolonising research means the discussion must be driven by reflections on ways in which one method excludes other forms of knowledge and in which the identities of the researchers and experiences are also transfigured. Another central tenet of decolonising research process involves using research to advance social justice.

Mpoe & Swartz (2019) framed the decolonising research under five headings which included power, trust, recognition of individual and community assets, culture and cultural competence and respectful and legimate research practices. Power is conceptualised as accepting the role of insider and outside and a status where the researcher is not the sole producer of knowledge, but allowing participants to make invaluable input. Power can be shown by class, gender, socio-economic status or even by a backdrop of colonialism. Colonialism has created a deficit of trust between the indigenous people and researchers. Decolonising research entails building an atmosphere of trust between the researcher and the researched. Culture and cultural competence has to be an integral to the research process taking into account cultural dyanmism. Finally, it is vital to provide coherent guidelines for ethics and practices that are deemed appropriate to indigenous people.

Parker et al.(2017) acknowledged the university space as a colonised space, which also holds prospects for radical transformation. Burgeoning research has called for the subaltern in the global south to rebut colonial histories. The hegemony of the colonial structures has perpetuated a situation where people of colour and women who enter academia as lecturers to retain the image of otherness in the academia. Parker et al.(2017) submitted that decolonising the classroom university classroom entails making three commitments which engaging in praxis which encourage openess in conducting research, listening to the voice of students, probing research conventions as areas of potential conflict that upholds colonial practices, facilitaing academic space for co-production of knowledge. In order to comprehend the necessity for decolonising academic space, it is important to examine the racial composition of lectures in countries like United States and the statistics reflected that 78% of the academic posts are taken by white males, by contrast blacks occupy 6% and 26% by white women and only 6% of professors are taken by people from the marginalised groups such as Latin Americans, African American, and Native American scholars (US Department of Education National Center for Education Statistics:, 2016). The demographic structure of university staff doesn't reflect the diversity of the United States where the people of colour constitute 40% of the population. Non-white in the university spaces in nations such as United States in academia, tend to experience a sense of alienation as their research is disregarded and lived experience taken for granted. In addition, they are treated as if they do not belong to academia, and they are expected to transform themselves into Anglo-Saxons. Marginalised groups in academia have to contend with hostile work environments, challenge of upending existing colonial methodologies and unfriendly work environment.

Research Ethics and Bias

Saunders et al.,(2019) indicated that ethical concerns have become a prominent issue in the contemporary university research. This alludes to the way one gains, access to a research site, how one deals with research subjects among a plethola of other issues. Moreover ethical issues include being fair and open- minded, respect for the participants, non- malfeasance, protecting the right to privacy for the participants, acknowledging the right of the participant to withdraw from the research, responsibility in data analysis, protecting the identity of participants and managing data in a responsible way. Universities have set up ethics commitees in order to ensure that there is compliance with the requisite procedures for conducting research. Research ethics do not have a long gealogy in university institution and they are traced to the end of the World War 2, particularly the denigrating experiments conducted by the Nazi on the Jews in concertration camps (Clinical Trials, 2009). In the aftermath of World War 2, trials to convict people accussed of human rights abuse were held at Nuremburg.A code was drafted to adjudicate on scientists and doctors who had participant in inhuman research in the concetration camps. This code provided the basis for ethical conduct when conducting research in conjunction with other declarations such as Helsinki Declaration of 1964 and Belmont Report of 1979.

The World Medical Association (WMA) was promulgated Helsinki declaration to provide ethics that inform research involving people (World Medical Association, 1964). The Belmont Report was compiled by the National Commission for the Protection of Human Subjects in research (Office for Human Research Protection, 1979).

This approach to research ethics is guided by a Eurocentric approach to research and ironically was not applied to research involving non-Caucasian people as will be demonstrated in this chapter. Lovo et al., (2021) expound that multiple indigenous people have been hurt due to researches that were not properly monitored, especially medical experiements. A number of studies which include postcolonial studies, feminist research and critical indigenous studies have disrupted the perspective that research can done without taking into account the context (Weheliye, 2014; Lugone, 2016). Decolonising methodologies entails also probing the notion that are taken for granted such as humanness, which in application in line with Eurocentric worldview tends to reject the humanity of black people (Weheliye, 2014; Wynter, 2003). Anderson (2002) averred that indigenous people were considered to be carriers of diseases which would jeopardise the well-being of the white race and for that reason miminal contacts were expected to be done with them. This claims were evident in South Africa, among the aborigines of Australia and also among the people from Asia. These various indigenous groups were perceived to be a people who were doomed. Research ethics were violated in the efforts to advance science using races that were different from the Caucasian race.

A case in point is the Henrietta Lacks, an Afro-American woman, who succumbed to cervical cancer. In the days leading up to the end of her life, the doctors had extracted cells from her body and gave them to a researcher without her permission and the Tuskegee Syphilis study (Brandt, 1978). The cells had some unusual properties that were latter used in research and also to treat people with cancer (Ahluwalia, 2020). In the Tuskegee Syphilis study the United States Public Health Service (USPHS) launched an experiment in Macony County, Alabama in order to assess the effects of latent untreated syphilis among the black population. The group constituted 400 black people suffering from syphilis and 200 who constituted a control group. This study was also driven by racial undercurrents of social Darwinism that were premised on the assumption that the black race was doomed and would eventually disappear from the country. In addition there was a fascination with the sexuality of black people, which when contrasted with that of white males, was perceived to be over -developed, while the brain was assumed to be underdeveloped (Brandt, 1978).

The experiments were run over 40 years with the black males receiving no treatment and those who succumbed to the disease had their bodies treated as objects for further research. Benatar (2002) not that burgeoning research in areas such as HIV/AIDS has led to renewed interests in research ethics, especially the asymmetrical relationship between the researcher and the subjects. Researchers tend to display a priviledged position, while the researched people display extreme conditions of deprivation and marginalisation. The same author added that the approach that undergird research from western scholars is a neoliberal mindset. This neoliberal worldview has permeated research at all levels in the global south due to their affiliation with the global north. Lovo et al (2021) asserted that Human Research Ethics (HRE) in nations such as Canada, Fiji, Australia and New Zealand seeks protect indigeous people. The inclusion of indigenous people in the management of research work in higher education has been neglected.

Gill,et al .,(2012) postulate that participatory action research ethics has the potential to challenge the hegemony of postivism, monolingualism, western epistemic dominance and colonialism.

Defining Participatory Action Research

le Grange (2001) opined that participatory action research is derived from two different etymologies. Bhana (1999) indicated that participatory research (PR) has a genesis in education and community health, while action research (AR) has its origins in efforts to effect organisational change in industry and other institutions. Overtime the differences have become inconsequential leading to a conflation, where the term participatory action research (PAR) has gained popular usage. Action research is constituted of multiple strands from its conception with numerous adherents and networks (Kindon et al.,2007). Action research is identified as being interdisiplinary and embraces multiple social movements which include feminism, critical race theory and neo-Marxism, and has major commitment to democratice involvement in decision making, wide participation in research and a commitment to praxis.

The origins of PAR is contested with some academics viewing it as an off-shot of cultural developments and others attributing to scholars in the United Sates in the 1920s. Kurt Lewin is in most cases acknowledged as the founder of PAR (Gomez et al., 2013). In addition the evolution is traced to the 1920s and opined to have grown intensely in the 1940s.

Newcastle University (2024) adduces that Participatory Action Research (PAR) connotes research that engages participants as partners to effect social change. Its emergence can be traced to the beginning of 1940s and gained prominence in the 1970s, during the cultural transition, PAR aims to completely undermine the power structure situated in traditional research approaches by appraising the engagement between the researcher and the participants, while placing value on the knowledge of ordinary people. Academics acknowledge that PAR does nt have its genesis from a single discipline, but subsumes the a cultural movement that places premium on experiences of those who who are subaltern, challenging inequality and embracing equality in terms of partnership in knowledge production.

Thiollent (2011) asserted that action research has its evolution from social psychology particularly the work of Kurt Lewin in the United States in the 1940s and also from the liberatory ideology emanating from Latin America in the 1950s and 60's under the auspices of Paulo Freire, in religious, social and political backgrounds. In addition, it is remarkable that these diverse political strands began to form a nexus in the 1980s and 90's and to coalesce into a coherent and systematic methodology that was contrary to the widely embraced methodologies which has etymology in positivism (Kindon et al.,2007). The practice of participatory action research has become orthodox among researcher from both the global north and south. Participatory action research promised a democratic approach to conducting research, but over the years the approach has faced a tumultuous path particularly in the 1990s and opposition from neoliberalism.

At a World Conference on Participatory Convergence in Knowledge in Catagena in 1997, multiple schools adding up to 32 were identified as affiliates of participtaory action research, making it a multi-disciplinary field, which had one common element of disastisafaction with positivism and it had closer links with radical thinking (Swantz, 2008). Cornish, et al., (2023) declared that action participatory action research is an approach that places premium on experiential knowledge for addressing challenges that are attributed to disparate and deleterious social systems and for conceptualising transformation and a different world. PAR incoporates the involvement of and leadership of communities that are experiencing the issues, who embark on a process of transformation, through doing rigorous and orderly research designed to effect change. PAR was a reaction to the shortcoming of postivism and interpretivism and has closer affirnity with critical philosophical leaning (le Grange, 2001).

However PAR converges with interpretivist's assumptions that reality is socially constructed (Au, 2007; Berger and Luckmann, 1966; Habermas,1985). Critical research foregrounds the premise that the contemporary world is unjust due to by design and this injustice include violation of the rights of women, people with a disability, elderly, minority groups among many other forms. The current system is also characterised by inequality in terms of distribution of resources.

PAR is conceptualised as placing at the centre the participants in the research and situating the university researcher from their privileged central position and in addition is committed to transformation of unequal power structures and challenges people to rethink the purpose of research and to whom university professional are answerable to. Gaskins et al., (2023) adduced that PAR is appropriate for marginalised and minoritised groups. In addition, PAR critiques disparate power relations and the systemic systems of oppression and privilege they generate. Furthermore, it raises political awareness of the minoritised groups for the purpose of transforming their lived reality (Greenwood and Levin, 2006). The research process is underpinned by the collaboration and efforts to confront injustice and inequality and seeking ways to change systemic injustice. PAR is rooted in the episteme and experience of the participants. Moreover PAR empowers participants, energises them and construct their self -esteem and build community collaboration (Barnes, 2018). It is for that reason that it has been associated with subversive elements as it challenges the status quo and seeks to replace it with more inclusive and democratic practices. PAR grants communities an opportunity to gain control over their own discourse, knowledge and experiences.

Participatory Action Research Methodologies

The methodologies of participatory action research includes the use of the following methods:

- Use of focus groups
- Semi-structured interviews
- Observations
- Document analysis
- Student work samples
- Use of pre and post test surveys

Ethics of Care in Participatory Action Research

Participatory action research is a broad term that subsumes emancipatory methodologies, decolonisation research and co-production methodologies. Participants engage in the process of creating knowledge in collaboration with the researcher (Edwards & Brannelly, 2017).As participatory research is expanding exponentially, it has been accompanied by a growth in ethics of care as vital response to neoliberal ideology and also as an approach to enhance social justice (Tronto, 1993; Tronto, 2013). In addition participatory action research is postulated to represent a democratic worldview (Abma, 2020). The relevance of the traditional approach to research has been questioned in Participatory Action Ressearch (Bussu, et al, 2020). The principles of informed consent, protected the identity of participants and confidentiality may not take into account the dynamic nature of the research context where decisions have to be made on daily basis. In order to address this fluid research context and the need for relevant research ethics, it is postulated that an ethic of care may provide an appropriate response. This means that a researcher has to take strenuous efforts to ascertain that appropriate moral framework is employed in a research context by building reflexivity and takes into account all prominent moral elements of a situation, relationship, collaboration and critical input from friends (Abma, 2020).

Groot, et al., (2019) opined that participatory action research has been noted to be riddled with ethical challenge issues. Some of the issues raised were collaboration with co-researchers, balancing power structures, building up of relations of trust, distribution of control and ownership (Kamlongera & Katenga-Kaunda, 2023; Omodan & Dastile, 2023; Brannelly & Boulton, 2017). Ethics of care has its pedigree in feminist approach (Held, 2006). The ethics of care is constructed on two pillars which include dealing with periphelisation, and respecting mutual relationships and an expanded conceptualisation of care and the centrality of care to inform and influence praxis. Centrality of care constitute elements which were expounded by Tronto (2013) as:

- Caring about-Attentiveness
- Caring for –responsibility
- Care giving-competence
- Care receiving-responsiveness
- Caring with-solidarity

An ethics of care to research places high premium on experiential knowledge and probing the political nuances in research and the associations that emerge in a research enterprise.Ethical engagements begin at the onset of research not at the end of long process of data extraction (Pickering & Kara, 2017). This creates a situation where the researcher is able to function with the participants or communities as opposed to treating the participants as subjects. Ethical decisions have to functions in a context. The way data is managed in participatory research runs contrary to ethics as practiced in universities using postivitic approach where the data becomes the property of the researcher. In participatory research data is treated as a form of representation.

Criticism of Participatory Research

Participatory research has been censured on the issue of contribution, for example, who contribute in research, concerns about participation based on dissimulation, power asymmetry (as shown by whose thoughts are valuable and who is paid to participant), which probe poor practices (Pickering and Kara, 2017). Others in a research may be conceptualised as participants or communities. Researchers have to actively take part in ethics at a variety of levels during participatory research which includes data analysis where decisions have to be made regarding the representation of the voices of the participants. Cornish, et al., (2023) opined that PAR projects have to contend with challenges such as lack of alignment with higher education resource infrustructure, dangers of being co-opted, and decentralisation of control. In order to address these challenges, PAR practitioners need to institutionalise PAR friendly support networks of people and institutions and urge the communities to probe their work and offer them honest feedback in order to promote accountability, employ reflexivity, share powers with other stakeholders and then have confidence in the approach.

Nel (2005) acknowledges that during the colonial era people were dislocated from their land, thus isolating them from the indigenous knowledge bases. This means that there has to be rigorous and extensive efforts to retrieve this knowledge and apply it for the benefit of the communities. In addition, the same author submitted that the implementation of indigenous systems in countries such as South Africa has been marked more by rhetoric than action, leading to very little action in integrating it in school curriculum and other platform of learning. Finally, there has been no clear delineation of the agenda of indigenous knowledge systems and the research methodology leading to it and it has become a catch -all phrase for all kinds of research practices, which maybe merely opportunistic. The next section explore ways in which methodologies may be decolonised.

Decolonising Methodologies

Parker et al.,(2017) explicated university research must create opportunities for students to experience tensions, ambivalences and contradiction in knowledge production. This begins with posing ontological questions on how knowledge is produced and interrogating the functions of research committees who have the capacity to either develop students' develop critical thinking skills or to denigrate them and crafting a research design that has potential to either dismantle colonial structures or to sustain them. Gill,et al.,(2012) the adherents of action research postulate that it is driven by a desire to redesign asymmetrical power relations and to get acquainted with the knowledge or power within disparate communities and to dispute inequality and advance the agenda for social justice.

Gill, et al., (2012) opines that participatory action research envision an attempt to dismantle skewed power relations in the research work, to acknowledge the value of knowledge and research in multiple and different communities and to tease inequalities and thereby advance social justice. A decolonial approach to action research aspires to unravel modernism, monolingualism, colonialism and Eurocentrism. The next section explores two cases studies of using case studies in Tanzania and South Africa.

Participatory Action Research in Tanzanian University

Swantz (2008) averred that action research in Tanzania was driven by the pragmatic demand to relate research at university level to national development aspirations, bridging the gap between workers and peasants and higher education. The avowed intention of PAR was to make research a tool for social transformation. The university of Dar es Salaam trained students in PAR and supported them during vacation as they helped to collect data. The approach had a major reception among people in rural areas who perceived the participatory elements with satisfaction. However it took time before the novel approach could be endorsed as legitimate by research leaders at the university. The first group of 12 students in Tanzania undertook to explore the income earning possibilities of of school leavers in five coastal villages in 1973.

Participatory Action Research in South Africa

Nel (2005) indicated that in IKS is acknowledged as vital although there has been limited progress on the ground. Key policy documents which includes Higher Education Quality Committee. However there is no foundation for teaching IKS at primary school level, which makes it difficult to implement at university level.

South African university have also been engaging in action research to varying degrees to address the challenges that communities and students contend with which include poverty, HIV/AIDS, lack of epistemological access to education, gender inequality and a host of other social injustices (Wood, et al., 2017). The concept of emancipatory action research in South Africa may be traced to the era of apartheid, when the teaching community responded to the Bantu education system by calling for education for the purpose of education in order to demonstrate the power of the people. The university of western Cape embraced PAR as a legitimate research inquiry. There is a growing practice in action research in South African universities including Nelson Mandela Metropolitan University, Walter Sisulu university, University of KwaZulu Natal among many others. PAR in South African universities has been employed as a tool for on-going professional development as well as a transformative mode of inquiry for addressing community issues. However, the utilisation of PAR in South African universities has not been without challenges. Some of the challenges include constrictive deadlines to research imposed on academy and students hinder researchers from development trust with communities in order to conduct ethical research. University committee responsible for research in some institution tend to place high value on traditional research above PAR (Wood, et al., 2017). The majority of rubrics employed in many higher education institutions in South Africa, do not accommodate the reflective nature of action research and have a propensity towards traditional research. People in communities are not used to exercising agency in community development leading to tendency to expect government to intiate interventions by teachers, administrators and school management. Other challenges that have been associated with PAR in South Africa is the question of how to make programmes sustainable once the university has withdrawn funding and the limitations of time impedes the involvement of post –graduate students from engaging in PAR. The next section explores Harbermas communicative theory and and the way it applies to PAR.

Habermas Communicative Action

Jurgen Habermas is perceived as a leading theorist and philosopher to emerge from Gemany after the second World War.He associated briefly with the Frankfurt school of Marxist philosophy before he was forced to leave by by Max Horkhimer, who perceived him to be an extreme Marxist (Stanford Encyclopedia of Philosophy, 2024). Later, Habermas returned to Franfurt to take -over the position left vacant with the departure of Horkhimer. Habermas contributed significantly to critical theory critical theory, communication theory and political philosophy (Philo-Notes, 2023).

One of his magnum opus is the theory of communicative action, a holistic theory, where he postulated that human communication shapes social reality. Habermas opined that everything that is encapsulated in the domain of social life is embraced as " the object area of social sciences" (Ozkul, 2020, p. 217). This observation foregrounds an important criticism of positivism which underestimated the value of social reality in an objective world. For, Habermas people are always engaging in dyanamic activities in a lifesphere that involves communication and diverse norms. In addition he submitted that positivism fell short in accounting for the interaction between human beings and life and positivism restricts human agency by applying laws of natural sphere to the social and communicative sphere. The social sphere is riddled with subjective reality and can not be governed by laws that apply to physical objects.

The theory sought to address the tension between individual and society and reason and subjectivity and between freedom and social control (Stanford Encyclopedia of Philosophy, 2024). Habermas postulated that the liberatory potential of a society is not located in a social class, but in communication, which is a universal attribute in all people and in spoken language which is spoken in by all people and which is the basis of communication (Ozkul, 2020). The fundamental tenets of communicative theory is that communication is the basic medium, through which people exchange ideas and construct and sustain social reality. Communication transcends the exchange of ideas and emotions and demonstrate the way in which individuals negotiate relationships with each other and the overarching social structures. In addition communication is the process of reciprocal understanding and this reciprocal understanding is what provides the basis for social order.

One of the central principle of communication theory is the concept of communicative rationality which connotes that when people communicate, they should undertake a dual process of interrogating ideas and assumptions in order to appropriate a common understanding of the issue. Communication should therefore be framed in a context that allows for debate and justification of claims made (Schaefer et al.,2013).

Habermas postulated that the use of language for the purpose of creating understanding is desirable, and its appropriate, while the use of language for strategic rationality is not morally unacceptable (Schaefer et al.,2013). Communicative rationality is the bedrock of prescriptive democracy. In principle in a democracy, communicative rationality must shape the engagements until a consensus is realised. Furthermore, Habermas averred that four general guidelines for argumentation:

- All parties with capacity to contribute to argumentation are included
- Equality of voice among participants
- They are guaranted the freedom to express their honest opinion without deception or self deception
- There are no tools of coercion that are structured to influence the process or structure of discourse

This resonates with the assumptions of PAR in terms of upholding democracy as a value by treating the participants as subjects as opposed to objects. Both views have a tenet that assumes that democracy is necessary and vital in social engagements and research. Moreover PAR seeks to offer pragmatic solutions to local challenges, to build human capabilities. PAR eschews the theory practical divide as it generates knowledge and at the same time is commited to praxis. In terms of ontology PAR holds that the social world we inhabit is co-constructed and shaped by a given context is intersubjective and positioned. In addition PAR acknowledges the agency of the people in being able to change their society, and seeks to raise awareness of the constraints that people find themselves in, in order to challenge the power structures. The next section explores Freire thinking on critical consciousness.

Freirean Thinking on Critical Consciousness

Paulo Freire (1921-1997) advocated a world where all students are able to develop critical consciousness through the practice of critical pedagogy (Freire,1973). Freire denounced education that sought to promote regurgitation of facts and information, an approach he termed banking concept of education (Freire,1973; Orlowski, 2019). There are commonalities between teaching PAR and teaching for social justice. In a traditional teaching context research is approached from a perspective where students are objects and the they are expected to locate the objective knowledge that is hidden within the learning institution and the world. The knowledge and culture of the dominant class is reproduced in every generation. The use of standardised tests and traditional assessments in the United States and Canada is a sign to the recalcitrance of traditional teaching methods, and they are being further buttressed by neoliberalism (Orlowski, 2019;Apple, 2013; Solomon & Singer, 2011). Freire underscored the importance of ideology when selecting a research methodology. At thc same time ideology is central to the selection of didactic approach for the oppressed.

Freire submitted that consciousness plays a central role in the liberation of mankind. Freire acknowledged that since human beings are intimately engaging with the world they develop a social consciousness, which is intentionality towards the world. Freire postulated that human beings can reflect critically on their world and act to change it or engage in praxis (Freire,1982). Praxis entails that human beings either as individuals or in groups engage in changing their material world which impacts on their manner of acting. Furthermore to treat human beings as objects is to degrade or dehumanise them (Shor & Freire, 1987). It is in this context that PAR aims to treat people as humans empowering them to act on their social reality in order to transform it as the seek to be more human. The next section explores ubuntu and indigenous knowledge system.

Ubuntu as a Research Theoretical Lens

Mpoe & Swartz (2019) has been postulated as the basis lens to frame research among indigenous people in Africa. Ubuntu alludes to an ethic among the people from Sub-Sahara which foregrounds relatedness and interdependence. Chakabwata & Mukazi (2023) averred that ubuntu is based on four variables which include reciprocal appropriation, respectful representation, rights and regulation and relational accountability. This means that all the procedures that are designed to enhance accountability in research are followed. Making research reciprocal alludes to the fact that all the benefits that accrue from university research and the knowledge must be shared with the community. This means that research enables the creation of spaces for indigenous people to articulate their world view on issues.

Ubuntu serves the needs of indigenous communities in Sub-Sahara Africa by placing the well-being of the people at the centre of the research discourse. Using Ubuntu in research at university in higher education enables university to realise the ambivalence in western research practices and the importance of allowing for plurality and hybridity in knowledge production (Mawere & Mubaya, 2016). Ubuntu lens holds that research in higher education in Africa must be grounded in the philosophy of Ubuntu, which places high premium on community relationships as opposed to individualistic accumulation of wealth. The next section discusses the chapter.

DISCUSSION

There is a growing interest in the use of action research in higher education as shown by initiatives such as South African Educational Research Association and Higher Education Learning and Teaching Association of Southern Africa, have Action Research Special Interest Group (Wood, et al., 2017). A number of conferences have been held in South Africa on action research including the one hosted by University of Pretoria in 2003. This is in line with the short coming of positivistic research paradigm and its association with colonialism and coloniality. It is apparent that the goals of PAR and social justice tend to converge (Orlowski, 2019). Researchers who function within PAR and Social Justice hold the same assumptions that the world is amenable to transformation that can promote a more egalitarian society. The teachers who function with the ambit of the two worldviews appreciate the social construction of knowledge. In addition, both positions acknowledge the tenet that knowledge is not isolated to certain individuals as in a positivistic framework, but is found in all people.

Dancis, et al., (2023) noted that PAR has capacity to energise students in learning once they develop knowledge that they can take part in knowledge production. Gradually PAR is becoming ubiquitous as it is also employed in the USA and other countries in spite of its leftist leanings. Critics have also underscored the weakness of formalising a weakened version of PAR in universities, thus depriving the framework of its liberatory potential. University spaces offer an incredible opportunity to engage in liberatory discourse. Ritterbusch (2019) postulated that a depoliticised version of PAR at university level would assume the guise of preoccupation with students experiences while downplaying institutional transformation. An approach that implements PAR as rubric of antipated behaviours can stunt the capacity of the programme at university level to develop the needed critical skills.

Davies (2022) noted that PAR has benefits and challenges. Literature encapsulated challenges such as PAR is to intricate and demands a lot of time, each member who partakes in a PAR project must be prepared to acquire new skills and the risks of reproducing inequalities (Baum et al., 2006; Davies, 2022). Furthermore the messy nature of PAR, at times makes it difficult to attract funding. Some of the benefits of PAR are the way it changes the role of the researched to be researchers on their own issues, its capacity to elicit agency from participants and the way it turns participants from being objects to subjects. PAR is being implemented globally due to the way it addresses some of the shortcomings related to the use of traditional research. It is noticeable that PAR promotes participatory democracy as espoused by Harbemas which asserted the discourse ethics as the basis for Communicative Action Theory. In addition Harbemas opined that mankind inability to create a just society is not a reflection of lack of reason, but inability to engage in interpersonal communication. Most of the engagements that occur are marred by domination where one person seeks to make their interests paramount. This observation characterises research that is conducted using a positivistic approach in universities and which tends to use strategic instrumental rationality. If we are going to value indigenous comunities, we have tupholds human rights by training graduates from higher education to engage in PAR, with communities without displaying disparate power dynamics. It is such democratic approaches to research that will empower communities and also lead to entrecnhment of democratic practices.

Higher education is in need of transformation from a postivistic worldview towards research to a more inclusive approach, that envisions societal transformation. This transformation is inline with Freire development of theory and praxis and also ubuntu philosophy.

The agenda for IKS has also been questioned as well as the methodology (Nel, 2005). In addition there has been more political rhetoric on incorporating IKS in school curriculum which has not been acted upon. In the light of the foregoing discussion the following recommendations are offered.

Recommendation

The foregoing discussion has explored the importance of engaging in PAR when conducting research at university level, particularly when working among indigenous communities that were denied a voice through positivistic research ideologies. In order to develop the use of PAR in higher education in Africa recommendations made by Wood, et al., (2017, p. 231) are relevant and may be encapsulated as:

- Setting up networks that can create a platform for exchange of ideas and collaboration on PAR
- Forming collaboration with global action researchers who have expansive experience in the area

- Universities must integrate PAR in their practices in order to ensure better students learning outcomes
- Student must be encouraged to develop a culture of inquiry that allows them to ask question regarding their material conditions and perceived social injustice and inequality
- IKS needs to be integrated into school systems as well to ensure that learners are exposed to it prior to higher education, as a legitimate way of knowing

CONCLUSION

In this chapter it was noted that indigenous knowledge system represent a new way of knowing from the positivistic ways that are associated with enlightenment and modernity. The convergence of IKS and PAR gives indigenous communities a voice in knowledge production in their own communities and also helps empower them to control their own destiny. However, in order to integrate PAR in higher education institution changes have to start at primary school level. It was also noted that PAR holds the key to democratic participation of communities in different ways of knowing, from the researcher based ways of conducting research.

REFERENCES

Abma, T. (2020). *Ethics for good for good participatory action research: Engagement in a ommitment to epistemic justice.* BeleidsonderzoekOnline., DOI: 10.5553/BO/221335502020000006001

Ahluwalia, M. (2020). Genetic privacy: We must learn from the story of Henrietta Lacks. *New Scientist.*

Allen, R. (1990). *Concise Oxford dictionary of current english.* Clarendon.

Anderson, W. (2002). The Cultivation of whiteness: Science, Health and racial destiny in Australia. Melborne: Melborne University Press.

Apple, M. (2013). *Can education change society?* Routledge.

Au, W. (2007). Epistemology of the oppressed: Dialectics of Paulo Freire's theory of knowledge. *The Journal for Critical Education Policy Studies, 5*(2), 1–14.

Barnes, B. (2018). Decolonising research methololodologies: Opportunity and. *South African Journal of Psychology. Suid-Afrikaanse Tydskrif vir Sielkunde, 48*(3), 379–387. DOI: 10.1177/0081246318798294

Baum, F., MacDougall, C., & Smith, D. (2006). Participatory action research. *Journal of Epidemiology and Community Health, 60*(10), 854–857. DOI: 10.1136/jech.2004.028662 PMID: 16973531

Benatar, S. (2002). Reflections and recommendations on research ethics in developing countries. *Social Science & Medicine, 54*(7), 1131–1141. DOI: 10.1016/S0277-9536(01)00327-6 PMID: 11999507

Berger, P., & Luckmann, T. (1966). *The social construction of reality.* Penguin Books.

Bhana, A. (1999). Participatory action research: A practical guide for realistic radicals. In Blanche, M., Durrheim, K., & Painter, D. (Eds.), *Research in practice: Applied methods for social sciences* (pp. 227–238). UTC.

Brandt, A. M. (1978). Racism and research: The case of the Tuskegee study. *The Hastings Center Report, 9*(6), 21–29. DOI: 10.2307/3561468 PMID: 721302

Brannelly, T., & Boulton, A. (2017). The ethics of care and transformation research practices in Aotearoa New Zealand. *Qualitative Research, 17*(3), 340–350. DOI: 10.1177/1468794117698916

Bussu, S., Lalani, M., Pattison, S., & Marshall, M. (2020). Engaging with ethical care: Ethical issues in participatory research. Retrieved 04 19, 2024, from http://discovery.ucl.ac.uk

Butler, C. (2002). *Postmodernism: A very short introduction.* Oxford University Press.

Chakabwata, W., & Mukazi, F. (2023). Ubuntu philosophy and online assessment in Higher Education Institutions. In J. Keengwee, *Transformative and innovative pedagogies in education* (pp. 257-275). Dakota: IGI Global Publishers.

Clinical Trials. (2009). The origins of the research ethics. Retrieved 04 19, 2024, from https://www.appliedclinicaltrialsonline.com/view/origins-research-ethics

Cornish, F., Breton, N., Moreno Tabarez, U., Delgado, J., Rua, M., & de-Graft Aikins, A. (2023). Participatory action research. *Nature Reviews. Methods Primers*, *3*(34), 1–14.

Cunningham, C., & Stanley, F. (2003). Indigenous by definition,experience or World view. *BMJ (Clinical Research Ed.)*, *327*(7412), 403–404. DOI: 10.1136/bmj.327.7412.403 PMID: 12933705

Dancis, J., Coleman, B., & Ellison, E. (2023). Participatory action research as pedagogy. *Journal of Participatory Action Research Methods*, *4*(2), 1–24.

Davies, S. (2022). Towards digital justice: Participatory action research in global digital health. *BMJ Global Health*, *7*, 1–4. PMID: 35580914

Edwards, R., & Brannelly, T. (2017). Approaches to democratising qualitative research methods. *Qualitative Research*, *17*(3), 271–277. DOI: 10.1177/1468794117706869

Fernandez, J., Gaston, J., Nguyen, M., Rovaris, J., Rhyann, L., & Anguilar, D. (2018). Documenting sociopolitical development via participatory action research (PAR) with women of colour student activists in the neoliberal university. *Journal of Social and Political Psychology*, *6*(2), 591–607. DOI: 10.5964/jspp.v6i2.900

Fine, M., Torre, M., Oswald, A. G., & Avery, S. (2021). Critical participatory action research: Methods and praxis for intersectional knowledge production. *Journal of Counseling Psychology*, *68*(3), 344–356. DOI: 10.1037/cou0000445 PMID: 34043378

Freire, P. (1973). *Education for critical consciousness*. Continuum.

Freire, P. (1974). *Pedagogy of the oppressed*. Seabury Press.

Freire, P. (1982). Education as the practice of freedom. In *M. Ramos, in education for critical consciousness* (pp. 1–84). Bergin & Garvey Publishers.

Gaffney, M. (2008). Participatory action research: An overview. *Kariraranga*, *9*(3), 9–15. DOI: 10.54322/kairaranga.v9i3.131

Gaskins, W., Batsheva, G., & Brittany, A. (2023). Reflections on implementing participatory action research in engineering. *Journal of Education and Development*, *7*(3), 18–24. DOI: 10.20849/jed.v7i3.1369

Gill, H., Puru, K., Lin, G., & Gill, H. (2012). In the midst of participatory action research practices moving towards decolonizing and decolonial praxis. *Reconceptualizing Educational Research Methodology*, *3*(1), 1–15. DOI: 10.7577/rerm.357

Gomez, A., Elbo, C., & Capllonch, M. (2013). Beyond action research: The communicative methodology of research. *International Journal of Qualitative Research*, *6*(2), 183–197. DOI: 10.1525/irqr.2013.6.2.183

Greenwood, D., & Levin, M. (2006). Introduction to action research: Social research for social change. Carlifornia: Sage Publications.

Groot, B., Vink, A., Haveman, A., Huberts, M., Schout, G., & Abma, T. (2019). Ethics of care in participatory research: Mutual responsibility in collaboration with co-researchers. *Educational Action Research*, *27*(2), 286–302. DOI: 10.1080/09650792.2018.1450771

Habermas, J. (1971). *Knowledge and human interests*. Bacon.

Habermas, J. (1985). *The theory of communicative action*. Beacon Press.

Harrison, G. (2005).. . *Science and the British Empire*, *96*, 13–28.

Held, V. (2006). *The ethics of care: Personal political and global*. Oxford University Press.

Hooks, B. (1994). *Teaching to transgress: Education as a freedom of practice*. Routledge.

Joona, T. (2012). *ILO Convention No 169 in a Nordic context with comparative analysis: An interdisciplinary approach.* Rovaniemi: Lapland Univiversity Press.

Joseph, G. (1987). Foundations of Eurocentrism in mathematics. *Race & Class*, *28*(3), 13–28. DOI: 10.1177/030639688702800302

Kalawole, O. (2014). Repositioning local knowledge in development initiatives. In Alvares, C. (Ed.), *Multicultural knowledge and the university* (pp. 251–258). Multiversity and Citizens International.

Kamlongera, M., & Katenga-Kaunda, M. (2023). Reflections reflections on ethics of care as decolonial research practice: Understanding Indigenous knowledge communication systems to navigate moments of ethical tension in in rural Malawi. *Research Ethics*, *19*(3), 312–324. DOI: 10.1177/17470161231169484

Kindon, S., & Elwood, S. (2009). IntroductioN: More than methods-reflections on participatory action research in geographic teaching, learning and research. *Journal of Geography in Higher Education*, *33*(1), 19–32. DOI: 10.1080/03098260802276474

Kindon, S., Pain, R., & Kesby, M. (2007). Participatory action research: Origins, approaches and methods. In Kindon, S., Pain, R., & Kesby, M. (Eds.), *Participatory Action Research Approaches and Methods Connecting People, Participation and Place* (pp. 9–18). Routledge. DOI: 10.4324/9780203933671-13

Kittay, E. (2015). Centering justice on dependency and recovering freedom. *Hypatia*, *30*(1), 285–291. DOI: 10.1111/hypa.12131

Kolawole, O. (2022). Is local knowledge peripheral? The future of indigenous knowledge in research and development. *Alternative*, *18*(1), 121–129. DOI: 10.1177/11771801221088667

Kuhn, T. (1970). *The structure of scientific revolutions*. Chicago Press: University of Chicago Press.

Lakso, L., & Adu, K. (2024). "The unofficial curriculum is where the real teaching takes":faculty experiences of decolonising the curriculum. *Higher Education*, *87*(1), 185–200. DOI: 10.1007/s10734-023-01000-4

le Grange, L. (2001). Challenges for participatory action research and indigneous knowledge in Africa. *Acta Academica*, *33*(3), 1–15.

Lovo, E., Woodward, L., Larkins, S., Preston, R., & Baba, U. (2021). Indigenous knowledge around the ethics of human research from the Oceania Region: A scoping Literature Review. *Philosophy, Ethics, and Humanities in Medicine; PEHM*, *16*(2), 1–14. DOI: 10.1186/s13010-021-00108-8 PMID: 34625108

Lugone, M. (2016). The coloniality of gender. In Harcourt, W. (Ed.), *The Palgrave Handbook Gender and Development* (pp. 13–33). Springer. DOI: 10.1007/978-1-137-38273-3_2

Lykes, M., Lloyd, C., & Nicholson, K. (2018). Participatory and action research within and beyond the academy: Contesting racism through decolonial praxis and teaching "against the grain". *American Journal of Community Psychology*, *62*(3-4), 406–418. DOI: 10.1002/ajcp.12290 PMID: 30479010

Magni, G. (2016). *Background paper prepared for the 2016 Global Education Monitoring Report*. UNESCO.

Mawere, M., & Mubaya. (2016). *African philosophies and thought systems: A search for a culture and philosophy of being*. Bameda: Langaa Research and Publishing CIG.

Medina, S., & Boharquez, Y. (2020). Modern and postmodern views of education that shaped EFL mentoring in the teaching practicum. *Colombian Applied Lingustic Journal Vo*, *22*(1), 1–5.

Mitrovic, L. (1999). New social paradigm Habermas's theory of communicative action. *The Scientific Journal of FACTA UNIVERSITATIS*, *6*(2), 217–223.

Molebatsi, M. (2006). The Africanisation of the university in Africa. *Alternation (Durban)*, *13*(1), 49–69.

Mpoe, J., & Swartz, L. (2019). Decolonising research methodologies: Lessons from a qualitative research project, Cape Town, South Africa. *Global Health Action*, *12*, 1–7. PMID: 30661464

Nel, P. (2005). Indigenous knowledge systems: Contestations rhetoric and space. *Indilinga*, *4*(1), 1–13.

Newcastle University. (2024, 05 2). *Participatory action research*. Retrieved from https://research.ncl.ac.uk/methodshub/methods/participatoryactionresearch/

Office for Huma Research Protection. (1979, 04 18). *The Belmont Report*. Retrieved from Ethical Principles and Guidelines for the Protection of Human Subjects of Research: https://www.hhs.gov/ohrp/regulations-and-policy/belmont-report/read-the-belmont-report/index.html

Omodan, B., & Dastile, N. (2023). Analysis of participatory action research as a decolonial. *Social Science*, *12*(507), 1–11.

Orlowski, P. (2019). Freirean conceptions of participatory action research and teaching for social justice -Same struggle different fronts. *Canadian Journal of Action Research*, *20*(1), 30–51. DOI: 10.33524/cjar.v20i1.445

Ozkul, O. (2020). Communication Action Theory: A system-lifeworld compatibility or incompatibility? *Journal of Social Sciences and Humanities*, •••, 215–237.

Parker, P., Smith, S., & Dennison, J. (2017). Decolonising the classroom creating and sustaining revolutionary spaces inside the academy. *TIJDSCHRIFT VOOR GENDERSTUDIES*, *20*(3), 233–247. DOI: 10.5117/TVGN2017.3.PARK

Parker, P., Smith, S., & Dennison, J. (2017). Decolonising the classroom Creating and sustaining revolutionary spaces inside the academy. *TIJDSCHRIFT VOOR GENDERSTUDIES VOL.*, *20*(3), 233–247. DOI: 10.5117/TVGN2017.3.PARK

Philo-Notes. (2023, 03 22). *Habermas's theory of communicative action: Key concepts*. Retrieved from Philo-Notes: http://philonotes.com/2023/03/habermass-theory-of-communicative-action-key-concepts

Pickering, L., & Kara, H. (2017). Presenting and representing others: Towards an ethics of engagement. *International Journal of Social Research Methodology, 20*(3), 1–15. DOI: 10.1080/13645579.2017.1287875

Planent Anomaly. (2023, 4 16). *Mapped: The World Indigenous Peoples*. Retrieved from Mapped: The world indigenous peoples: https://www.visualcapitalist.com/cp/mapped-the-worlds-indigenous-peoples/

Ritterbusch, A. E. (2019). Empathy at knifepoint: The dangers of research and;its pedagogies for social justice movements. *Antipode, 51*(4), 1296–1317. DOI: 10.1111/anti.12530

Roy, R. (2014). *The untold story of modern science is one of emoire and colonial exploitation*. Quartz India.

Sarivaara, E., Maatta, K., & Uusiautti, S. (2013). Who is indigenous? Definitions of indegeneity. *European Scientific Journal, 1*, 369–378.

Sarivana, E., Maatta, K., & Uusiautti, S. (2013). Who is indigenous? Defintions of indigeneity. *European Scientific Journal, 1*, 369–378.

Sarkhel, J. (2016). Strategies of indigenous knowledge in libraries. *Qualitative and Quantitative Methods in Libraries, 5*, 427–439.

Saunders, M., Lewis, P., & Thornhill, A. (2019). *Research methods for business students*. Pearson Professional Ltd.

Schaefer, M., Hans-Jachen, H., & Denke, C. (2013). Communuicative versus strategic rationality: Habermas Theory of Communicative Action and social brain. *PLoS One*, •••, 1–7.

Shava, S., & Manyike, T. (2018). The decolonial role of African indigenous languages and indigenous knowledges. *Indilinga, 17*(1), 36–52.

Shor, I., & Freire, P. (1987). *A pedagogy for liberation: Dialogues on transforming education*. Bergin & Garvey Publishers.

Siambombe, A., Mutale, Q., & Muzingili, T. (2018). Indigenous knowledge systems: A synthesis of Batonga people's traditional knowledge on weather dynamism. *African Journal of Social Work, 8*(2), 46–54.

Smith, L. (1999). *Decolonizing methodologies research and indegenous peoples*. Zed Books Ltd.

Solomon, P., & Singer, J. (2011). School reform as inequity: The casc of standards, standadization, and accountability. In Solomon, P., Campbell, S. J., Campbell, A., & Allen, A. (Eds.), *Brave new teachers doing social justice work in neoliberal times* (pp. 69–110). Canadian Scholras' Press.

South African San Institute. (2017). *San code of research ethics*. Retrieved 4 13, 2024, from https://trust-project.eu/wp-content/uploads

Stanford Encyclopedia of Philosophy. (2024). *Stanford encyclopedia of philosophy*. Retrieved from Jurgen Habermas: https://plato.stanford.edu/entries/habermas/#Biog

Stuart, C. (1998). Care and Concern: An ethical journey in participatory action research. *Canadian Journal of Counselling/Revue canadienne de counseling, Vol. 32:4*, 298-314.

Swantz, M. (2008). Participatory action research as practice. In Reason, P., & Bradbury, H. (Eds.), *The Sage Handbook of Action Research Participative Inquiry and Practice* (pp. 31–48). Sage Publications. DOI: 10.4135/9781848607934.n8

Tegama, N., & Fox, A. (2023). Ethics, the university, and society: Toward a decolonial approach to research ethics. *Alliance for African Partnership Perspectives*, *2*, 75–91.

Thiollent, M. (2011). Action research and participatory research: An overview. *International Journal of Action Research*, *7*(2), 160–174.

Tronto, J. (1993). *Moral Boundaries, a political argument for an ethic of care*. New York University Press.

Tronto, J. (2013). *Caring democracy*. New York University Press.

US Department of Education National Center for Education Statistics. (2016). *The condition of education 2016: Characteristics of postsecondary faculty*. USA: US Department of Education National Center for Education Statistics. Retrieved 4 15, 2024, from https://nces.ed.gov/programs/coe/indicator_csc.asp

Weheliye, A. (2014). *Habeas viscus: Racializing assemblages, biopolitics and black feminist theories of the human*. Duke University Press.

Wilmer, F. (1996). Domination and Resistance, exlusion and inclusion: Indigenous people's quest for peace and justice. *Peace Conflict Studies Vol 3 Article 4*, 1-9.

Wood, L., Dean, C., du Toit, P., Esau, O., James, A., Mokhele, P., & Pillay, A. (2017). A snapshot of action research in and by higher education in southern Africa. The Palgrave international handbook of action research, 221-234.

World Medical Association. (1964). *WMI Declaration of Helsinki-Ethical Principles for medical research involving human subjects*. Retrieved from WMI Declaration of Helsinki-Ethical Principles for medical Research involving human subjects: https://www.wma.net/policies-post/wma-declaration-of-helsinki-ethical-principles-for-medical-research-involving-human-subjects/

Wynter, S. (2003). Unsettling the coloniality of being/power/truth/freedom: Towards the human after. *CR (East Lansing, Mich.)*, *3*(3), 257–337. DOI: 10.1353/ncr.2004.0015

Yates, M. (2011). Postmetaphysical thinking. In Fultner, B. (Ed.), *Jurgen Habimas: Key concepts* (pp. 35–111). Routledge. DOI: 10.1017/UPO9781844654741.003

Zhang, Q. (2019). Postmodern view of knowledge and reorientation of teachers' role. *Advances in Social Science, Education and Humanities Research*, *350*, 189–191. DOI: 10.2991/aermt-19.2019.47

Zima, P. (2015). *Subjectivity and identity in between modernity and postmodernity*. Bloomsbury Academic.

KEY TERMS AND DEFINITIONS

Indigenous: People who were inhabitants to an area for about 1000 years
Auchthonous: Something that can be obtained from the local community
EMIC: Comparison of culture

Globalization: Interconnectedness of the world through information, technology and internet

Local Knowledge: Knowledge developed by indigenous communities

Modernity: Development of industrialization and enlightenment and accompanying inventions

Consciousness: Knowledge or awareness people develop which can lead them to improve their conditions

Chapter 18

Indigenous vs. Eurocentric Teaching Methods:

The Place of Afro-Centric Games in South African Primary School Mathematics Curriculum

Austin Musundire
https://orcid.org/0000-0001-8784-0616
University of Limpopo, South Africa

ABSTRACT

This paper addresses disparities in Mathematics teaching between Indigenous and Eurocentric methods in South African primary schools, aiming to improve mathematical performance in underprivileged settings. Through systematic literature review, action research, and action learning, it identifies Afro-centric games like nhodo, tsoro, and pada as effective tools for teaching basic math concepts. The study reflects the potential of integrating these games into the South African curriculum to enhance learning outcomes.

1. INTRODUCTION

Consultation of findings, recommendations and conclusions of published research articles reflect a continues debate that seem widening the controversy of Afro-centric and Eurocent methods of teaching mathematics without proposing theoretical and practical ways of closing it. This probably the reason why proponents of ethno mathematical theoretical perspective encourages researchers to identify cultural groups within the community in order to collect ideas ranging from distinct numerical and mathematical systems to multicultural mathematics education. In the context of this study, this would lead curriculum designers have an understanding of mathematics, and appreciate inclusion connections between different cultural groups at stake. Taking this into consideration, this study finds the place of African games and application of the Eurocentric Vygotsky's Social Constructivist Theory in the African context in South African primary school Mathematics in order to fulfil the principles of the ethno mathematic principles. The findings from the study are expected to contribute to theoretical and practical strategies of closing gaps between the controversies of n Eurocentric and Afrocentric methods of teaching. Special focus is

DOI: 10.4018/979-8-3693-3542-0.ch018

on addressing challenges associated with poor mathematical calculations in the four basic operations, addition, subtraction, multiplication and division specifically in underprivileged South African communities. Studied is in the field of mathematics agree that all the mathematical concepts at primary, secondary and tertiary are built upon these 4 operations which must be grasped at grassroots, the primary school curriculum.

1.1 Background to the Study

The integration of Indigenous knowledge systems within the frameworks of action research and action learning represents a vital connection of traditional wisdom and contemporary methodologies. Indigenous knowledge, rooted in centuries-old practices and deeply connected to local ecosystems and cultural heritage, offers unique perspectives on societal challenges and sustainable solutions. Action research and action learning, on the other hand, are dynamic approaches that emphasize collaboration, reflection, and transformative change. By bridging these two domains, this research seeks to explore how Indigenous knowledge can enrich and inform action-oriented research and learning processes, fostering greater cultural understanding, community empowerment, and holistic problem-solving. Other studies concluded that most African countries including South Africa still depend on European systems of designing, development and implementing of their national curriculum (Gumbo and Masila, 2015). Literature also shows that the South African curriculum is still influenced by Eurocentric methods of teaching and learning and research while the African content and context is being neglected (Watson-Vandiver & Wiggan, 2021). Within its narrow perspective, and focusing on the socio-history of the Black race, it serves as a "radical critique of the Eurocentric ideology that masquerades as a universal view ... ", and thus provides "an alternative perspective on phenomena" (Asante, 1987, p. 3). In view of this, the dependency theory in which put more focus on global in-equalities calls for substantive amount of research to be done in the art of socio-economic and political disparities (Romaniuk, 2017). It is perhaps unfortunate that during globalization, it has become necessary to come up with an intercultural understanding of knowledge from across civilizations and cultures in order to avoid a clash of societies (Gundara 2014). There is also a concern that where Eurocentric based curriculum is obligatory to the African continent, western curriculum philosophy is mistakenly perceived as more special than other curriculum ideologies (Nnama-Okechukwu & McLaughlin, 2022). This has a negative perspective towards the African people's culture (Hickling-Hudson & Ahlquist 2003, Battiste, 2019, Macedo, 2019). Despite the constitutional call for democracy after the apartheid error, the privileges of white education are prevalent in the South African education system. In fact, the largely black, indigenous people of South Africa prefer their children to attend former white dominated schools and then their township schools. This promotes and prolongs Eurocentric values of education in the country.

In an attempt to solve the current curriculum development and implementation challenges in the context of the South African Education system, there is a growing demand of eradicating the African Curriculum Dependent syndrome within the African Education systems.

This article will make use of the analysis of the South African educational numeracy strategies in order to come up with what Olivet and Jagusah (2001, p.113) termed "... the problematic of education and socio-cultural development in Africa." In line with the same scholarly focus, the intention of this study is also to challenge the perception that "...practices that are taken for granted as best practices elsewhere in education have little or no place in the educational policy and processes in Africa" (Olivet & Jagusah, 2001, p.113). "Issues such as the role of culture, the role of languages of instruction, the role of indigenous philosophy, and the role of the community in the education of its youth and citizens, are generally given little consideration" (Olivet & Jagusah, 2001, p.113). These aspects are usually dismissed as not necessary for the education of African (Nnama-Okechukwu & McLaughlin, 2022).

Battiste (2009) points out that Indigenous Knowledge (IK) is part of the combined genius of humanity of Indigenous societies which takes form when the people are learning and knowing the places where they have hunted, lived, journeyed, explored, tilled, built communities, looked after their families, and survived from time immemorial notwithstanding the inexorable bombardment on the culture and languages of such indigenous people.

1.1.1 Indigenous Versus Eurocentric Methods of Teaching Mathematics

This section refers to a collection of findings from researchers who used the qualitative research methodology. The findings indicate that Eurocentric thoughts have tended to regard indigenous knowledge as obsolete and/or inferior (Maweu, 2011; Shizha, 2013; Mehta & Henriksen 2022). Curriculum materials and mathematics textbooks included are written in English with the language of instruction being English as well from the fourth grade upwards. This is a second language and a challenge to most indigenous or rural learners and even to some teachers (Chikiwa, Westaway, & Graven, 2019; Barcelos & Ruohotie-Lyhty, 2018). This also often calls for teachers to engage in language switching or code switching. Additionally, most rural learners are unfamiliar with the examples used in these textbooks. At the same time, there are a variety of traditional African games that children/learners can use as indigenous ways of learning mathematics in their socio-cultural environment. In support of the use of indigenous games in learning (Bayeck, 2018, Mosimege 2020) maintains that games can be tools for cognitive development, social learning, as an alternative teaching strategy and as motivation in the mathematics classrooms. In these games, the scaffolding comes from the more skilled peers or members of society (Nyota & Mapara, 2008; Kajee & Sibanda, 2019; Zippert, Marshall, & Ramani). Through indigenous games such as "nhodo" in the Shona context, learners get the chance to practice cognitive skills such as counting, adding, subtracting multiplication and dividing until they master these skills (Madondo & Tsikira 2021; Siamonga, 2017; Tatira, 2014). The Euro-centric methods make use of modern electronic gadgets instead. Maweu (2011) argues the integration of indigenous knowledge into the mainstream knowledge systems as a way of showing that the Western/modern system is not the only viable one. Many current researchers seem to be focusing much on improvement mathematical skills in basic operations in terms of subtraction, addition, multiplication and division by intergarating modern technology games and other technological actives (Hoadley, 2012; Reddy, Zuze, Visser, Winnaar & Juan, 2015; Räsänen, 2019). They seem to forget that majority of primary schools are located in poor communities such as farms, locations, rural and semi-urban where technological resources are hardly accessed (Hoadley, 2012; Reddy, Zuze, Visser, Winnaar & Juan, 2015; Tandika & Ndijuye, 2019).

1.1.2 The Ethno Mathematics Theoretical Perspective in the Context of Eradicating the Eurocentric Curriculum Dependent Syndrome

These section comprise of findings from published articles that attempted to establish The Ethno mathematics theoretical perspective in the context of eradicating the Eurocentric Curriculum Dependent Syndrome using the qualitative research methods. The findings seem to recommend that the mentality of the Eurocentric dependency curriculum syndrome must be eradicated by African Education systems considering that the platform and power to that perception is already at their disposal. Although Eurocentric approaches regard indigenous knowledge as obsolete and/or inferior (Maweu, 2011; Shizha, 2013; Battiste, 2019; Zaslavsky, 1999), critical analysis of their characteristics from a different perspective clearly indicate full evidence of how they recognize the value of the African indigenous knowledge in curriculum development and implementation. For instance, ethnomathematics is a theoretical perspective developed in the last two decades that have shown the extent to which Eurocentric curriculum theorists acknowledge how powerful the African knowledge is through mathematical sciences and cross-cultural analysis (Kaino, 2017, p. 97; Mchombo, 2019; Rogers, 2003; Adama & Englas, 2002).

Despite efforts to undermine African capabilities of using mathematics in their lives, Zaslavsky (1999), points out that in Chinua Achebe's book, *Arrow of God* the author narrates that, the medicine man counted cowrie shells, "...carefully on the ground as a woman would before she bought or sold in the market in groups of six. There were four groups and he nodded his head (Zaslavsky, 1999, p. 147). From a historical perspective, traditional studies conducted by researchers like (Zaslavsky, 1999) came up with the following findings to signify the significance of Afrocentric methods in mathematical adopting basic numeracy skills:

- The Igbo are referred to as an ethnic group that was able to apply indigenous knowledge in developing numerical skills such as counting, addition, subtraction, multiplication and division using cowrie shells. six and four were important to their counting systems
- Some African groups were able to observe and reproduce patterns both geometrically and numerically in a mode of surviving. For example, the so-called Bushmmen of the Kalahari Desert in both South Africa and Namibia are able to locate a berry which they would have identified many weeks before without the aid of geographical systems.
- cattle herders in Africa were able to identify their cattle of markings on the hides to make distinction of them and also making reference to the shapes of their horns.
- Many numerical patterns were made when playing games which are similar to tic-tac-toe and the African board game which uses pebbles and seeds are captured (Zaslavsky, 1999).
- Even carvings and patterns on cloth, wood and pottery have some meaning.

Ethnomathematics, which is expressed through most indigenous games raises the need to transform the curriculum and teaching of Mathematics by way of respecting traditional knowledge (Kaino, 2017: 97; Nur, Waluya, Rochmad, & Wardono, 2020; Rogers, 2003; Adama & Englas, 2002). Many recent researchers agree that the important role of ehtnomathematics in the value of African knowledge is expressed through the use of African games (Kaino, 2017, p. 97; Rogers, 2003; Adama & Englas, 2002; Supiyati & Hanum, 2019). Many studies indicate that ethnomathematics is a theoretical perspective that deals with reflecting the important of tribal knowledge in Mathematics curricular (Mchombo, 2019). To support the a-fore mentioned assumption, Vygotsky (1978)'s social constructivist and socio-cultural from a Eurocentric perspective is one of the theories that is seen biased to Africanist way of teaching and learning. The theory accept that culture has an effect on how people live and learn. Every culture has a wide range of activities or practices that can be used to explain the mathematical concepts that are taught in schools (Nyota & Mapara, 2008; Howard, 2019). The African games reflect a society's cultural values (Siamonga, 2017; Tatira, 2014; Ejuu, 2019). In explaining the socio-cultural theory of learning Vygotsky (1978) stated that human cognition and learning are more social/cultural than individual phenomena. An individual makes socio-cultural meanings from the activities and actions of others, using signs, symbols and tools. In one way or the other, Vygotsky's theory acknowledges how significant the African indigenous is during curriculum development and implementation. Vygosky further claims that social interaction in a variety of settings is an important component in the development of knowledge – and this includes mathematical knowledge. This social interaction occurs between learners, teachers, with other learners and with other learning materials and the whole community at large. Vygotsky further argued that the concepts discussed in these social settings can be internalized or added to what an individual already knows and that this 'transfer' from the external to the personal can only take place in each learner's zone of proximal development (ZPD). The ZPD is 'the distance between what one can achieve alone and what one can achieve with help' (Eun, 2019; Peer and McClendon 2002: 137; Sellberg, Lundin & Säljö, 2021). Vygotsky then coined the term 'scaffolding' to refer to a pulling or raising of a learner to a higher level of cognition which occurs in the ZPD. This connection-making process is achieved through help from the teacher or more capable peers as they explain or demonstrate the process or task.

Analysis of the above literature indicates that African Ethonomathamatics theoretical is a possible tool that can be used to eradicating the Eurocentric Curriculum Dependent Syndrome. It also gives room for Africanising the teaching and learning theories in the African way with relevant African terminologies. There is evidence from other Eurocentric teaching and learning theories such as Vygotsky's theory of teaching and learning that seem to fully support the power of the African indigenous knowledge. This also seem a privilege of Africanising curriculum theories and curriculum content.

1.1.3 Mathematical Instructional Challenges in South African Under Privileged Schools

This sections discusses findings from a collection of studies that used the quantitative research methodology to establish Mathematical instructional challenges in South African under privileged Schools by way of statistical compassions. In 2011, the Southern and Eastern African Consortium for Monitoring Education Quality (SACMEQ III) conducted numeracy tests whereby South African average performance in primary schools was 4th out of 15 African countries (Spaull, 2011; Rule & land, 2017: 2). The South African Annual National Assessment (ANA) was established as a strategy to improve mathematical performance. In 2014, the results were far below 50% across all the primary school grades because of lack of knowledge and understanding of basic mathematical operations particularly in rural and farm schools (Spaull, 2011; Rule & land, 2017: 2; Department of Basic Education, 2014).

Form a provincial perspective, the Gauteng Department of Education started by initiating the Gauteng Primary Language Strategy in 2010 (DoBE, 2014) focusing on improving the Annual National Assessment (ANA) results by addressing the poor language performance. By April 2012, the term "Mathematics" was added to read as "Gauteng Primary Language and Mathematics Strategy (GPLMS)" after research findings established that acquisition of linguistic skills is pre requisite of grasping mathematical language in one way or the other (DoBE, 2014, Graven, 2015; Hoadley, 2012; Maharajh & Chilli 2021; Reddy, Zuze, Visser, Winnaar & Juan, 2015; Stott & Graven, 2013; Venkat & Naidoo, 2012). The drop in the pass rate is said to be due to the change in language of instruction (Hoadley, 2012; Reddy, Zuze, Visser, Winnaar & Juan, 2015). Thus, to researchers, the major barrier to understanding of the basic operations comprising of addition, subtraction, multiplication and division was the use of English as a language of instruction from indigenous languages (Hoadley, 2012; Maharajh & Chilli 2021; Reddy, Zuze, Visser, Winnaar & Juan, 2015). It can still be argued that the decline in learners' performance was not only caused by the introduction of English but failure to value teaching mathematics concepts from an African perspective using indigenous languages (Graven, 2015; Hoadley, 2012; Mbirimi-Hungwe, 2019). Hence, it can be argued that the Eurocentric language, English, used to teach the learners is not all about what they experience in their socio-cultural environments (Chikiwa, 2019). The failure to appropriately use the idea of developing lessons from the African indigenous knowledge by means of interactive activities appears to have caused the drop in performance, hence the need to use African traditional games as a way of scaffolding and helping the learners to get into their zone of proximal development (Hoadley, 2012; Reddy, Zuze, Visser, Winnaar & Juan, 2015). African Games are interactive, language rich and draw heavily from Montessori's strategies and social cognitivist theory by leading scholars in this subjacent area (Aberšek, 2019; Vygotsky, 1978; & Grabowski, 2007; Antony, 2002, p. 179). There is also consensus among other researches that children enjoy social games that involve music, dance and movement as well as those that involve playing with concrete objects (Ke & Grabowski, 2007; Harris, 2009; Nabie, 2015; Nyota and Mapara, 2008).

Basing on the synthesis of the above findings, there is an assumption that there is a link between mathematical calculations and the integration of these indigenous African games in the curriculum (Ke & Grabowski, 2007; Harris, 2009; Peng, Lin, Ünal, Lee, Namkung, Chow & Sales, 2020). Hence, it is important to help learners to realize that indigenous games play a role in their everyday lives as they apply it to solve their everyday mathematics problems (Antony, 2002, p.179; Setiyadi, Zaenuri, & Mulyono, 2018). The process of addition, subtraction and multiplication and division develop out of social activities that are developed through social interaction (Vygotsky, 1978).

The GPLMS comprised 830 underperforming primary schools in Gauteng province which after through research and statistical comparison were identified among those underperforming schools located in poorly resourced environments such as townships, farms and other informal settlements. These schools did not have enough provision in terms of electricity, computers, internet connections and other Eurocentric mathematical gadgets to improve their basic mathematical skills. The GPLMS strategy was designed to ensure that at least 60% of learners in the province perform at 50% and above by 2014 (DoBE, 2014). The GPLMS coaches focused on direct training and support for teachers in the 830 under performing schools where one of the strategies was the traditional CRC approach characterised by use of teacher and child made concrete objects during the teaching and learning process (DoBE, 2014). Although the approach to a certain extent improved the language and mathematical skills in primary schools, there was still a demand for continuous implementation of the programme that suited the context of rural, farm schools as well as other disadvantaged communities where the Eurocentric technological devices were a scarce resource. The argument is that lessons involving song, dance, play and manipulation of concrete material will produce greater mathematical achievement than where lessons in which manipulative materials are not used particularly where computer-based manipulatives are lacking (Ke & Grabowski, 2007; Pangrazi & Beighle 2019; Suryaratri, Prayitno, & Wuryani, 2019). Even at home children learn through observation, interaction and play (Gomes & Fleer, 2019; Ke & Grabowski, 2007; Harris, 2009). This is not only part of the African tradition but part of every other culture in the world (Ke & Grabowski, 2007; Taylor & Leung 2020). This is one of the reasons why this study examines the role played by African traditional games in developing concepts of addition, subtraction, multiplication and division. This approach is based on the analysis which was carried out by the International Mathematics Union (IMU) which found out that mathematical performance is weak in most African countries which in turn affects student academic future particularly at universities (IMU, 2020).

1.1.4 The Principles and Phases of Action Research in the Context of Systematic Literature Research

This study applies the Principles and Phases of Action Research in the context of systematic literature research. Action research is characterized by its participatory nature, iterative cycles of planning, action, observation, and reflection. Recent contributions by Creswell and Creswell (2021) have highlighted key principles such as democratic collaboration, self-reflection, and collective problem-solving within action research frameworks. The phases typically include identifying a research question or problem, planning and implementing interventions collaboratively, observing outcomes, reflecting on the process, and iteratively refining strategies based on feedback and learning. Action learning emphasizes individual and collective learning through action and reflection. Scholars like Marquardt and Waddill (2020) have contributed to the core concepts of action learning, including the set problem, action, reflection, and learning. Participants engage in real-world projects or challenges, take action, reflect on their experiences and insights, and apply new knowledge to improve practice. Action learning groups foster peer support, knowledge sharing, and continuous improvement, enhancing individual and organizational learning capacities.

There are Linkages between Action-Oriented Methodologies and Indigenous Knowledge Systems. The principles and processes of action research and action learning resonate with Indigenous knowledge systems in several ways. Both emphasize community collaboration, holistic perspectives, and experiential learning. Recent research by McGregor, Restoule, and Johnston (2020) explores the synergies between action-oriented methodologies and Indigenous knowledge, highlighting the value of integrating Indigenous ways of knowing into research and learning processes. Indigenous knowledge systems, with their emphasis on reciprocity, interconnection, and respect for the land, offer insights into sustainable practices, community empowerment, and cultural revitalization, aligning with the transformative aims of action-oriented approaches. Below is the discussion and analysis and interpretation of the study findings in the context of systematic literature study, action –oriented methodologies and indigenous and Eurocentric approaches to teaching and learning of Mathematical concepts.

1.2 Types of African games and Application of the Vygotsky's Social Constructivist Theory in the African Context

Three African games will be discussed and analysed in line with the current implementation of the South African Curriculum Assessment Policy Statements (CAPS) (2012) for the primary school mathematics subject. As highlighted before. Africans had their ways of making use of mathematics training which help learners to assimilate learning in class. The identified games are nhodo, tsoro and pada. Nhodo is the Shona name, and it is known as diketo in Tshwana and jacks in English (Mogege, 2017; Mosimege, 2012). Below are synthesised findings from qualitative research methods by means of Action research, a method used for improving practice. It involves action, evaluation, and critical reflection and – based on the evidence gathered – changes in practice are

2. ANALYSIS, INTERPRETATION AND DISCUSSION OF THE FINDINGS IN COLLABORATION WITH

The analysis, interpretation and discussion of the findings were done in the following criteria per each identified game:

- Description of the game
- Rules of the game
- Mathematical implications
- Application of Vygotsky's Social Constructivist Theory
- incorporating use of Afro-centric games into the South African Math Curriculum and Assessment Policy Statement (CAPS).

2.1 Explanation of the African Traditional Game "Nhodo"

Figure 1. Illustration of how to play the game "Nhodo"

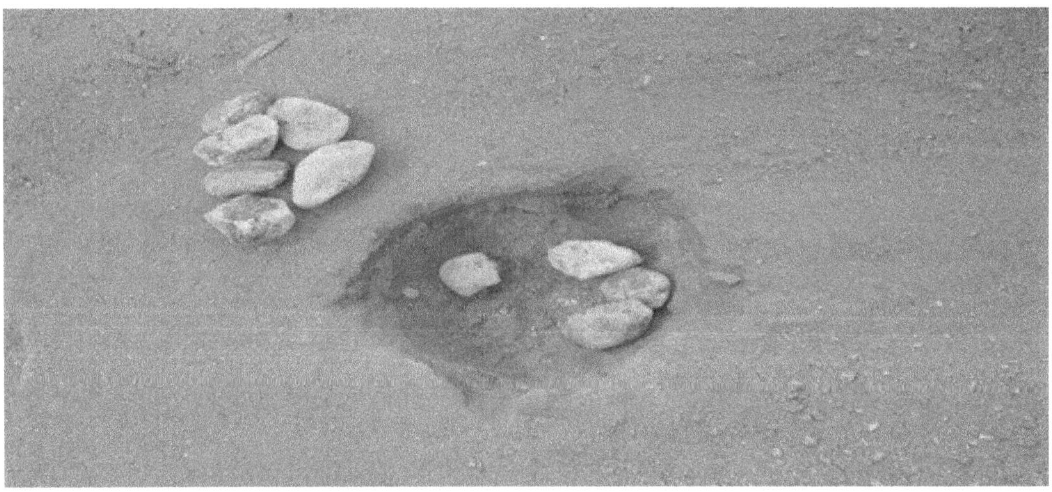

Figure 1 shows how children play diteko (Tshwana) or nhodo (Shona) or jacks (English). What one needs in order to play nhodo is just to dig a shallow hole on the ground about 5 – 10cm deep and the children gather 6-10 pebbles or seeds (Mugari & Matemera, 2016).

2.1.1 Description

The equivalent or approximate term for nhodo (Shona) is diketo in Tswana and Jacks in English (Mogege, 2017; Mosimege, 2012). Nhodo/diketo/jacks is mostly played by children who live in a resource-poor environment (Burnett, 2018; Ketsitlile, 2011; Tshube, & Hanrahan, 2016). Such children are exposed to minimal materials that they can play with unlike wealthy environments that can make use of technology to play electronic games (Mugari & Matemera, 2016). Nhodo/diketo/jacks is a game played mostly by young girls although nowadays boys have joined in (Mugari & Matemera 2016). Those playing the game have to sit in a circular formation with their legs sometimes crossed (Mugari & Matemera 2016). In the centre will be a hole with at least 6 to 30 small stones or seeds (Mugari & Matemera 2016). The one playing will have a much bigger stone (Nyota & Mapara, 2008). Play involves throwing the main big stone up, scooping all the stones from the hole and catching the main big stone before it hits the ground (Nyota & Mapara, 2008, Mugari & Matemera, 2016). This stone is thrown up again, and the player returns one stone at a time into the hole until all stones are back in the groove (Tatira, 2014). The player must not let the main big stone hit the ground (Nyota & Mapara, 2008). The second round will involve scooping the stones out, then returning 2 stones at a time until they are all in the hole. In round three all stones are scooped out, then return 3 at a time, and so on (Nyota & Mapara, 2008). A player can lose the game in one of three ways: 1) when the player drops the main big stone. 2) When the player fails to scoop all the stones out of the hole in one scoop. 3) The player fails to return the correct number of stones into the shallow holes (Nyota & Mapara, 2008).

2.1.2 Rules

Participants take turns to play the game. Every player is supposed to throw the big stone up, remove the small stones from the groove and catch the thrown stone again with the same hand without dropping the stone to the ground (Mugari & Matemera 2016). One the sequence of activities is violated, the next person takes over while the rest are watching and strictly observing if the rules are being followed (Mugari & Matemera 2016). There is need for coordinating skills between the eye sight and hand.

2.1.3 Mathematical Implications

Observation of children playing Nhodo clearly indicates a natural development of mathematical concepts and application of counting, addition, subtraction, and multiplication (Nyota and Mapara, 2008). The players sit around the groove. Thus the game helps its players to develop their numeracy/cognitive and psycho-motor skills while also entertaining themselves. In support of the use of games in mathematics teaching, Rutherford (2015) contends that learners can discern new patterns and relationships, develop their computational fluency, and build a deeper understanding of concepts. For example, a player will eventually know that if you scoop six stones out of the hole and you have to return 2 at a time then you must do it three times so that all stones are back in the hole. Mathematical concepts and skills involve counting in sets and count on (addition), sharing and repeated subtraction (Division), grouping in 2s, 3s, 4s, 5s and etc. (multiplication).

2.1.4 Application of Vygotsky's Social Constructivist Theory

The implication of literature analysis is that teachers can integrate game "nhodo" in their Mathematics curriculum by way of applying the principles of Vygotsky's Social Constructivist Theory whereby learners are expected to learn mathematical concepts through interactions and communications with others whilst they play the game (Vygotsky, (1962). Vygotsky (1962) argues that our social environments influence the learning process. The educators may also use the game to teach such concepts as addition, subtraction, multiplication and addition in accordance to the social learning theory which suggests that learning takes place through the interactions students have with their peers, teachers, and other experts as highlighted before (Vygotsky, 1962). Consequently, teachers can create a learning environment that maximizes the learner's ability to interact with each other through discussion, collaboration, and feedback Vygotsky (1962). Moreover, Vygotsky (1962) argues that culture is the primary determining factor for knowledge construction. We learn through this cultural lens by interacting with others and following the rules, skills, and abilities shaped by our culture (Madondo, & Tsikira, 2021; Vygotsky, 1962). Social Learning Theory initiates development of knowledge of language and mathematical skills together with mathematical habits of the mind.

2.1.5 Incorporating Use of the "Nhodo" Game Into the South African Math Curriculum and Assessment Policy Statement (CAPS)

Table 1. The place of "nhodo" game into the South African Curriculum and Assessment Policy Statement (CAPS)

Name of African traditional game	Topic according to CAPS	Skills	Questions based on the CAPS Document
Nhodo	Numbers, Operations and Relationships	repeated addition and subtraction	45÷9 =45-9=36 =36-9=27 =27-9=18 =18-9=9 =9-9 =0 How many times has 9 being subtracted

From an ethno mathematical perspective. incorporating use of the "nhodo" game into the South African Math Curriculum and Assessment Policy Statement (CAPS) seem to support Vygotsky (1978)'s social constructivist and socio-cultural by way of incorporating a Eurocentric perspectives of addition, subtraction and division with the Africanised Nhodo Method. The theory accept that culture has an effect on how people live and learn. It is therefore important to note that every culture has a wide range of activities or practices that can be used to explain the mathematical concepts that are taught in schools

2.2 Playing Tsoro

Figure 2. Illustration of how to play the game "Tsoro"

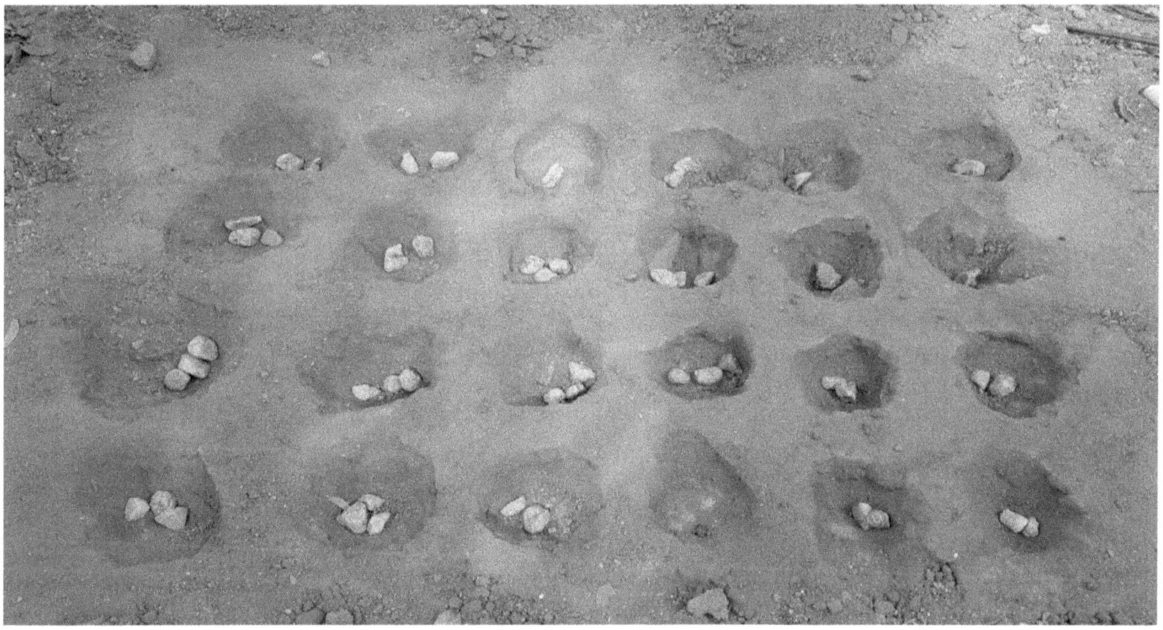

Apparatus to play tsoro are 84 stones or pebbles: (Kumari, Ncube & Munapo, 2003; Mugari & Matemera 2016)

2.2.1 Players

The game is played by two people at a time.

2.2.2 Description

From a traditional point of view, tsoro was played by adults long ago particularly warriors as a strategy of improving their prowess, marauding and seizing techniques of their enemies (Kumar, Ncube & Munapo, 2003; Mugari & Matemera 2016). This has been converted into a leisure game by both children and adults (Kumari, Ncube & Munapo, 2003; Mugari & Matemera 2016). Under normal circumstances, each of the two players occupy two rows of holes (Mugari & Matemera 2016). Before the start of the game, players ensure that all holes are filled with the same number of pebbles or seeds while they sit opposite each other. These pebbles are referred to as vana in Shona. Vana means children. Each player takes charge of the two nearest holes to him or her (Kumari, Ncube & Munapo, 2003. The game is flexible as the number of holes can be been currently modified according to the needs of the players. Below is a representation of a 10 holed tsoro board with 2 seeds in each shallow hole at the beginning of a game (Kumari, Ncube & Munapo, 2003; Mugari & Matemera 2016). "The players put two stones in each of the twenty-four ruts" (Jesuit Missions, 2014. p .3; Mawere, 2012). The groves are dug into the ground and should be shallow (Mugari & Matemera 2016). The first player picks up stones from a groove of choice. Then s/he goes clockwise dropping one stone in each one of the following grooves (Mugari & Matemera 2016). If the last stone or pebble is dropped in a groove in the front line, the player collects the stones in the grooves opposite hers/his, of the other player. If player A's stones end in the rear line, then it is the other player's turn. If the stones end in front line in a groove which is empty, then player B continues (Jesuit Missions, 2014. p .3; Kumari, Ncube & Munapo, 2003; Mugari & Matemera 2016).

2.2.3 Rules of Playing Tsoro

The player stops playing if the last stone is dropped in an empty groove (Kumari, Ncube & Munapo, 2003; Mugari & Matemera 2016). The game is over when all seeds or pebbles are collected from the opponent's groove (Kumari, Ncube & Munapo, 2003; Mugari & Matemera 2016). That is mean that the opponent loses once his or her opponent has collected their "children" and taken to them to become part of theirs.

2.2.4 Mathematical Implications

Just like nhodo above, mathematical concepts and skills involve counting in sets and count on (addition), sharing (Division), grouping in 2s, 3s, 4s, 5s and etc. (multiplication). This is related to the example of the counting and grouping in sets by the Igbo people when they count cowrie shells in Nigeria.

2.2.5 Application of Vygotsky's Social Learning Theory

The implication of the game "Tsoro" in the mathematical school curriculum is that it creates a social context that allows for collaboration on mathematical combinations that making it possible to solve problems and tasks that build a child's skills in language, mathematics and experience shaped by their individual cultures as advocated by Vygotsky (1978). Vygotsky also argued that language is the main tool which is used to develop reasoning, promote thinking, and support educational activities like reading and writing and performing numerical calculations (Vygotsky 1978). As highlighted above, Vygotsky further argued that the concepts discussed in these social settings can be internalized or added to what individuals already know and that this 'transfer' from the external to the personal can only take place in each learner's zone of proximal development (ZPD). The ZPD is 'the distance between what one can achieve alone and what one can achieve with help' (Peer & McClendon 2002, p. 137; Sellberg, Lundin & Säljö, 2021). Therefore, the application of the principles of Vygotsky's Social Learning Theory initiates development of knowledge of language and mathematical skills together with mathematical habits of the mind.

Below is a synthesis analysis of the findings of the following researchers who used the qualitative research methods (Kumari, Ncube & Munapo, 2003; Mugari & Matemera 2016):

2.2.6 Incorporating Use of the "Tsoro" Game Into the South African Math Curriculum and Assessment Policy Statement (CAPS)

Table 2. The place of "tsoro" into the South African Curriculum and Assessment Policy (CAPS)

Name of African traditional game	Topic	Skills	Questions that can be asked based on caps document
Tsoro/ draft	Numbers, operations and relationships	Dividing, sharing and sets	1.How many sets of 4 can be created from 20 2.20÷ 5 3.Share 20 items among 5 people 4.Write a number sentence to describe the game

From an ethno mathematical perspective. incorporating use of the "nhodo" game into the South African Math Curriculum and Assessment Policy Statement (CAPS) seem to support Vygotsky (1978)'s social constructivist by way of incorporating a Eurocentric perspectives of division with the Africanised Nhodo Method.

2.3 Explanation of the African Traditional Game "Pada"

In playing pada, the apparatus used is just a flat object which can be a stone, a piece of asbestos o any object really. It can be circular, oval or rectangular. It must be flat so that it can be dragged on the ground using feet. (Jesuit Missions, 2014. p .3)

2.3.1 Players

The game can be played by a combination of a number of players. Two teams of three to five players each game (Jesuit Missions, 2014. p .3). But the number of teams can be increased up to seven; however, the number of players in each team should become less. It is not uncommon to see two children playing this game (Jesuit Missions, 2014. p .3).

Figure 3. Illustration of how to play the game "Pada"

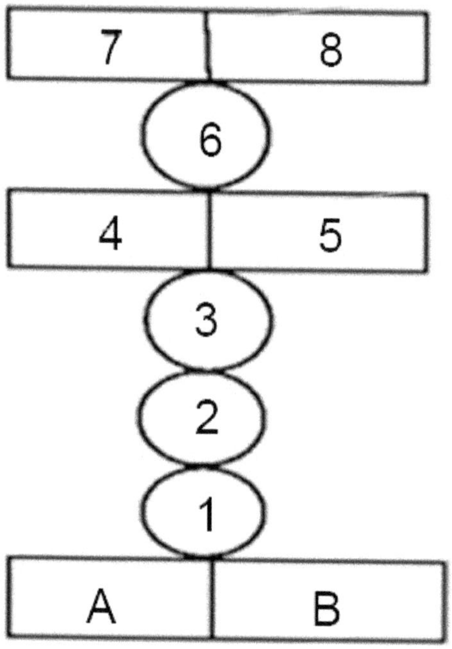

2.3.2 Playing Sequence

A case study conducted by Jesuit Missions (2014) at St. Rupert Mayer Primary School in Zimbabwe gave the following illustration of the playing sequence of "Pada" supported by other researchers like Mugari and Matemera (2015) as linked to figure 3 above:

- The first player stands in the first two boxes A and B with both feet astride facing the circles in front of them in the following sequence and throws the flat stone in the circle marked 1.
- The player hops over the circle into circle 2 with one leg as well as in circle 3
- The player hops into boxes 4 and 5 legs astride at the same time
- The player then hops into circle 6 with one leg, then with two in rectangles 7 and 8 astride.
- The player hops turning around in 7 and 8 with feet astride and hops back to the starting boxes A and B.
- When the player reaches circle 2 he/she balances and stands still on one leg while bending down to pick up the flat stone in box and hops over circles 1 and steps on two feet astride into the starting boxes A and B.
- The player turns and faces the boxes as before and throws the stone in circle 2 and repeats the process.

- After throwing in circle 3 and so forth up to box 8, he/she hops through all the boxes, picks the stone in 8 and returns to the starting boxes A and B. Here the game ends for the player (Jesuit Missions, 2014:3; Kumari, Ncube & Munapo, 2003; Mugari & Matemera 2015).

2.3.3 Rules

- The player is not allowed to hop into the box which contains the stone. If that is done, the player automatically loses the game (Kumari, Ncube & Munapo, 2003; Mugari & Matemera 2015).
- Secondly, the thrown stone must not land on the marked line of the box or go outside the box. When that happens, the player is penalized and loses the chance to the next player who then takes over (Kumari, Ncube & Munapo, 2003; Mugari & Matemera 2015).

Stepping on a marking is a penalty. If a player is out, s/he has to wait his turn. The team, whose players finish first, has won.

2.3.4 Mathematical Implications

It can be noted that as the child through a stone in the first circle to the second up to the 8th, there is counting that happens. As the child skips back to the original box from 8, there is count down. This is all about addition and subtraction.

2.3.5 Application of Vygotsky's Theory

As highlighted in other games above, the educators may use the pada game to teach concepts like addition, subtraction, by way of following the principles of the social learning theory. This involves the educator creating a social environment whereby students interact with their peers, during game playing. Consequently, teachers can create a learning environment that maximizes the learner's ability to interact with each other through conversation, collaboration, and feedback. Vygotsky (1978) has already confirmed that learners interact using language at the level of their peers to share ideas and create knowledge which they will use to solve problems (Van de Walle, 2007:22). When learners interact with more knowledgeable peers they go into the ZPD. This is the distance between what they know and what they would acquire with the help of more knowledgeable peers or adults

2.3.6 Incorporating Use of the "Pada" Game Into the South African Math Curriculum and Assessment Policy Statement (CAPS)

Table 3. The place of "pada" into the Siuth African Curricullum and Assrssment Policy (CAPS)

Name of African traditional game	Topic according to CAPS	Skills	Questions based on the CAPS Document
Pada	Numbers, Operations and Relationships	Counting in sets in relationship to addition, subtraction and multiplication	1.Counting backwards and forwards in sets of 4,5,3or 6 2.Find the missing number 3, 4 X 5 4.5 sets of 4 5.Draw a number line to show the movement

Results of table 3 confirm that from an ethno mathematical perspective. incorporating use of the "Pada" game into the South African Math Curriculum and Assessment Policy Statement (CAPS) seem to support Vygotsky (1978)'s social constructivist and socio-cultural by way of incorporating a Eurocentric perspectives of multiplication with the Africanised Pada Method.

3. SUMMARY OF THE FINDINGS

Research synthesis methods used for qualitative research studies were employed in order to identify three broad themes and related subthemes across the studies. These themes comprise the following: (1) Mathematical instructional challenges in South African under privileged Schools (2) Indigenous Africa games versus Euro-centric Methods of Math teaching (3) Implications of African games on application of the Vygotsky's Social Constructivist. The Study found that lack of inclusion of African games as a teaching methodology of the 4 basic operations is one of the challenges affecting mathematical performance in South African underprivileged primary schools. Curriculum imbalances between Afrocentric and Eurocentric approaches to curriculum development and implementation of the primary school Math subject were identified. Eurocentric methods are dominating and suppressing indigenous knowledge. The study revealed a close connection between Indigenous Africa games, Vygotsky's social constructivist and learners' readiness of grasping addition, subtraction, multiplication, and division skills in line with the South African Curriculum Assessment Policy (CAPS).

4 CONCLUSION

The article has shown that the integration of indigenous knowledge into the mainstream knowledge systems can assist in helping learners acquire skills to tackle mathematical problems. The case study has demonstrated that the four basic operations subtraction, addition, multiplication and division can be enriched by integrating modern technology games and other technological activities which come with

African traditional games. Afro-centric games are an effective way of teaching basic concept from an indigenous knowledge systems perspective in South African underprivileged primary schools

5 RECOMMENDATIONS

5.1 African games can be used as a tool to improve learner performance in the four basic mathematical calculations and operations. As such, they should be incorporated into the teacher training curriculum to assist learners who struggle to understand concepts such as division, multiplication, addition and subtraction.
5.2 Contexts from African games can be used to explain mathematical terminology to enhance transition of learners from the use of the vernacular languages to the use of English as a medium of instruction.
5.3 African games promote collaborative learning and sharing of ideas among learners and should be adopted for use in the mathematics curriculum. By so doing we will be integrating African indigenous knowledge and Western/Euro-centric knowledge systems in a mutually beneficial way
5.4 Traditional African games can also be used to enhance interactive learning if they are harnessed and computerized. Those learners with access to computers or other forms of technology can play and consolidate their understanding of mathematics.

6 CONTRIBUTION OF THE STUDY

This contributes to the existing body of knowledge in Ethno mathematics perspective's designed to make a balance between Afrocentric and Eurocentric methods of teaching Mathematics.

REFERENCES

Aberšek, B. (2019). *The impact of innovative ICT education and AI on the pedagogical paradigm*. Cambridge Scholars Publishing.

Aderajew, A. (2019). African systems of thought: Whether they fit scientific knowledge. *Journal of Philosophy and Culture*, *7*(2), 14–25.

Asante, M. K. (1991). The Afrocentric idea in education. *The Journal of Negro Education*, *60*(2), 170–180. DOI: 10.2307/2295608

Barcelos, A. M. F., & Ruohotie-Lyhty, M. (2018). *Teachers' Emotions and Beliefs in Second Language Teaching:* Implications for Teacher Education. In Agudo, J. D. D. M. (Ed.), *Emotions in Second Language Teaching* (pp. 109–124). Springer International Publishing. DOI: 10.1007/978-3-319-75438-3_7

Battiste, M. (2009). Naturalizing Indigenous knowledge in Eurocentric education. *Canadian Journal of Native Education*, *32*(1).

Battiste, M. (2019). *Decolonizing education: Nourishing the learning spirit*. UBC press. DOI: 10.59962/9781895830972

Bayeck, R. Y. (2018). A review of five African board games: Is there any educational potential? *Cambridge Journal of Education*, *48*(5), 533–552. DOI: 10.1080/0305764X.2017.1371671

Burnett, C. (2018). Traditional sports and games in Eastern, Central and Southern Africa. In *Sport in the African World* (pp. 121–145). Routledge. DOI: 10.4324/9781351212755-8

CAPS. (2014). *Foundations phase*. Department of Basic Education.

Chikiwa, S., Westaway, L., & Graven, M. (2019). What mathematics knowledge for teaching is used by a Grade 2 teacher when teaching counting. *South African Journal of Childhood Education*, *9*(1), 567. DOI: 10.4102/sajce.v9i1.567

Creswell, J. D., & Creswell, J. W. (2021). Key principles of action research: democratic collaboration, self-reflection, and collective problem-solving. In *Action Research in Systematic Literature Research* (pp. 25–46). Sage Publications.

Department of Basic Education. (2014). *Report on the annual national assessment of 2014: Grades 1–6 and 9*. Department of Basic Education.

Ejuu, G. (2019). African indigenous games: Using Bame Nsamenang's Africentric thoughts to reflect on our heritage, pedagogy, and practice in a global village. *Journal of Psychology in Africa*, *29*(4), 319–327. DOI: 10.1080/14330237.2019.1647496

Eun, B. (2019). The zone of proximal development as an overarching concept: A framework for synthesizing Vygotsky's theories. *Educational Philosophy and Theory*, *51*(1), 18–30. DOI: 10.1080/00131857.2017.1421941

Gomes, J., & Fleer, M. (2019). The development of a scientific motive: How preschool Science and home play reciprocally contribute to Science learning. *Research in Science Education*, *49*(2), 613–634. DOI: 10.1007/s11165-017-9631-5

Graven, M., (2015), *Strengthening math's learning dispositions through math clubs*. South African Journal of Childhood Education, VOL 5.

Gundara, J. S. (2014). Global and civilisational knowledge: Eurocentrism, intercultural education and civic engagements. *Intercultural Education*, *25*(2), 114–127. DOI: 10.1080/14675986.2014.888802

Hickling-Hudson, A., & Ahlquist, R. (2003). Contesting the curriculum in the schooling of Indigenous children in Australia and the United States: From Eurocentrism to culturally powerful pedagogies. *Comparative Education Review*, *47*(1), 64–89. DOI: 10.1086/345837

Hoadley, U. (2012). What do we know about teaching and learning in South African primary schools? *Education as Change*, *16*(2), 187–202. DOI: 10.1080/16823206.2012.745725

Howard, T. C. (2019). *Why race and culture matter in schools: Closing the achievement gap in America's classrooms*. Teachers College Press.

International Mathematics Union (IMU). (2020). International non-governmental and non-profit scientific organization. Retrieved from http://www.mathunion.org

Jagusah, O. I. W. (2001). Educational policy in Africa and the issue(s) of context: The case of Nigeria and South Africa, International Education Journal, 2(5). *WCCES Commission*, *6*(special), 2001.

Jakovljevic, M., Bjedov, S., Jaksic, N., & Jakovljevic, I. (2020). COVID-19 pandemia and public and global mental health from the perspective of global health security. *Psychiatria Danubina*, *32*(1), 6–14. DOI: 10.24869/psyd.2020.6 PMID: 32303023

Jesuit Missions. (2014) Hwishu Pada, Tsoro Nhodo and Bvutu Rakaraka, Zimbabwean Children Games Played By The children of the St. Rupert Mayer Primary School. https://jesuitmissions.org.uk

Kajee, L., & Sibanda, R. (2019). Home as a primary space: Exploring out-of-school literacy practices in early childhood education in a township in South Africa. *South African Journal of Childhood Education*, *9*(1), 1–10.

Ke, F., & Grabowski, B. (2007). Game playing for maths learning: Cooperative or not? *British Journal of Educational Technology*, *38*(2), 249–259. DOI: 10.1111/j.1467-8535.2006.00593.x

Ketsitlile, L. E. (2011). San Junior Secondary Students' Home–School Literacy Disconnection: A Case Study of a Remote Area Dweller School in Botswana. *Diaspora, Indigenous, and Minority Education*, *5*(2), 88–99. DOI: 10.1080/15595692.2011.559773

Kimberly Sue Peer and Ronald C McClendon. (2002). Sociocultural Learning Theory in Practice: Implications for Athletic Training Educators. *Journal of Athletic Training*. PMID: 12937534

Kumari, S., Ncube, O., & Munapo, E. (2003). Tsoro and Hungarian Approaches: A Hybrid Algorithm for An Assignment Problem. *Asia-Pacific Journal of Operational Research*, *20*, 41–56.

Macedo, D. (2019). *Decolonizing foreign language education. The Misteaching of English and Other Colonial Languages*. Routledge.

Madondo, F., & Tsikira, J. (2021). Traditional children's games: Their relevance on skills development among rural zimbabwean children age 3–8 years. *Journal of Research in Childhood Education*, •••, 1–15.

Maharajh, L. R., & Chilli, B. (2021). English as an Alien Language: Teachers' Experiences of the English Across the Curriculum Strategy in South Africa.

Marquardt, M. J., & Waddill, D. (2020). Core concepts of action learning: set problem, action, reflection, and learning. In *Action Learning for Improving Practice* (pp. 67–89). Daves –Black Publishers.

Marquardt, M. J., & Waddill, D. (2020). Core concepts of action learning: set problem, action, reflection, and learning. In *Action Learning for Improving Practice* (pp. 67–89). Daves –Black Publishers.

Maweu, J. M. (2011, December). *Indigenous Ecological Knowledge and Modern Western Ecological Knowledge: Complementary, not Contradictory*, A Journal of the Philosophical Association of Kenya (PAK). *New Series*, *3*(2), 35–47.

Mbirimi-Hungwe, V. (2019). Stepping beyond linguistic boundaries in multilingual Science education: Lecturer's perceptions of the use of translanguaging. *Southern African Linguistics and Applied Language Studies*, *37*(1), 15–26. DOI: 10.2989/16073614.2019.1598877

Mchombo, S. (2019). Verbal arts as culturally relevant pedagogical tools in math/science education. In *Promoting language and STEAM as human rights in education* (pp. 17–38). Springer. DOI: 10.1007/978-981-13-2880-0_2

Mehta, R., & Henriksen, D. A. (2022). To Democratize, First Decolonize: Approaches Beyond Eurocentric and Colonial Epistemologies in Creativity. *Review of Research in Education*, *46*(1), 105–133. DOI: 10.3102/0091732X221084324

Mogege, M. (2017). Mathematical concepts from community elders: Exploring the connection between ethnomathematical contexts and classroom practices. *ETD-Educação Temática Digital*, *19*(3), 667–686. DOI: 10.20396/etd.v19i3.8648368

Mosimege, M. (2012). Methodological challenges in doing ethnomathematical research. *International Journal of African Renaissance Studies-Multi-. Inter-and Transdisciplinarity*, *7*(2), 59–78.

Mosimege, M. (2020). The use of indigenous games in the teaching and learning of Mathematics. *Revemop*, *2*, e202009–e202009.

Gumbo, M. T., & Msila, V. (Eds.). (2017). *African voices on indigenisation of the curriculum: Insights from practice*. Reach Publisher.

Mugari A, Matemera N (2015). *Demystifying Mathematical Operations: Leveraging On The Power Of Traditional Games,* European Journal of Physical Education and Sport Science, ISSN: 2501 - 1235

Nnama-Okechukwu, C. U., & McLaughlin, H. (2022). Indigenous knowledge and social work education in Nigeria: Made in Nigeria or made in the West? *Social Work Education*, •••, 1–18.

Nur, A. S., Waluya, S. B., Rochmad, R., & Wardono, W. (2020). Contextual Learning with Ethnomathematics in Enhancing the Problem Solving Based on Thinking Levels. *Journal of Research and Advances in Mathematics Education*, *5*(3), 331–344. DOI: 10.23917/jramathedu.v5i3.11679

Nyota, S., & Mapara, J. (2008). Shona traditional children's games and play: Songs as indigenous ways of knowing. *The Journal of Pan African Studies*, *2*(4).

Pangrazi, R. P., & Beighle, A. (2019). *Dynamic physical education for elementary school children*. Human Kinetics Publishers.

Peng, P., Lin, X., Ünal, Z. E., Lee, K., Namkung, J., Chow, J., & Sales, A. (2020). Examining the mutual relations between language and mathematics: A meta-analysis. *Psychological Bulletin*, *146*(7), 595–634. DOI: 10.1037/bul0000231 PMID: 32297751

Räsänen, P., Laurillard, D., Käser, T., & Aster, M. V. (2019). Perspectives to technology-enhanced learning and teaching in mathematical learning difficulties. In *International handbook of mathematical learning difficulties* (pp. 733–754). Springer. DOI: 10.1007/978-3-319-97148-3_42

Romaniuk, S. N. (2017). Dependency: "Dependency Theory" SAGE Publications, Inc. City: Thousand Oaks, ISBN: 9781483359892 pages: 482-483

Rule, P., and Land, S (2017). *Finding the plot in South African reading education* Reading & Writing - Journal of the Reading Association of South Africa, ISSN: (2308-1422, p2079-8245)

Sellberg, C., Lundin, M., & Säljö, R. (2021). Assessment in the zone of proximal development: simulator-based competence tests and the dynamic evaluation of knowledge-in-action. *Classroom Discourse*, 1-21.

Setiyadi, D., Zaenuri, Z., & Mulyono, M. (2018). The Problem Based Learning Model with Etnomatematics Nuance by Using Traditional Games to Improve Problem Solving Ability. *Journal of Primary Education*, *7*(2), 176–186.

Siamonga, E. (2017). *Children"s Traditional Games and Promotion of Social Cohesion*. The Patriot.

Spaull, N. (2011). A Preliminary Analysis of SACMEQ III South Africa. Stellenbosch Economic Working Papers

Suryaratri, R. D., Prayitno, E. H., & Wuryani, W. (2019). The implementation of multi-sensory learning at elementary schools in Jakarta. *Jurnal Pendidikan Usia Dini*, *13*(1), 100–113. DOI: 10.21009/10.21009/JPUD.131.08

Tandika, P. B., & Ndijuye, L. G. (2019). Pre-primary teachers' preparedness in integrating information and communication technology in teaching and learning in Tanzania. *Information and Learning Science*, *121*(1/2), 79–94. DOI: 10.1108/ILS-01-2019-0009

Taylor, S. V., & Leung, C. B. (2020). Multimodal literacy and social interaction: Young children's literacy learning. *Early Childhood Education Journal*, *48*(1), 1–10. DOI: 10.1007/s10643-019-00974-0

Tshube, T., & Hanrahan, S. J. (2016). Sport psychology in Botswana: A prime breeding ground. *International Journal of Sport and Exercise Psychology*, *14*(2), 126–134. DOI: 10.1080/1612197X.2016.1142462

Van de Walle, J. A. (2007). *Elementary and middle school mathematics education: Teaching developmentally* (6th ed.). Pearson education.

Venkat, H., & Naidoo, D. (2012). Analyzing coherence for conceptual learning in a Grade 2 numeracy lesson. *Education as Change*, *16*(1), 21–33. DOI: 10.1080/16823206.2012.691686

Vygotsky, L. S. (1978). *Mind in society: the development of higher psychological processes*. Harvard University Press.

Watson-Vandiver, M. J., & Wiggan, G. (2021). *The healing power of education: Afrocentric pedagogy as a tool for restoration and liberation*. Teachers College Press.

Zaslavsky, C. (1999). *Africa counts: Number and pattern in African cultures*. Chicago Review Press.

Zippert, E. L., Eason, S. H., Marshall, S., & Ramani, G. B. (2019). Preschool children's math exploration during play with peers. *Journal of Applied Developmental Psychology*, *65*, 101072. DOI: 10.1016/j.appdev.2019.101072

Compilation of References

Abah, J., Mashebe, P., & Denuga, D. D. (2015). Prospect of integrating African Indigenous knowledge Systems into the teaching of sciences in Africa. *American Journal of Educational Research*, *3*(6), 668–673.

Aberšek, B. (2019). *The impact of innovative ICT education and AI on the pedagogical paradigm*. Cambridge Scholars Publishing.

Abma, T. (2020). *Ethics for good for good participatory action research: Engagement in a ommitment to epistemic justice*. BeleidsonderzoekOnline., DOI: 10.5553/BO/221335502020000006001

Abu Moghli, M., & Kadiwal, L. (2021). Decolonising the curriculum beyond the surge: Conceptualisation, positionality and conduct. *London Review of Education*, *19*(1), 1–16. DOI: 10.14324/LRE.19.1.23

Adam, K., & Apou, K. (2013). Indigenous traditional knowledge (ITK)-an useful tools in pests and diseases management. *Indian Agriculturist*, *57*(3), 185–190.

Adelman, C. (1993). Kurt Lewin and the Origins of Action Research. *Educational Action Research*, *1*(1), 7–24. DOI: 10.1080/0965079930010102

Aderajew, A. (2019). African systems of thought: Whether they fit scientific knowledge. *Journal of Philosophy and Culture*, *7*(2), 14–25.

Adom, D., Asare-Aboagye, I., Thompson, J., & Kquofi, S. (2024). Integration of Indigenous Knowledge Systems in African Higher Education in Ghana: A Post-CESA (2016-2025) Review. What Next for Africa's Higher Education?: What Should Follow the Continental Education Strategy for Africa (CESA 2016-2025), 161.

Adom, D. (2016). *Asante IKS: Repositories of Conservation Ethics for Ghana's Biodiversity. Sub-Sahara indigenous knowledge and new discoveries in the new millennium interdisciplinary approach: Uthman Danfodio University, Sokoto, PTF I Hall, Sokoto, Nigeria. 7. Sokoto State*. Uthman Danfodio University.

Agherdien, N., Pillay, R., Dube, N., & Masinga, P. (2022). What does decolonising education mean to us? Educator reflections. *Scholarship of Teaching and Learning in the South*, *6*(1), 55–78. DOI: 10.36615/sotls.v6i1.204

Ahluwalia, M. (2020). Genetic privacy: We must learn from the story of Henrietta Lacks. *New Scientist*.

Ajaps, S. O. (2021). Decolonising education for environmental conservation: A participatory action research (Doctoral dissertation, New York University).

Ali, A. D. (2020). Implementing Action Research in EFL/ESL Classrooms: A Systematic Review of Literature 2010-2019. *Systemic Practice and Action Research, 33*(3), 341–362. DOI: 10.1007/s11213-020-09523-y

Aliverti, A., Carvalho, H., Chamberlen, A., & Sozzo, M. (2021). Decolonizing the criminal question. *Punishment & Society, 23*(3), 297–316. DOI: 10.1177/14624745211020585

Alkan, F. (2015). The role of technology in science teaching activities: Web based teaching applications. *Journal for the Education of Gifted Young Scientists, 3*(2), 1–7. DOI: 10.17478/JEGYS.2015213531

Allan, C., Nguyen, T., Seddaiu, G., Wilson, B., & Roggero, P. (2013). Integrating local knowledge with experimental research: Case studies on managing cropping systems in italy and australia. *Italian Journal of Agronomy, 8*(2), 15. DOI: 10.4081/ija.2013.e15

Allen, R. (1990). *Concise Oxford dictionary of current english.* Clarendon.

Almond, A. (2021). *We Are All Related:(Re).* Storying With Augmented Reality to Build Indigenous-Settler Relations.

Altrichter, H., Posch, P., & Somekh, B. (1993). Professors investigate their work: An introduction to methods of action research. Research With Children: Perspectives and Practices, 241-257.

Altrichter, H., & Posch, P. (1989). Does the 'Grounded Theory' Approach Offer a Guiding Paradigm for Teacher Research? *Cambridge Journal of Education, 19*(1), 21–32. DOI: 10.1080/0305764890190104

Amoyedo-Peter, C. (2023). Albert Bandura's Social Learning Theory: The significance. *The Prodigious Musicologist: Celebrating Professor Agatha Ijeoma Onwuekwe*, 194-204.

Anderson, W. (2002). The Cultivation of whiteness: Science, Health and racial destiny in Australia. Melborne: Melborne University Press.

Angadi, S. C., Mansur, C. P., & Kammar, M. R. (2021). Indigenous technical knowledge on soil and water conservation, soil health management and tillage techniques in Bagalkot district.

Ankiewicz, P. (2016). The relevance of indigenous technology knowledge systems (ITKS) for the 21st century classroom. In PATT 32 *conference proceedings* (pp. 22-34).

Antons, C. (2009). The international debate about traditional knowledge and approaches in the Asia-Pacific region. In Antons, C. (Ed.), *Traditional knowledge, traditional cultural expressions and intellectual property law in the Asia-Pacific region* (pp. 40–51). Wolters Kluwer.

Apple, M. (2013). *Can education change society?* Routledge.

Argryis, C., Putnam, R., & McLain Smith, D. (2019) *Action Science: Concepts, Methods, and Skills for Research and Intervention.* New York: Jossey Bass (Out of print: http://www.actiondesign.com/action_science/)

Argyris, C., Putnam, R., & Smith, D. M. (1985). *Argyris, Chris, Robert Putnam, and Diane Maclain Smith, Action Science: Concepts, Methods, and Skills for Research and Intervention.* Jossey-Bass.

Argyris, C., & Schön, D. A. (1976). *Theory in Practice: increasing professional effectiveness.* Jossey-Bass.

Arlin, P. (1990). Wisdom: the art of problem-finding. In Sternberg, R. J. (Ed.), *Wisdom: its nature, origins and development*. Cambridge University Press. DOI: 10.1017/CBO9781139173704.012

Arnalte-Mur, L., Ortiz-Miranda, D., Cerrada-Serra, P., Martinez-Gómez, V., Moreno-Pérez, O., Barbu, R., & Šūmane, S. (2020). The drivers of change for the contribution of small farms to regional food security in Europe. *Global Food Security*, *26*, 100395. Advance online publication. DOI: 10.1016/j.gfs.2020.100395

Arnold, J., Edwards, T., Hooley, N., & Williams, J. (2012). Conceptualising teacher education and research as 'critical praxis'. *Critical Studies in Education*, *53*(3), 281–295. DOI: 10.1080/17508487.2012.703140

Asante, M. K. (1991). The Afrocentric idea in education. *The Journal of Negro Education*, *60*(2), 170–180. DOI: 10.2307/2295608

Ashwath, M. N., Bodiga, D., Wagmare, B., Dechamma, N. L., Hosur, S. R., Toushif, P. K., & Dinesha, S. (2023). Threatened and near threatened underutilized edible fruit species of southern India for food security and diversifying agroecology. *Indian Journal of Ecology*, *50*(1), 19–31. DOI: 10.55362/IJE/2023/3847

Atkin, J. M. (1993). Developments in the Philosophy/Sociology of Science and Action Research. *Educational Action Research*, *1*(1), 187–188. DOI: 10.1080/0965079930010111

Ausubel, D. P. (1968). *Educational psychology: A cognitive view*. Holt, Rinehart & Winston, Inc.

Au, W. (2007). Epistemology of the oppressed: Dialectics of Paulo Freire's theory of knowledge. *The Journal for Critical Education Policy Studies*, *5*(2), 1–14.

Azizan, N., Smith, R., Cooper, V., & Abu Bakar, N. (2018). Qualitative Content Analysis: A Case Study of a Malaysian Government Education Website. *Malaysian Journal of Science Health & Technology*, *1*(1). Advance online publication. DOI: 10.33102/mjosht.v1i1.14

Baan, J., Gaikhorst, L., & Volman, M. (2019). Stimulating teachers' inquiring attitude in academic and professional teacher education programmes. *European Journal of Teacher Education*, *43*(3), 352–367. DOI: 10.1080/02619768.2019.1693994

Bada, S. O., & Olusegun, S. (2015). Constructivism learning theory: A paradigm for teaching and learning. *Journal of Research and Method in Education*, *5*(6), 66–70. DOI: 10.9790/7388-05616670

Balakrishnan, V., & Claiborne, L. (2017). Participatory action research in culturally complex societies: Opportunities and challenges. *Educational Action Research*, *25*(2), 185–202. DOI: 10.1080/09650792.2016.1206480

Baloyi, T. M., Maphanga, T., Madonsela, B. S., Mongwe, Q. G., Malakane, K. C., Grangxabe, X. S., & Gqomfa, B. (2024). Indigenous Strategies for Managing Bush Encroachment in Rural Areas of South Africa. *Challenges*, *15*(3), 33. DOI: 10.3390/challe15030033

Banes, G. G., & Cruz, K. B. D. (2021). The integration of Indigenous knowledge systems in the tertiary level curriculum of Benguet State University La Trinidad Campus. [formerly Benguet State University Research Journal]. *Mountain Journal of Science and Interdisciplinary Research*, *81*(1), 19–36. DOI: 10.70884/mjsir.v81i1.279

Barcelos, A. M. F., & Ruohotie-Lyhty, M. (2018). *Teachers' Emotions and Beliefs in Second Language Teaching:* Implications for Teacher Education. In Agudo, J. D. D. M. (Ed.), *Emotions in Second Language Teaching* (pp. 109–124). Springer International Publishing. DOI: 10.1007/978-3-319-75438-3_7

Barnes, B. (2018). Decolonising research metholodologies: Opportunity and. *South African Journal of Psychology. Suid-Afrikaanse Tydskrif vir Sielkunde, 48*(3), 379–387. DOI: 10.1177/0081246318798294

Barnhardt, R., & Kawagley, A. O. (2005). Indigenous Knowledge Systems and Alaska Native Ways of Knowing. *Anthropology & Education Quarterly, 36*(1), 8–23. DOI: 10.1525/aeq.2005.36.1.008

Bartlett, J. G., Iwasaki, Y., Gottlieb, B., Hall, D., & Mannell, R. (2007). Framework for Aboriginal-guided decolonizing research involving Métis and First Nations persons with diabetes. *Social Science & Medicine, 65*(11), 2371–2382. DOI: 10.1016/j.socscimed.2007.06.011 PMID: 17689163

Basole, A., & Basu, D. (2015). Fueling calorie intake decline: Household-level evidence from rural India. *World Development, 68*, 82–95. DOI: 10.1016/j.worlddev.2014.11.020

Basu, R. (2018). Constructivism Based Active Learning in Higher Education. *Shodh Sarita, 5*(19), 24–28.

Bates, D. (2014). Traditional ecological knowledge and natural resource management. *Local Environment, 19*(7), 826–841. https://www.nebraskapress.unl.edu/nebraska/9780803232464

Battiste, M. (2009). Naturalizing Indigenous knowledge in Eurocentric education. *Canadian Journal of Native Education, 32*(1).

Battiste, M. (2013). *Decolonizing Education: Nourishing the Learning Spirit.* Purich Publishing.

Battiste, M. (2019). *Decolonizing education: Nourishing the learning spirit.* UBC press. DOI: 10.59962/9781895830972

Battiste, M., & Henderson, J. Y. (2000). *Protecting Indigenous Knowledge and Heritage: A Global Challenge.* Purich Publishing. DOI: 10.59962/9781895830439

Baum, F., MacDougall, C., & Smith, D. (2006). Participatory action research. *Journal of Epidemiology and Community Health, 60*(10), 854–857. DOI: 10.1136/jech.2004.028662 PMID: 16973531

Bayeck, R. Y. (2018). A review of five African board games: Is there any educational potential? *Cambridge Journal of Education, 48*(5), 533–552. DOI: 10.1080/0305764X.2017.1371671

Bele, B., Nielsen, V. K. S., Fairclough, G., & Herring, P. (2024). A Heritage Without Boundaries: Nature-Culture Synergies in and around Norwegian Protected Landscapes. *Human Ecology: an Interdisciplinary Journal, 52*(1), 1–18. DOI: 10.1007/s10745-023-00468-z

Benatar, S. (2002). Reflections and recommendations on research ethics in developing countries. *Social Science & Medicine, 54*(7), 1131–1141. DOI: 10.1016/S0277-9536(01)00327-6 PMID: 11999507

Bennett, M. (2019). A Review of the Literature on the Benefits and Drawbacks of Participatory Action Research. *First Peoples Child & Family Review, 14*(1), 109–122. DOI: 10.7202/1071290ar

Berger, P. L., & Luckmann, T. (1966). *The Social Construction of Reality: A Treatise in the Sociology of Knowledge.* Anchor Books.

Berger, P., & Luckmann, T. (1966). *The social construction of reality*. Penguin Books.

Bergmark, U. (2020). Teachers' professional learning when building a research-based education: Context-specific, collaborative and teacher-driven professional development. *Professional Development in Education*, *49*(2), 210–224. DOI: 10.1080/19415257.2020.1827011

Bertram, C., & Christiansen, I. (2020). *Understanding research: an introduction to reading research* (2nd ed.). Van Schaik.

Bezemer, J., & Cowan, K. (2022). We are exploring reading in social semiotics: Theory and methods. In *Reading* (pp. 107–118). Routledge. DOI: 10.4324/9781003208648-9

Bezuidenhout, R.-M., & Cronjé, F. (2014). Qualitative data analysis. In du Plooy, F., Davis, C., & Bezuidenhout, R.-M. (Eds.), *Research matters* (pp. 228–251). Juta.

Bhana, A. (1999). Participatory action research: A practical guide for realistic radicals. In Blanche, M., Durrheim, K., & Painter, D. (Eds.), *Research in practice: Applied methods for social sciences* (pp. 227–238). UTC.

Bhan, S., Bharti, V. K., Bharti, S., & Bharti, S. (2016). Impactful Indigenous Technical Knowledge (ITK) for sustainable agriculture. *Journal of Soil and Water Conservation*, *15*(4), 349–355. DOI: 10.5958/2455-7145.2016.00023.0

Bheemappa, A., Shruthi, S. M., Maheshwari, K. D., & Biradar, N. (2022). Indigenous technical knowledge based rainfall prediction: A review. *Agricultural Reviews (Karnal)*, *43*(2), 205–210. DOI: 10.18805/ag.R-2347

Bhuda, M. T. (2024). Decolonising and indigenising African Indigenous research: Why Indigenous ways of knowing, being and living matter in research. In *Higher Education Transformation in Africa* (pp. 50–69). Routledge. DOI: 10.4324/9781003528319-4

Bhuda, M. T., & Marumo, P. (2022). Ubuntu Philosophy and African Indigenous Knowledge Systems: Insights from decolonization and indigenization of research. *Gender & Behaviour*, *20*(1), 19133–19151.

Bilal, A., & Chand, A. (2023). Decolonising education and research: An Indianised approach to pedagogy. *The Arab Journal of the Social Sciences*, *22*(2), 23–40.

Bion, W. R. (1961). *Experiences in Groups*. Tavistock Publications.

Blaikie, N. (2000). *Designing social research: The logic of anticipation* (1st ed.). Polity Press.

Blose, P. (2019). *The integration of indigenous graphics knowledge and skills to enhance Grade 9 learners' understanding of graphic designs in Technology Education*. MEd Dissertation: University of South Africa, Pretoria.

Blose, P. (2023). *The incorporation of education for sustainable development in Technology Education for an indigenous context* (Doctoral Thesis). https://uir.unisa.ac.za/handle/10500/30606

Blose, P., & Gumbo, M. T (2024). Developing a framework for integrating Indigenous knowledge Systems in Technology Education with sustainable development principles. *African Journal of Research in Mathematics, Science and Technology Education*, 1-13. https://doi.org/DOI: 10.1080/18117295.2024.2356416

Blumenthal, D. S. (2011). Is community-based participatory research possible? *American Journal of Preventive Medicine*, *40*(3), 386–389. DOI: 10.1016/j.amepre.2010.11.011 PMID: 21335275

Boboc, R. G., Băutu, E., Gîrbacia, F., Popovici, N., & Popovici, D. M. (2022). Augmented reality in cultural heritage: An overview of the last decade of applications. *Applied Sciences (Basel, Switzerland)*, *12*(19), 9859. DOI: 10.3390/app12199859

Bonk, C. J., & Graham, C. R. (2006). *Blended learning: In theory and practice*. Jossey-Bass.

Bonk, C. J., & Graham, C. R. (2012). *The hybrid learning imperative: Seven principles for implementing blended learning successfully*. Jossey-Bass.

Boomer, G. (1985). Fair Dinkum Teaching and Learning: Reflections on Literacy and Power. Boynton/Cook Publishers, Inc., 52 Upper Montclair Plaza, PO Box 860, Upper Montclair, NJ 07043.

Borah, N., Bey, B. S., & Deka, N. (2023). Indigenous technical knowledge (ITK) used in agriculture by selected ethnic communities of Assam. [IJTK]. *Indian Journal of Traditional Knowledge*, *22*(2), 264–272. DOI: 10.56042/ijtk.v22i2.46568

Borgen Project. (2021). Education and Improving Sanitation in Developing Countries.

Borthakur, A., & Singh, P. (2020). Indigenous knowledge systems in sustainable water conservation and management. In *Water conservation and wastewater treatment in BRICS Nations* (pp. 321–328). Elsevier., DOI: 10.1016/B978-0-12-818339-7.00016-3

Bourgeois, R., Karuri-Sebina, G., & Feukeu, K. E. (2024). The future as a public good: decolonising the future through anticipatory participatory action research. *foresight, 26*(4), 533-549.

Bourgeois, R., Karuri-Sebina, G., & Feukeu, K. E. (2024). The future as a public good: Decolonising the future through anticipatory participatory action research. *foresight, 26*(4), 533-549.

Bourgeois, R., Karuri-Sebina, G., &Feukeu, K. E. (2024). The future as a public good: decolonising the future through anticipatory participatory action research. *foresight, 26*(4), 533-549.

Bowrey, K., Cochrane, T., Hadley, M., McKeough, J., Pappalardo, K., & Weatherall, K. (2024). Managing ownership of copyright in research publications to increase the public benefits from research. *Federal Law Review*, *52*(1), 3–33. DOI: 10.1177/0067205X231213676

Brandt, A. M. (1978). Racism and research: The case of the Tuskegee study. *The Hastings Center Report*, *9*(6), 21–29. DOI: 10.2307/3561468 PMID: 721302

Brannelly, T., & Boulton, A. (2017). The ethics of care and transformation research practices in Aotearoa New Zealand. *Qualitative Research*, *17*(3), 340–350. DOI: 10.1177/1468794117698916

Braun, V., & Clarke, V. (2006). Using thematic analysis in psychology. *Qualitative Research in Psychology*, *3*(2), 77–101. DOI: 10.1191/1478088706qp063oa

Brayboy, B. M. J. (2005). Toward a Tribal Critical Race Theory in Education. *The Urban Review*, *37*(5), 425–446. DOI: 10.1007/s11256-005-0018-y

Brondízio, E. S., Aumeeruddy-Thomas, Y., Bates, P., Carino, J., Fernández-Llamazares, Á., Ferrari, M. F., Galvin, K., Reyes-García, V., McElwee, P., Molnár, Z., Samakov, A., & Shrestha, U. B. (2021). Locally based, regionally manifested, and globally relevant: Indigenous and local knowledge, values, and practices for nature. *Annual Review of Environment and Resources*, *46*(1), 481–509. DOI: 10.1146/annurev-environ-012220-012127

Brook, R. K., & McLachlan, S. M. (2005). On using expert-based science to "test" local ecological knowledge. *Ecology and Society 10*(2): r3. [online] URL: https://www.ecologyandsociety.org/vol10/iss2/resp3/

Brown, A., & Dowling, P. (2001). *Doing research/reading research: A mode of interrogation for teaching*. Routledge Falmer.

Bruner, J. S. (1966). *Toward a theory of instruction*. Harvard University Press.

Brydon-Miller, M., Kral, M., & Ortiz Aragón, A. (2020). Participatory action research: International perspectives and practices. *International Journal of Qualitative Research, 13*(2), 103–111. DOI: 10.1177/1940844720933225

Buckingham, B. R. (2018). *Research for teachers*. New York: Silver, Burdett and Co Berge, B-M. with H. Ve (2000) *Action research for gender equity*. Buckingham: Open University Press.

Burgess, C., Bishop, M., & Lowe, K. (2022). Decolonising Indigenous education: The case for cultural mentoring in supporting indigenous knowledge reproduction. *Discourse (Abingdon)*, *43*(1), 1–14. DOI: 10.1080/01596306.2020.1774513

Burnett, C. (2018). Traditional sports and games in Eastern, Central and Southern Africa. In *Sport in the African World* (pp. 121–145). Routledge. DOI: 10.4324/9781351212755-8

Bussu, S., Lalani, M., Pattison, S., & Marshall, M. (2020). Engaging with ethical care: Ethical issues in participatory research. Retrieved 04 19, 2024, from http://discovery.ucl.ac.uk

Butler, C. (2002). *Postmodernism: A very short introduction*. Oxford University Press.

Bybee, R. W. (2010). Advancing STEM education: A 2020 vision. *Technology and Engineering Teacher*, *70*(1), 3035.

Cajete, G. A. (1994). *Look to the Mountain: An Ecology of Indigenous Education*. Kivaki Press.

Cajete, G. A. (1999). *Ignite the Sparkle: An Indigenous Science Education Model*. Kivaki Press.

Calhoun, E. F. (1994). *How to Use Action Research in the Self-Renewing School*. Association for Supervision and Curriculum Development.

Calicioglu, O., Alessandro, F., Bracco, S., Lorenzo, B., & Ralph, S. (2019). The Future Challenges of Food and Agriculture: An Integrated Analysis of Trends and Solutions. *Sustainability (Basel)*, *11*(1), 222. DOI: 10.3390/su11010222

Campbell, J. Y. (2019). No Sustainable Development Without Indigenous Peoples https://sdg.iisd.org/commentary/guest-articles/no-sustainable-development-without-indigenous-peoples/

CAPS. (2014). *Foundations phase*. Department of Basic Education.

Carr, W. (1993). Whatever Happened to Action Research? *Educational Action Research*, *2*(3), 427–436. DOI: 10.1080/0965079940020310

Carr, W., & Kemmis, S. (1986). *Becoming Critical: education, knowledge and action research*. Falmer Press.

Carr, W., & Kemmis, S. (2018). *Becoming critical: knowing through action research*. Falmer Press.

Caspersen, J., & Smeby, J.-C. (2023). Research-based teacher education in Norway – a longitudinal perspective. *International Journal of Educational Research*, *119*, 102177. Advance online publication. DOI: 10.1016/j.ijer.2023.102177

Castro, B. M. (2017). Ijiam: Augmented Reality Technologies for the Adaptation of Ecuadorian Intangible Cultural Heritage (Doctoral dissertation, OCAD University).

Caudill, C., Pulsifer, P. L., Thumbadoo, R. V., & Taylor, D. R. F. (2024). Meeting the challenges of the un sustainable development goals through holistic systems thinking and applied geospatial ethics. *ISPRS International Journal of Geo-Information*, *13*(4), 110. DOI: 10.3390/ijgi13040110

Caxaj, C. S. (2015). Indigenous storytelling and participatory action research: Allies toward decolonization? Reflections from the peoples' international health tribunal. *Global Qualitative Nursing Research*, *2*, 2333393615580764. DOI: 10.1177/2333393615580764 PMID: 28462305

Cedervall, Y., & Åberg, A. C. (2010). Physical activity and implications on well-being in mild Alzheimer's disease: A qualitative case study on two men with dementia and their spouses. *Physiotherapy Theory and Practice*, *26*(4), 226–239. DOI: 10.3109/09593980903423012 PMID: 20397857

Čepienė, A., & Vaitiekienė, N. (2019). Studentų profesinės praktikos organizavimo proceso kolegijoje analizė. *Mokslo Taikomieji Tyrimai Lietuvos Kolegijose*, (14), 53–62. Retrieved from https://ojs.kaunokolegija.lt/index.php/mttlk/article/view/256

Chahine, I. C. (2020). Towards African Humanicity: Re-mythogolising ubuntu through reflections on the ethnomathematics of African cultures. [CriSTal]. *Critical Studies in Teaching and Learning*, *8*(2), 95–111. DOI: 10.14426/cristal.v8i2.251

Chaka, C. (2020). Translanguaging, decoloniality, and the Global South: An integrative review study. *Scrutiny2, 25*(1), 6-42.

Chakabwata, W., & Mukazi, F. (2023). Ubuntu philosophy and online assessment in Higher Education Institutions. In J. Keengwee, *Transformative and innovative pedagogies in education* (pp. 257-275). Dakota: IGI Global Publishers.

Chambers, R. (1983). *Rural Development: Putting the Last First*. Longman Press.

Chanda, T., & Sharma, A. (2024). *Shivarapatna stone craft in Karnataka - Exploring the transformative role of innovation in traditional handicraft and cultural identity*. ResearchGate. DOI: 10.29121/shodhkosh.v4.i2CDSDAD.2023.595

Chan, M. (2018). *Ten years in public health 2007-2017: report by dr margaret chan director-general world health organization*. World Health Organization., DOI: 10.1093/ww/9780199540884.013.249876

Chauke, O. R., Balotyi, T., Mapindani, A., Chauke, W. S., & Motlhaka, H. A. (2021). Rain petitioning as an indigenous agenda: Fusing ecological traditions with the modern IKS philosophy. *Turkish Journal of Computer and Mathematics Education, 12*(6), 4946–4950. DOI: 10.17762/turcomat.v12i6.8741

Checkland, P., & Holwell, S. (1998). *Information, Systems, and Information Systems: Making Sense of the Field*. Wiley.

Chein, S., Cook, S. W., & Harding, J. (1948). The field of action research. *The American Psychologist, 3*(2), 33–44. DOI: 10.1037/h0053515 PMID: 18911932

Chen, P. G., Diaz, N., Lucas, G., & Rosenthal, M. S. (2010). Dissemination of results in community-based participatory research. *American Journal of Preventive Medicine, 39*(4), 372–378. DOI: 10.1016/j.amepre.2010.05.021 PMID: 20837290

Chernela, J. (2004, December 3). The rights of indigenous people. Retrieved from http://janetchernela.outfoxing.com/rights-of-indigenous-peoples.html

Chikiwa, S., Westaway, L., & Graven, M. (2019). What mathematics knowledge for teaching is used by a Grade 2 teacher when teaching counting. *South African Journal of Childhood Education, 9*(1), 567. DOI: 10.4102/sajce.v9i1.567

Cilliers, E. J. (2021). Reflecting on Social Learning Tools to Enhance the Teaching-Learning Experience of Generation Z Learners. *Frontiers in Education, 5*(606533), 606533. Advance online publication. DOI: 10.3389/feduc.2020.606533

Cindi, L. (2021). *Incorporating African IKS Indigenous knowledge into systems into the Basic Education curriculum: Experiences from two schools in the Gauteng and KwaZulu-Natal Provinces, South Africa* (Dissertation).

Clandinin, D. J. (1989). Developing Rhythm in Teaching: The narrative study of a beginning teacher's personal practical knowledge of classrooms. *Curriculum Inquiry, 19*(2), 121–141. DOI: 10.1080/03626784.1989.11075320

Clarke, V., & Braun, V. (2017). Thematic analysis. *The Journal of Positive Psychology, 12*(3), 297–298. DOI: 10.1080/17439760.2016.1262613

Clinical Trials. (2009). The origins of the research ethics. Retrieved 04 19, 2024, from https://www.appliedclinicaltrialsonline.com/view/origins-research-ethics

Cochran-Smith, M., Barnatt, J., Friedman, A., & Pine, G. (2009). Inquiry on Inquiry: Practitioner Research and Student Learning. *Action in Teacher Education, 31*(2), 17–32. DOI: 10.1080/01626620.2009.10463515

Coghlan, D. (2010). Seeking Common Ground in the Diversity and Diffusion of Action and Collaborative Management Research Methodologies: The Value of a General Empirical Method. In W. Pasmore, A.B.(Rami) Shani, & R. Woodman (Eds.), *Research in Organizational Change and Development*, 18:149–181. Bingley: Emerald

Coghlan, D., Brady, V., O'Leary, D., & Hynes, G. (2024). Exploring vulnerability and risk in an action research writing group: A cooperative inquiry. *Educational Action Research*, *32*(1), 25–42. DOI: 10.1080/09650792.2022.2084133

Collins, P. H. (2013). Critical intervention in western social theory: Reflections on power and southern theory. *Decentering Social Theory. Political Power and Social Theory*, *25*, 173–182.

Collis, J., & Hussey, R. (2003). *Business research: A practical guide for undergraduate and postgraduate students* (2nd ed.). Palgrave Macmillan.

Connell, R. (2007). *Southern Theory: The Global Dynamics of Knowledge in Social Science*. Polity Press.

Cooke, W. (Bill) (undated) *A Foundation correspondence on action research: Ronald Lippitt and John Collier*. Produced by Working Paper No.6/2002 of the Institute for Development Policy and Management.

Corbett, M., & Gereluk, D. (Eds.). (2020). Introduction: Why do rural matters in Canadian education? In M. Corbett & D. Gereluk (Eds.), *Rural teacher education: connecting land and people*. Singapore: Springer Nature. [Online]. Available at: https://doi.org/ (Accessed on 10 April 2020).DOI: 10.1007/978-981-15-2560-1

Cornish, F., Breton, N., Moreno-Tabarez, U., Delgado, J., Rua, M., de-Graft Aikins, A., & Hodgetts, D. (2023). Participatory action research. *Nature Reviews. Methods Primers*, *3*(1), 34. DOI: 10.1038/s43586-023-00214-1

Cox, R., Heykoop, C., Fletcher, S., Hill, T., Scannell, L., Wright, L., & Plush, T. (2021). Creative action research. *Educational Action Research*, *29*(4), 569–587. DOI: 10.1080/09650792.2021.1925569

Crawford-Garrett, K., Anderson, S., Grayson, A., & Suter, C. (2015). Transformational practice: Critical teacher research in pre-service teacher education. *Educational Action Research*, *23*(4), 479–496. DOI: 10.1080/09650792.2015.1019902

Creswell, J. W. (1994). *Research design: Quantitative and qualitative approaches*. California: Sage publication, Inc.

Creswell, J. D., & Creswell, J. W. (2021). Key principles of action research: democratic collaboration, self-reflection, and collective problem-solving. In *Action Research in Systematic Literature Research* (pp. 25–46). Sage Publications.

Creswell, J. W. (2012). *Educational Research: Planning, Conducting, and Evaluating Quantitative and Qualitative Research* (4th ed.). Pearson.

Crotty, M. (1998). *The foundations of social research: Meaning and perspective in the research process* (1st ed.). Sage Publication, Inc.

Crush, J. (2001). Twentieth-century Swaziland When the Sleeping Grass Awakens: Land and Power in Swaziland. By Richard Levin. Johannesburg: University of Witwatersrand Press, 1997. Pp. 290. SA Rand 98.94 (ISBN 1-86814-301-5). *Journal of African History*, *42*(2), 307–352. DOI: 10.1017/S0021853701457938

Cruzes, D. S., Dybå, T., Runeson, P., & Höst, M. (2015). Case studies synthesis: A thematic, cross-case, and narrative synthesis worked example. *Empirical Software Engineering, 20*(6), 1634–1665. DOI: 10.1007/s10664-014-9326-8

Cummings, S., Munthali, N., & Shapland, P. (2021). A systemic approach to the decolonisation of knowledge. In *The politics of knowledge in inclusive development and innovation* (pp. 65-79). DOI: 10.4324/9781003112525-7

Cunningham, C., & Stanley, F. (2003). Indigenous by definition,experience or World view. *BMJ (Clinical Research Ed.), 327*(7412), 403–404. DOI: 10.1136/bmj.327.7412.403 PMID: 12933705

da Silva, C., Pereira, F., & Amorim, J. P. (2023). The integration of Indigenous knowledge in school: A systematic review. *Compare: A Journal of Comparative Education, 54*(7), 1210–1228. DOI: 10.1080/03057925.2023.2184200

Dadich, A., Moore, L., & Eapen, V. (2019). What does it mean to conduct participatory research with Indigenous peoples? A lexical review. *BMC Public Health, 19*(1), 1388. DOI: 10.1186/s12889-019-7494-6 PMID: 31660911

Dana, N. F., & Yendol-Hoppey, D. (2019). *The Reflective Educator's Guide to Classroom Research: Learning to Teach and Teaching to Learn Through Practitioner Inquiry*. Corwin.

Dancis, J., Coleman, B., & Ellison, E. (2023). Participatory action research as pedagogy. *Journal of Participatory Action Research Methods, 4*(2), 1–24.

Darwin, S., & Barahona, M. (2019). Can an outsider become an insider? Analysing the effect of action research in initial EFL teacher education programs. *Educational Action Research, 27*(5), 709–725. DOI: 10.1080/09650792.2018.1494616

Das, A., Ramkrushna, G. I., Choudhury, B. U., Munda, G. C., Patel, D. P., Ngachan, S. V., ... & Kumar, M. (2012). Natural resource conservation through indigenous farming systems: Wisdom alive in North East India.

Datta, R. (2018). Decolonizing both researcher and research and its effectiveness in Indigenous research. *Research Ethics, 14*(2), 1–24. DOI: 10.1177/1747016117733296

Datta, R., Khyang, N. U., Prue Khyang, H. K., Prue Kheyang, H. A., Ching Khyang, M., & Chapola, J. (2015, November 02). Hla Kray Prue Khyang, Hla Aung Prue Kheyang,Mathui Ching Khyang&Jeb-unnessaChapola (2014): Participatory action research and researcher's responsibilities: an experience with an Indigenous community. *International Journal of Social Research Methodology, 18*(6), 581–599. Advance online publication. DOI: 10.1080/13645579.2014.927492

Davies, S. (2022). Towards digital justice: Participatory action research in global digital health. *BMJ Global Health, 7*, 1–4. PMID: 35580914

De Oliveira, B. (2023). Participatory action research as a research approach: Advantages, limitations and criticisms. *Qualitative Research Journal, 23*(3), 287–297. DOI: 10.1108/QRJ-08-2022-0101

De Sousa Santos, B., Nunes, J., & Meneses, M. (2007). *Another knowledge is possible: Beyond Northern epistemologies*. Verso.

Deaton, A., & Eze, J. (2009). Food and nutrition in India: Facts and interpretations. *Economic and Political Weekly, 44*(7).

Debisa, N. G. (2022). Building peace by peaceful approach: The role of Oromo Gadaa system in peace-building. *Cogent Social Sciences, 8*(1), 1–17. DOI: 10.1080/23311886.2021.2023254

Dei, G. J. S. (1996). *Anti-Racism Education: Theory and Practice*. Fernwood Publishing.

Deka, S., Nath, R. K., Sehgal, M., Ahuja, D. B., Kakoti, R. K., & Barbora, A. C. (2017). Indigenous technological knowledge (ITK) and practices in pest management of Assam. *Annals of Plant Protection Sciences, 25*(1), 119–125.

De, L. C. (2021). Traditional knowledge practices of North East India for sustainable agriculture. *Journal of Pharmacognosy and Phytochemistry, 10*(1S), 549–556.

Deloria, V., & Wildcat, D. (2001). *Power and Place: Indian Education in America*. Fulcrum Publishing.

Denzin, N. K., Lincoln, Y. S., & Smith, L. T. (2008). *Handbook of Critical and Indigenous Methodologies*. Sage Publications. DOI: 10.4135/9781483385686

Deo, M. R. C., Deo, C., Wangchu, L., Sarma, P., Singh, A. K., & Dobhal, S. (2021). Ethnobotany of indigenous (Traditional) vegetables in Adi tribe of East Siang region of Arunachal Pradesh, India. *Environment Conservation Journal, 22*(3), 299–307. DOI: 10.36953/ECJ.2021.22335

Department of Basic Education (DBE). (2010). Curriculum and Assessment Policy Statement (CAPS) Technology, grades 7–9. Final draft [Online]. Available at: https://www.education.gov.za/CAPS.asp (Accessed on 15 November 2011).

Department of Basic Education. (2011). *National Curriculum Statement: Curriculum and Assessment Policy Statement (CAPS) Technology Grades 7 – 9*. Government Printers.

Department of Basic Education. (2014). *Report on the annual national assessment of 2014: Grades 1–6 and 9*. Department of Basic Education.

Dhivya, C, Monika, A., Jayashree, V. and Kamali, S P. (2024). The Role of Indigenous Knowledge in Sustainable Farming Practices. GreenariA- An Agricultural Monthly e-Magazine. 2 (10), 2584-153X

Dönmez, D., Isak, M. A., İzgü, T., & Şimşek, Ö. (2024). Green horizons: Navigating the future of agriculture through sustainable practices. *Sustainability (Basel), 16*(8), 3505. DOI: 10.3390/su16083505

Doré, T., Makowski, D., Malézieux, É., Munier-Jolain, N., Tchamitchian, M., & Tittonell, P. (2011). Facing up to the paradigm of ecological intensification in agronomy: Revisiting methods, concepts and knowledge. *European Journal of Agronomy, 34*(4), 197–210. DOI: 10.1016/j.eja.2011.02.006

Dreze, J., & Khera, R. (2015). Understanding leakages in the Public Distribution System. *Economic and Political Weekly, 50*(7), 39–42.

Du Plessis, P. (2021). Decolonisation of education in South Africa: Challenges to decolonise the University curriculum. *South African Journal of Higher Education, 35*(1), 54–69. DOI: 10.20853/35-1-4426

Dua, P., Dhir, A., Naranyan, A., & Allamraju, A. (2023). A Holistic Approach to Achieving SDGs: A Case Study of Dayalbagh, India. https://uwspace.uwaterloo.ca/ handle/10012/19181

Dube, E., & Munsaka, E. (2018). The Contribution of Indigenous knowledge to disaster risk reduction activities in Zimbabwe: A big call to practitioners. *Jàambá, 10*(1), 1–8. DOI: 10.4102/jamba.v10i1.493 PMID: 29955265

Dudgeon, P., Bray, A., Darlaston-Jones, D. and Walker, R., (2020). Aboriginal participatory action research: an indigenous research methodology strengthening decolonisation and social and emotional wellbeing.

Dudgeon, P., Bray, A., Darlaston-Jones, D., & Walker, R. (2020). Aboriginal participatory action research: An indigenous research methodology strengthening decolonisation and social and emotional wellbeing.

Dudgeon, P., Bray, A., Walker, R., Wright, M., & Sutherland, S. (2024). Caring for Country: Indigenous Well-being, Law, and Environmental Justice. In *Climate Change and Mental Health Equity* (pp. 383–407). Springer International Publishing. DOI: 10.1007/978-3-031-56736-0_15

Easterby-Smith, M., Thorpe, R., & Lowe, A. (1999). *Management research: An introduction* (1st ed.). SAGE Publications.

Edinyang, S. D. (2016). The significance of social learning theories in the teaching of social studies education. *International Journal of Sociology and Anthropology Research, 2*(1), 40–45.

Edwards, R., & Brannelly, T. (2017). Approaches to democratising qualitative research methods. *Qualitative Research, 17*(3), 271–277. DOI: 10.1177/1468794117706869

Ejuu, G. (2019). African indigenous games: Using Bame Nsamenang's Africentric thoughts to reflect on our heritage, pedagogy, and practice in a global village. *Journal of Psychology in Africa, 29*(4), 319–327. DOI: 10.1080/14330237.2019.1647496

Elliott, J. (2009). Building educational theory through action research. In *The SAGE Handbook of Educational Action Research* (pp. 28–38). SAGE Publications Ltd., DOI: 10.4135/9780857021021

Emeagwali, G., & Dei, G. J. S. (2014). *African Indigenous Knowledge and the Disciplines*. Sense Publishers. DOI: 10.1007/978-94-6209-770-4

Errico, S. (2021). Women's Right to Land Between Collective and Individual Dimensions. Some Insights From Sub-Saharan Africa. *Frontiers in Sustainable Food Systems, 5*, 690321. DOI: 10.3389/fsufs.2021.690321

Ertmer, P. A. (2009). Blended learning in K-12 schools: A review of the literature. *Journal of Educational Technology & Society, 12*(1), 25–35.

Escobar, A. (2018). *Design for the Pluriverse: Radical Interdependence, Autonomy, and the Making of Worlds. United State*. Duke University Press. DOI: 10.1215/9780822371816

Eun, B. (2019). The zone of proximal development as an overarching concept: A framework for synthesizing Vygotsky's theories. *Educational Philosophy and Theory, 51*(1), 18–30. DOI: 10.1080/00131857.2017.1421941

Evangelista, R. D., Corrêa, D. H., Nunes, M. J., Figueiredo, J. S., Mascarenhas, S. W., Reis, M. da S., & Ferreira da Silva, D. (2024). Environmental Education in Promoting Sustainable Practices: Use of Indigenous Games and Play in Raising Awareness in Elementary School Students. *Journal of Social and Environmental Management*, *18*(12), e010376. DOI: 10.24857/rgsa.v18n12-099

Fade, S. (2009). *Learning and assessing through reflection: a practical guide.* [Online]. Available from: from www.practicebasedlearning.org [2009, June 12]

Fahlberg, A. (2023, March). Decolonizing sociology through collaboration, co-learning and action: A case for participatory action research. *Sociological Forum*, *38*(1), 95–120. DOI: 10.1111/socf.12867

FAO, IFAD, UNICEF, WFP and WHO. (2024). The State of Food Security and Nutrition in the World 2024 – Financing to end hunger, food insecurity, and malnutrition in all its forms. Rome.DOI: 10.4060/cd1254en

FAO. (2002). State of Food Insecurity in the World. Retrieved from: www.fao.org

FAO. (2018). *Transforming Food and agriculture to achieve the SDGs*. Food and Agriculture Organization of the United Nations.

FAO. (2021). Indigenous Peoples' Food Systems: Insights on Sustainability and resilience from the front line of climate change.

Feekery, A. (2023). The 7 C's framework for participatory action research: Inducting novice participant-researchers. *Educational Action Research*, *32*(3), 332–347. DOI: 10.1080/09650792.2023.2234417

Feri, M., & Maemonah, N. H. (2022). Analysis of social attitudes of elementary school learners with the application of Albert Bandura's learning theory. *Jurnal Pendidikan Dasar*, *14*(2), 149–157. DOI: 10.17509/eh.v14i2.41081

Fernández, J. S. (2022). Decolonising participatory action research in community psychology. *Decolonial enactments in community psychology*, 29-51.

Fernandez, J., Gaston, J., Nguyen, M., Rovaris, J., Rhyann, L., & Anguilar, D. (2018). Documenting sociopolitical development via participatory action research (PAR) with women of colour student activists in the neoliberal university. *Journal of Social and Political Psychology*, *6*(2), 591–607. DOI: 10.5964/jspp.v6i2.900

Fine, M., Torre, M., Oswald, A. G., & Avery, S. (2021). Critical participatory action research: Methods and praxis for intersectional knowledge production. *Journal of Counseling Psychology*, *68*(3), 344–356. DOI: 10.1037/cou0000445 PMID: 34043378

Firdaussy, U. F., Ningsih, S., & Asrawijaya, E. (2024). Basic education for indigenous peoples in Indonesia: Limiting children's cultural alienation and loss of identity. *Issues in Educational Research*, *34*(3), 995–1015.

Firmansyah, D., & Saepuloh, D. (2022). Social learning theory: Cognitive and behavioral approaches. *Jurnal Ilmiah Pendidikan Holistik*, *1*(3), 297–324. DOI: 10.55927/jiph.v1i3.2317

Fonua, S. (2021). Enabling Sustainable Development by Embedding Tongan Knowledge into University Science Curricula. *Journal of Comparative & International Higher Education, 13*(3), 157–179. DOI: 10.32674/jcihe.v13iSummer.3164

Food and Agriculture Organization. (2001). *International treaty on plant and genetic resources for food and agriculture*. Retrieved from ftp://ftp.fao.org/docrep/fao/011/i0510e/i0510e.pdf

Ford, J. D., King, N., Galappaththi, E. R., Pearce, T., McDowell, G., & Harper, S. L. (2020). The resilience of indigenous peoples to environmental change. *One Earth, 2*(6), 532–543. DOI: 10.1016/j.oneear.2020.05.014

Frater, L. (2009). Protecting indigenous knowledge: The role of the convention on biodiversity and intellectual property rights (Working Paper Series No. 52). The Centre for Business Relationships, Accountability, Sustainability and Society.

Freese, A. R. (1999). The role of reflection on preservice teachers' development in the context of a professional development school. *Teaching and Teacher Education, 15*(8), 895–909. DOI: 10.1016/S0742-051X(99)00029-3

Freire, P. (1970). *Pedagogy of the Oppressed*. Herder and Herder.

Freire, P. (1973). *Education for critical consciousness*. Continuum.

Freire, P. (1974). *Pedagogy of the oppressed*. Seabury Press.

Freire, P. (1982). Education as the practice of freedom. In *M. Ramos, in education for critical consciousness* (pp. 1–84). Bergin & Garvey Publishers.

Gaffney, M. (2008). Participatory action research: An overview. *Kariraranga, 9*(3), 9–15. DOI: 10.54322/kairaranga.v9i3.131

Galloway, F. B., Wynberg, R. P., & Nott, K. (2016). Commercialising a perfume plant, Commiphora wildii: Livelihood implications for indigenous Himba in north-west Namibia. *International Forestry Review, 18*(4), 429–443. DOI: 10.1505/146554816820127541

Garrison, D. R. (1993). A Cognitive Constructivist View of Distance Education: An Analysis of Teaching-Learning Assumptions. *Distance Education, 14*(2), 199–211. DOI: 10.1080/0158791930140204

Garrison, D. R., & Cleveland-Innes, S. (2005). *Blended learning in online education: Theory, practice, and policy*. Jossey-Bass.

Garrison, D. R., & Kanuka, H. A. (2004). Blended learning in a K-12 school: One school's experience. *Journal of Asynchronous Learning Networks, 8*(1), 1–17.

Gartaula, H., Patel, K., Shukla, S., & Devkota, R. (2020). Indigenous knowledge of traditional foods and food literacy among youth: Insights from rural Nepal, Journal of Rural Studies, Volume 73, Pages 77-86, ISSN 0743-0167, DOI: 10.1016/j.jrurstud.2019.12.001

Gaskins, W., Batsheva, G., & Brittany, A. (2023). Reflections on implementing participatory action research in engineering. *Journal of Education and Development, 7*(3), 18–24. DOI: 10.20849/jed.v7i3.1369

Gates, P. (2019). Developing consciousness and pedagogical knowledge through mutual observation. In Woods, P. (Ed.), *Working for Teacher Development.*

Gay, G. (2010). *Culturally Responsive Teaching: Theory, Research, and Practice.* Teachers College Press.

Geographical Indications Registry. (n.d.). GI Search Version 2.0. Government of India. https://search.ipindia.gov.in/GIRPublic/Application/Details/710

George, N.A. and McKay, F.H. (2019). The Public Distribution System and Food Security in India. International Journal of Environmental Research and Public Health, 3;16(17):3221. . PMID: 31484374; PMCID: PMC6747310.DOI: 10.3390/ijerph16173221

Gilchrist, G., Mallory, M., & Merkel, F. (2005). Can local ecological knowledge contribute to wildlife management? Case studies of migratory birds. *Ecology and Society, 10*(1).

Gill, H., Purru, K., & Lin, G. (2012). In the midst of participatory action research practices: Moving towards decolonizing and decolonial praxis. *Reconceptualizing Educational Research Methodology, 3*(1).

Gillespie, S., & Kadiyala, S. (2012). *Exploring the agriculture nutrition disconnect in India.* International Food Policy Research Institute.

Gill, H., Puru, K., Lin, G., & Gill, H. (2012). In the midst of participatory action research practices moving towards decolonizing and decolonial praxis. *Reconceptualizing Educational Research Methodology, 3*(1), 1–15. DOI: 10.7577/rerm.357

Gilstrap, D. L., & Dupree, J. (2008). A regression model of predictor variables on critical reflection in the classroom: Integration of the critical incident questionnaire and the framework for reflective thinking. *Journal of Academic Librarianship, 34*(6), 469–481. DOI: 10.1016/j.acalib.2008.09.019

Ginn, C. S., & Kulig, J. C. (2015). Participatory Action Research with a Group of Urban First Nations Grandmothers. *International Indigenous Policy Journal, 6*(1), 1–16. DOI: 10.18584/iipj.2015.6.1.4

Global Hunger Index. (2023). *The power of youth in shaping food systems.* Welt Hunger Hilfe.

Glowach, T., Mitchell, R., Bennett, T., Donaldson, L., Jefferson, J., Panford, L., & Hemmings, E. (2023). Making spaces for collaborative action and learning: Reflections on teacher-led decolonising initiatives from a professional learning network in England. *Curriculum Journal, 34*(1), 100–117. DOI: 10.1002/curj.186

Gluck, A. (2015). De-Westernisation: Key concept paper. United Kingdom" the University of Leeds.

Gobena, E. B., Hean, S., Heaslip, V., & Studsrød, I. (2024). The challenge of western-influenced notions of knowledge and research training: Lessons for decolonizing the research process and researcher education. *Journal of Ethnic & Cultural Diversity in Social Work, 33*(5), 276–287.

Gogia, J. and Sikarwar, N. (2019). Agriculture and Dairy Sectors contribution to Food Security in India, International Journal of Basic and Applied Research, Vol. 9, No. 4

Gomes, J., & Fleer, M. (2019). The development of a scientific motive: How preschool Science and home play reciprocally contribute to Science learning. *Research in Science Education, 49*(2), 613–634. DOI: 10.1007/s11165-017-9631-5

Gomez, A., Elbo, C., & Capllonch, M. (2013). Beyond action research: The communicative methodology of research. *International Journal of Qualitative Research, 6*(2), 183–197. DOI: 10.1525/irqr.2013.6.2.183

Gonçalves, C. B. Q., Schlindwein, M. M., & Martinelli, G. C. (2021). Agroforestry Systems: A Systematic Review Focusing on Traditional Indigenous Practices, Food and Nutrition Security, Economic Viability, and the Role of Women. *Sustainability (Basel), 13*(20), 11397. DOI: 10.3390/su132011397

Gopi, R., Avasthe, R. K., Kalita, H., Kapoor, C., Yadav, A., Babu, S., & Das, S. K. (2016). Traditional pest and disease management practices in Sikkim Himalayan region. https://doi.org/DOI: 10.23910/IJBSM/2016.7.3.1543

Gough, K. V., Veléz-Torres, I., Ruette-Orihuela, K., Fayad, J., Bueno, B., Corredor, G., & van Grootheest, S. (2023). Engaged pedagogic research: Transforming societies through co-learning and social action. *Environment and Planning C. Politics and Space, 41*(1), 109–129.

Govender, N., & Mutendera, G. (2020). Teachers' and custodians' views and dilemmas arising thereof regarding the integration of indigenous knowledge in the primary school. *Alternative, 16*(4), 356–368. DOI: 10.1177/1177180120970935

Government of India. (2012). *Indian minerals yearbook 2011 (Part-II): Marble* (50th ed.). Ministry of Mines, Indian Bureau of Mines.

Government of India. (n.d.). Application details of Jabalpur stone craft. Retrieved from https://search.ipindia.gov.in/GIRPublic/Application/Details/710

Graham, C. R. (2006). Blended learning: An effective model for K-12 education? *Educational Technology Review, 52*(4), 233–255.

Grande, S. (2004). *Red Pedagogy: Native American Social and Political Thought*. Rowman & Littlefield Publishers.

Grandin, T. (2022). Grazing cattle, sheep, and goats are important parts of a sustainable agricultural future. *Animals (Basel), 12*(16), 2092. DOI: 10.3390/ani12162092 PMID: 36009682

Graven, M., (2015), *Strengthening math's learning dispositions through math clubs*. South African Journal of Childhood Education, VOL 5.

Greenwood, D., & Levin, M. (2006). Introduction to action research: Social research for social change. Carlifornia: Sage Publications.

Greiner, C. (2012). Unexpected consequences: Wildlife conservation and territorial conflict in northern Kenya. *Human Ecology: an Interdisciplinary Journal, 40*(3), 415–425. DOI: 10.1007/s10745-012-9491-6

Groot, B., Vink, A., Haveman, A., Huberts, M., Schout, G., & Abma, T. (2019). Ethics of care in participatory research: Mutual responsibility in collaboration with co-researchers. *Educational Action Research, 27*(2), 286–302. DOI: 10.1080/09650792.2018.1450771

Guillemin, M., Gillam, L., Barnard, E., Stewart, P., Walker, H., & Rosenthal, D. (2016). "We're checking them out": Indigenous and non-Indigenous research participants' accounts of deciding to be involved in research. *International Journal for Equity in Health, 15*(1), 1–10. DOI: 10.1186/s12939-016-0301-4 PMID: 26772174

Gullion, J. S., & Tilton, A. (2020). *Researching with: A decolonizing approach to community-based action research* (Vol. 6). Brill. DOI: 10.1163/9789004424852

Gumbo, M. T. (2017). Indigenous knowledge in the teaching of technology: a quest for transforming Technology education. In M. T. Gumbo & V. Msila (Eds.), *African voices on indigenization of the curriculum: insights from practice*, pp. 119–154. Wandsbeck: Reach.

Gumbo, M. T., & Williams, J. P. (Eds.). (2023). An Introduction to Indigenous Technology Knowledge Systems. In M. T. Gumbo & J. P. Williams (Eds.), *Indigenous Technology Knowledge Systems: Decolonizing the Technology Education Curriculum*, Part I, pp (v)-(xvii). Singapore: Springer.

Gumbo, M. T. (2015). Indigenous technology in Technology Education curricula and teaching. In Williams, P. J., Jones, A., & Bunting, C. (Eds.), *Contemporary issues in Technology Education: The future of Technology Education* (pp. 57–74). Springer.

Gumbo, M. T. (2015). Indigenous technology in Technology Education curricula and teaching. In Williams, P. J., Jones, A., & Buntting, C. (Eds.), *Contemporary Issues in Technology Education: The future of Technology Education* (pp. 57–75). Springer.

Gumbo, M. T. (2019). Integrating forms of knowledge and technology. In Gumbo, M. T. (Ed.), *Teaching Technology: Intermediate to Senior Phase* (pp. 20–39). Oxford University Press.

Gumbo, M. T. (2019). What is technology? In Gumbo, M. T. (Ed.), *Teaching technology: Intermediate to senior phase* (pp. 2–19). Oxford University Press.

Gumbo, M. T. (2024). In defence of African philosophy. *South African Journal of Higher Education, 38*(2), 120–140. DOI: 10.20853/38-2-6003

Gumbo, M. T., & Msila, V. (Eds.). (2017). *African voices on indigenisation of the curriculum: Insights from practice*. Reach Publisher.

Gumbo, M. T., & Williams, J. P. (2014). Investigating Grade 8 Technology teachers' pedagogical content knowledge in the Tshwane district of Gauteng province of South Africa. *International Journal of Educational Sciences, 6*(3), 479–488. DOI: 10.1080/09751122.2014.11890159

Gumbo, M. T., & Williams, P. J. (Eds.). (2023). *Indigenous Technology Knowledge Systems: Decolonizing the Technology Education Curriculum*. Springer Nature. DOI: 10.1007/978-981-99-1396-1

Gundara, J. S. (2014). Global and civilisational knowledge: Eurocentrism, intercultural education and civic engagements. *Intercultural Education, 25*(2), 114–127. DOI: 10.1080/14675986.2014.888802

Gupta, A. D. (2011). Does Indigenous Knowledge have anything to deal with Sustainable Development? Antrocom Online Journal of Anthropology. 7 (1). https://antrocom.net/wp/wp-content/uploads/2024/05/gupta-does-indigenous-knowledge-deal-sustainable-development.pdf

Gupta, V. K. (2011, June). Protecting India's traditional knowledge. *WIPO Magazine*. Retrieved from https://www.wipo.int/wipo_magazine/en/2011/03/article_0002.html

Gupta, A. D. (2012). Way to study indigenous knowledge and indigenous knowledge system. *Cultural Anthropology*, *373*, 393.

Gupta, A., & Mishra, D. K. (2018). Public Distribution System in Uttar Pradesh: Access, Utilization and Impact. *Indian Journal of Human Development.*, *12*(1), 20–36. DOI: 10.1177/0973703018774668

Gupta, P., & Singh, U. (2016). Efficiency of Public Distribution System. *Business Review*, *17*, 79–95.

Guru, R. (2024). A study on challenges to the revival of handicrafts in India. *ResearchGate*. Retrieved from https://www.researchgate.net/publication/383661483_A_Study_on_Challenges_to_Revival_of_Handicrafts_in_India

Guskey, T. R. (2002). Professional Development and Teacher Change. *Teachers and Teaching*, *8*(3), 381–391. DOI: 10.1080/135406002100000512

Gustafson, K., (1999). Issue and difficulties in promoting learner reflection: results from a three – year study. *Critical reflection*.

Gyamerah, K. (2024). The Role of African Indigenous Knowledge Systems and Pedagogies in Decolonizing and Transforming Mathematics and Science in Ghana (Doctoral dissertation).

Habermas, J. (1971). *Knowledge and human interests*. Bacon.

Habermas, J. (1985). *The theory of communicative action*. Beacon Press.

Halewood, M. (1999). Indigenous and local knowledge in international law: A preface to sui generis intellectual property protection. *McGill Law Journal. Revue de Droit de McGill*, *44*, 953–996.

Halliday, A. J., Kern, M. L., Garrett, D. K., & Turnbull, D. A. (2019). The student voice in well-being: A case study of participatory action research in positive education. *Educational Action Research*, *27*(2), 173–196. DOI: 10.1080/09650792.2018.1436079

Hamlyn, D. W. (1995). History of epistemology. In Honderich, T. (Ed.), *The Oxford companion to philosophy* (pp. 242–245). Oxford University Press.

Hanglem, A., Thokchom, R., Singh, S. S., & Thokchom, J. (2017). Harnessing the indigenous knowledge base associated with horticultural enterprise for sustainable development of north eastern state, Manipur, India. *International Journal of Current Microbiology and Applied Sciences*, *6*(10), 3266–3274. DOI: 10.20546/ijcmas.2017.610.383

Harding, S. (1987). *Feminism and Method: Incorporating Standpoint Theory*. Open University Press.

Hare, J. (2021). Indigenous Synergies for Decolonizing Action Research. *The Canadian Journal of Action Research*, *21*(3), 4–16. DOI: 10.33524/cjar.v21i3.561

Harrison, G. (2005).. . *Science and the British Empire*, *96*, 13–28.

Harrison, N., & Clarke, I. (2022). Decolonising curriculum practice: Developing the indigenous cultural capability of university graduates. *Higher Education*, *83*(1), 183–197. DOI: 10.1007/s10734-020-00648-6

Hart, M. A. (2010). Indigenous worldviews, knowledge, and research: The development of an Indigenous research paradigm. *Journal of Indigenous Social Development, 1*(1A).

Hayes, A., Luckett, K., & Misiaszek, G. (2021). Possibilities and complexities of decolonising higher education: Critical perspectives on praxis. *Teaching in Higher Education, 26*(7-8), 887–901. DOI: 10.1080/13562517.2021.1971384

Held, V. (2006). *The ethics of care: Personal political and global.* Oxford University Press.

Heleta, S. (2018). Decolonizing knowledge in South Africa: Dismantling the 'pedagogy of big lies'. *Ufahamu. Journal of African Studies, 40*(2).

Hensler, L., Alatorre Frenk, G., & Merçon, J. (2023). Participatory action research. *Handbook Transdisciplinary Learning, 6*, 257–266. Advance online publication. DOI: 10.1515/9783839463475-027

Heron, J., & Reason, P. (2008). Extending Epistemology within a Cooperative Inquiry. In Reason, P., & Bradbury, H. (Eds.), *The SAGE Handbook of Action Research* (2nd ed., pp. 367–380). Sage. DOI: 10.4135/9781848607934.n32

Hickling-Hudson, A., & Ahlquist, R. (2003). Contesting the curriculum in the schooling of Indigenous children in Australia and the United States: From Eurocentrism to culturally powerful pedagogies. *Comparative Education Review, 47*(1), 64–89. DOI: 10.1086/345837

Higgs, P. (2012). African philosophy and the decolonisation of education in Africa: Some critical reflections. *Educational Philosophy and Theory, 44*(2), 22–34. DOI: 10.1111/j.1469-5812.2011.00794.x

Higgs, P. (2016). The African Renaissance and decolonization of the curriculum. In Msila, V., & Gumbo, M. T. (Eds.), *Africanising the curriculum: indigenous perspective and theories* (pp. 1–15). Sun Media Metro.

Hill, R., Adem, ., Alangui, W. V., Molnár, Z., Aumeeruddy-Thomas, Y., Bridgewater, P., Teng , M., Thaman, R., Adou Yao, C. Y., Berkes, F., Carino, J., & Xue, D. (2020). Working with indigenous, local, and scientific knowledge in assessments nature and nature's linkage with people. In *Current opinion in environmental sustainability* (Vol. 43, pp. 8–20). Elsevier B. V., DOI: 10.1016/j.cosust.2019.12.006

Hincapié, M., Díaz, C., Zapata-Cárdenas, M. I., Rios, H. D. J. T., Valencia, D., & Güemes-Castorena, D. (2021). Augmented reality mobile apps for cultural heritage reactivation. *Computers & Electrical Engineering, 93*, 107281. DOI: 10.1016/j.compeleceng.2021.107281

Hoadley, U. (2012). What do we know about teaching and learning in South African primary schools? *Education as Change, 16*(2), 187–202. DOI: 10.1080/16823206.2012.745725

Hoban, G., & Hastings, G. (2006). Developing different forms of student feedback to promote teacher reflection: A 10-year collaboration. *Teaching and Teacher Education, 22*(8), 1006–1019. DOI: 10.1016/j.tate.2006.04.006

Hodges, C., Moore, S., Lockee, B., Trust, T., & Bond, A. (2020). The difference between emergency remote teaching and online learning. *EDUCAUSE Review, 27*(1), 1–9.

Holmberg, B. (2005). *Distance education: Culture and creativity.* Routledge.

Holst, J. D. (2020). Fifty Years of Pedagogy of the Oppressed: A Book Review Essay.

Hooks, B. (1994). *Teaching to transgress: Education as a freedom of practice*. Routledge.

Horbačauskienė, J., Ratkevičienė, M., Kasperė, R., Staniškienė, E., Simonaitienė, B., Stankevičiūtė, Ž., & Ramanauskaitė, J. (2024). How to enhance EdTech integration in secondary education: teachers' and managers' experiences // *Navigating teaching in an era of social distancing and global conflict* / Shalin Hai-Jew (ed.). Hershey, PA: IGI Global. ISBN 9798369333822. eISBN 9798369333839. p. 127-154. DOI: DOI: 10.4018/979-8-3693-3382-2.ch006

House, E. (1992). Response to 'Notes on pragmatism and scientific realism'. *Educational Researcher*, *21*(6), 18–19. DOI: 10.3102/0013189X021006018

Howard, T. C. (2019). *Why race and culture matter in schools: Closing the achievement gap in America's classrooms*. Teachers College Press.

Huang, Z., Nya, E. L., Rahman, M. A., Mwamila, T. B., Cao, V., Gwenzi, W., & Noubactep, C. (2021). Integrated water resource management: Rethinking the contribution of rainwater harvesting. *Sustainability (Basel)*, *13*(15), 8338. DOI: 10.3390/su13158338

Hungwe, J. P., & Mkhize, T. R. (2022). The Africanisation of universities in Africa: Reclamation of humanity and rationality. *Critical Studies in Teaching and Learning*, *10*(1), 136–152. DOI: 10.14426/cristal.v10i1.487

Hutchins, R. M. (1954). *Great Books: The foundation of liberal education*. Simon & Schuster.

Igwe, P. A., Madichie, N. O., & Rugara, D. G. (2022). Decolonising research approaches towards non-extractive research. *Qualitative Market Research*, *25*(4), 453–468. DOI: 10.1108/QMR-11-2021-0135

Ihcjirika, C. (2024). Harnessing African Indigenous Knowledge Systems for Knowledge Production: A Redefinition of a Culture-Centric Epistemology. *Journal of Contemporary Philosophical and Anthropological Studies*, *2*(1). Advance online publication. DOI: 10.59652/jcpas.v2i1.103

Indian Bureau of Mines. (2012, October). *Indian minerals yearbook 2011: Marble (Part-II, 50th ed.)*. Ministry of Mines, Government of India. Government of India. Geographical Indications Registry. Application details of Jabalpur stone craft (Application No. 710).

International Council for Science. (2002, March). *Science and traditional knowledge: Report from the ICSU Study Group on Science and Traditional Knowledge*.

International Labour Organization. (1989). *Convention No. 169 - Indigenous and Tribal Peoples Convention*. Retrieved from https://normlex.ilo.org/dyn/normlex/en/f?p=NORMLEXPUB:12100:0:NO:P12100_ILO_CODE:C169

International Mathematics Union (IMU). (2020). International non-governmental and non-profit scientific organization. Retrieved from http://www.mathunion.org

Irvine, A. (2023). First steps in qualitative secondary analysis: Experiences of engaging with the primary research team. *International Journal of Social Research Methodology*, •••, 1–13. DOI: 10.1080/13645579.2023.2267836

Israel, B. A., Eng, E., Schulz, A. J., & Parker, E. A. (2005). Introduction to methods in community-based participatory research for health. *Methods in community-based participatory research for health, 3*, 26.

Itwaru, S. (2024). Decolonizing higher education pedagogy: From theory to practice. *Collected Essays on Learning and Teaching, 15*(1).

Jacobs, K. R. (2015). *The classroom implementation of indigenous knowledge in the science curriculum by science teachers in the Western Cape province*. DEd Thesis: University of Cape Town, Cape Town.

Jagosh, J., Macaulay, A. C., Pluye, P., Salsberg, J. O. N., Bush, P. L., Henderson, J. I. M., Sirett, E., Wong, G., Cargo, M., Herbert, C. P., Seifer, S. D., Green, L. W., & Greenhalgh, T. (2012). Uncovering the benefits of participatory research: Implications of a realist review for health research and practice. *The Milbank Quarterly*, *90*(2), 311–346. DOI: 10.1111/j.1468-0009.2012.00665.x PMID: 22709390

Jagusah, O. I. W. (2001). Educational policy in Africa and the issue(s) of context: The case of Nigeria and South Africa, International Education Journal, 2(5). *WCCES Commission*, *6*(special), 2001.

Jakavonytė-Staškuvienė, D., Jašinauskas, L., & Kudinovienė, J. (2023). An analysis of prospective primary school teachers' action research: The case of Lithuania. *Educational Action Research*, •••, 1–16. DOI: 10.1080/09650792.2023.2298744

Jakovljevic, M., Bjedov, S., Jaksic, N., & Jakovljevic, I. (2020). COVID-19 pandemia and public and global mental health from the perspective of global health security. *Psychiatria Danubina*, *32*(1), 6–14. DOI: 10.24869/psyd.2020.6 PMID: 32303023

Jegede, O. J., & Aikenhead, G. S. (1999). Transcending cultural borders: Implication for Science teaching. *Research in Science & Technological Education*, *17*(1), 45–66. DOI: 10.1080/0263514990170104

Jesuit Missions. (2014) Hwishu Pada, Tsoro Nhodo and Bvutu Rakaraka, Zimbabwean Children Games Played By The children of the St. Rupert Mayer Primary School. https://jesuitmissions.org.uk

Jha, A. K., Verma, V. K., Deshmukh, N. A., Rymbai, H., Assumi, S. R., Devi, M. B., & Talang, H. D. (2017). Spices for Income Enhancement in NE Region: Needs and Focus. *Training Manual on Entrepreneurship and Skill Development on Market Driven Production and Processing of High Value Crops of NEH Region*, 32.

Jima, A. O. (2022). Significance and restraint of indigenous knowledge inclusion in Ethiopian higher education curriculum: In focus Gadaa system. *Cogent Education*, *9*(1), 1–16. DOI: 10.1080/2331186X.2022.2046241

John, R. M. (2008). Crowding out effect of tobacco expenditure and its implications on household resource allocation in India. *Social Science & Medicine*, *66*(6), 1356–1367. DOI: 10.1016/j.socscimed.2007.11.020 PMID: 18187245

Johnson, D. W., Johnson, R. T., & Smith, K. A. (2011). Cooperative learning: Returning to the classroom. *Educational Leadership*, *68*(7), 26–31.

Johnson, L. R. (2010). The Cultural Implications of Navajo Language Shift. *International Journal of the Sociology of Language*, *2010*(201), 149–158.

Jonassen, D., Spector, M. J., Driscoll, M., Merrill, M. D., van Merrienboer, J., & Driscoll, M. P. (2008). *Handbook of research on educational communications and technology: a project of the association for educational communications and technology*. Routledge.

Joona, T. (2012). *ILO Convention No 169 in a Nordic context with comparative analysis: An interdisciplinary approach.* Rovaniemi: Lapland Univiversity Press.

Joseph, G. (1987). Foundations of Eurocentrism in mathematics. *Race & Class*, *28*(3), 13–28. DOI: 10.1177/030639688702800302

Kajee, L., & Sibanda, R. (2019). Home as a primary space: Exploring out-of-school literacy practices in early childhood education in a township in South Africa. *South African Journal of Childhood Education*, *9*(1), 1–10.

Kalawole, O. (2014). Repositioning local knowledge in development initiatives. In Alvares, C. (Ed.), *Multicultural knowledge and the university* (pp. 251–258). Multiversity and Citizens International.

Kamakaula, Y. (2024). Ethnoecology and climate change adaptation in agriculture. *Global*, *2*(2), 473–485. DOI: 10.59613/global.v2i2.99

Kamlongera, M., & Katenga-Kaunda, M. (2023). Reflections reflections on ethics of care as decolonial research practice: Understanding Indigenous knowledge communication systems to navigate moments of ethical tension in in rural Malawi. *Research Ethics*, *19*(3), 312–324. DOI: 10.1177/17470161231169484

Kamwendo, G. H. (2017). Foreword. In Ngulube, P. (Ed.), *Social, cultural, and educational considerations of indigenous knowledge in developing countries* (p. xix). IGI Global.

Kanene, K. M. (2016). 'Indigenous practices of environmental sustainability in the Tonga community of southern Zambia'. *Jàambá*, *8*(1), 1–7. DOI: 10.4102/jamba.v8i1.331 PMID: 29955300

Kanter, R., Kennedy, G., & Boza, S. (2023). Editorial: Local, traditional, and indigenous food systems in the 21st century to combat obesity, undernutrition, and climate change. *Frontiers in Sustainable Food Systems*, *7*, 1195741. DOI: 10.3389/fsufs.2023.1195741

Kanu, I. A. (2017). The Philosophical Canons of Africa Indigenous Education. *Hummingbird: Journal of African Sustainable Development, 9*(8), 78-90. http://www.academicexcellencesociety.com/the_philosophical_canons_of_african_indigenous_education.pdf

Kasulwe, A. (2014). *African Indigenous Education*. https://patient.over-blog.fr/2014/07/african-indigenous-education-by-apollinaire-kaselwe.html

Kavanagh, K. (Ed.). (1978). *The South African concise Oxford dictionary*. Rhodes University.

Kawagley, A. O. (1995). *A Yupiaq Worldview: A Pathway to Ecology and Spirit.* Waveland Press.

Kayira, J., Lobdell, S., Gagnon, N., Healy, J., Hertz, S., McHone, E., & Schuttenberg, E. (2022). Responsibilities to decolonize environmental education: A co-learning journey for graduate students and instructors. *Societies (Basel, Switzerland)*, *12*(4), 96. DOI: 10.3390/soc12040096

Keane, M., Khupe, C., & Seehawer, M. (2017). Decolonising methodology: Who benefits from indigenous knowledge research?. *Educational Research for social change, 6*(1), 12-24.

Keane, M. (2008). Science education and worldview. *Cultural Studies of Science Education, 3*(3), 587–621. DOI: 10.1007/s11422-007-9086-5

Keerthana, V., & Divakaran, A. (2021). A Study on Adoption of Indigenous Technical Knowledge Practices among Tribal Farmers in Kolli Hills. https://doi.org/DOI: 10.9734/ajaees/2021/v39i1130783

Ke, F., & Grabowski, B. (2007). Game playing for maths learning: Cooperative or not? *British Journal of Educational Technology, 38*(2), 249–259. DOI: 10.1111/j.1467-8535.2006.00593.x

Kemmis, S., McTaggart, R., Nixon, R., Kemmis, S., McTaggart, R., & Nixon, R. (2014). Introducing critical participatory action research. *The action research planner: Doing critical participatory action research*, 1-31.

Kemmis, S. (2009). Action research as a practice-based practice. *Educational Action Research, 17*(3), 463–474. DOI: 10.1080/09650790903093284

Kemmis, S., McTaggart, R., & Nixon, R. (2014). *The action research planner: Doing critical participatory action research.* Springer. DOI: 10.1007/978-981-4560-67-2

Ketsitlile, L. E. (2011). San Junior Secondary Students' Home–School Literacy Disconnection: A Case Study of a Remote Area Dweller School in Botswana. *Diaspora, Indigenous, and Minority Education, 5*(2), 88–99. DOI: 10.1080/15595692.2011.559773

Khasinah, S. (2013). Clasroom action research. *Jurnal Pionir, 1*(1), 107–114.

Khera, R., & Somanchi, A. (2020). A review of the coverage of PDS. Ideas for India: https://www.ideasforindia.in/topics/poverty-inequality/a-review-of-the-coverage-of-pds.html

Khupe, C. (2014). *Indigenous knowledge and school science: Possibilities for integration* (Doctoral dissertation, University of the Witwatersrand, Faculty of Science, School of Science Education).

Khupe, C., & Keane, M. (2017). Towards an African education research methodology: Decolonising new knowledge. *Educational Research for Social Change, 6*(1), 25–37. DOI: 10.17159/2221-4070/2017/v6i1a3

Kidd, S. A., & Kral, M. J. (2005). Practicing participatory action research. [REMOVED HYPERLINK FIELD]. *Journal of Counseling Psychology, 52*(2), 187–195. DOI: 10.1037/0022-0167.52.2.187

Kimbell, R. (2008). Indigenous knowledge, know-how, and design & technology. *Design and Technology Education: An International Journal, 10*(3), 8-10. https://openjournals.ljmu.ac.uk/DesignTechnologyEducation/article/view/1130

Kimberly Sue Peer and Ronald C McClendon. (2002). Sociocultural Learning Theory in Practice: Implications for Athletic Training Educators. *Journal of Athletic Training*. PMID: 12937534

Kindon, S., & Elwood, S. (2009). IntroductioN: More than methods-reflections on participatory action research in geographic teaching, learning and research. *Journal of Geography in Higher Education, 33*(1), 19–32. DOI: 10.1080/03098260802276474

Kindon, S., Pain, R., & Kesby, M. (2007). Participatory action research: Origins, approaches and methods. In Kindon, S., Pain, R., & Kesby, M. (Eds.), *Participatory Action Research Approaches and Methods Connecting People, Participation and Place* (pp. 9–18). Routledge. DOI: 10.4324/9780203933671-13

King, T. (2003). *The Truth About Stories: A Native Narrative*. House of Anansi Press.

Kirkpatrick, S. I., McIntyre, L., & Potestio, M. L. (2010). Child hunger and long-term adverse consequences for health. *Archives of Pediatrics & Adolescent Medicine*, *164*(8), 754–762. DOI: 10.1001/archpediatrics.2010.117 PMID: 20679167

Kirshner, J., & Kamberelis, G. (2021). *Decolonizing transcultural teacher education through participatory action research: Dialogue, culture, and identity*. Routledge. DOI: 10.4324/9781003111573

Kishore, K. (2018). Geographical indications in horticulture: An Indian perspective. *Journal of Intellectual Property Rights*, 159-166.

Kittay, E. (2015). Centering justice on dependency and recovering freedom. *Hypatia*, *30*(1), 285–291. DOI: 10.1111/hypa.12131

Kizilcec, R., Sarı, H., & Uyar, D. A. (2013). The effectiveness of blended learning: A meta-analysis of literature. *Computers & Education*, *62*, 210–221.

Klein, E. S., & Merritt, E. (1994). Environmental Education as a Model for Constructivist Teaching. *The Journal of Environmental Education*, *25*(3), 14–21. DOI: 10.1080/00958964.1994.9941953

Kohl, H., & Kohl, S. (1990). *The Long Haul - an autobiography of Myles Horton*. Doubleday.

Kolawole, O. (2022). Is local knowledge peripheral? The future of indigenous knowledge in research and development. *Alternative*, *18*(1), 121–129. DOI: 10.1177/11771801221088667

Kolb, D. A. (1984). *Experiential Learning: Experience as the Source of Learning and Development*. Prentice Hall.

Kolb, D. A. (1984). *Experiential learning: Experiences as a source of learning and development*. Prentice-Hall.

Koro-Ljungberg, M., Yendol-Hoppey, D., Smith, J. J., & Hayes, S. (2009). (E)pistemological awareness, instantiation of methods, and uninformed methodological ambiguity in qualitative research projects. *Educational Researcher*, *38*(9), 687–699. DOI: 10.3102/0013189X09351980

Koshy, V. (2010). Action research for improving educational practice: A step-by-step guide. *Sage (Atlanta, Ga.)*.

Kosoe, E. A., Adjei, P. O. W., & Diawuo, F. (2020). From sacrilege to sustainability: The role of IKS indigenous knowledge insystems in biodiversity conservation in the Upper West Region of Ghana. *GeoJournal*, *85*(4), 1057–1074. DOI: 10.1007/s10708-019-10010-8

Kovach, M. (2009). *Indigenous Methodologies: Characteristics, Conversations, and Contexts*. University of Toronto Press.

Kpan, M. A. (2020). *Indigenous knowledge systems and the curriculum of basic schools in the Nadowli district of the upper west region*. MA Dissertation: University for Development Studies, Tamale.

Kreber, C. (2005). Reflection on teaching and the scholarship of teaching: Focus on science instructors. *Higher Education*, *50*(2), 323–359. DOI: 10.1007/s10734-004-6360-2

Krishna Panicker, L., Nair, A. S., & Oommen, O. V. (2024). ABS-Bioprospecting Research and Benefit Sharing. In *Biodiversity and Business* (pp. 201–215). Springer. DOI: 10.1007/978-3-031-71674-4

Kruger, C., & Vass, G. (2025). Indigenous knowledges and learners. In *Teaching Middle Years* (4th ed., pp. 36–49). Routledge., https://www.taylorfrancis.com/chapters/edit/10.4324/9781003458586-4/indigenous-knowledges-learners-candace-kruger-greg-vass

Kugara, S. L., Kugedera, A. T. T., Sakadzo, N., Chivhenge, E., & Museva, T. (2022). The role of Indigenous Knowledge Systems in climate change. In Tshifhumulo, R., & Makhanikhe, T. J. (Eds.), *Handbook of Research on Protecting and Managing Global Indigenous Knowledge Systems* (pp. 1–23). IGI Global. DOI: 10.4018/978-1-7998-7492-8.ch001

Kuhnlein, H. V., & Chotiboriboon, S. (2022). Why and How to Strengthen Indigenous Peoples' Food Systems with examples from two unique Indigenous communities. *Frontiers in Sustainable Food Systems*, *6*, 808670. DOI: 10.3389/fsufs.2022.808670

Kuhn, T. (1970). *The structure of scientific revolutions*. Chicago Press: University of Chicago Press.

Kumari, S., Ncube, O., & Munapo, E. (2003). Tsoro and Hungarian Approaches: A Hybrid Algorithm for An Assignment Problem. *Asia-Pacific Journal of Operational Research*, *20*, 41–56.

Kumar, M. J. (2024). Forging Connections: Integrating Indian Knowledge Systems in Higher Education. *IETE Technical Review*, *41*(3), 271–273. DOI: 10.1080/02564602.2024.2342625

Kumbhar, S., & Kharwar, P. (2023). Indigenous Knowledge System and Food Security. *Contemporary Social Science*, *32*(4).

Labonne, J. (2024). *Higher Education for Climate Change Mitigation: Sámi and Indigenous Knowledge Integration in Norwegian University Curricula* (Masters dissertation). Oslo: Oslo Metropolitan University.

Ladson-Billings, G. (2021). *Critical race theory in education: A scholar's journey*. Teachers College Press.

Lakso, L., & Adu, K. (2024). "The unofficial curriculum is where the real teaching takes":faculty experiences of decolonising the curriculum. *Higher Education*, *87*(1), 185–200. DOI: 10.1007/s10734-023-01000-4

Lal, C., & Verma, L. R. (2008). Indigenous technological knowledge on soil and water management from Himachal Himalaya.

Landsberger, H. A. (2021). *Hawthorne Revisited: management and the worker, its critics and developments in human relations in industry*. Cornell University Press.

Lattimer, H. (2012). Action Research in Pre-service Teacher Education: Is There Value Added*? i.e.Inquiry in Education*, *3*(1), 5.

le Grange, L. (2001). Challenges for participatory action research and indigneous knowledge in Africa. *Acta Academica*, *33*(3), 1–15.

Lepore, W., Hall, B. L., & Tandon, R. (2021). The Knowledge for Change Consortium: A decolonising approach to international collaboration in capacity-building in community-based participatory research. *Canadian Journal of Development Studies. Canadian Journal of Development Studies*, *42*(3), 347–370. DOI: 10.1080/02255189.2020.1838887

Leung, M. W., Yen, I. H., & Minkler, M. (2004). Community based participatory research: A promising approach for increasing epidemiology's relevance in the 21st century. *International Journal of Epidemiology*, *33*(3), 499–506. DOI: 10.1093/ije/dyh010 PMID: 15155709

Lewin, K.(2021) Die Sozialisierung des Taylorsystems, Praktischer Sozialismus, No.4.

Lewin, K. (1946). Action research and minority problems. *The Journal of Social Issues*, *2*(1), 34–46. DOI: 10.1111/j.1540-4560.1946.tb02295.x

Lin, J., Stoltz, A., Aruch, M., & Rappeport, A. (2021). Decolonization and transformation of higher education for sustainability: Integrating indigenous knowledge into policy, teaching, research, and practice. *Journal of Comparative & International Higher Education*, *13*(3), 134–156. DOI: 10.32674/jcihe.v13iSummer.3255

Li, Y. (2023). The Influence of Top-Down Mode and Bottom-up Mode to National Innovation and Entrepreneurship. *Journal of Innovation and Development*, *2*(3), 158–161. DOI: 10.54097/jid.v2i3.7640

Loughran, J. J. (2002). Effective Reflective Practice: In Search of Meaning in Learning about Teaching. *Journal of Teacher Education*, *53*(1), 33–43. DOI: 10.1177/0022487102053001004

Louis, R. P. (2007). Can you hear us now? Voices from the margin: Using indigenous methodologies in geographic research. *Geographical Research*, *45*(2), 130–139. DOI: 10.1111/j.1745-5871.2007.00443.x

Lovo, E., Woodward, L., Larkins, S., Preston, R., & Baba, U. (2021). Indigenous knowledge around the ethics of human research from the Oceania Region: A scoping Literature Review. *Philosophy, Ethics, and Humanities in Medicine; PEHM*, *16*(2), 1–14. DOI: 10.1186/s13010-021-00108-8 PMID: 34625108

Ludwig, D., & Poliseli, L. (2018). Relating traditional and academic ecological knowledge: Mechanistic and holistic epistemologies across cultures. *Biology & Philosophy*, *33*(5-6), 43. Advance online publication. DOI: 10.1007/s10539-018-9655-x PMID: 30595613

Lugo-Morin, D. (2020). Indigenous communities and their food systems: A contribution to the current debate. *Journal of Ethnic Foods*, *7*(1), 6. DOI: 10.1186/s42779-019-0043-1

Lugone, M. (2016). The coloniality of gender. In Harcourt, W. (Ed.), *The Palgrave Handbook Gender and Development* (pp. 13–33). Springer. DOI: 10.1007/978-1-137-38273-3_2

Lykes, M., Lloyd, C., & Nicholson, K. (2018). Participatory and action research within and beyond the academy: Contesting racism through decolonial praxis and teaching "against the grain". *American Journal of Community Psychology*, *62*(3-4), 406–418. DOI: 10.1002/ajcp.12290 PMID: 30479010

Maathai, W. (2009). *The challenge for Africa: a new vision*. Random House.

MacDonald, B. (1978). *The Experience of Innovation*. CARE,University of East Anglia.

Macedo, D. (2019). *Decolonizing foreign language education. The Misteaching of English and Other Colonial Languages*. Routledge.

Madaus, G. F. (2022). *Educational Evaluation: classic works of Ralph W. Tyler*. Kluwer.

Maditsi, M. E., & Bhuda, M. T. (2023). Decolonising and Humanising Pedagogies in South African Postgraduate Education: Lessons From Indigenous Knowledge Systems. In *Digital Preservation and Documentation of Global Indigenous Knowledge Systems* (pp. 282-302). IGI Global.

Madondo, F., & Tsikira, J. (2021). Traditional children's games: Their relevance on skills development among rural zimbabwean children age 3–8 years. *Journal of Research in Childhood Education*, •••, 1–15.

Maelene, M. (2002). Implications of indigenous African beliefs and customs for health education and prevention of HIV/AIDS. *Nutrition Update*, *26*(8), 14–17.

Magni, G. (2016). *Background paper prepared for the 2016 Global Education Monitoring Report*. UNESCO.

Maguire, P. (1987). Doing Participatory Research: A Feminist Approach.

Mahadew, A., & Hlalele, D. J. (2022). Understanding inclusion in early childhood care and education: A participatory action learning and action research study. *South African Journal of Childhood Education*, *12*(1), 1073. DOI: 10.4102/sajce.v12i1.1073

Mahamallik, M., & Sahu, G. B. (2011). Identification of the poor: Errors of exclusion and inclusion. *Economic and Political Weekly*, *46*(9), 71–77.

Maharajh, L. R., & Chilli, B. (2021). English as an Alien Language: Teachers' Experiences of the English Across the Curriculum Strategy in South Africa.

Maine, M. K. (2022). Towards a decolonised research curriculum in psychology: A participatory action research exploration of students' perspectives (Doctoral dissertation, University of Pretoria).

Maira, J. N., & Kauka, E. O. (2018). Analysis of Ecopedagogy: Lessons from African Indigenous Education. *International Journal of Social Science and Human Research*, *6*(2), 108–114.

Malkani, J., & Allen, J. (2005). *Cases in teacher education: beyond reflection into practice.*

Maluleke, R. (2019). *A pedagogy for technology education: An Indigenous perspective* (Doctoral thesis). Pretoria: University of South Africa.

Maluleke, R. (2019). *A pedagogy for technology education: an indigenous perspective*. DEd Thesis: University of South Africa, Pretoria.

Manathunga, C. (2020). Decolonising higher education: Creating space for Southern knowledge systems. *Scholarship of Teaching and Learning in the South*, *4*(1), 4–25. DOI: 10.36615/sotls.v4i1.138

Mapara, J. (2022). Indigenous Knowledge across the curriculum: An epistemic and cognitive shifting process. In Muyambo, T., Mapangisana-Hlatywayo, A., & Sithole, P. (Eds.), *Re-imagining Indigenous Knowledge and Practices in 21st Century Africa: Debunking Myths and Misconceptions for Conviviality and Sustainability* (pp. 41–68). Langaa Publishers.

Mapotse, T. A. (2012). The teaching practice of senior phase Technology Education teachers in selected schools of Limpopo Province: An Action Research Study. DEd Thesis: University of South Africa, Pretoria.

Mapotse, T. (2014). Technology Education and Action Research: Two Complementary Processes to Capacitate Teachers. *Mediterranean Journal of Social Sciences*, •••, 5. DOI: 10.5901/mjss.2014.v5n3p338

Mapotse, T. A. (2018). Development of a technology education cascading theory through community engagement site-based support. *International Journal of Technology and Design Education*, *28*(3), 685–699. DOI: 10.1007/s10798-017-9411-6

Mapotse, T. A. (2020). Developing Technology teachers through Action Research: An African perspective. In Gumbo, M. T. (Ed.), *Decolonization of Technology Education: African indigenous perspectives* (pp. 331–350). Peter Lang.

Marquardt, M. J., & Waddill, D. (2020). Core concepts of action learning: set problem, action, reflection, and learning. In *Action Learning for Improving Practice* (pp. 67–89). Daves –Black Publishers.

Martinez-Vargas, C. (2020). Decolonising higher education research: From a uni-versity to a pluri-versity of approaches. *South African Journal of Higher Education*, *34*(2), 112–128. DOI: 10.20853/34-2-3530

Martin, S., Nicolas, C., Burq, H., & Rives, A. (2019). Enhancing cultural heritage education with augmented reality technologies. *Journal of Computer Assisted Learning*, *35*(1), 74–87. https://link.springer.com/book/10.1007/978-3-030-42156-4

Matos, J. F., Piedade, J., Freitas, A., Pedro, N., Dorotea, N., Pedro, A., & Galego, C. (2023). Teaching and Learning Research Methodologies in Education: A Systematic Literature Review. *Education Sciences*, *13*(2), 173. DOI: 10.3390/educsci13020173

Matriano, E. A. (2020). Ensuring student-centred constructivist and project-based experiential learning applying the Exploration, Research, Interaction and Creation Learning Model. *International Online Journal of Education & Teaching*, *7*(1), 214–227. https://iojet.org/index.php/IOJET/article/view/727

Matsekoleng, T. K. (2022). *Challenges Confronting Technology Education Teachers in Raising learners' Awareness on littering in Gauteng Province* (DEd thesis). Pretoria: University of South Africa.

Matsekoleng, T. K., Mapotse, T. A., & Gumbo, M. T. (2024). The role of indigenous games in education: A technology and environmental education perspective. *Diaspora, Indigenous, and Minority Education*, *18*(1), 68–82. DOI: 10.1080/15595692.2022.2160436

Mattu, V. K. (2017). Perspectives and challenges of apiculture for sustainable agriculture under changing climatic conditions. *Journal of Research: THE BEDE ATHENAEUM*, *8*(1), 130–142. DOI: 10.5958/0976-1748.2017.00018.2

Mawere, M., & Mubaya. (2016). *African philosophies and thought systems: A search for a culture and philosophy of being*. Bameda: Langaa Research and Publishing CIG.

Mawere, M. (2014). *Culture, indigenous knowledge and development in Africa: Reviving interconnections for sustainable development*. Langaa Rpcig.

Maweu, J. M. (2011, December). *Indigenous Ecological Knowledge and Modern Western Ecological Knowledge: Complementary, not Contradictory*, A Journal of the Philosophical Association of Kenya (PAK). *New Series*, *3*(2), 35–47.

Mbah, M. F., & Bailey, M. (2022). Decolonisation of research methodologies for sustainable development in Indigenous settings. In *Indigenous Methodologies, Research and Practices for Sustainable Development* (pp. 21–48). Springer International Publishing. DOI: 10.1007/978-3-031-12326-9_2

Mbah, M., Ajaps, S., & Molthan-Hill, P. (2021). A systematic review of the deployment of indigenous knowledge systems towards climate change adaptation in developing world contexts: Implications for climate change education. *Sustainability (Basel)*, *13*(9), 4811. DOI: 10.3390/su13094811

Mbasalaki, P. K. (2023). *14 Building Coalitions Across Structural Borders As A Form Of Radical Intimacy And Kinship1*. The Sage Handbook of Global Sociology.

Mbeki, T. (1998). The African Renaissance, South Africa, and the world. South African Deputy President Thabo Mbeki speaks at the United Nations University 9 April 1998.

Mbembe, A. J. (2016). Decolonizing the university: New directions. *Arts and Humanities in Higher Education*, *15*(1), 29–45. DOI: 10.1177/1474022215618513

Mbirimi-Hungwe, V. (2019). Stepping beyond linguistic boundaries in multilingual Science education: Lecturer's perceptions of the use of translanguaging. *Southern African Linguistics and Applied Language Studies*, *37*(1), 15–26. DOI: 10.2989/16073614.2019.1598877

McConaghy, C. (2000). *Rethinking Indigenous Education: Culturalism, Colonialism and the Politics of Knowing*. Post Pressed.

McGowan, K., Kennedy, A., El-Hussein, M., & Bear Chief, R. (2020). Decolonization, social innovation and rigidity in higher education. *Social Enterprise Journal*, *16*(3), 299–316. DOI: 10.1108/SEJ-10-2019-0074

Mchombo, S. (2019). Verbal arts as culturally relevant pedagogical tools in math/science education. In *Promoting language and STEAM as human rights in education* (pp. 17–38). Springer. DOI: 10.1007/978-981-13-2880-0_2

McHugh, T. L. F., & Kowalski, K. C. (2009). Lessons learned: Participatory action research with young Aboriginal women. *Pimatisiwin*, *7*(1), 117–131.

McKee, A. (2003). Textual analysis: A beginner's guide. *Metro Magazine*, *127/128*, 138–149.

McKinley, E., & Smith, L. T. (2019). *Handbook of Indigenous Education*. Springer. DOI: 10.1007/978-981-10-3899-0

McLeod, S. (2024). Albert Bandura's social learning theory. *Simply Psychology. London*. https://www.simplypsychology.org/bandura.html

Means, B., Toyama, Z., & Vavrus, S. (2009). Implementing blended learning: Where the rubber meets the road. *The International Journal of Learning and Teaching in Educational Settings*, *1*(1), 3–24.

Medina, S., & Boharquez, Y. (2020). Modern and postmodern views of education that shaped EFL mentoring in the teaching practicum. *Colombian Applied Lingustic Journal Vo*, *22*(1), 1–5.

Mehta, R., & Henriksen, D. A. (2022). To Democratize, First Decolonize: Approaches Beyond Eurocentric and Colonial Epistemologies in Creativity. *Review of Research in Education*, *46*(1), 105–133. DOI: 10.3102/0091732X221084324

Melash, A. A., Bogale, A. A., Migbaru, A. T., Chakilu, G. G., Percze, A., Ábrahám, É. B., & Mengistu, D. K. (2023) Indigenous agricultural knowledge: A neglected human based resource for sustainable crop protection and production. Heliyon 9(1): e12978 9 p. ISSN: 2405-8440

Mentz, E., & De Beer, J. (2019). The use of Cultural-Historical Activity Theory in researching the affordances of indigenous knowledge for self-directed learning. In J. De Beer (Ed.), *The decolonisation of the curriculum project: The affordances of indigenous knowledge for self-directed learning* (NWU Self-directed Learning Series Volume 2), pp. 49-86. Cape Town: AOSIS. https://doi.org/DOI: 10.4102/aosis.2019.BK133.03

Merriam-Webster. (n.d.). Didactic Definition and Meaning. In *Merriam-Webster.com dictionary*. Retrieved September 10, 2024, from https://www.merriam-webster.com/dictionary/didactic [Didactic Definition & Meaning - Merriam-Webster].

Mertler, C. (2024). *Action Research: Improving Schools and Empowering Educators*. SAGE Publications, Inc.

Meyer, M. A. (2008). *Indigenous and Authentic: Hawaiian Epistemology and the Triangulation of Meaning*. AltaMira Press.

Meyiwa, T., Letsekha, T., & Wiebesiek, L. (2013). Masihambisane, Lessons Learnt Using Participatory Indigenous Knowledge Research Approaches in a School-Based Collaborative Project of the Eastern Cape. *South African Journal of Education*, *33*(4), 1–15. DOI: 10.15700/201412171329

Michie, M., Hogue, M., & Rioux, J. (2023). Two-Ways thinking and Two-Eyed Seeing as ways of implementing Indigenous perspectives in the science education curriculum. *Disciplinary and Interdisciplinary Science Education Research*, *5*(23), 1–14. DOI: 10.1186/s43031-023-00084-3

Mignolo, W. D. (2020). On decoloniality: Second thoughts. *Postcolonial Studies*, *23*(4), 612–618. DOI: 10.1080/13688790.2020.1751436

Mihesuah, D. A. (2004). *Indigenous American Women: Decolonization, Empowerment, Activism*. University of Nebraska Press.

Miller, C. A. (2007). "Action Research: Making Sense of Data." On-line article recovered 11/05/2007, www.coe.fau.edu/sfcel/sensdata.htm

Mishra, A., Singh, S. R. K., & Raut, A. A. (2020). Traditional Knowledge in Agriculture. Division of Agricultural Extension, ICAR-Agricultural Technology Application Research Institute, New Delhi. https://icar.org.in/sites/default/files/2022-06/IITKA_Book_Traditional-Knowledge-in-Agriculture-English_0_0.pdf

Misra, V., & Misra, P. K. (2020). Implications of constructivist approaches in the classrooms: The role of the teachers. *Asian Journal of Education and Social Studies*, *11*, 17–25. DOI: 10.9734/ajess/2020/v7i430205

Mitrovic, L. (1999). New social paradigm Habermas's theory of communicative action. *The Scientific Journal of FACTA UNIVERSITATIS*, *6*(2), 217–223.

Mji, G. (2012). The health knowledge utilised by rural older Xhosa women in the management of health problems in their home situation, with a special focus on indigenous knowledge, DEd Thesis: University of Stellenbosch, Cape Town.

Mnguni, J. N. C., & Mapotse, T. T. (2024). Decolonizing Postgraduate Studies Embracing African Spirituality in South African Higher Education. In Global Perspectives on Decolonizing Postgraduate Education (pp. 55-81). IGI Global. DOI: 10.4018/979-8-3693-1289-6.ch005

Mnguni, J. N. C. (2013). *Some challenges in teaching natural sciences in the context of National Curriculum Statement, MEd.* University of South Africa.

Mnguni, J. N. C., & Mapotse, T. A. (2023). Examining the Technological Divide Between Africa and the Western World: A Case of South Africa. In Gumbo, M. T., & Williams, P. J. (Eds.), *Indigenous Technology Knowledge Systems. Contemporary Issues in Technology Education*. Springer., DOI: 10.1007/978-981-99-1396-1_14

Mogege, M. (2017). Mathematical concepts from community elders: Exploring the connection between ethnomathematical contexts and classroom practices. *ETD-Educação Temática Digital*, *19*(3), 667–686. DOI: 10.20396/etd.v19i3.8648368

Molebatsi, M. (2006). The Africanisation of the university in Africa. *Alternation (Durban)*, *13*(1), 49–69.

Mooney-Somers, J., Perz, J., & Ussher, J. M. (2008). A complex negotiation: Women's experiences of naming and not naming premenstrual distress in couple relationships. *Women & Health*, *47*(3), 57–77. DOI: 10.1080/03630240802134134 PMID: 18714712

Moore, M. G. (1983). The individual adult learner. In Tight, M. (Ed.), *Adult Learning and Education* (pp. 153–168). Croom Helm.

Moran, M. J. (2007). Collaborative action research and project work: Promising practices for developing collaborative inquiry among early childhood preservice teachers. *Teaching and Teacher Education*, *23*(4), 418–431. DOI: 10.1016/j.tate.2006.12.008

Mosimege, M. (2012). Methodological challenges in doing ethnomathematical research. *International Journal of African Renaissance Studies-Multi-. Inter-and Transdisciplinarity*, *7*(2), 59–78.

Mosimege, M. (2020). The use of indigenous games in the teaching and learning of Mathematics. *Revemop*, *2*, e202009–e202009.

Mpoe, J., & Swartz, L. (2019). Decolonising research methodologies: Lessons from a qualitative research project, Cape Town, South Africa. *Global Health Action*, *12*, 1–7. PMID: 30661464

Mpofu, V., Otulaja, F. S., & Mushayikwa, E. (2014). Towards culturally relevant classroom Science: A theoretical framework focusing on traditional plant healing. *Cultural Studies of Science Education*, *9*(9), 221–242. DOI: 10.1007/s11422-013-9508-5

Mpuangnan, K. N., & Ntombela, S. (2024). Community voices in curriculum development. *Curriculum Perspectives*, *44*(1), 49–60. DOI: 10.1007/s41297-023-00223-w

Msila, V., & Gumbo, M. T. (Eds.). (2016). *Africanising the curriculum: Indigenous perspectives and theories*. African Sun Media.

Muchenje, F., Gora, R. B., & Makhuvaza, N. (2016). Interrogating the concept of time among the Shona: a post-colonial discourse. In Emeagwali, G., & Shizha, E. (Eds.), *African Indegenous Knowledge and the Sciences: Journeys into the past and present* (pp. 79–92). Sense. DOI: 10.1007/978-94-6300-515-9_6

Mugari A, Matemera N (2015). *Demystifying Mathematical Operations: Leveraging On The Power Of Traditional Games,* European Journal of Physical Education and Sport Science, ISSN: 2501 - 1235

Muigua, P. D. (2021). Revisiting the Place of Indigenous Knowledge in the Sustainable Development Agenda in Kenya. Journal of cmsd, 7(2).

Mwangi, R., Maina, S. M., & Munene, M. (2023). Participatory Action Research (Par) for Community Based Conservation in Kenya. *Africa Design Review Journal*, *1*(4), 480–496.

My Courses Editor. (2024, January). Applying Indigenous Knowledge Systems (IKS) in Design Project. https://mycourses.co.za/applying-indigenous-knowledge-systems-iks-in-design-projects/

NABARD. (2020). Impact Assessment of COVID-19 on Indian Agriculture and Rural Economy. Department of Economic Analysis and Research. National Bank for Agriculture and Rural Development. https://www.nabard.org/auth/writereaddata/tender/1211203145Impact%20Assessment%20of%20COVID.pdf

Nadasdy, P. (2003). *Hunters and Bureaucrats: Power, Knowledge, and Aboriginal-State Relations in the Southwest Yukon*. UBC Press.

Nadeau, L., Gaulin, D., Johnson-Lafleur, J., Levesque, C., & Fraser, S. (2022, December). The challenges of decolonising participatory research in indigenous contexts: The Atautsikut community of practice experience in Nunavik. *International Journal of Circumpolar Health*, *81*(1), 2087846. DOI: 10.1080/22423982.2022.2087846 PMID: 35979584

Nair, K. P. (2023). *Biodiversity in Agriculture: Sustainability of Soil, Soil Fauna and Soil Flora*. Springer Nature., DOI: 10.1007/978-3-031-44252-0

Nakata, M. (2007). *Disciplining the Savages: Savaging the Disciplines*. Aboriginal Studies Press.

Narayanan, S. (2015). Food Security in India: The Imperative and Its Challenges. *Asia & the Pacific Policy Studies*, *2*(1), 197–209. DOI: 10.1002/app5.62

Ndlovu, E. C. (2023). *Senior phase technology teachers' integration of indigenous creative design processes in systems and control* (Doctoral Thesis). https://hdl.handle.net/10500/30901

Ndlovu, E. C., & Gumbo, M. T. (2018). Technology teachers' integration of technology-society-environment in teaching-learning activities. *Paper presented at the9th Annual UNISA ISTE Conference on Mathematics, Science and Technology Education.* Skhukhuza, South Africa

Ndlovu, E. C., & Gumbo, M. T. (2024). Decolonising technology education: Integrating indigenous knowledge for sustainable development in electrical and mechanical systems and control. African Journal of Research in Mathematics. *Science and Technology Education*, *28*(3), 362–372.

Ndlovu-Gatsheni, S. J. (2015). Decoloniality in Africa: A continuing search for a new world order. *Australasian Review of African Studies. The*, *36*(2), 22–50.

Ndonye, H. N., Mulwa, A., & Kyalo, D. N. (2021). Stakeholder engagement and performance of community based conservation projects: A case of Laikipia Conservation Region Conservancies. *Journal of Economics and Sustainable Development*, *12*(4), 41–50.

Nel, P. (2005). Indigenous knowledge systems: Contestations rhetoric and space. *Indilinga*, *4*(1), 1–13.

Nelson, M. K., & Shilling, D. (Eds.). (2018). *Traditional ecological knowledge: Learning from Indigenous practices for environmental sustainability*. Cambridge University Press., DOI: 10.1017/9781108552998

Newcastle University. (2024, 05 2). *Participatory action research*. Retrieved from https://research.ncl.ac.uk/methodshub/methods/participatoryactionresearch/

Ngcaweni, B. (2024, March). BRICS building a "pluriversal" world. *UBUNTU: Stronger Together, 31*, 34-35, 110. South Africa's Public Diplomacy in action: Department of International Relationship and Cooperation (DIRCO).

Ngulube, P. (Ed.). (2016). *Handbook of research on social, cultural, and educational considerations of indigenous knowledge in developing countries*. IGI global.

Nicoline, D. H. (2024). Transforming Food Systems Must Include Indigenous Peoples' Rights, Voices and Knowledge. Food Security and Nutrition. Farming First. https://farmingfirst.org/2024/08/transforming-food-systems-must-include-indigenous-peoples-rights-voices-and-knowledge/

Nkambule, B., Ngubane, S. A., & Mashiane-Nkabinde, B. (2024). Management of Context-Specific Knowledge Creation and Application through Collective Social Responsibility in School? Community Partnerships. International Journal of Learning. *Teaching and Educational Research*, *23*(7), 499–524.

Nkhata, B., Mkandawire, S. B., Nachiyunde, K., Phiri-Nalube, P., Kaani, B., Mulenga, I. M., & Chipindi, F. M. (2019). Exploring selected Theories applicable to Educational Disciples and Social Sciences Research. *International Journal of Humanities Social Sciences and Education*, *6*(12), 97–116. http://dx.org/10.20431/2349-0381.0612008

Nnama-Okechukwu, C. U., & McLaughlin, H. (2022). Indigenous knowledge and social work education in Nigeria: Made in Nigeria or made in the West? *Social Work Education*, •••, 1–18.

Novak, J. D., & Gowin, B. D. (1984). *Learning How To Learn*. Cambridge University Press. DOI: 10.1017/CBO9781139173469

Nur, A. S., Waluya, S. B., Rochmad, R., & Wardono, W. (2020). Contextual Learning with Ethnomathematics in Enhancing the Problem Solving Based on Thinking Levels. *Journal of Research and Advances in Mathematics Education*, *5*(3), 331–344. DOI: 10.23917/jramathedu.v5i3.11679

Nyerere, J. (1968). *Ujamaa: essay on socialism*. Oxford University Press.

Nyota, S., & Mapara, J. (2008). Shona traditional children's games and play: Songs as indigenous ways of knowing. *The Journal of Pan African Studies*, *2*(4).

O'Donoghue, R., van Staden, W., Bhurekeni, J., Snow-Macleod, J., & Ndlamlenze, L. (2024). A Formative Study towards the inclusion of Indigenous Technologies and Knowledge practices in Science, Technology, Engineering, Arts, and Mathematics (STEAM) Curriculum settings. *Educational Research for Social Change*, *13*(1), 34–47. DOI: 10.17159/2221-4070/2024/v13i1a3

Obiora, A. C., & Emeka, E. E. (2015). African Indigenous knowledge system and environmental sustainability. *International Journal of Environmental Protection and Policy*, *3*(4), 88–96. DOI: 10.11648/j.ijepp.20150304.12

O'Donoghue, R. K. (1993). Environment, development, and environmental education: An environmental policy and curriculum initiative in formal education. Howick: Share-net.

Odora-Hoppers, C. A. (2001). Poverty, power, and partnerships in educational development: Post-victimology perspective. *Compare: A Journal of Comparative Education*, *31*(1), 21–38. DOI: 10.1080/03057920020030144

Office for Huma Research Protection. (1979, 04 18). *The Belmont Report*. Retrieved from Ethical Principles and Guidelines for the Protection of Human Subjects of Research: https://www.hhs.gov/ohrp/regulations-and-policy/belmont-report/read-the-belmont-report/index.html

Office of Controller General of Patents. (2022). Designs and Trademarks. *Geographical indications journal No. 166*. Department for Promotion of Industry and Internal Trade, Ministry of Commerce and Industry, Government of India.

Office of Controller General of Patents. (2022). Designs and Trademarks. *Registered geographical indications*. Department for Promotion of Industry and Internal Trade, Ministry of Commerce and Industry, Government of India. https://ipindia.gov.in/registered-gls.htm

Oliver, R. (2013). The Role of Indigenous Knowledges in Science Education: Learning from Two-Way Science. *Australian Journal of Indigenous Education*, *42*(2), 179–189.

Omodan, B. I., & Dastile, N. P. (2023). Analysis of participatory action research as a decolonial research methodology. *Social Sciences (Basel, Switzerland)*, *12*(9), 507. DOI: 10.3390/socsci12090507

Omodan, B., & Dastile, N. (2023). Analysis of participatory action research as a decolonial. *Social Science*, *12*(507), 1–11.

Opoku, M., & James, A. (2020). Challenges of teaching Akan (Ghana) culturally specific environmental ethics in senior high schools: Voices of Akan Indigenous Knowledge -holders and biology teachers. *Southern African Journal of Environmental Education*, *36*, 33–50. DOI: 10.4314/sajee.v36i1.4

Orlowski, P. (2019). Freirean conceptions of participatory action research and teaching for social justice -Same struggle different fronts. *Canadian Journal of Action Research*, *20*(1), 30–51. DOI: 10.33524/cjar.v20i1.445

Owusu-Ansah, F. E., & Mji, G. (2013). African indigenous knowledge and research. *African Journal of Disability*, *2*(1), 1–5. DOI: 10.4102/ajod.v2i1.30 PMID: 28729984

Owusu, S., & Asante, R. (2020). Rainwater harvesting and primary uses among rural communities in Ghana. *Journal of Water, Sanitation, and Hygiene for Development : a Journal of the International Water Association*, *10*(3), 502–511. DOI: 10.2166/washdev.2020.059

Ozkul, O. (2020). Communication Action Theory: A system-lifeworld compatibility or incompatibility? *Journal of Social Sciences and Humanities*, •••, 215–237.

Pandey, N. K., Somvanshi, S. P. S., Prakash, O., & Kumar, S. (2020). Indigenous knowledge of agriculture and animal husbandry practiced by monpa tribes of tawang Arunachal Pradesh. *The Pharma Innovation Journal*, *9*(4), 1012–1016.

Pandey, V., Mittal, R., & Sharma, P. (2017). Documentation and application of indigenous traditional knowledge (itk) for sustainable agricultural development. *Asian Journal of Agricultural Extension Economics & Sociology*, *15*(3), 1–9. DOI: 10.9734/AJAEES/2017/31481

Pangrazi, R. P., & Beighle, A. (2019). *Dynamic physical education for elementary school children*. Human Kinetics Publishers.

Pant, R. (2015). Protecting and promoting traditional knowledge in India: What role for geographical indications? *IIED*. Retrieved from https://www.iied.org/16576iied#:~:text=Sustainable%20markets-,Protecting%20and%20promoting%20traditional%20knowledge%20in,what%20role%20for%20geographical%20indications%3F&text=Geographical%20indications%20(GIs)%20can%20help,and%20promotion%20of%20traditional%20knowledge

Paris, D., & Alim, H. S. (2017). *Culturally Sustaining Pedagogies: Teaching and Learning for Justice in a Changing World*. Teachers College Press.

Parker, P., Smith, S., & Dennison, J. (2017). Decolonising the classroom creating and sustaining revolutionary spaces inside the academy. *TIJDSCHRIFT VOOR GENDERSTUDIES*, *20*(3), 233–247. DOI: 10.5117/TVGN2017.3.PARK

Parsons, R. D., & Brown, K. S. (2002). *Teacher as Reflective Practitioner and Action Researcher*. Wadsworth/Thomson Learning.

Patel, S. K., Sharma, A., & Singh, G. S. (2020). Traditional agricultural practices in India: An approach for environmental sustainability and food security. *Energy, Ecology & Environment*, *5*(4), 253–271. DOI: 10.1007/s40974-020-00158-2

Peltier, C. (2018). An application of two-eyed seeing: Indigenous research methods with participatory action research. *International Journal of Qualitative Methods*, *17*(1), 1609406918812346. DOI: 10.1177/1609406918812346

Peng, P., Lin, X., Ünal, Z. E., Lee, K., Namkung, J., Chow, J., & Sales, A. (2020). Examining the mutual relations between language and mathematics: A meta-analysis. *Psychological Bulletin, 146*(7), 595–634. DOI: 10.1037/bul0000231 PMID: 32297751

Phillips, A. (2017). *Afrophobia, a result of colonialism and global capitalism*. University of South Africa.

Phillips, D. K., & Carr, K. (2010). *Becoming a Teacher through Action Research: Process, Context, and Self-Study* (2nd ed.). Routledge., DOI: 10.4324/9780203861776

Philo-Notes. (2023, 03 22). *Habermas's theory of communicative action: Key concepts*. Retrieved from Philo-Notes: http://philonotes.com/2023/03/habermass-theory-of-communicative-action-key-concepts

Pickering, L., & Kara, H. (2017). Presenting and representing others: Towards an ethics of engagement. *International Journal of Social Research Methodology, 20*(3), 1–15. DOI: 10.1080/13645579.2017.1287875

Pihama, L., Smith, L. T., Taki, M., & Lee, J. (2004). A Literature Review on Kaupapa Maori and Maori Education Pedagogy. *International Journal of Learning*, *11*, 181–188.

Pine, G. J. (2009). *Teacher Action Research: Building Knowledge Democracies*. SAGE Publications. DOI: 10.4135/9781452275079

Planent Anomaly. (2023, 4 16). *Mapped: The World Indigenous Peoples*. Retrieved from Mapped: The world indigenous peoples: https://www.visualcapitalist.com/cp/mapped-the-worlds-indigenous-peoples/

Pooley, S. (2021). Fire in African landscapes. In *Oxford Research Encyclopedia of African History*. http://dx.doi.org/DOI: 10.1093/acrefore/9780190277734.013.984

Pophiwa, N., & Saidi, U. (2022). Approaches to embedding IKS in Made in Africa Evaluations. *African Evaluation Journal, 10*(1), 623. DOI: 10.4102/aej.v10i1.623

Prensky, M. (2001). Digital Natives, Digital Immigrants Part 1. *On the Horizon*, *9*(5), 1–6. DOI: 10.1108/10748120110424816

Prior, D. (2007). Decolonising research: A shift toward reconciliation. *Nursing Inquiry*, *14*(2), 162–168. DOI: 10.1111/j.1440-1800.2007.00361.x PMID: 17518828

Puustinen, M., Säntti, J., Koski, A., & Tammi, T. (2018). Teaching: A practical or research-based profession? Teacher candidates' approaches to research-based teacher education. *Teaching and Teacher Education*, *74*, 170–179. DOI: 10.1016/j.tate.2018.05.004

Rajan, S., Sharma, M., & Mishra, P. (•••). Geographical Indication for Horticulture Commodities in Northeastern India: Opportunities and Challenges. *Indian Horticulture*, *68*(6), 21–24.

Ramose, M. B. (2003). I doubt, therefore African philosophy exists. *South African Journal of Philosophy*, *22*(2), 110–121. DOI: 10.4314/sajpem.v22i2.31364

Randle, H., & Tilander, K. (2007). Reflection: A method for organizational and individual development. *US-China Education Review.*, *4*(9), 1–10.

Rangana, N., Geduld, D., & Sathorar, H. (2023). Multiple stakeholder collaborations to co-develop school improvement plans in Community schools: A Participatory Action Learning and Action Research (PALAR) study. *South African Journal of Educational Leadership and Management*, *3*(1), 5–25.

Rannikmäe, M., Holbrook, J., & Soobard, R. (2020). In Akpan, B., & Kennedy, T. J. (Eds.), *Science education in theory and practice: An introductory guide to learning theory* (pp. 259–275). Springer. DOI: 10.1007/978-3-030-43620-9_18

Rao, C. H. H. (2000). Declining demand for food grains in rural India causes and implications. *Economic and Political Weekly*, *35*(4), 112–123.

Räsänen, P., Laurillard, D., Käser, T., & Aster, M. V. (2019). Perspectives to technology-enhanced learning and teaching in mathematical learning difficulties. In *International handbook of mathematical learning difficulties* (pp. 733–754). Springer. DOI: 10.1007/978-3-319-97148-3_42

Rasool, S., Cerchione, R., Centobelli, P., Oropallo, E., & Salo, J. (2024). Sustainability and socially responsible food consumption: An empirical investigation based on self-awareness and self-congruity theories. *Asia Pacific Journal of Marketing and Logistics*, *36*(4), 993–1023. DOI: 10.1108/APJML-07-2022-0559

Rasool, S., & Harms-Smith, L. (2021). Towards decoloniality in a social work programme: A process of dialogue, reflexivity, action and change. *Critical African Studies*, *13*(1), 56–72. DOI: 10.1080/21681392.2021.1886136

Rawat, D. S., & Kharwal, A. D. (2016). Folk knowledge on indigenous practices in North-West Himalaya with special reference to Himachal Pradesh (HP), India. *Folk (Kobenhavn)*, *1*(8). Advance online publication. DOI: 10.1007/s10745-011-9396-9

Rawat, V. S., Rawat, Y. S., & Shah, S. (2010). Indigenous knowledge and sustainable development in the Tones Valley of Garhwal Himalaya. *Journal of Medicinal Plants Research*, *4*(19), 2043–2047. DOI: 10.5897/JMPR10.191

Ray, L. (2021). Decolonizing action research through Two-Eyed Seeing: The Indigenous quality assurance project. *The Canadian Journal of Action Research*, *21*(3), 95–114. DOI: 10.33524/cjar.v21i3.510

Reason, P. (1999). Integrating Action and Reflection through Cooperative Inquiry. *Management Learning*, *30*(2), 207–226. DOI: 10.1177/1350507699302007

Redmond, B. (2004). *Reflection in action: developing reflective practice in health and social services*. Ashgate.

Reinhardt, K. S., Robertson, P. M., & Johnson, R. D. (2021). Connecting inquiry and Universal Design for Learning (UDL) to teacher candidates' emerging practice: Development of a signature pedagogy. *Educational Action Research*, *31*(3), 437–454. DOI: 10.1080/09650792.2021.1978303

Remenyi, D., Williams, B., Money, A., & Swartz, E. (2005). *Doing research in business and management: An introduction to process and method* (1st ed.). SAGE Publication.

Revathi, T. (2020). Cooperative learning as a constructivist teaching-learning strategy. *International Journal of Pedagogical Studies*, *8*(1), 84–91.

Ritchie, J., & Rau, C. (2006). Enacting a Te Whāriki Curriculum: The Complexities of Bicultural Involvement. *Early Childhood Folio*, *10*, 20–24.

Ritterbusch, A. E. (2019). Empathy at knifepoint: The dangers of research and;its pedagogies for social justice movements. *Antipode*, *51*(4), 1296–1317. DOI: 10.1111/anti.12530

Rivero-Romero, A., Moreno-Calles, A., Casas, A., Melgoza-Castillo, A., & Camou-Guerrero, A. (2016). Traditional climate knowledge: A case study in a peasant community of tlaxcala, mexico. *Journal of Ethnobiology and Ethnomedicine*, *12*(1), 33. Advance online publication. DOI: 10.1186/s13002-016-0105-z PMID: 27539307

Robson, C. (2002). *Real world research* (2nd ed.). Blackwell Publishing Ltd.

Rogers, C. R. (1969). *Freedom to learn*. Charles E. Merrill Publishing.

Romaniuk, S. N. (2017). Dependency: "Dependency Theory" SAGE Publications, Inc. City: Thousand Oaks, ISBN: 9781483359892 pages: 482-483

Roy, R. (2014). *The untold story of modern science is one of emoire and colonial exploitation*. Quartz India.

Roy, S., Rathod, A., Sarkar, S., & Roy, K. (2015). Use of ITK in plant protection. *Popular Kheti*, *3*(2), 75–78.

Ruggiano, N., & Perry, T. E. (2019). Conducting secondary analysis of qualitative data: Should we, can we, and how? *Qualitative Social Work: Research and Practice*, *18*(1), 81–97. DOI: 10.1177/1473325017700701 PMID: 30906228

Rule, P., and Land, S (2017). *Finding the plot in South African reading education* Reading & Writing - Journal of the Reading Association of South Africa, ISSN: (2308-1422, p2079-8245)

Rwela, A. G. (2018). Indigenous Knowledge Use in Horticulture for Sustainable Agriculture Development in Tanzania: Evidence from Nyandira Village in Morogoro.

Saldana, J. (2013). *The Coding Manual for Qualitative Researchers*. Sage.

Sambrook, S., & Stewart, J. (2008). Developing critical reflection in professional focused doctorates: A facilitator's perspective. *Journal of European Industrial Training*, *32*(5), 359–373. DOI: 10.1108/03090590810877085

Samuels, K. G. (2003). *Perceptions and knowledge regarding indigenous knowledge and Environmental Education in the Natural Science curriculum*. DEd Thesis: University of South Africa, Pretoria.

Sanders, D. (1980). *Background information on the formation of the World Council of Indigenous Peoples*. Centre for World Indigenous Studies.

Sannigrahi, K. (2016). Dietary Transition from Traditional to Modernization increase Non-Communicable Diseases. *Indian Journal of Applied Research*, *6*(8).

Sarivaara, E., Maatta, K., & Uusiautti, S. (2013). Who is indigenous? Definitions of indegeneity. *European Scientific Journal*, *1*, 369–378.

Sarivana, E., Maatta, K., & Uusiautti, S. (2013). Who is indigenous? Defintions of indigeneity. *European Scientific Journal, 1*, 369–378.

Sarkar, A. K. (2019). Impact of geographical indication (GI) on craftsmen of Varanasi in Uttar Pradesh. *NBS*. Retrieved from https://www.nbsc.in/MediaGallery/Impact%20of%20Geographical%20Indication%20(GI)%20on%20Craftsmen%20of%20Varanasi%20in%20Uttar%20Pradesh.pdf

Sarkar, S., Padaria, R. N., Vijayragavan, K., Pathak, H., Kumar, P., & Jha, G. K. (2015). Assessing the potential of Indigenous Technological Knowledge (ITK) for adaptation to climate change in the Himalayan and arid ecosystems.

Sarkhel, J. (2016). Strategies of indigenous knowledge in libraries. *Qualitative and Quantitative Methods in Libraries, 5*, 427–439.

Saruchera, M., & Xaba, M. B. (2024). Embracing African Traditional Knowledge Health Systems in a HIV/AIDS and COVID-19 Pandemic Era. *Journal of Black Studies, 55*(2), 95–116. DOI: 10.1177/00219347231213373

Satsangi, P. S., & Dhir, A. (2024). *Role of Communities in Achieving Sustainable Development*. Academic Foundation India.

Saunders, M., Lewis, P., & Thornhill, A. (2003). *Research methods for business students* (3rd ed.). Pearson Education Limited.

Scarborough, P., Appleby, P. N., Mizdrak, A., Briggs, A. D., Travis, R. C., Bradbury, K. E., & Key, T. J. (2014). Dietary greenhouse gas emissions of meat-eaters, fish-eaters, vegetarians, and vegans in the UK. *Climatic Change, 125*(2), 179–192. DOI: 10.1007/s10584-014-1169-1 PMID: 25834298

Schaefer, M., Hans-Jachen, H., & Denke, C. (2013). Communuicative versus strategic rationality: Habermas Theory of Communicative Action and social brain. *PLoS One*, •••, 1–7.

Schmuck, R. A. (2006). *Practical Action Research for Change*. Corwin Press, Inc.

Seale, C. (2012). Researching Society and Culture. *Sage (Atlanta, Ga.)*.

Sebata, S., Dhlamini, N., & Ndlovu, M. (2024). Indigenous Weather Forecast Systems: A Case Study of Weather Forecast Indicators for Chapo Village, Ward 17, in Matopo Zimbabwe. *Journal of the Kenya National Commission for UNESCO, 4*(2), 1–21. https://www.ajol.info/index.php/jkncu/article/view/281841. DOI: 10.62049/jkncu.v4i2.106

Seda, O. (2024). *Sho Madjozi: The pop star using traditional culture to shape a fresh identity for young South Africans. UBUNTU: Stronger Together, 31, 106-107, 110. South Africa's Public Diplomacy in action: Department of International Relationship and Cooperation*. DIRCO.

Sefa-Dei, G. J., & Cacciavillani, A. (2024). ActualisingDecolonisation: A Case for Anti-Colonising and Indigenising the Curriculum. *Journal of Philosophy of Education*, •••, qhae036.

Sekaran, U. (2003). *Research methods for business: A skill building approach* (4th ed.). John Wiley & Sons, Inc.

Sekhar, M., Rastogi, M., Saikanth, D. R. K., Rout, S., Kumar, S., & Patel, A. K.Rajesh C M. (2024). Exploring traditional agricultural techniques integrated with modern farming for a sustainable future: A review. *Journal of Scientific Research and Reports*, *30*(3), 185–198. DOI: 10.9734/jsrr/2024/v30i31871

Sellberg, C., Lundin, M., & Säljö, R. (2021). Assessment in the zone of proximal development: simulator-based competence tests and the dynamic evaluation of knowledge-in-action. *Classroom Discourse*, 1-21.

Semali, L. M., & Kincheloe, J. L. (1999). *What is Indigenous Knowledge? Voices from the Academy*. Falmer Press.

Semali, L., & Kincheloe, J. L. (1999). *What is indigenous knowledge?: Voices from the academy*. Routledge.

Seng, T. W. (2008). *Measuring practicum student teachers' reflectivity: the reflective pedagogical thinking scale*.

Serote, M. W. (1998). Initiatives for the protection of rights of holders of traditional knowledge, indigenous peoples, and local communities. *Proceedings of the World Intellectual Property Organisation conference*, Geneva, July 23 & 24.

Seroto, J. (2011). Indigenous education in the pre-colonial era. In Booyse, J. J., le Roux, C. S., Seroto, J., & Wolhuter, C. C. (Eds.), *A history of schooling in South Africa: Method and context* (2nd ed., pp. 37–56). Van Schaik.

Seroto, J. (2020). Indigenous history and education before colonial times. In Seroto, J., Davids, M. N., & Wolhuter, C. C. (Eds.), *Decolonising education in the global south: Historical and comparative international perspectives* (pp. 40–58). Pearson.

Setiyadi, D., Zaenuri, Z., & Mulyono, M. (2018). The Problem Based Learning Model with Etnomatematics Nuance by Using Traditional Games to Improve Problem Solving Ability. *Journal of Primary Education*, *7*(2), 176–186.

Sharma, I. P., Kanta, C., Dwivedi, T., & Rani, R. (2020). Indigenous agricultural practices: A supreme key to maintaining biodiversity. *Microbiological advancements for higher altitude agro-ecosystems & sustainability*, 91-112. https://doi.org/DOI: 10.1007/978-981-15-1902-4_6

Sharma, A., Manpoong, C., Pandey, H., Gupta, C. K., Baja, Y., Singh, M. S., & Mounglang, C. C. (2023). A comprehensive update on traditional agricultural knowledge of farmers in India. In *Wild food plants for zero hunger and resilient agriculture* (pp. 331–386). Springer Nature Singapore., DOI: 10.1007/978-981-19-6502-9_14

Sharma, I. P., Kanta, C., Dwivedi, T., & Rani, R. (2020). *Indigenous Agricultural Practices: ASupreme Key to Maintaining Biodiversity* (Goel, R., Eds.). Microbiological Advancements for Higher Altitude Agro-Ecosystems & Sustainability, Rhizosphere Biology, Springer Nature Singapore Pte Ltd., DOI: 10.1007/978-981-15-1902-4_691

Sharma, P. K., Srivastava, S., & Chandauriya, M. (2023). Indigenous Knowledge and Traditional Practices for Water Resource Management in Rajasthan, India. In *Traditional Ecological Knowledge of Resource Management in Asia* (pp. 137–157). Springer International Publishing., DOI: 10.1007/978-3-031-16840-6_9

Sharma, R., & Mishra, S. (2015). Blended learning in higher education: A review of the literature. *International Journal of Learning and Teaching in Higher Education*, *36*(2), 181–201.

Shava, G. N., Hleza, S., Tlou, F., Shonhiwa, S., & Mathonsi, E. (2021). Qualitative Content Analysis, Utility, Usability and Processes in Educational Research. *International Journal of Research and Innovation in Social Science* [IJRISS]. *International Journal of Research and Innovation in Social Science*, *5*(07), 553–558.

Shava, S. (2013). The representation of indigenous knowledges. In Stevenson, R. B., Brody, M., Dillon, J., & Wals, A. (Eds.), *International handbook of research on Environmental Education*. Routledge.

Shava, S., & Manyike, T. (2018). The decolonial role of African indigenous languages and indigenous knowledges. *Indilinga*, *17*(1), 36–52.

Shava, S., & Manyike, T. V. (2018). The decolonial role of African indigenous languages and indigenous knowledges in formal education process. *Indilinga*, *17*(1), 36–52.

Shava, S., & Manyike, T. V. (2018). The decolonial role of African Indigenous languages and Indigenous knowledges in formal education processes. *Indilinga*, *17*(1), 36–52. https://hdl.handle.net/10520/EJC-fe647aa43

Sheffield, J. (2005). Systemic knowledge and the v-model. *International Journal of Business Information Systems*, *1*(1/2), 83. DOI: 10.1504/IJBIS.2005.007401

Sheya, E. (2014). *Indigenous knowledge and Environmental Education: A case study of selected schools in Namibia* (Masters dissertation). Stellenbosch: University of Stellenbosch.

Shor, I., & Freire, P. (1987). *A pedagogy for liberation: Dialogues on transforming education*. Bergin & Garvey Publishers.

Shrivastava, P., Smith, M. S., O'Brien, K., & Zsolnai, L. (2020). Transforming sustainability science to generate positive social and environmental change globally. *One Earth*, *2*(4), 329–340. DOI: 10.1016/j.oneear.2020.04.010 PMID: 33501419

Siambombe, A., Mutale, Q., & Muzingili, T. (2018). Indigenous knowledge systems: A synthesis of Batonga people's traditional knowledge on weather dynamism. *African Journal of Social Work*, *8*(2), 46–54.

Siamonga, E. (2017). *Children"s Traditional Games and Promotion of Social Cohesion*. The Patriot.

Sikarwar, N., & Gogia, J. (2023). Linkages between food systems and food security. *Research Journal of Humanities and Social Sciences.*, *14*(4), 179–186.

Silva, C. M. D., Sauerbronn, F. F., & Thiollent, M. (2022). Decolonial studies, non-extractive methods, and participatory action research in Accounting. *RAC. Revista de Administração Contemporânea*, *26*(04), e210014. DOI: 10.1590/1982-7849rac2022210014.en

Silverman, D. (2010). *Doing qualitative research* (3rd ed.). Sage publications, Inc.

Simpson, D. J., & Jackson, M. J. (2021). *Educational reform: A Deweyan perspective*. Routledge. DOI: 10.4324/9780203825617

Simpson, J. E., & Mendenhall, T. J. (2022). Community-based participatory research with Indigenous youth: A critical review of literature. *Alternative*, *18*(1), 192–202. DOI: 10.1177/11771801221089033

Singh, B., Shirsath, P.B., Jat, M.L., McDonald, A.J., Srivastava, A.K., Craufurd, P., Rana, D.S., Singh, A.K., Chaudhari, S.K., Sharma, P.C., Singh, R., Jat, H.S., Sidhu, H.S., Gerard, B. and Braun, H. (2020). Agricultural labour, COVID-19, and potential implications for food security and air quality in the breadbasket of India. Agricultural Systems, 185, 0308-521X, .DOI: 10.1016/j.agsy.2020.102954

Singh, A. K., Janakiram, T., & Singh, J. (2020). Status of indigenous and minor vegetables Research "Way forward. *Indian Horticulture*, *65*(3).

Sirpurkar, R. J., & Gurpur, S. (2014). Protection of traditional knowledge through geographical indications: A case study of marble craft of Bhedaghat, Jabalpur. *American International Journal of Research in Humanities, Arts and Social Sciences*, 6(1, 2, 3). Retrieved from https://www.academia.edu/7371551/ISSN_Print_2328_3734_ISSN_Online_2328_3696_ISSN_CD_ROM_2328_3688

Skatkin, M. N., & Lerner, I. A. (Eds.). (1970-1979). *The great Soviet encyclopaedia* (3rd ed) [Online]. Retrieved from: https://encyclopedia2.thefreedictionary.com/Didactics (Accessed on 9 March 2020).

Smith, G. (1987). Whatever happened to educational priority areas? *Oxford Review of Education*, *13*(1), 23–39. DOI: 10.1080/0305498870130103

Smith, K., & Flores, M. A. (2019). The Janus faced teacher educator. *European Journal of Teacher Education*, *42*(4), 433–446. DOI: 10.1080/02619768.2019.1646242

Smith, L. (1999). *Decolonizing methodologies research and indegenous peoples*. Zed Books Ltd.

Smith, L. T. (1999). *Decolonizing Methodologies: Research and Indigenous Peoples*. Zed Books.

Smith, T. L. (2012). *Decolonizing methodologies: Research and Indigenous peoples*. Zed Books.

Snively, G., & Corsiglia, J. (2001). Discovering Indigenous Science: Implications for Science Education. *Science Education*, *85*(1), 6–34. DOI: 10.1002/1098-237X(200101)85:1<6::AID-SCE3>3.0.CO;2-R

Sobel, D. (2004). *Place-Based Education: Connecting Classrooms & Communities*. The Orion Society.

Sofowora, A. (1996). Research on medicinal plants and traditional medicine in Africa. *Journal of Alternative and Complementary Medicine (New York, N.Y.)*, *2*(3), 365–372. DOI: 10.1089/acm.1996.2.365 PMID: 9395671

Solomon, P., & Singer, J. (2011). School reform as inequity: The case of standards, standadization, and accountability. In Solomon, P., Campbell, S. J., Campbell, A., & Allen, A. (Eds.), *Brave new teachers doing social justice work in neoliberal times* (pp. 69–110). Canadian Scholras' Press.

Somekh, B., & Zeichner, K. (2009). Action research for educational reform: Remodelling action research theories and practices in local contexts. *Educational Action Research*, *17*(1), 5–21. DOI: 10.1080/09650790802667402

Souravi, K., & Rajasekharan, P. E. (2019). Traditional Knowledge in Horticulture. *Conservation and Utilization of Horticultural Genetic Resources*, 621-636. https://doi.org/DOI: 10.1007/978-981-13-3669-0_21

South African San Institute. (2017). *San code of research ethics.* Retrieved 4 13, 2024, from https://trust-project.eu/wp-content/uploads

Spaull, N. (2011). A Preliminary Analysis of SACMEQ III South Africa. Stellenbosch Economic Working Papers

Spencer Clark, J., Porath, S., Thiele, J., & Jobe, M. (2020). *Action Research.* NPP eBooks. 34. Retrieved from https://newprairiepress.org/ebooks/34

Sridhar, S., Alizadeh, F., Ratner, L., Russ, C. M., Sun, S. W., Sundberg, M. A., & Rosman, S. L. (2023). Learning to walk the walk: Incorporating praxis for decolonization in global health education. *Global Public Health: An International Journal for Research, Policy and Practice*, *18*(1), 2193834. DOI: 10.1080/17441692.2023.2193834 PMID: 36989128

Stanford Encyclopedia of Philosophy. (2024). *Stanford encyclopedia of philosophy*. Retrieved from Jurgen Habermas: https://plato.stanford.edu/entries/habermas/#Biog

Stenhouse, L. (Ed.). (1975). *An Introduction to Curriculum Research and Development London*. Heinemann.

Sterenberg, G. (2013). Considering Indigenous Knowledges and Mathematics Curriculum. *Canadian Journal of Science, Mathematics and Technology Education = Revue Canadienne de l'Enseignement des Sciences, des Mathématiques et de la Technologie*, *13*(1), 18–32. DOI: 10.1080/14926156.2013.758325

Stern, T. (2019). Participatory action research and the challenges of knowledge democracy. *Educational Action Research*, *27*(3), 435–451. DOI: 10.1080/09650792.2019.1618722

Stewart, S. (2002). *Native American Education: A Reference Handbook.* ABC-CLIO.

Stoll, P. T., & von Hahn, A. (2008). Indigenous peoples, indigenous knowledge and indigenous resources in international law. In von Lewinski, S. (Ed.), *Indigenous heritage and intellectual property: Genetic resources, traditional knowledge and folklore* (2nd ed.). Kluwer Law International.

Street, R. A., & Prinsloo, G. (2013). Commercially important medicinal plants of South Africa: A review. *Journal of Chemistry*, *2013*(1), 205048. DOI: 10.1155/2013/205048

Stuart, C. (1998). Care and Concern: An ethical journey in participatory action research. *Canadian Journal of Counselling/Revue canadienne de counseling, Vol. 32:4*, 298-314.

Suarta, I. M., Noortyani, R., Yarsama, K., & Adhiti, I. A. I. (2022). The role of teachers' Indigenous Knowledge and Cultural Competencies in enhancing students' engagement and learning outcomes. *Journal of Ethnic and Cultural Studies*, *9*(1), 244–264. DOI: 10.29333/ejecs/1025

Sumaiya, B. (2024). Indigenizing Education: Integrating Indigenous Knowledge Systems for Cultural Relevance, Cognitive Justice, and Student Engagement. In Waris, A., & Tiwari, D. (Eds.), *Integration of Indian Knowledge System in Higher. Education, 16(1), 93-105.*

Suryaratri, R. D., Prayitno, E. H., & Wuryani, W. (2019). The implementation of multi-sensory learning at elementary schools in Jakarta. *Jurnal Pendidikan Usia Dini*, *13*(1), 100–113. DOI: 10.21009/10.21009/JPUD.131.08

Swaminathan, M.S. (2001). A future attack on the PDS, Frontline. 18 (2).

Swantz, M. (2008). Participatory action research as practice. In Reason, P., & Bradbury, H. (Eds.), *The Sage Handbook of Action Research Participative Inquiry and Practice* (pp. 31–48). Sage Publications. DOI: 10.4135/9781848607934.n8

Swift, D. (1992). Indigenous Knowledge in the Service of Science and Technology in Developing Countries. *Studies in Science Education, 20*(1), 1–28. DOI: 10.1080/03057269208560002

Swisher, K., & Tippeconnic, J. (1999). *Next Steps: Research and Practice to Advance Indian Education.* ERIC Clearinghouse on Rural Education and Small Schools.

Tagliari, M., Bogoni, J., Blanco, G., Cruz, A., & Peroni, N. (2023). Disrupting a socio-ecological system: Could traditional ecological knowledge be the key to preserving the araucaria forest in brazil under climate change? *Climatic Change, 176*(2), 2. Advance online publication. DOI: 10.1007/s10584-022-03477-x

Tandika, P. B., & Ndijuye, L. G. (2019). Pre-primary teachers' preparedness in integrating information and communication technology in teaching and learning in Tanzania. *Information and Learning Science, 121*(1/2), 79–94. DOI: 10.1108/ILS-01-2019-0009

Tanksale, A., & Jha, A. K. (2015). Implementing National Food Security Act in India: Issues and challenges. *British Food Journal, 117*(4), 1315–1335. Advance online publication. DOI: 10.1108/BFJ-07-2014-0239

Tariq, M. U. (2024). Harnessing Persuasive Technologies for Enhanced Learner Engagement and Motivation. In M. Sanmugam, D. Lim, N. Mohd Barkhaya, W. Wan Yahaya, & Z. Khlaif (Eds.), *Power of Persuasive Educational Technologies in Enhancing Learning* (pp. 30-62). IGI Global. https://doi.org/ DOI: 10.4018/979-8-3693-6397-3.ch002

Tariq, M. U. (2024). Generative AI in Curriculum Development in Higher Education. In Fields, Z. (Ed.), *Impacts of Generative AI on Creativity in Higher Education* (pp. 227–258). IGI Global., DOI: 10.4018/979-8-3693-2418-9.ch009

Tariq, M. U. (2024). Navigating the Personalization Pathway: Implementing Adaptive Learning Technologies in Higher Education. In Minh Tung, T. (Ed.), *Adaptive Learning Technologies for Higher Education* (pp. 265–291). IGI Global., DOI: 10.4018/979-8-3693-3641-0.ch012

Tariq, M. U. (2025). Beyond the Classroom: The Sustained Influence of Educators on Student Success. In West, E. (Ed.), *Former Educators' Roles in Supporting Student Growth and Learning* (pp. 321–346). IGI Global Scientific Publishing., DOI: 10.4018/979-8-3693-6427-7.ch013

Tariq, M. U. (2025). Gamification and Virtual Reality: Transforming Early Childhood Language Learning. In Manditereza, B. (Ed.), *Digital Pedagogy in Early Childhood Language Development* (pp. 25–52). IGI Global Scientific Publishing., DOI: 10.4018/979-8-3693-7437-5.ch002

Tariq, M. U. (2025). Integrating Theory and Practice: Transformative Approaches in Practice-Led Research for Professional Development. In Kwok, S., Yu, P., & Le Corre, J. (Eds.), *Exploring Practice-Led Research for Professional Development* (pp. 419–436). IGI Global Scientific Publishing., DOI: 10.4018/979-8-3693-6376-8.ch016

Tariq, M. U. (2025a). Transforming Pedagogy: Best Practices for SoTL in Online Learning Environments. In Rahimi, R., & Soares, L. (Eds.), *Enhancing the Scholarship of Teaching and Learning in Online Learning Environments* (pp. 21–46). IGI Global Scientific Publishing., DOI: 10.4018/979-8-3693-7595-2.ch002

Tariq, M. U. (2025d). Empowering Learning Through Networked and Connected Education: Case Studies in Digital Engagement. In Gierhart, A. (Ed.), *Cases on Enhancing P-16 Student Engagement With Digital Technologies* (pp. 169–198). IGI Global Scientific Publishing., DOI: 10.4018/979-8-3693-5633-3.ch007

Tariq, M. U. (2025f). The Belmont Report: Guiding Ethical Principles in Human Research. In Throne, R. (Ed.), *IRB, Human Research Protections, and Data Ethics for Researchers* (pp. 245–268). IGI Global Scientific Publishing., DOI: 10.4018/979-8-3693-3848-3.ch010

Tashakkori, A., & Teddlie, C. *Handbook of mixed methods in social & behavioral research*. Califorina: Sage publications, Inc. 2003.

Tavernaro-Haidarian, L. (2019). Decolonization and development: Reimagining key concepts in education. *Research in Education*, *103*(1), 19–33. DOI: 10.1177/0034523719839750

Taylor. (2010). *Transformative learning theory-An overview.*

Taylor, S. V., & Leung, C. B. (2020). Multimodal literacy and social interaction: Young children's literacy learning. *Early Childhood Education Journal*, *48*(1), 1–10. DOI: 10.1007/s10643-019-00974-0

Tegama, N., & Fox, A. (2023). Ethics, the university, and society: Toward a decolonial approach to research ethics. *Alliance for African Partnership Perspectives*, *2*, 75–91.

Thaman, K. H. (2003). Decolonizing Pacific Studies: Indigenous Perspectives, Knowledge, and Wisdom in Higher Education. *The Contemporary Pacific*, *15*(1), 1–17. DOI: 10.1353/cp.2003.0032

The Economic Times. (2012, December 18). Bhedaghat may make it to UNESCO list soon.

Thelma, C. C., & Madoda, D. (2024). Exploring the relationship between cultural identity and educational equity: A Zambian analysis. *International Journal of Research Publication and Reviews*, *5*(5), 11440–11448.

Thiollent, M. (2011). Action research and participatory research: An overview. *International Journal of Action Research*, *7*(2), 160–174.

Trapido, A. (2018, December 21). 'Meet the tea makers of Sekhukhune who honour traditions while building a modern business. City Press, 21 December, viewed 8 February 2022. https://www.news24.com/citypress/news/meet-the-tea-makers-of-sekhukhune-who-honour-traditions-while-building-a-modern-business-20181221

Triyanto & Handayani, R.D. (2020). Prospect of integrating indigenous knowledge in the teacher learning community. *Diaspora, Indigenous, and Minority Education*, •••, 1–14.

Tronto, J. (1993). *Moral Boundaries, a political argument for an ethic of care*. New York University Press.

Tronto, J. (2013). *Caring democracy*. New York University Press.

Tshube, T., & Hanrahan, S. J. (2016). Sport psychology in Botswana: A prime breeding ground. *International Journal of Sport and Exercise Psychology*, *14*(2), 126–134. DOI: 10.1080/1612197X.2016.1142462

Tuck, E., & Yang, K. W. (2014). Unbecoming claims: Pedagogies of refusal in qualitative research. *Qualitative Inquiry*, *20*(6), 811–818. DOI: 10.1177/1077800414530265

Ulvik, M. (2014). Student-teachers doing action research in their practicum: Why and how? *Educational Action Research*, *22*(4), 518–533. DOI: 10.1080/09650792.2014.918901

Ulvik, M., Riese, H., & Roness, D. (2017). Action research – connecting practice and theory. *Educational Action Research*, *26*(2), 273–287. DOI: 10.1080/09650792.2017.1323657

UNESCO. (2003). *Convention for the safeguarding of the intangible cultural heritage*. Retrieved from http://unesdoc.unesco.org

Unisa [University of South Africa]. (n.d). Indigenous knowledge systems & education. Retrieved from https://www.unisa.ac.za/sites/corporate/default/Unisa-History-and-Memory-Project/Themes/All-themes/Indigenous-knowledge-systems-&-education [Indigenous knowledge systems & education] (Accessed on 8 August 2024).

United Nations Conference on Trade and Development. (2014). The Convention on Biodiversity and the Nagoya Protocol: Intellectual Property Implications, Chapter 5 Protection of Traditional Knowledge. https://unctad.org/system/files/official-document/diaepcb2014d3_ch5_en.pdf

United Nations Conference on Trade and Development. (n.d.). The Convention on Biodiversity and the Nagoya Protocol: Intellectual Property Implications, Chapter 5 Protection of Traditional Knowledge. Retrieved from https://unctad.org/system/files/official-document/diaepcb2014d3_ch5_en.pdf

United Nations. (1945). *Charter of the United Nations.*

United Nations. (1948). *Universal declaration of human rights*. Retrieved from https://www.un.org/en/about-us/universal-declaration-of-human-rights

United Nations. (1992). *Convention on biological diversity*. Retrieved from www.cbd.int

United Nations. (1994). *Convention to combat desertification.*

US Department of Education National Center for Education Statistics. (2016). *The condition of education 2016: Characteristics of postsecondary faculty.* USA: US Department of Education National Center for Education Statistics. Retrieved 4 15, 2024, from https://nces.ed.gov/programs/coe/indicator_csc.asp

Utter, A., White, A., Mendez, V. E., & Morris, K. (2021). Co-creation of knowledge in agroecology. *Elementa*, *9*(1), 1. DOI: 10.1525/elementa.2021.00026

Van de Walle, J. A. (2007). *Elementary and middle school mathematics education: Teaching developmentally* (6th ed.). Pearson education.

Varadharajan, K. S., Thomas, T., and Kurpad, A. V. (2013). Poverty and the state of nutrition in India. Asia Pacific Journal of Clinical Nutrition, 22(3), 326–339. https://doi.org/. 3.19.DOI: 10.6133/apjcn.2013.22

Vaughan, D. (2007). Blended learning: A framework for the effective integration of online and face-to-face learning. *International Journal of Learning and Teaching in Educational Settings*, *1*(1), 3–30.

Vaughan, D. (2007). Blended learning: An effective model for K-12 education. *Journal of Asynchronous Learning Networks*, *11*(1), 1–10.

Vaughn, L. M., & Jacquez, F. (2020). Participatory research methods–choice points in the research process. *Journal of Participatory Research Methods*, *1*(1). Advance online publication. DOI: 10.35844/001c.13244

Venkat, H., & Naidoo, D. (2012). Analyzing coherence for conceptual learning in a Grade 2 numeracy lesson. *Education as Change*, *16*(1), 21–33. DOI: 10.1080/16823206.2012.691686

Vijayan, D., Ludwig, D., Rybak, C., Kaechele, H., Hoffmann, H., Schönfeldt, H. C., Mbwana, H. A., Rivero, C. V., & Löhr, K. (2022). Indigenous knowledge in food system transformations. *Communications Earth & Environment*, *3*(1), 213. DOI: 10.1038/s43247-022-00543-1

Viswanath, S., & Lubina, P. A. (2017). Traditional Agroforestry Systems. *Agroforestry: Anecdotal to Modern Science*, 91-119. http://dx.doi.org/DOI: 10.1007/978-981-10-7650-3_3

Volk, K. S. (2010). Action research as a sustainable endeavor for teachers: Does initial training lead to further action? *Action Research*, *8*(3), 315–332. DOI: 10.1177/1476750309351358

Vozoris, N. T., & Tarasuk, V. S. (2003). Household food insufficiency is associated with poorer health. *The Journal of Nutrition*, *133*(1), 120–126. DOI: 10.1093/jn/133.1.120 PMID: 12514278

Vygotsky, L. S. (1978). *Mind in society: the development of higher psychological processes*. Harvard University Press.

Wahyudi, R. (2014). 'Democratic' online courses as 'globalized communication' in English language teaching and applied linguistics: A proposal. *Journal of Global Literacies, Technologies, and Emerging Pedagogies*, *2*(3), 248–260.

Wardhani, D. F., Arisanty, D., Nugroho, A., & Utami, U. B. L. (2024). Environmental Education Model Based on Local Wisdom of the Dayak Paramasan Tribe Indonesia. *Nature Environment and Pollution Technology*, *23*(4), 2259–2272. DOI: 10.46488/NEPT.2024.v23i04.030

Watkins, K. E., Marsick, V. J., & Wasserman, I. (2019). Action research, action learning, and appreciative inquiry: Interventions that build individual and group capacity for EBOCD. In *Evidence-based initiatives for organizational change and development* (pp. 76–92). IGI Global. DOI: 10.4018/978-1-5225-6155-2.ch004

Watson-Vandiver, M. J., & Wiggan, G. (2021). *The healing power of education: Afrocentric pedagogy as a tool for restoration and liberation*. Teachers College Press.

Watts, M., & Lawson, M. (2009). Using a meta-analysis activity to make critical reflection explicit in teacher education. *Teaching and Teacher Education*, *25*(5), 609–616. DOI: 10.1016/j.tate.2008.11.019

Weheliye, A. (2014). *Habeas viscus: Racializing assemblages, biopolitics and black feminist theories of the human*. Duke University Press.

White, R. E., & Cooper, K. (2022). Action Research. In *Qualitative Research in the Post-Modern Era: Critical Approaches and Selected Methodologies* (pp. 387–439). Springer International Publishing. DOI: 10.1007/978-3-030-85124-8_10

Whiteside, K., & Lodge, J. M. (2019). Blended learning in higher education: A review of the literature. *International Journal on Teaching and Learning in Higher Education*, *30*(2), 228–242.

WHO. (2020). Food safety. https://www.who.int/news-room/fact-sheets/detail/food-safety

WHO. (2024). Malnutrition. https://www.who.int/news-room/fact-sheets/detail/malnutrition

Wilmer, F. (1996). Domination and Resistance, exlusion and inclusion: Indigenous people's quest for peace and justice. *Peace Conflict Studies Vol 3 Article 4*, 1-9.

Wilson, C., Heinrich, L., Heidari, P., & Adams, K. (2020). Action research to implement an Indigenous health curriculum framework. *Nurse Education Today*, *91*, 104464. DOI: 10.1016/j.nedt.2020.104464 PMID: 32526618

Wilson, S. (2003). Progressing Towards an Indigenous Research Paradigm in Canada and Australia. *Canadian Journal of Native Education*, *27*(2), 161–178.

Wilson, S. (2008). *Research is Ceremony: Indigenous Research Methods*. Fernwood Publishing.

Wiratchai, N. (2009). *Research and statistics: questions and answers*. Bangkok: Icon printing.

Woldegiorgis, E. T. (2021). Decolonising a higher education system which has never been colonised. *Educational Philosophy and Theory*, *53*(9), 894–906. DOI: 10.1080/00131857.2020.1835643

Wood, L., Dean, C., du Toit, P., Esau, O., James, A., Mokhele, P., & Pillay, A. (2017). A snapshot of action research in and by higher education in southern Africa. The Palgrave international handbook of action research, 221-234.

World Bank. (2018). Food-borne Illnesses Cost US$ 110 Billion Per Year in Low- and Middle-Income Countries. https://www.worldbank.org/en/news/press-release/2018/10/23/food-borne-illnesses-cost-us-110-billion-per-year-in-low-and-middle-income-countries

World Intellectual Property Organization. (1967). *Convention establishing the World Intellectual Property Organization*. Retrieved from https://www.wipo.int/treaties/en/text.jsp?file_id=283854

World Intellectual Property Organization. (1996). *WIPO performances and phonograms treaty*. Retrieved from https://www.wipo.int/treaties/en/ip/wppt/

World Intellectual Property Organization. (2001). *Intellectual property needs and expectations of traditional knowledge holders: WIPO report on fact-finding missions on intellectual property and traditional knowledge (1998-1999)*. WIPO.

World Intellectual Property Organization. (2012). *The protection of traditional cultural expressions: Draft articles* (WIPO/GRTKF/IC/22/4). Retrieved from https://www.wipo.int/edocs/mdocs/tk/en/wipo_grtkf_ic_22/wipo_grtkf_ic_22_ref_facilitators_text.pdf

World Intellectual Property Organization. (2012). *The protection of traditional knowledge: Draft articles* (WIPO/GRTKF/IC/21/4). Retrieved from https://www.wipo.int/edocs/mdocs/tk/en/wipo_grtkf_ic_21/wipo_grtkf_ic_21_ref_facilitators_text.pdf

World Intellectual Property Organization. (2019, August 9). WIPO Director General quote on Indigenous Day [Tweet]. X (formerly Twitter).

World Intellectual Property Organization. (2023). *Assemblies of member states of WIPO, 64th series of meetings, agenda item 15(v) report on intergovernmental committee on intellectual property, genetic resources, traditional knowledge and folklore (IGC), decision*. Retrieved from https://www.wipo.int/export/sites/www/tk/en/igc/docs/igc-mandate-2024-2025.pdf

World Intellectual Property Organization. (n.d.). *Traditional cultural expressions*. Retrieved from https://www.wipo.int/tk/en/folklore/

World Medical Association. (1964). *WMI Declaration of Helsinki-Ethical Principles for medical research involving human subjects*. Retrieved from WMI Declaration of Helsinki-Ethical Principles for medical Research involving human subjects: https://www.wma.net/policies-post/wma-declaration-of-helsinki-ethical-principles-for-medical-research-involving-human-subjects/

World Trade Organization. (1994). *Agreement on trade-related aspects of intellectual property rights*. Retrieved from https://www.wto.org/english/docs_e/legal_e/27-trips_01_e.htm

World Trade Organization. (2001). *Doha ministerial declaration*. Retrieved from https://www.wto.org/english/tratop_e/dda_e/dohaexplained_e.htm

Wotschack, P., Vladova, G., de Paiva Lareiro, P., & Thim, C. (2023). Learning via assistance systems in industrial manufacturing. an experimental study in an industry 4.0 environment. *Journal of Workplace Learning*, *35*(9), 235–258. DOI: 10.1108/JWL-09-2022-0119

Wynter, S. (2003). Unsettling the coloniality of being/power/truth/freedom: Towards the human after. *CR (East Lansing, Mich.)*, *3*(3), 257–337. DOI: 10.1353/ncr.2004.0015

Yamunarani, K., Sangeetha, S., Dhanushkodi, V., Chitra, K., & Dhanalakshmi, K. (2023). Application of Indigenous Technological Knowledge (ITK) for Plant Disease Management in Tamil Nadu, India. *Asian Journal of Agricultural Extension. Economia e Sociologia*, *41*(12), 310–315. DOI: 10.9734/ajaees/2023/v41i122334

Yates, M. (2011). Postmetaphysical thinking. In Fultner, B. (Ed.), *Jurgen Habimas: Key concepts* (pp. 35–111). Routledge. DOI: 10.1017/UPO9781844654741.003

Yeseraw, A. (2023). *Indigenous Knowledge Inclusion in the New Curriculum Materials and Teachers 'Classroom Practices in the Amhara Region Primary and Middle Schools* (Doctoral dissertation).

Yishak, D., & Gumbo, M. T. (2017). The blending of indigenisation and standardisation in the Ethopian Education and Training Policy: A critical inquiry. In Gumbo, M.T. & Msila, V. (Eds.). African voices on indigenisation of the curriculum: Insights from practice, pp. 246-284. Wandsbeck: Reach Publishers.

Yishak, D., & Gumbo, M. (2015). A stand-alone, blended or restructured indigenisation approach to curriculum? A critical perspective. International Journal of African Renaissance Studies-Multi-. *Inter- and Transdisciplinarity*, *10*(1), 60–82.

Yosief, A., Sulieman, M. S., & Biede, T. (2024). Improving the practices of teacher educators through collaborative action research: Challenges and hopes. *Educational Action Research*, *32*(2), 204–221. DOI: 10.1080/09650792.2022.2066147

Yucel, A. S. (2006). E-learning approach in teacher training. *Turkish Online Journal of Distance Education*, *7*(4), 123–131.

Zappalaglio, A. (2013). Traditional knowledge: Emergence and history of the concept at international level. *University of Oxford*. Retrieved from https://www.researchgate.net/publication/271207052_Traditional_Knowledge_Emergence_and_History_of_the_Concept_at_International_Level

Zaslavsky, C. (1999). *Africa counts: Number and pattern in African cultures*. Chicago Review Press.

Zhang, Q. (2019). Postmodern view of knowledge and reorientation of teachers' role. *Advances in Social Science, Education and Humanities Research*, *350*, 189–191. DOI: 10.2991/aermt-19.2019.47

Ziai, A., Bendix, D., & Müller, F. (2020). Decolonizing knowledge orders, research methodology and the academia: An introduction. In Bendix, D., Müller, F., & Ziai, A. (Eds.), *Beyond the Master's Tool* (pp. 1–16).

Zidny, R., Sj str m, J., & Eilks, I. (2021). A multi-perspective reflection on how indigenous knowledge and related ideas can improve Science Education for sustainability. *Science & Education*, *29*(1), 145–185. DOI: 10.1007/s11191-019-00100-x

Zikmund, W. G. (2003). *Business research methods* (7th ed.). Thomson/South-Western.

Zima, P. (2015). *Subjectivity and identity in between modernity and postmodernity*. Bloomsbury Academic.

Zippert, E. L., Eason, S. H., Marshall, S., & Ramani, G. B. (2019). Preschool children's math exploration during play with peers. *Journal of Applied Developmental Psychology*, *65*, 101072. DOI: 10.1016/j.appdev.2019.101072

Zlatanović, D., & Mulej, M. (2015). Soft-systems approaches to knowledge-cum-values management as innovation drivers. *Baltic Journal of Management*, *10*(4), 497–518. DOI: 10.1108/BJM-01-2015-0015

Zuber-Skerritt, O. (2015). Participatory action learning and action research (PALAR) for community engagement: A theoretical framework. *Educational research for social change, 4*(1), 5-25.

Zuber-Skerritt, O. (2002). A model for designing action learning and action research programs. *The Learning Organization*, *9*(4), 143–149. DOI: 10.1108/09696470210428868

Zuber-Skerritt, O. (2015). Participatory action learning and action research for community engagement: A theoretical framework. *Educational Research for Social Change*, *4*(1), 5–25. DOI: 10.1108/978-1-78769-537-520191017

Zuber-Skerritt, O. (2018). An educational framework for participatory action learning and action research. *Educational Action Research*, *26*(4), 513–532. DOI: 10.1080/09650792.2018.1464939

Zuber-Skerritt, O. (Ed.). (2021). *Action research for change and development*. Routledge. DOI: 10.4324/9781003248491

Zuckermann, G., et al. (2015). ENGAGING - A guide to interacting respectfully and reciprocally with Aboriginal and Torres Strait Islander people, and their arts practices and intellectual property [PDF]. Australian Government: Indigenous Culture Support.

About the Contributors

Tomé Awshar Mapotse was formerly a Technical Education teacher; Departmental Head of Mathematics, Science and Technology Education; School Principal for Leaners with Special Needs; Technology Education (TE): teacher, lecturer, and senior lecturer. Currently a Full Professor at the University of South Africa (Unisa) where he is designing online open distance modules that are Africanized and incorporate Indigenous Knowledge Systems (IKS). Passionate about community engagement, Action Research (AR), IKS, and Open Distance Learning (ODL) with TE teachers. He is a pan-Africanist. Prof Mapotse supervises students registered for Masters and Doctoral (M&D) degrees within Unisa and have respectively examined other M & D dissertations and thesis from other universities. Purpose driven by the vision of Unisa, Prof Mapotse has therefore upheld the high standard of research and reporting in ODL, where he has shared his findings in journals and conferences both nationally and globally. He has contributed book chapters on ODL, TE, IKS, AR and Environmental Education (EE) in other books. Having couple of years at the contact university, this fact alone put him in a better position as an author to write effectively and efficiently about both distance and contacts modes of delivery. Proud to lift the African flag in any field.

Emmanuel Tetteh is a Contributing Doctoral Research Faculty Mentor at Walden University, a Research Faculty at Saybrook University, a Dissertation Faculty at California Baptist University, a Senior Lecturer at Mercy University, & an Adjunct Professor at Metropolitan College of New York (MCNY). He was a Senior Lecturer at Norwich University, a faculty member at Grand Canyon University and Capella University, and mentored several doctoral candidates. He holds BPS & MS degrees from MCNY and a Ph.D. from Walden University. Through his master's degree research program that led to a grounded theory, Dr. Tetteh coined the metaphor of "Communal Photosynthesis (CP)," developed the "creative-reflective methodology," unraveling the Identify, Act, Reflect, Evaluate, and Produce (IAREP) model for participatory action research, and an Internet of Things (IoT) action-learning solution model of big data policy-analytic epistemology. He serves as International Vice President of Action Learning, Action Research Association.

Tsebo Kgoto Matsekoleng holds a PhD in Environmental Education from University of South Africa. The co-editor published 10 papers, including 7 research articles, two book chapters and one conference paper. He is an action research practitioner in Environmental Education with a research focus on environmental issues, action research, sustainable development, sustainability, curriculum, and Indigenous Knowledge Systems and teaching.

***

Canan Koçak Altundağ is currently an Associate Professor in the Department of Mathematics and Science Education at Hacettepe University, Ankara, Turkey. Dr. Altundağ received B.Sc., M.Sc., and Ph.D. degree from the Faculty of Education at Hacettepe Univer-sity. Her research interests include daily life chemistry, context based learning and chemistry education. She has teaching experience of more than 10 years in the area of chemistry education. Dr. Altundağ published a number of papers in preferred Journals and participated in a range of conferences.

Monicca Thulisile Bhuda is an indigenous scholar, children's book author, award winning academic and a cultural heritage expert. At the University of Mpumalanga, Dr. Bhuda coordinates a course and lectures on culture and heritage studies. She graduated from North-West University with a Bachelor's degree (hons) in Indigenous Knowledge Systems, a Master's degree also in Indigenous Knowledge Systems and a PhD studies in Social Sciences with Indigenous Knowledge Systems (IKS). She has been featured on numerous media platforms such as TV, Radio, Magazines and Newspapers that were based on her interests as an indigenous scholar and on Indigenous Knowledge Systems. She has also written a number of publications on various subjects related to Indigenous Knowledge Systems, which is a multidisciplinary discipline.

Princess Blose is the Lecturer for Technology Education at the University of South Africa. Her true passion is in teaching Technology Education, Engineering Graphics, and Design. But recently, she has developed a keen interest in researching and learning more about indigenous knowledge systems and education for sustainable development in the context of teaching the subject of Technology Education. As a young scholar, she is interested also in the incorporation of indigenous knowledge in graphic design teachings. To maintain the standard of living in the modern world, her future endeavors include researching the teaching methodologies of Open Distance eLearning (ODeL), digitalization in Technology Education, and Technology and Vocational Education.

Chakabwata William holds a PhD in Curriculum and Instruction obtained from University of South Africa on 27 September 2023.He also holds a master degree in Curriculum and arts education which he earned from University of Zimbabwe. He has taught at Midlands State University in Namibia and also at Namibia University of Science and Technology(NUST).

Jyoti Gogia holds the position of Professor of Economics at the Dayalbagh Educational Institute (Deemed to be University) in Agra, India. Her expertise lies in microeconomics, statistics, and welfare economics. Driven by her passion for the subject, she has expanded the horizons of economics by integrating it with other disciplines, leading to numerous publications in esteemed journals, edited books, and magazines. Her research focuses on gender issues, food systems, food security, conscious economic systems, and sustainable development. Dr. Gogia also plays pivotal roles as the coordinator of the Dayalbagh Knowledge for Change (K4C) Hub and as the Nodal Officer for the Association of Indian Universities- Dayalbagh Educational Institute- Academic & Administrative Development Centre (AIU-DEI-AADC). As a dedicated educator and mentor, she has inspired numerous students to pursue careers in economics.

Anette Hay is a senior lecturer and a researcher in the School of Mathematics, Sciences and Technology Education at North-West University, South Africa and research niche Environmental

Education for Sustainable Development in communities. She has an interest in environmental education and sustainable development with a focus on the implementation of different teaching and learning strategies, and indigenous knowledge systems in the subject Geography. Community-Based Educational Research (COMBER).

Jolita Horbačauskienė has expertise in the fields of foreign language teaching and learning in various settings. Her main research interests lay in discipline literacy, intercultural communication, in the use of educational technologies, and experimental research methodology.

Sapna Jarial is an Associate Professor, Department of Agricultural Economics and Extension, Lovely Professional University.

Motheo Koitsiwe is currently the Director of Indigenous Knowledge Systems Centre (IKS Center), Faculty of Natural and Agricultural Sciences (FNAS), North - West University (NWU) in South Africa. He is responsible for the Teaching, Learning, Research, Innovation, and Community Engagement in Indigenous Knowledge Systems at the North West University. Dr Koitsiwe is mainly driving the decolonial and transformation agenda at the North West University and using IKS as a strategy to make impact not only among the academia, students but also to the communities, government and industry. Dr Motheo Koitsiwe is the first African indigenous scholar to have completed successfully the IKS PhD thesis in African indigenous astronomy (bolepa - dinaledi) of Batswana in Botswana and South Africa. He has done and continues to conduct indigenous research with indigenous astronomy experts, host conferences, seminars, community-based engagement (Maitiso) sessions in indigenous astronomy. Dr Koitsiwe was the first coordinator of the Bachelor of Indigenous Knowledge Systems (BIKS) which is a four (04) years professional degree registered with the South African Qualification Authority (SAQA) and approved by the Council for Higher Education (CHE). He is currently working on the Programme Qualification Mix (PQM) Review of the Bachelor of Indigenous Knowledge Systems. to revise the curriculum, make it relevant, responsive to society, economy, industry and to be futuristic. He also established reputation with engaging and working with African indigenous health practitioners, traditional leaders in the Recognition of Prior Learning which is a project sponsored by the Department of Science and Innovation. Dr Koitsiwe is also running the Coachlab Entrepreneurship Programme in partnership with the Innovation Hub sponsored by the Department of Science and Innovation. He is also the project Leader for the IKS Cannabis Sandbox Project in partnership with the Rastafari Nation, Traditional Health Practitioners, and the Traditional Leaders. Dr Koitsiwe also runs the project for the Indigenous Knowledge Systems Documentation Center in the North West province funded by the Department of Science and Innovation. Dr Koitsiwe is a distinguished decolonial researcher using indigenous epistemologies, ontologies, cosmologies, and philosophies in doing research with the students, scientists, community-based researchers in South Africa and other countries such as Botswana, Namibia, Uganda, and Lesotho. At an international level, Dr Koitsiwe is looking forward to establishing collaborations, partnerships with international universities in countries such as Colombia, Canada, Australia with the purposes of promoting, protecting, developing, open science, data sovereignty, decolonization of IKS. He organized the first International Indigenous Astronomy Conference sponsored by the National Convention Bereau and Department of Science and Innovation including the Conference on Recognition of Prior Learning (RPL) on Male and Female Rites of Passage (Bojale and Bogwera) in partnership with the Department of Science and Innovation.

Joseph Nkosana Chitja Mnguni is a prominent educator and academic committed to teaching and learning. His educational journey includes a University Diploma in Education (UDEP) and an Advanced Certificate in Education (ACE) from North-West University, as well as a Bachelor of Education Honours (BEd Hons) and a Master of Education (MEd) from the University of South Africa (UNISA). I am pursuing a PhD in Technology Education at UNISA, reflecting my dedication to academic excellence. Throughout my career, I have focused on promoting quality education, specializing in curriculum development, teacher development, and educational research. I have presented at conferences and contributed to two book chapters on technology education. Known for my passion for teaching and lifelong learning, I actively engage in community service and value relationships. My contributions have established me as a respected figure in the field of education, allowing me to make a significant impact on my learners and the broader educational community.

Elliot Charles Ndlovu is a Lecturer at the University of South Africa in the Department of Science and Technology Education. He specialises in Science and Technology Education and recently has developed an interest in the teachings of Indigenous Knowledge Systems in educational settings particularly Technology Education.

Chethan Patil N D is Scientist-B,Scientist-B (Technology Transfer and Rural Development) atCentre for Aromatic Plants, Dehradun, Uttarakhand, India

Milda Ratkevičienė has a long experience in higher education, both as a practitioner and as a researcher. Currently, Milda Ratkevičienė works as a Researcher at Kaunas University of Technology. She has more than ten years of experience in study programs quality assurance. Currently she is involved in the development of bi-and multilingual disciplinary literacies across educational levels addressing developmental trajectories of learner groups, focusing on the transition across educational levels (primary-secondary-tertiary education).

Mamokete Venolia Setshego is a dedicated researcher and academic specialising in indigenous knowledge systems. Currently she is pursuing her Doctoral degree in Indigenous Knowledge Systems in North West University (South Africa). She earned her Master's degree in Indigenous Knowledge Systems at North-West University (South Africa) and has contributed significantly to understanding the role of natural resources in traditional cosmeceuticals. Her research highlights the use of plant-based remedies for skin conditions and their cultural significance within rural communities, particularly in South Africa's Vhembe district. Setshego's work emphasises the preservation of ethnobotanical knowledge, exploring sustainable and biodegradable alternatives to synthetic skincare. By integrating both qualitative and quantitative methods, her research bridges scientific inquiry and traditional practices to address global challenges in food security, nutrition, and natural product development.

Neha Sikarwar obtained her PhD from Dayalbagh Educational Institute (Deemed to be University) in Agra, India, and holds an MSc in Business Analytics and Management Sciences from the University of Southampton, United Kingdom, which is a member of the prestigious Russell Group. She is currently serving as a research and data analyst at Fortade Limited in the UK, focusing on development economics, sustainability, and financial institutions and markets. Before this, she gained experience as a data science intern at the University of Moratuwa in Sri Lanka. Dr. Sikarwar specializes in creating robust and scalable analytical solutions for economic and financial organizations. Her work combines academic expertise with operational acumen to enhance business performance and facilitate evidence-based policymaking.

Berita Simonaitienė has been working in the KTU Educational Research Group since 2001. She is one of the authors of the landmark document "Good School Concept", approved by the Minister of Education and Science in Lithuanian (2015), an expert in teacher education and external evaluator of study programmes in the field of education and teacher training. Her research interests include: the learning school, learning modes, curriculum management, learning strategies and environments. Research experience gained through participation in projects: 2020-2023 „Disparities in school achievement from a person and variable-oriented perspective: A prototype of a learning analytics tool (NO-GAP)", 2019-2022 „The development of adolescent personal goals in the context of social inequality (PaNel)", 2021 „Solutions for the use of artificial intelligence and digital technologies to improve the quality of education in response to COVID-19 (TRANSFORM-ED)", 2017-2018 "The Impact of Learning Strategy and Learning Environment on Student Achievement", „Investigation of the effectiveness and feasibility of applying the theoretical model of learning strategy and learning environment factors in real con

Muhammad Usman Tariq has more than 16+ year's experience in industry and academia. He has authored more than 200+ research articles, 100+ case studies, 130+ book chapters and several books other than 4 patents. He is founder and CEO of The Case HQ, a unique repository for courses, narrative and video case studies. He has been working as a consultant and trainer for industries representing six sigma, quality, health and safety, environmental systems, project management, and information security standards. His work has encompassed sectors in aviation, manufacturing, food, hospitality, education, finance, research, software and transportation. He has diverse and significant experience working with accreditation agencies of ABET, ACBSP, AACSB, WASC, CAA, EFQM and NCEAC. Additionally, Dr. Tariq has operational experience in incubators, research labs, government research projects, private sector startups, program creation and management at various industrial and academic levels. He is Certified Higher Education Teacher from Harvard University, USA, Certified Online Educator from HMBSU, Certified Six Sigma Master Black Belt, Lead Auditor ISO 9001 Certified, ISO 14001, IOSH MS, OSHA 30,

Sadık Taşkıran has been working as a science teacher in the Ministry of National Education of the Republic of Turkey since 2010. He began his education at Muğla University, graduating from the Department of Science Teaching in 2009. In 2024, he successfully completed his master's degree at Hacettepe University. Taşkıran is married and has one child, and he is 37 years old.

Index

A

B

C

D

E

P

R

S

T

W

www.ingramcontent.com/pod-product-compliance
Ingram Content Group UK Ltd.
Pitfield, Milton Keynes, MK11 3LW, UK
UKHW050952290926
15470UKWH00005B/1779

9 798369 335420